AF132960

Markov and Semi-markov Chains, Processes, Systems and Emerging Related Fields

Markov and Semi-markov Chains, Processes, Systems and Emerging Related Fields

Editors

Panagiotis-Christos Vassiliou
Andreas C. Georgiou

MDPI • Basel • Beijing • Wuhan • Barcelona • Belgrade • Manchester • Tokyo • Cluj • Tianjin

Editors
Panagiotis-Christos Vassiliou
University College London
UK

Andreas C. Georgiou
University of Macedonia
Greece

Editorial Office
MDPI
St. Alban-Anlage 66
4052 Basel, Switzerland

This is a reprint of articles from the Special Issue published online in the open access journal *Mathematics* (ISSN 2227-7390) (available at: https://www.mdpi.com/journal/mathematics/special_issues/Markov_Semi-markov_Chains).

For citation purposes, cite each article independently as indicated on the article page online and as indicated below:

LastName, A.A.; LastName, B.B.; LastName, C.C. Article Title. *Journal Name* **Year**, *Volume Number*, Page Range.

ISBN 978-3-0365-2398-9 (Hbk)
ISBN 978-3-0365-2399-6 (PDF)

© 2021 by the authors. Articles in this book are Open Access and distributed under the Creative Commons Attribution (CC BY) license, which allows users to download, copy and build upon published articles, as long as the author and publisher are properly credited, which ensures maximum dissemination and a wider impact of our publications.
The book as a whole is distributed by MDPI under the terms and conditions of the Creative Commons license CC BY-NC-ND.

Contents

Andreas C. Georgiou, Alexandra Papadopoulou, Pavlos Kolias, Haris Palikrousis and Evanthia Farmakioti
On State Occupancies, First Passage Times and Duration in Non-Homogeneous Semi-Markov Chains
Reprinted from: *Mathematics* **2021**, *9*, 1745, doi:10.3390/math9151745 . . . . **191**

Vlad Stefan Barbu, Guglielmo D'Amico and Thomas Gkelsinis
Sequential Interval Reliability for Discrete-Time Homogeneous Semi-Markov Repairable Systems
Reprinted from: *Mathematics* **2021**, *9*, 1997, doi:10.3390/math9161997 . . . . **209**

Mantas Dirma, Saulius Paukštys and Jonas Šiaulys
Tails of the Moments for Sums with Dominatedly Varying Random Summands
Reprinted from: *Mathematics* **2021**, *9*, 824, doi:10.3390/math9080824 . . . . **227**

Rodi Lykou and George Tsaklidis
Particle Filtering: A Priori Estimation of Observational Errors of a State-Space Model with Linear Observation Equation †
Reprinted from: *Mathematics* **2021**, *9*, 1445, doi:10.3390/math9121445 . . . . **253**

Ourania Theodosiadou and George Tsaklidis
State Space Modeling with Non-Negativity Constraints Using Quadratic Forms
Reprinted from: *Mathematics* **2021**, *9*, 1908, doi:10.3390/math9161908 . . . . **269**

About the Editors

Panagiotis-Christos Vassiliou is an Honorary Professor at the Department of Statistical Sciences, University College London and an Emeritus Professor at the Department of Mathematics, Aristotle University of Thessaloniki. He has spent research and teaching time at Ulster University, Ioannina University, Aristotle University of Thessaloniki, University College London. He has also spent research time at Imperial College London. He taught at the London Taught Course Center (LTCC), for 3rd and 4th year Ph.D. students of UCL, LSE, King's, Brunel and Kent. He has published more than 53 papers, which are indexed in the web of science, mainly in the areas of non-homogeneous Markov systems, non-homogeneous semi-Markov systems, and stochastic finance. His publication timespan is 47 years and the two journals with the highest frequency were Journal of Applied Probability and Linear Algebra and its Applications. He has also written four books in English (J. Wiley, Chapman and Hall, UCL, MDPI) and 13 books in Greek. He has supervised nine completed PhD theses and Refereed more than 25 scientific journals.

Andreas C. Georgiou is a Professor of Operational Research and Director of the Quantitative Methods and Decision Analytics Lab at the Department of Business Administration of the University of Macedonia. Formerly Director of the MBA program and of the MSc in Business Analytics and Data Science. His research includes Markov Processes, manpower planning, Business Analytics applications in operations, DES, MCDM, AHP and DEA. He teaches undergraduate and postgraduate courses at the Aristotle University of Thessaloniki, University of Macedonia and the Greek Open University. He is a member of several International Societies such as INFORMS and OR Society. His publications include more than 80 papers in international conferences proceedings and top-tier journals. He has also co-authored several textbooks in Greek, served as a reviewer for more than 50 International Journals and has more than 850 citations to his work.

Preface to "Markov and Semi-markov Chains, Processes, Systems and Emerging Related Fields"

The evolution of the theory and applications of Markov chains has continued for over a century since 1907 when A. Markov stated the fundamental notion. Since then, it has evolved into one of the most important areas of stochastic processes and its applications have an immense diversity in various scientific fields. Miscellaneous generalizations and/or specializations have also given rise to emerging related fields and even support technologies such as artificial intelligence, or foster critical research in genetics. The present volume includes the papers published in the Journal of Mathematics in the Special Issue entitled "Markov and Semi-Markov chains, processes, systems and emerging related fields". The motivation was to assemble a collection of papers by well-known researchers in the field, aiming to provide advanced theoretical and applicable material for discussion, application and further research. In this respect, the volume covers a broad range of theory and applications on Markov and Semi-Markov systems, as well as the related stochastic processes. All papers underwent the peer-review process according to the standards of the Journal of Mathematics in the section of Probability and Statistics during the period from January to August 2021. A complete list of all articles, a short description, and useful introductory information about their particular research area can be found in the editorial of the book. Acknowledgments for this Special Issue and volume are due to many people, including the Chief Editors, the Editorial Board, the Authors, the Editorial Manager and Editorial stuff and are provided in detail in the editorial of the Special Issue.

Panagiotis-Christos Vassiliou, Andreas C. Georgiou
Editors

mathematics

MDPI

Editorial

Markov and Semi-Markov Chains, Processes, Systems, and Emerging Related Fields

P.-C.G. Vassiliou [1,*] and Andreas C. Georgiou [2,*]

1 Department of Statistical Science, University College London, Gower St, London WC1E 6BT, UK
2 Quantitative Methods and Decision Analysis Lab, Department of Business Administration, University of Macedonia, 54636 Thessaloniki, Greece
* Correspondence: vasiliou@math.auth.gr (P.-C.G.V.); acg@uom.edu.gr (A.C.G.)

Citation: Vassiliou, P.-C.G.; Georgiou, A.C. Markov and Semi-Markov Chains, Processes, Systems, and Emerging Related Fields. *Mathematics* **2021**, *9*, 2490. https://doi.org/10.3390/math9192490

Received: 25 September 2021
Accepted: 26 September 2021
Published: 4 October 2021

Publisher's Note: MDPI stays neutral with regard to jurisdictional claims in published maps and institutional affiliations.

Copyright: © 2021 by the authors. Licensee MDPI, Basel, Switzerland. This article is an open access article distributed under the terms and conditions of the Creative Commons Attribution (CC BY) license (https://creativecommons.org/licenses/by/4.0/).

Probability resembles the ancient Roman God Janus since, like Janus, probability also has a face with two different sides, which correspond to the metaphorical gateways and transitions between the past and the future. Probability can be seen as a limbo state between abstractness and concreteness. This inherent duality renders one side the closest possible to a branch of pure mathematics, derived from certain axioms in classical areas of algebra.

Nonetheless, with its other side, probability is, indeed, an applied or applicable mathematics discipline, most commonly known as applied probability, although, in our opinion, the common distinction between pure and applied mathematics is, all too often, merely artificial and, at times, fuzzy. This side, without being less demanding in mathematical or stochastic terms, gives birth to valuable models for studying everyday phenomena of the real world. Stochastic processes are, by now, well established as an extension of probability theory. In the area of stochastic processes, Markov and semi-Markov processes play a vital role as an independent area of study, generating important and novel applications and new mathematical results.

The special issue with the title Markov and Semi-Markov Chains, Processes, Systems, and Emerging Related Fields includes fourteen articles published in the journal of "Mathematics" in the section of "Probability and Statistics", in the period from January–August 2021. The authors of this issue acted as Academic Editors to all the papers except their own papers for which the Editorial Board appointed Academic Editors that were unknown to the authors and became known after the publication and after they agreed to their names being published. We hope that this volume provides opportunities for future research ideas and that the interested reader will discover these paths between the lines and the mathematical formulas of the published papers.

The Guest Editors would like to thank the Chief Editors and the Editorial Board of the Journal of *Mathematics* for their invitation to edit the present volume. We cordially thank the authors for contributing to the publication of the volume by submitting their significant research articles and addressing all comments and suggestions with diligence and enthusiasm. We also pay our respects to the anonymous reviewers of the volume since, without their valuable assistance, this venture could have not been completed.

We would also like to express our gratitude to the Editorial Manager, Dr. Syna Mu, for his continuous efforts to facilitate the workflow of this issue, for the excellent collaboration with the Guest Editors, and for arranging for the partial funding of the publication of the present volume. The Guest Editors would also like to thank the Professors Andras Telecs and Alexander Zeifman for acting as Academic Editors for our own contributed articles. Last, but not least, many thanks are due to the numerous Editorial Assistants who successfully undertook the tedious tasks of managing the large number of submissions in the present volume.

We now proceed to a brief presentation of the articles by categorizing them in three sub-areas and also provide the reader with some useful references that might introduce them easily to the mathematical background of the papers. The order of the sub-areas (sections) generally follows the title of the special issue, and the articles within each section are sorted by the dates of publication.

(i) *Markov Chains, Processes, and Markov Systems.*

Markov processes are stochastic processes that exhibit the Markov property, while Markov Chains are their discrete time and discrete state space counterpart. That is, the probabilistic dependence on the past is only through the present state, which contains all the necessary information for the evolution of the process. Useful introductory texts on homogeneous and non-homogeneous Markov chains and processes are [1–3] and ([4], Chapter 3). For Markov Chains on general state space, an excellent reference is [5]. For Markov systems or Open Markov models, which are generalizations of the Markov chain, an introductory review paper is [6]. We now provide a brief description of the articles of the special issue that could be included in this category:

($i1$) *Geometric Ergodicity of the Random Walk Metropolis with Position-Dependent Proposal Covariance,* by Samuel Livingston [7]. In this paper, the ergodic behaviour of a Markov Chain Monte Carlo (MCMC) method is analysed and specifically the Metropolis–Hastings method. MCMC methods are used for estimating the expectations of a probability measure $\pi(.)$, which need not be normalized. This is done by sampling a Markov chain, which has asymptotic distribution $\pi(.)$, and computing the empirical averages. It is vital for the quality of the estimators to have conditions on $\pi(.)$ that will produce in a Markov chain, which will converge asymptotically at a geometric rate. In the present work with proposal $\mathcal{N}\left(x, hG(x)^{-1}\right)$ a Metropolis–Hastings algorithm is studied where x is its current state and the ergodicity properties are investigated. It is shown that appropriate selections of $G(x)$ severely influence the ergodicity properties in comparison to the respective Random Walk Metropolis.

($i2$) *Non-Homogeneous Markov Set Systems,* by P.-C.G. Vassiliou [8]. The class of stochastic processes defined as Non-Homogeneous Markov systems are, in effect, a generalization of a Markov chain. This provides a general framework for the many stochastic models used to model populations of different kinds of entities with a large diversity. In the present study, for the first time, the basic parameters of a NHMS are in intervals and not point estimates. It is proven that, under certain conditions of convexity of the intervals, the set of the relative expected population structure of memberships is compact and convex. A series of theorems are provided and proved on the asymptotic behaviour and the limit set of the expected relative population structure. Finally, an application for a geriatric and stroke patients in a hospital is presented, and, through which, solutions are provided for the problems that are usually surface in such applications.

($i3$) *Period-Life of a Branching Process with Migration and Continuous Time,* by Prysyazhnyk, K., Bazylevych, I., Mitkvova, L. and Ivanochko, I. [9]. Branching processes are a common tool for the mathematical representation of real processes, such as chemical, biological, demographic, and so on. The reason is that BP can easily describe the population dynamics of entities under different contexts (from physics and chemistry to biology and information technology). There exists a large number of variants of of BPs, and, in this article, the authors investigate the Markov branching process model with migration in continuous time.

The distribution of the period-life is the length of the time interval between the moment when the process is initiated by a positive number of particles and the moment when there are no individuals in the population for the first time. The form of the differential equation and the probability generating function of the random process that describes the behaviour of the process within its period-life is presented. In addition, the limit theorem for the period-life of the subcritical and critical BPMCT was found.

($i4$) *Optimizing a Multi-State Cold-Standby System With Multiple Vacations in the Repair and Loss of Units,* by Ruiz-Castro, J.E. [10]. This article focuses on redundant systems and preventive maintenance as fundamental pylons in ensuring systems reliability, minimizing failures, and reducing costs. In particular, the author studies a complex multi-state system subject to multiple events such as several types of internally or externally induced failures. The analysis takes into account the loss of units due to non-repairable failures, and it is assumed that the system can still operate with one less unit. There is also a repair person whose behaviour is determined by the number of units within the repair facility and the vacation policy applied. This system is is modelled via Markovian Arrival Processes with marked arrivals. The author presents the stationary distributions and multiple measures related to system and financial performance.

(*i*5) *Using Markov Models to Characterize and Predict Process Target Compliance,* by McClean, S. [11]. A general phase-type Markov model is presented to predict the process target compliance. The Markov model has several absorbing states with different targets and Poisson arrivals. Several theoretical results are provided, and several close analytic formulas are founded, which provide useful characterizations and predictions in a sufficient lead time of various target compliance. The results are illustrated using data from a stroke patient unit, where there are multiple discharge destinations for patients, namely death, private nursing home, or the patient's own home, where different discharge destinations may require disparate targets. Key performance indicators are also established, which are important and common place in health care, business, and industrial processes.

(*i*6) *Open Markov Type Population Models: From Discrete to Continuous Time,* by Esquivel, M.L., Krasii, N.P. and Guerreiro, G.R. [12]. The study of homogeneous open Markov population models in discrete and continuous time and discrete state space has a long and important history of seventy five years. Over the last forty years, attention has been shifted to the study of non-homogeneous Markov systems or equivalently to non-homogeneous open Markov population models in discrete and continuous time and discrete state space. Lately, there are also studies of non-homogeneous Markov systems in discrete time and general state space. The main contribution of the present work are to extend the results on open Markov chains in discrete time to some continuous time processes of Markov type using different methods of associating a continuous process to an observed process in discrete time.

(*i*7) *Partial Diffusion Markov Model of Heterogeneous TCP Link: Optimization With Incomplete Information,* by Borisov, A., Bosov, A., Miller, G., and Sokolov, I. [13]. This paper deals with an old acquaintance that still is an object of perpetual investigation and evolution: the Transmission Control Protocol (TCP). The authors present a new mathematical model of TCP using partially observable controllable Markov jump process (MJP) in a finite state space. The observations of the stochastic dynamic system are formed by low-frequency counting processes of packet losses and timeouts and a high-frequency compound Poisson process of packet acknowledgements.

In this respect, the entire information transmission process is considered as a stochastic control problem with incomplete information. The first aim of the paper is to present of a new mathematical model of the TCP link operation based on the heterogeneous (wired/wireless) channel, and the second aim is the presentation of a new TCP prototype version based on the solution of the optimal MJP state control under complete information as well as the solution to the optimal MJP state filtering given the diffusion and counting observations. The performance of the proposed model is demonstrated with numerical experiments.

(*i*8) *Evaluating the Efficiency of Off-Ball Screens in Elite Basketball Teams via Second-Order Markov Modelling,* by Stavropoulos, N., Papadopoulou, A.A., and Kolias, P. [14]. This paper falls in the area of sport oriented stochastic modelling, including sports performance analytics and mathematical optimization. The systematic use of performance indicators in the strategy orientation of sports teams has been the subject of extended research in recent years, and basketball is not an exception. The authors employ second-order, partially non-homogeneous, Markov models to gain insight into the behaviours and interactions of the players using the screens and the final attempt of the shots on the weak side. More specifically, they develop a second-order Markov modelling framework to evaluate the characteristics of off-ball screens that affect the finishing move and the outcome of the offensive movement. In addition, they examine how time, expressed either as the quarter of play or as the time clock (0–24 s), could influence the transition probabilities from screens and finishing moves to outcomes.The authors used a sample of 1170 possessions of the FIBA Basketball Champions League 2018–2019, and the particular variables of interest were the type of screen on the weak side, the finishing move, and the outcome of the shot. The proposed model provides useful information for coaches who can use it in both individual and group training programs as a part of their strategic planning for performance improvement.

(*ii*) *Semi-Markov Chains, Processes, and Semi-Markov Systems.*

Semi-Markov chains are generalizations of Markov chains where the time of transition from each state to another is now a random variable. The same applies for semi-Markov processes, except that now the time is continuous. A very good text on Semi-Markov chains and processes for the interested reader is [15]. For semi-Markov systems or open semi-Markov models, which are, again, generalizations of Markov chains, the first paper that introduced them was [16], and this is a good place to start. We provide below a brief description of the articles that could be categorized in this section:

(*ii*1) *Discrete Time Hybrid Semi-Markov Models in Manpower Planning,* by Verbeken, B. and Guerry, M-A. [17]. The present work is on non-homogeneous semi-Markov systems and, in particular, their traditional roots on manpower planning. Non-homogeneous semi-Markov systems have found important applications in a large variety of areas, such as biological phenomena, ecological modelling, DNA analysis, credit risk in mathematical finance, reliability and survival analysis, disability insurance problems, and wind tornado problems. The paper argues that, in a semi-Markov model for manpower problems, there is an advantage in considering some of the transitions as being purely in a Markov model. That is, the joint distribution in these states is a constant. There is also a section where solutions are provided for the problems that surface when applying such models to manpower systems.

(*ii*2) *On State Occupancies, First Passage Times and Duration in Non-Homogeneous Semi-Markov Chains,* by Georgiou, A.C., Papadopoulou, A.A., Kolias, P., Palikrousis, H., and Farmakioti, E. [18]. A basic aspect of Semi-Markov processes (SMC) is the utilization of general sojourn time distributions. This paper offers insights for three classes of relevant probabilities of a semi-Markov process and, more specifically, on the first passage time, the occupancy and the duration probabilities. The paper provides closed forms for the three classes of probabilities using the basic parameters of the process and initiating from the recursive relations of the aforementioned probabilities. The analytical results are accompanied with illustrations on the human genome DNA strands, which are often studied using Markovian models. There exist several algorithmic approaches analysing the occupancy and appearance of words in DNA sequences; however, the results suggest that the proposed modelling framework can be also used to investigate the structure of genome sequences.

(*ii*3) *Sequential Interval Reliability for Discrete-Time Homogeneous Semi-Markov Repairable Systems,* by Barbu, V.S., D'Amico, G. Gkelsinis, T. [19]. This is another paper of the special issue concerned with reliability indicators; however, this one is dedicated to semi-Markov systems. The authors introduce a new reliability measure, namely the sequential interval reliability (SIR), for homogeneous semi-Markov repairable systems in discrete time. This measure generalises the notion of interval reliability, and takes into account the dependence on what is called the final backward.

As mentioned by the authors, interval reliability was first introduced and studied for continuous-time semi-Markov systems. This measure computes the probability that a system is in a working state during a sequence of non-overlapping intervals, and this is important in applications where a system performs during consequent time periods, in the cases of extreme events over several time periods, in electricity consumption where certain thresholds are exceeded, or in financial modelling and relevant credit scoring models. The article introduces the sequential interval reliability measure from both aspects: transient analysis, providing a recurrence formula and its asymptotic result as time tends to infinity. The paper includes a numerical example.

(*iii*) *Related Stochastic Processes.*

The subject of introduction in the theory of stochastic processes is well established, and there are many useful introductory or advanced texts that could help any interested reader. Any text with a medium mathematical level will suffice as a first useful tool for the articles that follow. Nevertheless, as a first course in stochastic processes, the book [20] may well serve the purpose. For the more advanced reader, the books [21–23] are excellent and may well suffice for further reading.

We provide a brief description of the articles that could be categorized in this section:

(*iii*1) *Tails of the Moments for Sums with Dominatedly Varying Random Summands,* by Dirma, M., Paukstys, S., and Siaulys, J. [24] This paper investigates the asymptotic behaviour of tails of the moments for randomly weighted sums with possibly dependent dominatedly varying summands. The findings improve and generalise other related results of the relevant literature. For example, the authors achieve sharper asymptotic bounds under pairwise quasi-asymptotic independence structure. In addition, the relaxation of the exponent condition allows for the possibility to be any fixed non-negative real number. In the case of randomly weighted sums, the boundedness condition on the random weights is substituted by the less restrictive moments condition. The authors illustrate and conform their asymptotic results with a Monte Carlo simulation with three specific cases of random sums from disjoint sub-classes of dominatedly varying distributions.

(*iii*2) *Particle Filtering: A Priori Estimation of Observational Errors of a State-Space Model with Linear Observation Equation,* by Lykou, R. and Tsaklidis, G. [25] This is the first of two papers related to observational errors of Particle Filtering. Particle Filter (PF) methodology that deals with the estimation of latent variables of stochastic processes taking into consideration noisy observations generated by the latent variables. The paper focuses on state-space models with linear observation equations and provides an estimation of the errors of missing observations (in cases of missing data) aiming at the approximation of weights under a Missing At Random (MAR) assumption.

In this article, the observational errors are estimated prior to the upcoming observations. This action is added to the basic algorithm of the filter as a new step for the acquisition of the state estimations. As mentioned above, this intervention is mainly useful in the presence of missing data problems, as well as in sample tracking for impoverishment issues. The linearity assumption permits sequential replacements of missing values with equal quantities of known distributions. The contribution of the a priori estimation step to the study of impoverishment phenomena is also exhibited through Markov System (MS) framework. A simulation example is provided, highlighting the advantages of the proposed algorithm to existing approaches.

(*iii*3) *State Space Modeling with Non-Negativity Constraints Using Quadratic Forms,* by Theodosiadou, O. and Tsaklidis, G. [26] This article is the second on state space modelling methods. It proposes a method in state space modelling representation, which deals with hidden components that are subject to non-negativity constraints. It is known that state space models are used for the estimation of hidden random variables when noisy observations are available; however, if the state vector is subject to constraints, the standard Kalman filtering algorithm can no longer be used since it assumes linearity.

The proposed model's state equation describing the dynamic evolution of the hidden states vector is expressed through non-negative definite quadratic forms and, in fact, represents a non-negative valued Markovian stochastic process of order one. The proposed method provides a constrained optimization problem for which stationary points are derived and conditions for feasibility are provided. The proposed methodology exhibits a lower computational load when compared to other nonlinear filtering methods.

References

1. Iosifescu, M. *Finite Markov Processes and Applications*; John Wiley & Son: New York, NY, USA, 1980.
2. Isaacson, D.; Madsen, R. *Markov Chains Theory and Applications*; John Wiley & Son: New York, NY, USA, 1976.
3. Seneta, E. *Non-Negative Matrices and Markov Chains*; Springer: Berlin/Heidelberg, Germany, 1981.
4. Vassiliou, P.-C.G. *Discrete-Time Asset Pricing Models in Applied Stochastic Finance*; John Wiley & Son: New York, NY, USA, 2010.
5. Meyn, S.; Tweedie, R. *Markov Chains and Stochastic Stability*; Cambridge University Press: Cambridge, UK, 2009.
6. Vassiliou, P.-C.G. The evolution of the theory of non-homogeneous Markov systems. *Appl. Stoch. Models Data Anal.* **1997**, *13*, 159–176. [CrossRef]
7. Livingston, S. Geometric ergodicity of the random walk Metropolis with position-independent proposal covariance. *Mathematics* **2021**, *9*, 341. [CrossRef]
8. Vassiliou, P.-C.G. Non-homogeneous Markov set systems. *Mathematics* **2021**, *9*, 471. [CrossRef]
9. Prysyazhnyk, K.; Bazylevych, I.; Mitkvova, L.; Ivanochko, I. Period-Life of a Branching Process with Migration and continuous time. *Mathematics* **2021**, *9*, 868. [CrossRef]
10. Ruiz-Castro, J. Optimizing a multi-state cold-standby system with multiple vacations in the repair and loss of units. *Mathematics* **2021**, *9*, 913. [CrossRef]
11. McClean, S. Using Markov models to characterize and predict process target compliance. *Mathematics* **2021**, *9*, 1187. [CrossRef]
12. Esquivel, M.; Krasii, N.; Guerreiro, G. Open Markov type population models: From discrete to continuous time. *Mathematics* **2021**, *9*, 1496. [CrossRef]
13. Borisov, A.; Bosov, A.; Miller, G.; Sokolov, I. Partial diffusion Markov model of heterogeneous TCP link: Optimization with incomplete information. *Mathematics* **2021**, *9*, 1632. [CrossRef]
14. Stavropoulos, N.; Papadopoulou, A.; Kolias, P. Evaluating the efficiency of off-ball screens in elite basketball teams via second order Markov modelling. *Mathematics* **2021**, *9*, 1991. [CrossRef]
15. Howard, R. *Dynamic Probabilistic Systems: Semi-Markov and Decision Processes*; Dover Publications: New York, NY, USA, 2007.
16. Vassiliou, P.-C.G.; Papadopoulou, A.A. Non-homogeneous semi-Markov systems and maintainability of the state sizesl teams via second order Markov modelling. *J. Appl. Probab.* **1992**, *29*, 519–534. [CrossRef]
17. Verbeken, B.; Guerry, M.A. Discrete time hybrid semi-Markov models in Manpower planning. *Mathematics* **2021**, *9*, 1681. [CrossRef]
18. Georgiou, A.; Papadopoulou, A.; Kolias, P.; Palikrousis, H.; Farmakioti, E. On state occupancies, first passage times and duration in non-homogeneous semi-Markov chains. *Mathematics* **2021**, *9*, 1745. [CrossRef]
19. Barbu, V.; D' Amico, G.; Gkelsinis, T. Sequential interval reliability for discrete-time homogeneous semi-Markov repairable systems. *Mathematics* **2021**, *9*, 1997. [CrossRef]
20. Karlin, S.; Taylor, H. *A First Course in Stochastic Processes*; Academic Press: New York, NY, USA, 1975.
21. Cox, D.; Miller, H. *The Theory of Stochastic Processes*; Chapman and Hall: London, UK, 1965.
22. Doob, J. *Stochastic Processes*; All Time Classic Series; John Wiley & Son: New York, NY, USA, 1965.
23. Grimmett, G.; Stirzaker, D. *Probability and Random Processes*, 3rd ed.; Oxford University Press: Oxford, UK, 2001.
24. Dirma, M.; Paukstys, S.; Siaulys, J. Tails of the Moments for sums with dominatedly varying random summands. *Mathematics* **2021**, *9*, 824. [CrossRef]
25. Lykoy, R.; Tsaklidis, G. Particle filtering: A priori estimation of observational errors of a state-space model with linear observation equations. *Mathematics* **2021**, *9*, 1445. [CrossRef]
26. Theodosiadou, O.; Tsaklidis, G. State space modelling with non-negativity constraints using quadratic forms. *Mathematics* **2021**, *9*, 1908. [CrossRef]

mathematics MDPI

Article

Geometric Ergodicity of the Random Walk Metropolis with Position-Dependent Proposal Covariance

Samuel Livingstone

Department of Statistical Science, University College London, London WC1E 6BT, UK; samuel.livingstone@ucl.ac.uk

Abstract: We consider a Metropolis–Hastings method with proposal $\mathcal{N}(x, hG(x)^{-1})$, where x is the current state, and study its ergodicity properties. We show that suitable choices of $G(x)$ can change these ergodicity properties compared to the Random Walk Metropolis case $\mathcal{N}(x, h\Sigma)$, either for better or worse. We find that if the proposal variance is allowed to grow unboundedly in the tails of the distribution then geometric ergodicity can be established when the target distribution for the algorithm has tails that are heavier than exponential, in contrast to the Random Walk Metropolis case, but that the growth rate must be carefully controlled to prevent the rejection rate approaching unity. We also illustrate that a judicious choice of $G(x)$ can result in a geometrically ergodic chain when probability concentrates on an ever narrower ridge in the tails, something that is again not true for the Random Walk Metropolis.

Keywords: Monte Carlo; MCMC; Markov chains; computational statistics; bayesian inference

Citation: Livingstone, S. Geometric Ergodicity of the Random Walk Metropolis with Position-Dependent Proposal Covariance. *Mathematics* **2021**, *9*, 341. https://doi.org/10.3390/math9040341

Academic Editor: Panagiotis-Christos Vassiliou
Received: 19 January 2021
Accepted: 4 February 2021
Published: 8 February 2021

Publisher's Note: MDPI stays neutral with regard to jurisdictional claims in published maps and institutional affiliations.

Copyright: © 2021 by the author. Licensee MDPI, Basel, Switzerland. This article is an open access article distributed under the terms and conditions of the Creative Commons Attribution (CC BY) license (https://creativecommons.org/licenses/by/4.0/).

1. Introduction

Markov chain Monte Carlo (MCMC) methods are techniques for estimating expectations with respect to some probability measure $\pi(\cdot)$, which need not be normalised. This is done by sampling a Markov chain which has limiting distribution $\pi(\cdot)$, and computing empirical averages. A popular form of MCMC is the Metropolis–Hastings algorithm [1,2], where at each time step a 'proposed' move is drawn from some candidate distribution, and then accepted with some probability, otherwise the chain stays at the current point. Interest lies in finding choices of candidate distribution that will produce sensible estimators for expectations with respect to $\pi(\cdot)$.

The quality of these estimators can be assessed in many different ways, but a common approach is to understand conditions on $\pi(\cdot)$ that will result in a chain which converges to its limiting distribution at a *geometric* rate. If such a rate can be established, then a Central Limit Theorem will exist for expectations of functionals with finite second absolute moment under $\pi(\cdot)$ if the chain is reversible.

A simple yet often effective choice is a symmetric candidate distribution centred at the current point in the chain (with a fixed variance), resulting in the *Random Walk Metropolis* (RWM) (e.g., [3]). The convergence properties of a chain produced by the RWM are well-studied. In one dimension, essentially convergence is geometric if $\pi(x)$ decays at an exponential or faster rate in the tails [4], while in higher dimensions an additional curvature condition is required [5]. Slower rates of convergence have also been established in the case of heavier tails [6].

Recently, some MCMC methods were proposed which generalise the RWM, whereby proposals are still centred at the current point x and symmetric, but the variance changes with x [7–11]. An extension to infinite-dimensional Hilbert spaces is also suggested in Reference [12]. The motivation is that the chain can become more 'local', perhaps making larger jumps when out in the tails, or mimicking the local dependence structure of $\pi(\cdot)$ to propose more intelligent moves. Designing MCMC methods of this nature is particularly relevant for modern Bayesian inference problems, where posterior distributions are often high dimensional and exhibit nonlinear correlations [13]. We term this approach the *Position-dependent Random Walk Metropolis* (PDRWM), although technically this is a misnomer, since proposals are no longer random walks. Other choices of candidate distribution designed with distributions that exhibit nonlinear correlations were introduced in Reference [13]. Although powerful, these require derivative information for $\log \pi(x)$, something which can be unavailable in modern inference problems (e.g., [14]). We note that no such information is required for the PDRWM, as shown by the particular cases suggested in References [7–11]. However, there are relations between the approaches, to the extent that understanding how the properties of the PDRWM differ from the standard RWM should also aid understanding of the methods introduced in Reference [13].

In this article, we consider the convergence rate of a Markov chain generated by the PDRWM to its limiting distribution. Our main interest lies in whether this generalisation can change these *ergodicity* properties compared to the standard RWM with fixed covariance. We focus on the case in which the candidate distribution is Gaussian, and illustrate that such changes can occur in several different ways, either for better or worse. Our aim is not to give a complete characterisation of the approach, but rather to illustrate the possibilities through carefully chosen examples, which are known to be indicative of more general behaviour.

In Section 2 necessary concepts about Markov chains are briefly reviewed, before the PDRWM is introduced in Section 3. Some results in the one-dimensional case are given in Section 4, before a higher-dimensional model problem is examined in Section 5. Throughout $\pi(\cdot)$ denotes a probability measure (we use the terms probability measure and distribution synonymously), and $\pi(x)$ its density with respect to Lebesgue measure dx.

Since an early version of this work appeared online, some contributions to the literature were made that are worthy of mention. A Markov kernel constructed as a state-dependent mixture is introduced in Reference [15] and its properties are studied in some cases that are similar in spirit to the model problem of Section 5. An algorithm called *Directional Metropolis–Hastings*, which encompasses a specific instance of the PDRWM, is introduced and studied in Reference [16], and a modification of the same idea is used to develop the *Hop* kernel within the *Hug and Hop* algorithm of Reference [17]. Kamatani considers an algorithm designed for the infinite-dimensional setting in Reference [18] of a similar design to that discussed in Reference [12] and studies the ergodicity properties.

2. Markov Chains and Geometric Ergodicity

We will work on the Borel space $(\mathcal{X}, \mathcal{B})$, with $\mathcal{X} \subset \mathbb{R}^d$ for some $d \geq 1$, so that each $X_t \in \mathcal{X}$ for a discrete-time Markov chain $\{X_t\}_{t\geq 0}$ with time-homogeneous transition kernel $P : \mathcal{X} \times \mathcal{B} \to [0,1]$, where $P(x, A) = \mathbb{P}[X_{i+1} \in A | X_i = x]$ and $P^n(x, A)$ is defined similarly for X_{i+n}. All chains we consider will have invariant distribution $\pi(\cdot)$, and be both π-irreducible and aperiodic, meaning $\pi(\cdot)$ is the limiting distribution from π-almost any starting point [19]. We use $|\cdot|$ to denote the Euclidean norm.

In Markov chain Monte Carlo the objective is to construct estimators of $\mathbb{E}_\pi[f]$, for some $f : \mathcal{X} \to \mathbb{R}$, by computing

$$\hat{f}_n = \frac{1}{n} \sum_{i=1}^{n} f(X_i), \quad X_i \sim P^i(x_0, \cdot).$$

If $\pi(\cdot)$ is the limiting distribution for the chain then P will be *ergodic*, meaning $\hat{f}_n \xrightarrow{a.s.} \mathbb{E}_\pi[f]$ from π-almost any starting point. For finite n the quality of $\hat{f}_n$ intuitively depends on how quickly $P^n(x, \cdot)$ approaches $\pi(\cdot)$. We call the chain *geometrically ergodic* if

$$\|P^n(x, \cdot) - \pi(\cdot)\|_{TV} \leq M(x)\rho^n, \tag{1}$$

from π-almost any $x \in \mathcal{X}$, for some $M > 0$ and $\rho < 1$, where $\|\mu(\cdot) - \nu(\cdot)\|_{TV} := \sup_{A \in \mathcal{B}} |\mu(A) - \nu(B)|$ is the total variation distance between distributions $\mu(\cdot)$ and $\nu(\cdot)$ [19].

For π-reversible Markov chains geometric ergodicity implies that if $\mathbb{E}_\pi[f^2] < \infty$ for some $f : \mathcal{X} \to \mathbb{R}$, then

$$\sqrt{n}\left(\hat{f}_n - \mathbb{E}_\pi[f]\right) \xrightarrow{d} \mathcal{N}(0, v(P,f)), \tag{2}$$

for some asymptotic variance $v(P,f)$ [20]. Equation (2) enables the construction of asymptotic confidence intervals for $\hat{f}_n$.

In practice, geometric ergodicity does not guarantee that $\hat{f}_n$ will be a sensible estimator, as $M(x)$ can be arbitrarily large if the chain is initialised far from the typical set under $\pi(\cdot)$, and ρ may be very close to 1. However, chains which are not geometrically ergodic can often either get 'stuck' for a long time in low-probability regions or fail to explore the entire distribution adequately, sometimes in ways that are difficult to diagnose using standard MCMC diagnostics.

Establishing Geometric Ergodicity

It is shown in Chapter 15 of Reference [21] that Equation (1) is equivalent to the condition that there exists a *Lyapunov* function $V : \mathcal{X} \to [1, \infty)$ and some $\lambda < 1, b < \infty$ such that

$$PV(x) \le \lambda V(x) + b\mathbb{I}_C(x), \tag{3}$$

where $PV(x) := \int V(y)P(x,dy)$. The set $C \subset \mathcal{X}$ must be *small*, meaning that for some $m \in \mathbb{N}, \varepsilon > 0$ and probability measure $\nu(\cdot)$

$$P^m(x, A) \ge \varepsilon\nu(A), \tag{4}$$

for any $x \in C$ and $A \in \mathcal{B}$. Equations (3) and (4) are referred to as *drift* and *minorisation* conditions. Intuitively, C can be thought of as the centre of the space, and Equation (3) ensures that some one dimensional projection of $\{X_t\}_{t\ge0}$ drifts towards C at a geometric rate when outside. In fact, Equation (3) is sufficient for the return time distribution to C to have geometric tails [21]. Once in C, (4) ensures that with some probability the chain forgets its past and hence *regenerates*. This regeneration allows the chain to couple with another initialised from $\pi(\cdot)$, giving a bound on the total variation distance through the *coupling inequality* (e.g., [19]). More intuition is given in Reference [22].

Transition kernels considered here will be of the *Metropolis–Hastings* type, given by

$$P(x, dy) = \alpha(x,y)Q(x,dy) + r(x)\delta_x(dy), \tag{5}$$

where $Q(x,dy) = q(y|x)dy$ is some candidate kernel, α is called the acceptance rate and $r(x) = 1 - \int \alpha(x,y)Q(x,dy)$. Here we choose

$$\alpha(x,y) = 1 \wedge \frac{\pi(y)q(x|y)}{\pi(x)q(y|x)}, \tag{6}$$

where $a \wedge b$ denotes the minimum of a and b. This choice implies that P satisfies detailed balance for $\pi(\cdot)$ [23], and hence the chain is π-reversible (note that other choices for α can result in non-reversible chains, see Reference[24] for details).

Roberts and Tweedie [5], following on from Reference[21], introduced the following regularity conditions.

Theorem 1. *(Roberts and Tweedie). Suppose that $\pi(x)$ is bounded away from 0 and ∞ on compact sets, and there exists $\delta_q > 0$ and $\varepsilon_q > 0$ such that for every x*

$$|x - y| \le \delta_q \Rightarrow q(y|x) \ge \varepsilon_q.$$

Then the chain with kernel (5) is μ^{Leb}-irreducible and aperiodic, and every nonempty compact set is small.

For the choices of Q considered in this article these conditions hold, and we will restrict ourselves to forms of $\pi(x)$ for which the same is true (apart from a specific case in Section 5). Under Theorem 1 then (1) only holds if a Lyapunov function $V : \mathcal{X} \to [1, \infty]$ with $\mathbb{E}_\pi[V] < \infty$ exists such that

$$\limsup_{|x|\to\infty} \frac{PV(x)}{V(x)} < 1. \tag{7}$$

when P is of the Metropolis–Hastings type, (7) can be written

$$\limsup_{|x|\to\infty} \int \left[\frac{V(y)}{V(x)} - 1\right] \alpha(x,y) Q(x,dy) < 0. \tag{8}$$

In this case, a simple criterion for lack of geometric ergodicity is

$$\limsup_{|x|\to\infty} r(x) = 1. \tag{9}$$

Intuitively this implies that the chain is likely to get 'stuck' in the tails of a distribution for large periods.

Jarner and Tweedie [25] introduce a necessary condition for geometric ergodicity through a *tightness* condition.

Theorem 2. *(Jarner and Tweedie). If for any $\varepsilon > 0$ there is a $\delta > 0$ such that for all $x \in \mathcal{X}$*

$$P(x, B_\delta(x)) > 1 - \varepsilon,$$

where $B_\delta(x) := \{y \in \mathcal{X} : d(x,y) < \delta\}$, then a necessary condition for P to produce a geometrically ergodic chain is that for some $s > 0$

$$\int e^{s|x|} \pi(dx) < \infty.$$

The result highlights that when $\pi(\cdot)$ is heavy-tailed the chain must be able to make very large moves and still be capable of returning to the centre quickly for (1) to hold.

3. Position-Dependent Random Walk Metropolis

In the RWM, $Q(x,dy) = q(y-x)dy$ with $q(y-x) = q(x-y)$, meaning (6) reduces to $\alpha(x,y) = 1 \wedge \pi(y)/\pi(x)$. A common choice is $Q(x,\cdot) = \mathcal{N}(x, h\Sigma)$, with Σ chosen to mimic the global covariance structure of $\pi(\cdot)$ [3]. Various results exist concerning the optimal choice of h in a given setting (e.g., [26]). It is straightforward to see that Theorem 2 holds here, so that the tails of $\pi(x)$ must be uniformly exponential or lighter for geometric ergodicity. In one dimension this is in fact a sufficient condition [4], while for higher dimensions additional conditions are required [5]. We return to this case in Section 5.

In the PDRWM $Q(x,\cdot) = \mathcal{N}(x, hG(x)^{-1})$, so (6) becomes

$$\alpha(x,y) = 1 \wedge \frac{\pi(y)|G(y)|^{\frac{1}{2}}}{\pi(x)|G(x)|^{\frac{1}{2}}} \exp\left(-\frac{1}{2}(x-y)^T[G(y) - G(x)](x-y)\right).$$

The motivation for designing such an algorithm is that proposals are more able to reflect the local dependence structure of $\pi(\cdot)$. In some cases this dependence may vary greatly in different parts of the state-space, making a global choice of Σ ineffective [9].

Readers familiar with differential geometry will recognise the volume element $|G(x)|^{1/2}dx$ and the linear approximations to the distance between x and y taken at each point through $G(x)$ and $G(y)$ if $\mathcal{X}$ is viewed as a Riemannian manifold with metric G.

We do not explore these observations further here, but the interested reader is referred to Reference [27] for more discussion.

The choice of $G(x)$ is an obvious question. In fact, specific variants of this method have appeared on many occasions in the literature, some of which we now summarise.

1. *Tempered Langevin diffusions* [8] $G(x) = \pi(x)I$. The authors highlight that the diffusion with dynamics $dX_t = \pi^{-\frac{1}{2}}(X_t)dW_t$ has invariant distribution $\pi(\cdot)$, motivating the choice. The method was shown to perform well for a bi-modal $\pi(x)$, as larger jumps are proposed in the low density region between the two modes.
2. *State-dependent Metropolis* [7] $G(x) = (1+|x|)^{-b}$. Here the intuition is simply that $b > 0$ means larger jumps will be made in the tails. In one dimension the authors compare the expected squared jumping distance $\mathbb{E}[(X_{i+1} - X_i)^2]$ empirically for chains exploring a $\mathcal{N}(0,1)$ target distribution, choosing b adaptively, and found $b \approx 1.6$ to be optimal.
3. *Regional adaptive Metropolis–Hastings* [7,11]. $G(x)^{-1} = \sum_{i=1}^{m} \mathbb{I}(x \in \mathcal{X}_i)\Sigma_i$. In this case the state-space is partitioned into $\mathcal{X}_1 \cup ... \cup \mathcal{X}_m$, and a different proposal covariance Σ_i is learned adaptively in each region $1 \leq i \leq m$. An extension which allows for some errors in choosing an appropriate partition is discussed in [11]
4. *Localised Random Walk Metropolis* [10]. $G(x)^{-1} = \sum_{k=1}^{m} \check{q}_\theta(k|x)\Sigma_k$. Here $\check{q}_\theta(k|x)$ are weights based on approximating $\pi(x)$ with some mixture of Normal/Student's t distributions, using the approach suggested in Reference [28]. At each iteration of the algorithm a mixture component k is sampled from $\check{q}_\theta(\cdot|x)$, and the covariance Σ_k is used for the proposal $Q(x, dy)$.
5. *Kernel adaptive Metropolis–Hastings* [9]. $G(x)^{-1} = \gamma^2 I + \nu^2 M_x H M_x^T$, where $M_x = 2[\nabla_x k(z_1, x), ..., \nabla_x k(z_n, x)]$ for some kernel function k and n past samples $\{z_1, ..., z_n\}$, $H = I - (1/n)\mathbf{1}_{n\times n}$ is a centering matrix (the $n \times n$ matrix $\mathbf{1}_{n\times n}$ has 1 as each element), and γ, ν are tuning parameters. The approach is based on performing nonlinear principal components analysis on past samples from the chain to learn a local covariance. Illustrative examples for the case of a Gaussian kernel show that $M_x H M_x^T$ acts as a weighted empirical covariance of samples z, with larger weights given to the z_i which are closer to x [9].

The latter cases also motivate any choice of the form

$$G(x)^{-1} = \sum_{i=1}^{n} w(x, z_i)(z_i - x)^T (z_i - x)$$

for some past samples $\{z_1, ..., z_n\}$ and weight function $w : \mathcal{X} \times \mathcal{X} \to [0, \infty)$ with $\sum_i w(x, z_i) = 1$ that decays as $|x - z_i|$ grows, which would also mimic the local curvature of $\pi(\cdot)$ (taking care to appropriately regularise and diminish adaptation so as to preserve ergodicity, as outlined in Reference [10]).

Some of the above schemes are examples of adaptive MCMC, in which a candidate from among a family of Markov kernels $\{P_\theta : \theta \in \Theta\}$ is selected by learning the parameter $\theta \in \Theta$ during the simulation [10]. Additional conditions on the adaptation process (i.e., the manner in which θ is learned) are required to establish ergodicity results for the resulting stochastic processes. We consider the decisions on how to learn θ appropriately to be a separate problem and beyond the scope of the present work, and instead focus attention on establishing geometric ergodicity of the base kernels P_θ for any fixed $\theta \in \Theta$. We note that this is typically a pre-requisite for establishing convergence properties of any adaptive MCMC method [10].

4. Results in One Dimension

Here we consider two different general scenarios as $|x| \to \infty$, i) $G(x)$ is bounded above and below, and ii) $G(x) \to 0$ at some specified rate. Of course there is also the possibility that $G(x) \to \infty$, though intuitively this would result in chains that spend a long time in the tails of a distribution, so we do not consider it (if $G(x) \to \infty$ then chains will in fact exhibit the *negligible moves* property studied in Reference [29]). Proofs to Propositions in Sections 4 and 5 can be found in Appendix A.

We begin with a result that emphasizes that a growing variance is a necessary requirement for geometric ergodicity in the heavy-tailed case.

Proposition 1. *If $G(x) \geq \sigma^{-2}$ for some $\sigma^{-2} > 0$, then unless $\int e^{\eta|x|}\pi(dx) < \infty$ for some $\eta > 0$ the PDRWM cannot produce a geometrically ergodic Markov chain.*

The above is a simple extension of a result that is well-known in the RWM case. Essentially the tails of the distribution should be exponential or lighter to ensure fast convergence. This motivates consideration of three different types of behaviour for the tails of $\pi(\cdot)$.

Assumption 1. *The density $\pi(x)$ satisfies one of the following tail conditions for all $y, x \in \mathcal{X}$ such that $|y| > |x| > t$, for some finite $t > 0$.*

1. $\pi(y)/\pi(x) \leq \exp\{-a(|y| - |x|)\}$ *for some $a > 0$*
2. $\pi(y)/\pi(x) \leq \exp\{-a(|y|^{\beta} - |x|^{\beta})\}$ *for some $a > 0$ and $\beta \in (0,1)$*
3. $\pi(y)/\pi(x) \leq (|x|/|y|)^{p}$ *for some $p > 1$.*

Naturally Assumption 1 implies 2 and Assumption 2 implies 3. If Assumption 1 is not satisfied then $\pi(\cdot)$ is generally called *heavy-tailed*. When $\pi(x)$ satisfies Assumption 2 or 3 but not 1, then the RWM typically fails to produce a geometrically ergodic chain [4]. We show in the sequel, however, that this is not always the case for the PDRWM. We assume the below assumptions for $G(x)$ to hold throughout this section.

Assumption 2. *The function $G : \mathcal{X} \to (0, \infty)$ is bounded above by some $\sigma_b^{-2} < \infty$ for all $x \in \mathcal{X}$, and bounded below for all $x \in \mathcal{X}$ with $|x| < t$, for some $t > 0$.*

The heavy-tailed case is known to be a challenging scenario, but the RWM will produce a geometrically ergodic Markov chain if $\pi(x)$ is log-concave. Next we extend this result to the case of sub-quadratic variance growth in the tails.

Proposition 2. *If $\exists r < \infty$ such that $G(x) \propto |x|^{-\gamma}$ whenever $|x| > r$, then the PDRWM will produce a geometrically ergodic chain in both of the following cases:*

1. *$\pi(x)$ satisfies Assumption 1 and $\gamma \in [0,2)$*
2. *$\pi(x)$ satisfies Assumption 2 for some $\beta \in (0,1)$ and $\gamma \in (2(1-\beta), 2)$*

The second part of Proposition 2 is not true for the RWM, for which Assumption 2 alone is not sufficient for geometric ergodicity [4].

We do not provide a complete proof that the PDRWM will not produce a geometrically ergodic chain when only Assumption 3 holds and $G(x) \propto |x|^{-\gamma}$ for some $\gamma < 2$, but do show informally that this will be the case. Assuming that in the tails $\pi(x) \propto |x|^{-p}$ for some $p > 1$ then for large x

$$\alpha(x, x + cx^{\gamma/2}) = 1 \wedge \left(\frac{x}{x + cx^{\gamma/2}}\right)^{p+\gamma/2} \exp\left(-\frac{c^2 x^{\gamma}}{2h}\left[\frac{1}{(x + cx^{\gamma/2})^{\gamma}} - \frac{1}{x^{\gamma}}\right]\right). \tag{10}$$

The first expression on the right hand side converges to 1 as $x \to \infty$, which is akin to the case of fixed proposal covariance. The second term will be larger than one for $c > 0$ and less than one for $c < 0$. So the algorithm will exhibit the same 'random walk in the tails' behaviour which is often characteristic of the RWM in this scenario, meaning that the acceptance rate fails to enforce a geometric drift back into the centre of the space.

When $\gamma = 2$ the above intuition will not necessarily hold, as the terms in Equation (10) will be roughly constant with x. When only Assumption 3 holds, it is, therefore, tempting to make the choice $G(x) = x^{-2}$ for $|x| > r$. Informally we can see that such behaviour may lead to a favourable algorithm if a small enough h is chosen. For any fixed $x > r$ a typical proposal will now take the form $y = (1 + \xi\sqrt{h})x$, where $\xi \sim N(0,1)$. It therefore holds that

$$y = e^{\xi\sqrt{h}}x + r(x,h,\xi), \tag{11}$$

where for any fixed x and ξ the term $r(x,h,\xi)/\sqrt{h} \to 0$ as $h \to 0$. The first term on the right-hand side of Equation (11) corresponds to the proposal of the *multiplicative Random Walk Metropolis*, which is known to be geometrically ergodic under Assumption 3 (e.g., [3]), as this equates to taking a logarithmic transformation of x, which 'lightens' the tails of the target density to the point where it becomes log-concave. So in practice we can expect good performance from this choice of $G(x)$. The above intuition does not, however, provide enough to establish geometric ergodicity, as the final term on the right-hand side of (11) grows unboundedly with x for any fixed choice of h. The difference between the acceptance rates of the multiplicative Random Walk Metropolis and the PDRWM with $G(x) = x^{-2}$ will be the exponential term in Equation (10). This will instead become polynomial by letting the proposal noise ξ follow a distribution with polynomial tails (e.g., student's t), which is known to be a favourable strategy for the RWM when only Assumption 3 holds [6]. One can see that if the heaviness of the proposal distribution is carefully chosen then the acceptance rate may well enforce a geometric drift into the centre of the space, though for brevity we restrict attention to Gaussian proposals in this article.

The final result of this section provides a note of warning that lack of care in choosing $G(x)$ can have severe consequences for the method.

Proposition 3. *If $G(x)x^2 \to 0$ as $|x| \to \infty$, then the PDRWM will not produce a geometrically ergodic Markov chain.*

The intuition for this result is straightforward when explained. In the tails, typically $|y - x|$ will be the same order of magnitude as $\sqrt{G(x)^{-1}}$, meaning $|y - x|/|x|$ grows arbitrarily large as $|x|$ grows. As such, proposals will 'overshoot' the typical set of the distribution, sending the sampler further out into the tails, and will therefore almost always be rejected. The result can be related superficially to a lack of geometric ergodicity for Metropolis–Hastings algorithms in which the proposal mean is comprised of the current state translated by a drift function (often based in $\nabla \log \pi(x)$) when this drift function grows faster than linearly with $|x|$ (e.g., [30,31]).

5. A Higher-Dimensional Case Study

An easy criticism of the above analysis is that the one-dimensional scenario is sometimes not indicative of the more general behaviour of a method. We note, however, that typically the geometric convergence properties of Metropolis–Hastings algorithms do carry over somewhat naturally to more than one dimension when $\pi(\cdot)$ is suitably regular (e.g., [5,32]). Because of this we expect that the growth conditions specified above could be supplanted onto the determinant of $G(x)$ when the dimension is greater than one (leaving the details of this argument for future work).

A key difference in the higher-dimensional setting is that $G(x)$ now dictates both the *size* and *direction* of proposals. In the case $G(x)^{-1} = \Sigma$, some additional regularity conditions on $\pi(x)$ are required for geometric ergodicity in more than one dimension, outlined in References [5,32]. An example is also given in Reference [5] of the simple two-dimensional density $\pi(x,y) \propto \exp(-x^2 - y^2 - x^2y^2)$, which fails to meet these criteria. The difficult models are those for which probability concentrates on a ridge in the tails, which becomes ever narrower as $|x|$ increases. In this instance, proposals from the RWM are less and less likely to be accepted as $|x|$ grows. Another well-known example of this phenomenon is the *funnel* distribution introduced in Reference [33].

To explore the behaviour of the PDRWM in this setting, we design a model problem, the *staircase* distribution, with density

$$\mathfrak{s}(x) \propto 3^{-\lfloor x_2 \rfloor} \mathbb{I}_R(x), \quad R := \{y \in \mathbb{R}^2; y_2 \geq 1, |y_1| \leq 3^{1-\lfloor y_2 \rfloor}\}, \tag{12}$$

where $\lfloor z \rfloor$ denotes the integer part of $z > 0$. Graphically the density is a sequence of cuboids on the upper-half plane of $\mathbb{R}^2$ (starting at $y_2 = 1$), each centred on the vertical axis, with each successive cuboid one third of the width and height of the previous. The density resembles an ever narrowing staircase, as shown in Figure 1.

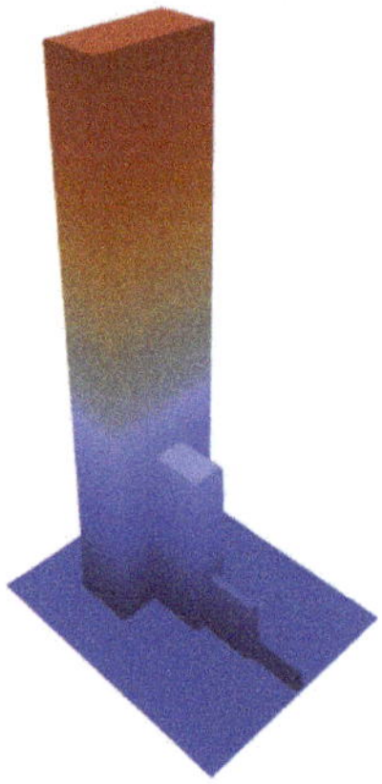

Figure 1. The staircase distribution, with density given by Equation (12).

We denote by Q_R the proposal kernel associated with the Random Walk Metropolis algorithm with fixed covariance $h\Sigma$. In fact, the specific choice of h and Σ does not matter provided that the result is positive-definite. For the PDRWM we denote by Q_P the proposal kernel with covariance matrix

$$hG(x)^{-1} = \begin{pmatrix} 3^{-2\lfloor x_2 \rfloor} & 0 \\ 0 & 1 \end{pmatrix},$$

which will naturally adapt the scale of the first coordinate to the width of the ridge.

Proposition 4. *The Metropolis–Hastings algorithm with proposal Q_R does not produce a geometrically ergodic Markov chain when $\pi(x) = \mathfrak{s}(x)$.*

The design of the PDRWM proposal kernel Q_P in this instance is such that the proposal covariance reduces at the same rate as the width of the stairs, therefore naturally adapting the proposal to the width of the ridge on which the density concentrates. This state-dependent adaptation results in a geometrically ergodic chain, as shown in the below result.

Proposition 5. *The Metropolis–Hastings algorithm with proposal Q_P produces a geometrically ergodic Markov chain when $\pi(x) = \mathfrak{s}(x)$.*

6. Discussion

In this paper we have analysed the ergodic behaviour of a Metropolis–Hastings method with proposal kernel $Q(x, \cdot) = \mathcal{N}(x, hG(x)^{-1})$. In one dimension we have characterised the behaviour in terms of growth conditions on $G(x)^{-1}$ and tail conditions on the target distribution, and in higher dimensions a carefully constructed model problem is discussed. The fundamental question of interest was whether generalising an existing Metropolis–Hastings method by allowing the proposal covariance to change with position can alter the ergodicity properties of the sampler. We can confirm that this is indeed possible, either for the better or worse, depending on the choice of covariance. The take home points for practitioners are (i) lack of sufficient care in the design of $G(x)$ can have severe consequences (as in Proposition 3), and (ii) careful choice of $G(x)$ can have much more beneficial ones, perhaps the most surprising of which are in the higher-dimensional setting, as shown in Section 5.

We feel that such results can also offer insight into similar generalisations of different Metropolis–Hastings algorithms (e.g., [13,34]). For example, it seems intuitive that any method in which the variance grows at a faster than quadratic rate in the tails is unlikely to produce a geometrically ergodic chain. There are connections between the PDRWM and some extensions of the Metropolis-adjusted Langevin algorithm [34], the ergodicity properties of which are discussed in Reference [35]. The key difference between the schemes is the inclusion of the drift term $G(x)^{-1}\nabla \log \pi(x)/2$ in the latter. It is this term which in the main governs the behaviour of the sampler, which is why the behaviour of the PDRWM is different to this scheme. Markov processes are also used in a wide variety of application areas beyond the design of Metropolis–Hastings algorithms (e.g., [36]), and we hope that some of the results established in the present work prove to be beneficial in some of these other settings.

We can apply these results to the specific variants discussed in Section 3. Provided that sensible choices of regions/weights are made and that an adaptation scheme which obeys the diminishing adaptation criterion is employed, the Regional adaptive Metropolis–Hastings, Locally weighted Metropolis and Kernel-adaptive Metropolis–Hastings samplers should all satisfy $G(x) \to \Sigma$ as $|x| \to \infty$, meaning they can be expected to inherit the ergodicity properties of the standard RWM (the behaviour in the centre of the space, however, will likely be different). In the State-dependent Metropolis method provided $b < 2$ the sampler should also behave reasonably. Whether or not a large enough value of b would be found by a particular adaptation rule is not entirely clear, and this could be an interesting direction of further study. The Tempered Langevin diffusion scheme, however, will fail to produce a geometrically ergodic Markov chain whenever the tails of $\pi(x)$ are lighter than that of a Cauchy distribution. To allow reasonable tail exploration when this is the case, two pragmatic options would be to upper bound $G(x)^{-1}$ manually or use this scheme in conjunction with another, as there is evidence that the sampler can perform favourably when exploring the centre of a distribution [8]. None of the specific variants discussed here are able to mimic the local curvature of the $\pi(x)$ in the tails, so as to enjoy the favourable behaviour exemplified in Proposition 5. This is possible using Hessian information as in Reference [13], but should also be possible in some cases using appropriate surrogates.

Funding: This research was supported by a UCL IMPACT PhD scholarship co-funded by Xerox Research Centre Europe and EPSRC.

Institutional Review Board Statement: Not applicable.

Informed Consent Statement: Not applicable.

Data Availability Statement: Not Applicable

Acknowledgments: The author thanks Alexandros Beskos, Krzysztof Łatuszyński and Gareth Roberts for several useful discussions, Michael Betancourt for proofreading the paper, and Mark Girolami for general supervision and guidance.

Conflicts of Interest: The author declares no conflict of interest.

Appendix A. Proofs

Proof of Proposition 1. In this case, for any choice of $\varepsilon > 0$ there is a $\delta > 0$ such that $Q(x, B_\delta(x)) > 1 - \varepsilon$. Noting that $P(x, B_\delta(x)) \geq Q(x, B_\delta(x))$ when P is of Metropolis–Hastings type, Theorem 2 can be applied directly. □

Proof of Proposition 2. For the log-concave case, take $V(x) = e^{s|x|}$ for some $s > 0$, and let B_A denote the integral (8) over the set A. We first break up $\mathcal{X}$ into $(-\infty, 0] \cup (0, x - cx^{\frac{\gamma}{2}}] \cup (x - cx^{\frac{\gamma}{2}}, x + cx^{\frac{\gamma}{2}}] \cup (x + cx^{\frac{\gamma}{2}}, x + cx^{\gamma}] \cup (x + cx^{\gamma}, \infty)$ for some $x > 0$ and fixed constant $c \in (0, \infty)$, and show that the integral is strictly negative on at least one of these sets, and can be made arbitrarily small as $x \to \infty$ on all others. The $-\infty$ case is analogous from the tail conditions on $\pi(x)$. From the conditions we can choose $x > r$ and therefore write $G(x)^{-1} = \eta x^{\gamma}$ for some fixed $\eta < \infty$.

On $(-\infty, 0]$, we have

$$\begin{aligned} B_{(-\infty,0]} &= e^{-sx} \int_{-\infty}^{0} e^{s|y|} \alpha(x,y) Q(x,dy) - \int_{-\infty}^{0} \alpha(x,y) Q(x,dy), \\ &\leq e^{-sx} \int_{0}^{\infty} e^{sy} Q(-x,dy). \end{aligned}$$

The integral is now proportional to the moment generating function of a truncated Gaussian distribution (see Appendix B), so is given by

$$e^{-sx + h\eta x^{\gamma} s^2/2} \Big[1 - \Phi\Big(x^{1-\gamma/2} / \sqrt{h\eta} - \sqrt{h\eta} s x^{\gamma/2} \Big) \Big].$$

A simple bound on the error function is $\sqrt{2\pi} x \Phi^c(x) < e^{-x^2/2}$ [37], so setting $\vartheta = x^{1-\gamma/2} / \sqrt{h\eta} - \sqrt{h\eta} s x^{\gamma/2}$ we have

$$\begin{aligned} B_{(-\infty,0]} &\leq \frac{1}{\sqrt{2\pi}} \exp\left(-2sx + \frac{h\eta s^2}{2} x^{\gamma} - \frac{1}{2}\left(\frac{1}{h\eta} x^{2-\gamma} - 2sx + h\eta s^2 x^{\gamma} \right) + \log \vartheta \right), \\ &= \frac{1}{\sqrt{2\pi}} \exp\left(-sx - \frac{1}{2h\eta} x^{2-\gamma} + \log \vartheta \right). \end{aligned}$$

which $\to 0$ as $x \to \infty$, so can be made arbitrarily small.

On $(0, x - cx^{\gamma/2}]$, note that $e^{s(|y| - |x|)} - 1$ is clearly negative throughout this region provided that $c < x^{1-\gamma/2}$, which can be enforced by choosing x large enough for any given $c < \infty$. So the integral is straightforwardly bounded as $B_{(0, x - cx^{\gamma/2}]} \leq 0$ for all $x \in \mathcal{X}$.

On $(x - cx^{\gamma/2}, x + cx^{\gamma/2}]$, provided $x - cx^{\gamma/2} > r$ then for any y in this region we can either upper or lower bound $\alpha(x,y)$ with the expression

$$\exp\left(-a(y-x) + \frac{\gamma}{2} \log\left| \frac{x}{y} \right| - \frac{1}{2h\eta} \left[(x-y)^2 y^{-\gamma} - (x-y)^2 x^{-\gamma} \right] \right).$$

A Taylor expansion of $y^{-\gamma}$ about x gives

$$y^{-\gamma} = x^{-\gamma} - \gamma x^{-\gamma-1}(y-x) + \gamma(\gamma+1) x^{-\gamma-2}(y-x)^2 + \ldots$$

and multiplying by $(y-x)^2$ gives

$$(y-x)^2 y^{-\gamma} = \frac{(y-x)^2}{x^{\gamma}} - \gamma \frac{(y-x)^3}{x^{\gamma+1}} + \gamma(\gamma+1) \frac{(y-x)^4}{x^{\gamma+2}} + \ldots$$

If $|y - x| = cx^{\gamma/2}$ then this is:

$$\frac{c^2 x^{\gamma}}{x^{\gamma}} - \gamma \frac{c^3 x^{3\gamma/2}}{x^{\gamma+1}} + \gamma(\gamma+1) \frac{c^4 x^{2\gamma}}{x^{\gamma+2}} + \ldots$$

As $\gamma < 2$ then $3\gamma/2 < \gamma + 1$, and similarly for successive terms, meaning each gets smaller as $|x| \to \infty$. So we have for large x, $y \in (x - cx^{\gamma/2}, x + cx^{\gamma/2})$ and any $\delta > 0$

$$(y-x)^2 y^{-\gamma} \geq \frac{(y-x)^2}{x^{\gamma}} - \gamma \frac{(y-x)^3}{x^{\gamma+1}} - 2h\eta\delta. \tag{A1}$$

So we can analyse how the acceptance rate behaves. First note that for fixed $\epsilon > 0$

$$\alpha(x, x+\epsilon) \le \exp\left(-a\epsilon + \frac{\gamma}{2}\log\left|\frac{x}{x+\epsilon}\right| + \frac{1}{2h}\gamma\frac{\epsilon^3}{x^{\gamma+1}} + \delta\right) \to \exp(-a\epsilon + \delta),$$

recalling that δ can be made arbitrarily small. In fact, it holds that the $e^{-a\epsilon}$ term will be dominant for any ϵ for which $\epsilon^3/x^{\gamma+1} \to 0$, i.e., any $\epsilon = o(x^{\gamma+1/3})$. If $\gamma < 2$ then $\epsilon = cx^{\gamma/2}$ satisfies this condition. So for any $y > x$ in this region we can choose an x such that

$$\alpha(x, y) \le \exp(-a(y-x) + \delta_x),$$

where $\delta_x \to 0$ as $x \to \infty$. Similarly we have (for any fixed $\epsilon > 0$)

$$\alpha(x, x-\epsilon) \ge \exp\left(a\epsilon + \frac{\gamma}{2}\log\left|\frac{x}{x-\epsilon}\right| - \frac{1}{2h}\gamma\frac{\epsilon^3}{x^{\gamma+1}} - \delta\right) \to \exp(a\epsilon - \delta).$$

So by a similar argument we have $\alpha(x, y) > 1$ here when $x \to \infty$. Combining gives

$$B_{(x-cx^{\gamma/2}, x+cx^{\gamma/2}]} \le \int_0^{cx^{\gamma/2}} \left[e^{(s-a)z+\delta_x} - e^{-az+\delta_x} + e^{-sz} - 1\right] q_x(dz),$$

where $q_x(\cdot)$ denotes a zero mean Gaussian distribution with the same variance as $Q(x, \cdot)$. Using the change of variables $z' = z/(h\eta x^{\gamma/2})$ we can write the above integral

$$\int_0^{\frac{c}{h\eta}} \left[e^{(s-a)h\eta x^{\gamma/2}z'+\delta_x} - e^{-ah\eta x^{\gamma/2}z+\delta_x} + e^{-sh\eta x^{\gamma/2}z'} - 1\right] \mu(dz)$$

where $\mu(\cdot)$ denotes a Gaussian distribution with zero mean and variance one. Provided $s < a$, then by dominated convergence as $x \to \infty$ this asymptotes to

$$-\int_0^{\frac{c}{h\eta}} \mu(dz) = -\frac{1}{2}\mathrm{erf}\left(\frac{c}{\sqrt{2}h\eta}\right) < 0,$$

where $\mathrm{erf}(z) := (2/\sqrt{\pi})\int_0^z e^{-t^2}dt$ is the Gaussian error function.

On $(x + cx^{\gamma/2}, x + cx^{\gamma}]$ we can upper bound the acceptance rate as

$$\alpha(x, y) \le \frac{\pi(y)}{\pi(x)} \exp\left(\frac{1}{2}\log\frac{|G(y)|}{|G(x)|} + \frac{G(x)}{2h}(x-y)^2\right)$$

If $y \ge x$ and $x > x_0$ we have

$$\alpha(x, y) \le \exp\left(-a(|y| - |x|) + \frac{1}{2h\eta}\frac{(x-y)^2}{x^\gamma}\right).$$

For $|y - x| = cx^\ell$ this becomes

$$\alpha(x, y) \le \exp\left(-acx^\ell + \frac{c^2}{2h\eta}x^{2\ell-\gamma}\right)$$

So provided $\gamma > \ell$ the first term inside the exponential will dominate the second for large enough x. In the equality case we have

$$\alpha(x, y) \le \exp\left(\left(\frac{c^2}{2h\eta} - a\right)cx^\gamma\right),$$

so provided we choose c such that $a > c^2/(2h\eta)$ then the acceptance rate will also decay exponentially. Because of this we have

$$B_{(x+cx^{\gamma/2},x+cx^{\gamma}]} \leq \int_{x+cx^{\gamma/2}}^{x+cx^{\gamma}} e^{s(y-x)}\alpha(x,y)Q(x,dy),$$
$$\leq e^{(c^2/(2h\eta)+s-a)cx^{\gamma/2}}Q(x,(x+cx^{\gamma/2},x+cx^{\gamma}]),$$

so provided $a > c^2/(2h\eta) + s$ then this term can be made arbitrarily small.

On $(x + cx^{\gamma}, \infty)$ using the same properties of truncated Gaussians we have

$$B_{(x+cx^{\gamma},\infty)} \leq e^{-sx}\int_{x+cx^{\gamma}}^{\infty} e^{sy}Q(x,dy),$$
$$= e^{s^2h\eta x^{\gamma}/2}\Phi^c\left(\left(\frac{c}{\sqrt{h\eta}} - \sqrt{h\eta}s\right)x^{\gamma}\right),$$

which can be made arbitrarily small provided that s is chosen to be small enough using the same simple bound on Φ^c as for the case of $B_{(-\infty,0]}$.

Combining gives that the integral (8) is bounded above by $-\text{erf}(c/\sqrt{2h^2\eta^2})/2$, which is strictly less than zero as c, h and η are all positive. This completes the proof under Assumption 1.

Under Assumption 2 the proof is similar. Take $V(x) = e^{s|x|^{\beta}}$, and divide $\mathcal{X}$ up into the same regions. Outside of $(x - cx^{\gamma/2}, x + cx^{\gamma/2}]$ the same arguments show that the integral can be made arbitrarily small. On this set, note that in the tails

$$(x + cx^{\ell})^{\beta} - x^{\beta} = \beta c x^{\ell+\beta-1} + \frac{\beta(\beta-1)}{2}c^2x^{2\ell+\beta-2} + ...$$

For $y - x = cx^{\ell}$, then for $\ell < 1 - \beta$ this becomes negligible. So in this case we further divide the typical set into $(x, x + cx^{1-\beta}] \cup (x + cx^{1-\beta}, x + cx^{\gamma/2})$. On $(x - cx^{1-\beta}, x + cx^{1-\beta})$ the integral is bounded above by $e^{-c_1}Q(x,(x - cx^{1-\beta}, x + cx^{1-\beta})) \to 0$, for some suitably chosen $c_1 > 0$. On $(x - cx^{\gamma/2}, x - cx^{1-\beta}] \cup (x + cx^{1-\beta}, x + cx^{\gamma/2}]$ then for $y > x$ we have $\alpha(x,y) \leq e^{-c_2(y^{\beta}-x^{\beta})}$, so we can use the same argument as in the the log-concave case to show that the integral will be strictly negative in the limit. □

Proof of Proposition 3. First note that in this case for any $g : \mathbb{R} \to (0,\infty)$ such that as $|x| \to \infty$ it holds that $g(x)/|x| \to \infty$ but $g(x)\sqrt{G(x)} \to 0$, then

$$Q(x,\{x - g(x), x + g(x)\}) = \Phi\left(g(x)\sqrt{G(x)}\right) - \Phi\left(-g(x)\sqrt{G(x)}\right) \to 0$$

as $|x| \to \infty$. The chain therefore has the property that $\mathbb{P}(\{|X_{i+1}| > g(X_i)/2\} \cup \{X_{i+1} = X_i\})$ can be made arbitrarily close to 1 as $|X_i|$ grows, which leads to two possible behaviours. If the form of $\pi(\cdot)$ enforces such large jumps to be rejected then $r(x) \to 1$ and lack of geometric ergodicity follows from (9). If this is not the case then the chain will be transient (this can be made rigorous using a standard Borel–Cantelli argument, see e.g., the proof of Theorem 12.2.2 on p. 299 of [21]). □

Proof of Proposition 4. It is sufficient to construct a sequence of points $x_p \in \mathbb{R}^2$ such that $|x_p| \to \infty$ as $p \to \infty$, and show that $r(x_p) \to 1$ in the same limit, then apply (9). Take $x_p = (0, p)$ for $p \in \mathbb{N}$. In this case

$$r(x_p) = 1 - \int \alpha(x_p, y)Q_R(x_p, dy)$$

Note that for every $\epsilon > 0$ there is a $\delta < \infty$ such that $Q(x_p, B_{\delta}^c(x_p)) < \epsilon$ for all x_p, where $B_{\delta}(x) := \{y \in \mathbb{R}^2 : |y - x| \leq \delta\}$. The set $A(x_p, \delta) := B_{\delta}(x_p) \cap R$ denotes the possible values of $y \in B_\delta(x)$ for which the acceptance rate is non-zero. Note that $A(x_p, \delta) \subset S(x_p, \delta) := \{y \in B_\delta(x_p) : |y_1| \le 3^{1-\lfloor p-\delta \rfloor}\}$, which is simply a strip that can be made arbitrarily narrow for any fixed δ by taking p large enough. Combining these ideas gives

$$\begin{aligned}\int \alpha(x_p, y) Q_R(x_p, dy) &\le \int_{A(x_p,\delta)} \alpha(x_p, y) Q_R(x_p, dy) + \epsilon \\ &\le Q_R(x_p, S(x_p, \delta)) + \epsilon.\end{aligned}$$

Both of the quantities on the last line can be made arbitrarily small by choosing p suitably large. Thus, $r(x_p) \to 1$ as $|x_p| \to \infty$, as required. □

Proof of Proposition 5. First note that $\inf_{x \in R} Q_P(x, R)$ is bounded away from zero, unlike in the case of Q_R, owing to the design of Q_P. The acceptance rate here simplifies, since for any $y \in R$

$$\frac{\mathfrak{s}(y)|G(y)|^{\frac{1}{2}}}{\mathfrak{s}(x)|G(x)|^{\frac{1}{2}}} = 1,$$

meaning only the expression $\exp\left(-\frac{1}{2}(y-x)^T[G(y) - G(x)](y-x)\right)$ needs to be considered. In this case the expression is simply

$$\exp\left(-\frac{1}{2}(3^{2\lfloor y_2 \rfloor} - 3^{2\lfloor x_2 \rfloor})(y_1 - x_1)^2\right).$$

Provided that $x_1 \neq y_1$, then when $1 \le \lfloor y_2 \rfloor < \lfloor x_2 \rfloor$ this expression is strictly greater than 1, whereas in the reverse case it is strictly less than one. The resulting Metropolis–Hastings kernel P using proposal kernel Q_P will therefore satisfy $\int y_2 P(x, dy) < x_2$ for large enough x_2, and hence geometric ergodicity follows by taking the Lyapunov function $V(x) = e^{s|x_2|}$ (which can be used here since the domain of x_1 is compact) and following an identical argument to that given on pages 404–405 of Reference [21] for the case of the proof of geometric ergodicity of the random walk on the half-line model for suitably small $s > 0$, taking the small set $C := [0,1] \times [1, r]$ for suitably large $r < \infty$ and $\nu(\cdot) = \int_\cdot \mathfrak{s}(x)dx$. □

Appendix B. Needed Facts about Truncated Gaussian Distributions

Here we collect some elementary facts used in the article. For more detail see e.g., [38]. If X follows a truncated Gaussian distribution $\mathcal{N}^T_{[a,b]}(\mu, \sigma^2)$ then it has density

$$f(x) = \frac{1}{\sigma Z_{a,b}} \phi\left(\frac{x-\mu}{\sigma}\right) \mathbb{I}_{[a,b]}(x),$$

where $\phi(x) = e^{-x^2/2}/\sqrt{2\pi}$, $\Phi(x) = \int_{-\infty}^x \phi(y)dy$ and $Z_{a,b} = \Phi((b-\mu)/\sigma) - \Phi((a-\mu)/\sigma)$. Defining $B = (b-\mu)/\sigma$ and $A = (a-\mu)/\sigma$, we have

$$\mathbb{E}[X] = \mu + \frac{\phi(A) - \phi(B)}{Z_{a,b}}\sigma$$

and

$$\mathbb{E}[e^{tX}] = e^{\mu t + \sigma^2 t^2/2}\left[\frac{\Phi(B - \sigma t) - \Phi(A - \sigma t)}{Z_{a,b}}\right].$$

In the special case $b = \infty$, $a = 0$ this becomes $e^{\mu t + \sigma^2 t^2/2}\Phi(\sigma t)/Z_{a,b}$.

References

1. Metropolis, N.; Rosenbluth, A.W.; Rosenbluth, M.N.; Teller, A.H.; Teller, E. Equation of state calculations by fast computing machines. *J. Chem. Phys.* **1953**, *21*, 1087–1092. [CrossRef]
2. Hastings, W.K. Monte Carlo sampling methods using Markov chains and their applications. *Biometrika* **1970**, *57*, 97–109. [CrossRef]

3. Sherlock, C.; Fearnhead, P.; Roberts, G.O. The random walk Metropolis: Linking theory and practice through a case study. *Stat. Sci.* **2010**, *25*, 172–190. [CrossRef]
4. Mengersen, K.L.; Tweedie, R.L. Rates of convergence of the Hastings and Metropolis algorithms. *Ann. Stat.* **1996**, *24*, 101–121. [CrossRef]
5. Roberts, G.O.; Tweedie, R.L. Geometric convergence and central limit theorems for multidimensional Hastings and Metropolis algorithms. *Biometrika* **1996**, *83*, 95–110. [CrossRef]
6. Jarner, S.F.; Roberts, G.O. Convergence of Heavy-tailed Monte Carlo Markov Chain Algorithms. *Scand. J. Stat.* **2007**, *34*, 781–815. [CrossRef]
7. Roberts, G.O.; Rosenthal, J.S. Examples of adaptive MCMC. *J. Comput. Graph. Stat.* **2009**, *18*, 349–367. [CrossRef]
8. Roberts, G.O.; Stramer, O. Langevin diffusions and Metropolis–Hastings algorithms. *Methodol. Comput. Appl. Probab.* **2002**, *4*, 337–357. [CrossRef]
9. Sejdinovic, D.; Strathmann, H.; Garcia, M.L.; Andrieu, C.; Gretton, A. Kernel Adaptive Metropolis-Hastings. In Proceedings of the 31st International Conference on Machine Learning, Beijing, China, 21–26 June 2014; Xing, E.P., Jebara, T., Eds.; PMLR: Beijing, China, 2014; Volume 32, pp. 1665–1673.
10. Andrieu, C.; Thoms, J. A tutorial on adaptive MCMC. *Stat. Comput.* **2008**, *18*, 343–373. [CrossRef]
11. Craiu, R.V.; Rosenthal, J.; Yang, C. Learn from thy neighbor: Parallel-chain and regional adaptive MCMC. *J. Am. Stat. Assoc.* **2009**, *104*, 1454–1466. [CrossRef]
12. Rudolf, D.; Sprungk, B. On a generalization of the preconditioned Crank–Nicolson Metropolis algorithm. *Found. Comput. Math.* **2018**, *18*, 309–343. [CrossRef]
13. Girolami, M.; Calderhead, B. Riemann manifold langevin and hamiltonian monte carlo methods. *J. R. Stat. Soc. Ser. B Stat. Methodol.* **2011**, *73*, 123–214. [CrossRef]
14. Brooks, S.; Gelman, A.; Jones, G.; Meng, X.L. *Handbook of Markov Chain Monte Carlo*; CRC Press: Boca Raton, FL, USA, 2011.
15. Maire, F.; Vandekerkhove, P. On Markov chain Monte Carlo for sparse and filamentary distributions. *arXiv* **2018**, arXiv:1806.09000.
16. Mallik, A.; Jones, G.L. Directional Metropolis-Hastings. *arXiv* **2017**, arXiv:1710.09759.
17. Ludkin, M.; Sherlock, C. Hug and Hop: A discrete-time, non-reversible Markov chain Monte Carlo algorithm. *arXiv* **2019**, arXiv:1907.13570.
18. Kamatani, K. Ergodicity of Markov chain Monte Carlo with reversible proposal. *J. Appl. Probab.* **2017**, 638–654. [CrossRef]
19. Roberts, G.O.; Rosenthal, J.S. General state space Markov chains and MCMC algorithms. *Probab. Surv.* **2004**, *1*, 20–71. [CrossRef]
20. Roberts, G.O.; Rosenthal, J.S. Geometric ergodicity and hybrid Markov chains. *Electron. Comm. Probab.* **1997**, *2*, 13–25. [CrossRef]
21. Meyn, S.P.; Tweedie, R.L. *Markov Chains and Stochastic Stability*; Cambridge University Press: Cambridge, UK, 2009.
22. Jones, G.L.; Hobert, J.P. Honest exploration of intractable probability distributions via Markov chain Monte Carlo. *Stat. Sci.* **2001**, *16*, 312–334. [CrossRef]
23. Tierney, L. Markov chains for exploring posterior distributions. *Annal. Stat.* **1994**, *22*, 1701–1728. [CrossRef]
24. Bierkens, J. Non-reversible Metropolis–Hastings. *Stat. Comput.* **2016**, *26*, 1213–1228. [CrossRef]
25. Jarner, S.F.; Tweedie, R.L. Necessary conditions for geometric and polynomial ergodicity of random-walk-type Markov chains. *Bernoulli* **2003**, *9*, 559–578. [CrossRef]
26. Roberts, G.O.; Rosenthal, J.S. Optimal scaling for various Metropolis-Hastings algorithms. *Stat. Sci.* **2001**, *16*, 351–367. [CrossRef]
27. Livingstone, S.; Girolami, M. Information-geometric Markov chain Monte Carlo methods using diffusions. *Entropy* **2014**, *16*, 3074–3102. [CrossRef]
28. Andrieu, C.; Moulines, É. On the ergodicity properties of some adaptive MCMC algorithms. *Ann. Appl. Probab.* **2006**, *16*, 1462–1505. [CrossRef]
29. Livingstone, S.; Faulkner, M.F.; Roberts, G.O. Kinetic energy choice in Hamiltonian/hybrid Monte Carlo. *Biometrika* **2019**, *106*, 303–319. [CrossRef]
30. Roberts, G.O.; Tweedie, R.L. Exponential convergence of Langevin distributions and their discrete approximations. *Bernoulli* **1996**, *2*, 341–363. [CrossRef]
31. Livingstone, S.; Betancourt, M.; Byrne, S.; Girolami, M. On the geometric ergodicity of Hamiltonian Monte Carlo. *Bernoulli* **2019**, *25*, 3109–3138. [CrossRef]
32. Jarner, S.F.; Hansen, E. Geometric ergodicity of Metropolis algorithms. *Stoch. Process. Their Appl.* **2000**, *85*, 341–361. [CrossRef]
33. Neal, R.M. Slice sampling. *Annal. Stat.* **2003**, 705–741. [CrossRef]
34. Xifara, T.; Sherlock, C.; Livingstone, S.; Byrne, S.; Girolami, M. Langevin diffusions and the Metropolis-adjusted Langevin algorithm. *Stat. Probab. Lett.* **2014**, *91*, 14–19. [CrossRef]
35. Łatuszyński, K.; Roberts, G.O.; Thiery, A.; Wolny, K. Discussion on 'Riemann manifold Langevin and Hamiltonian Monte Carlo methods' (by Girolami, M. and Calderhead, B.). *J. R. Stat. Soc. Ser. B Statist. Methodol.* **2011**, *73*, 188–189.
36. Chen, S.; Tao, Y.; Yu, D.; Li, F.; Gong, B. Distributed learning dynamics of Multi-Armed Bandits for edge intelligence. *J. Syst. Archit.* **2020**, 101919. Available online: https://www.sciencedirect.com/science/article/abs/pii/S1383762120301806 (accessed on 29 May 2015). [CrossRef]

37. Cook, J.D. Upper and Lower Bounds on the Normal Distribution Function; Technical Report. 2009. Available online: http://www.johndcook.com/normalbounds.pdf (accessed on 29 May 2015).
38. Johnson, N.L.; Kotz, S. *Distributions in Statistics: Continuous Univariate Distributions*; Houghton Mifflin: Boston, MA, USA, 1970; Volume 1.

mathematics

MDPI

Article

Non-Homogeneous Markov Set Systems

P.-C.G. Vassiliou

Department of Statistical Science, University College London, Gower St, London WC1E 6BT, UK; vasiliou@math.auth.gr

Abstract: A more realistic way to describe a model is the use of intervals which contain the required values of the parameters. In practice we estimate the parameters from a set of data and it is natural that they will be in confidence intervals. In the present study, we study Non-Homogeneous Markov Systems (NHMS) processes for which the required basic parameters are in intervals. We call such processes Non-Homogeneous Markov Set Systems (NHMSS). First we study the set of the relative expected population structure of memberships and we prove that under certain conditions of convexity of the intervals of the parameters the set is compact and convex. Next, we establish that if the NHMSS starts with two different initial distributions sets and allocation probability sets under certain conditions, asymptotically the two expected relative population structures coincide geometrically fast. We continue proving a series of theorems on the asymptotic behavior of the expected relative population structure of a NHMSS and the properties of their limit set. Finally, we present an application for geriatric and stroke patients in a hospital and through it we solve problems that surface in an application.

Keywords: Non-Homogeneous Markov Systems; Markov Set Systems; limiting set

Citation: Vassiliou, P.-C.G. Non-Homogeneous Markov Set Systems. *Mathematics* **2021**, *9*, 471. https://doi.org/10.3390/math9050471

Academic Editor: Andras Telcs

Received: 24 January 2021
Accepted: 19 February 2021
Published: 25 February 2021

Publisher's Note: MDPI stays neutral with regard to jurisdictional claims in published maps and institutional affiliations.

Copyright: © 2021 by the authors. Licensee MDPI, Basel, Switzerland. This article is an open access article distributed under the terms and conditions of the Creative Commons Attribution (CC BY) license (https://creativecommons.org/licenses/by/4.0/).

1. Introduction

The class of stochastic processes called Non-Homogeneous Markov Systems (NHMS) was first defined in [1] The class of NHMS provided a general framework for many applied probabilities models used to model populations of a wide diversity of entities. The primary motive was to provide a general framework for a wide class of stochastic models in social processes ([2]). They also include as special cases non-homogeneous Markov chain models in manpower systems such as [3–5]. The literature on NHMS has flourished since then to a large extent and presently exist a large volume of theoretical results as well a variety of applications. In Section 2 of the present we provide a definition and a concise description of a NHMS. As we will discuss in Section 2, it is important for the reader to have in mind that actually the well-known non-homogeneous Markov chain is a special case of a NHMS.

In many stochastic processes and so naturally in Markov chains and specifically in non-homogenous Markov chains and NHMS, the values of the various parameters are assumed to be exact while in practice these are estimated from the data. Therefore, actually the values of the parameters is more realistic to be viewed as being contained in intervals with the desired probability confidence. This approach has been used in systems of linear equations and in this case the solutions are given as the set of all possible solutions. Two books have been written on this topic by [6,7]. For the analogous problem for differential equations a book was written by [8]. For homogeneous Markov chains with this approach a book was written see [9].

In Section 3 of the present we will now add some additional assumptions on a NHMS in our way to define a non-homogeneous Markov set system (NHMSS). In this way now a NHMSS will be a NHMS whose basic parameters will be assumed to be in compact convex intervals.

The NHMSS is a stochastic system which has a population of members which increases at every point in time. I addition the initial members need not to be the same entities at different time points since there is wastage from the system. The members of the population move among the different states, exit from the system (population) and new members are coming into the population (system) as replacements or to expand the system. In the case of non-homogeneous Markov set chains we have only one particle in the population, which never leaves the system and no procedure to replace this particle exists. Mathematically now the NHMSS has more elements in the one step in time equation with more parameters introduced in the stochastic difference equation. As the equation is applied recursively we end up with series of components which interact together with more parameters being in an interval. Hence, the problems to be solved are a lot harder, and new strategies and tools must be used, than the simple case of the Markov set chain. The introduction of the concept of a membership is crucial in dealing with the different individual members as time progress. The tool of Minkowski sum of vectors and its properties for convex combination of compact sets will play a vital role which was not needed in the case of Markov set chains. One of the hard problems which we encounter which does not exist in the Markov set chains is finding the range of infinite series. The Hausdorff metric for compact sets and the coefficient of ergodicity together with properties of appropriate norms introduced and the manipulation of infinite series will help to provide the following:

In Section 4, we establish in the form of a theorem, using the Minkowski sum of two sets, under which conditions in a NHMSS the set of all possible expected relative population structures at a certain point in time is a convex set. Also, we establish a Theorem where we provide conditions under which the set of all expected relative population structures at a certain point in time is a convex polygon.

In Section 5 we study the asymptotic behavior of an NHMSS, a problem that has been of central importance for homogeneous Markov chains, non-homogeneous Markov chains, NHMS and homogeneous Markov set chains. In Theorem 4, with the use of the coefficient of ergodicity and the Hausdorff metric we prove the following: Let that in an NHMSS the sets of initial structures are different but compact and convex; also, the sets of allocation probabilities of the memberships are different but convex and compact; the inherent non-homogeneous Markov set chain is common; then the Hausdorff metric of the two different sets of all possible expected relative structures asymptotically goes to zero geometrically fast. This is equivalent with concluding that the two sets asymptotically coincide geometrically fast. In Theorem 5 we prove that in an NHMSS if the total population of memberships converges in a finite number geometrically fast, and the sets of initial structures and allocation probabilities of memberships are compact and convex, then the set of all possible expected relative population structure converges to a limit set geometrically fast. These two theorems have important consequences for a NHMS process also. The first is Theorem 6 which relaxes important assumptions of the basic asymptotic theorem for NHMS which is provided as Theorem 7. The second labeled as Theorem 8 answers a novel question for NHMS, i.e., provides conditions under which two different NHMS, with the same number of states and population, but different initial states and different allocation probabilities of memberships if they have the same transition probabilities sequence of memberships, they converge in the same relative population structure geometrically fast.

In Section 6 we study properties of the limit set of expected relative population structures. In Theorem 10 we prove the first property, that under some mild conditions the limit set of the expected relative population structures of an NHMSS remains invariant if any selected transition probability matrix of the inherent non-homogeneous Markov chain from the respective interval is multiplied by it from the right. We also prove that the limit set is the only set with this property if the interval of selection of transition probabilities of the inherent non-homogeneous Markov chain is product scrambling. In Theorem 11 the second property is established, i.e., let two different NHMSS in the sense that they have different sets of selecting initial distributions, different sets of selecting allocation probabilities and different intervals of selecting the transition probabilities of the inherent non-homogeneous Markov chains. What they have in common is that their respective intervals are uniformly scrambling with a common bound and they have the same total population of memberships. We prove that the Hausdorff metric of the limit sets of the expected relative population structures of the two NHMSS is bounded by the multiplication of a function of the Hausdorff metric of the two tight intervals of selection of the stochastic matrices of the inherent non-homogeneous Markov set chains and the bound of their uniform coefficients of ergodicity.

In Section 7 we present a representative application for geriatric and stroke patients in a hospital. Through this application we provide solutions in problems arising in an application by providing respective Lemmas and a general Algorithm with computational geometry procedures which are applicable to any population system.

2. The Non-Homogeneous Markov System

Consider a population which has $T(t)$ memberships at time t. These memberships could be held by any kind of entities, i.e., human beings, animals, T-cells in a biological entity, fish in an organized area in the sea, cars on a highway etc. We assume that $T(t)$ is known for every t, for example in a hospital the memberships are the beds for patients and from the management of the hospital's planning the number of beds are known. Let that the population is stratified into classes which we call states and let that there are a finite number of states, i.e., the state space is $\mathbb{S} = \{1, 2, \ldots, k\}$. We assume that the evolution of the population is in discrete time, i.e., $t = 1, 2, ...$ and we call the vector of random variables $\boldsymbol{N}(t) = [N_1(t), N_2(t), \ldots, N_k(t)]$ where $N_i(t)$ is the number of memberships in state i at time t, the *population structure* of the NHMS. Define by $\boldsymbol{q}(t) = \boldsymbol{N}(t)/T(t)$ to be the *relative population structure*. At every time instant $t = 1, 2, ...$, we have internal transitions of members among the states in $\mathbb{S}$ with probabilities which we collect in the $k \times k$ matrix $\boldsymbol{P}(t)$; we have wastage from all the states with probabilities which we collect in the $1 \times k$ vector $\boldsymbol{p}_{k+1}(t) = [p_{1,k+1}(t), p_{2,k+1}(t), \ldots, p_{k,k+1}(t)]$; finally, we have recruitment or allocation probabilities of replacements or new entrants in the various states at time t which we collect in the $1 \times k$ stochastic vector $\boldsymbol{p}_0(t) = [p_{01}(t), p_{02}(t), \ldots, p_{0k}(t)]$. We assume that the system is expanding, i.e., $\Delta T(t) = T(t) - T(t-1) \geq 0$. During the time interval $(t-1, t]$ a member of the system in state i either moves internally to another state j of the system with probability $p_{ij}(t)$ or leave the system and his membership remains at the exit of the system. New entrants to the system are of two types, those to replace leavers and those needed to be added in the system to meet the target of $T(t)$ total memberships. The new entrant gets his membership at the entrance and he is being allocated or recruited at state j with probability $p_{0j}(t)$. Hence, the probability of movement of a membership from state i to state j at time t is $q_{ij}(t) = p_{ij}(t) + p_{0j}(t)p_{i,k+1}(t)$. We collect these probabilities in the $k \times k$ matrix $\boldsymbol{Q}(t) = \boldsymbol{P}(t) + \boldsymbol{p}_{k+1}^{\top}(t)\boldsymbol{p}_0(t)$ which apparently is a stochastic matrix. We call the Markov chain defined by the sequence of matrices $\{\boldsymbol{Q}(t)\}_{t=0}^{\infty}$ the *imbedded or inherent non-homogeneous Markov chain* of the NHMS.

It is of interest the *expected relative population structure*. Let X_t be the random variable representing the state that a membership of the system is at time t. Define by

$$\boldsymbol{q}(s,t) = [q_1(s,t), q_2(s,t), \ldots, q_2(s,t)],$$

where

$$q_j(s,t) = \mathbb{P}[X_t = j \mid \boldsymbol{q}(s)] \text{ for } s \leq t, \tag{1}$$

then from ([10] p. 140) we get that

$$\mathbb{E}[\boldsymbol{q}(s,t)] = a(t-1)\mathbb{E}[\boldsymbol{q}(t-1)]\boldsymbol{Q}(t) + b(t-1)\boldsymbol{p}_0(t), \tag{2}$$

where

$$a(t-1) = \frac{T(t-1)}{T(t)} \quad \text{and} \quad b(t-1) = \frac{T(t) - T(t-1)}{T(t)}, \tag{3}$$

from which we get that

$$\mathbb{E}[\boldsymbol{q}(0,t)] = \frac{T(0)}{T(t)}\boldsymbol{q}(0)\boldsymbol{Q}(0,t)+$$

$$\frac{1}{T(t)}\sum_{\tau=1}^{t}\Delta T(\tau)\boldsymbol{p}_0(\tau)\boldsymbol{Q}(\tau,t), \tag{4}$$

where $\boldsymbol{Q}(s,t) = \boldsymbol{Q}(s+1)\boldsymbol{Q}(s+2)\ldots\boldsymbol{Q}(t)$ for $(s \leq t)$. We set $\boldsymbol{Q}(s,t) = \boldsymbol{I}$ the identity matrix for $s \geq t$. Please note that we set $\boldsymbol{q}(s,t) = \boldsymbol{0}$ for $s > t$. We call any such process as described above a *Non-homogeneous Markov process* in discrete time and discrete state space. It is important for the reader to realize that the well-known ordinary Markov chain is a very special case of a NHMS with $T(t) = 1$, $\boldsymbol{p}_0(t) = \boldsymbol{0}$, $p_{k+1}(t) = 0$ and $\boldsymbol{Q}(t) = \boldsymbol{P}$.

As we mentioned in the Introduction the stochastic process NHMS was first introduced in [1] as a discrete time, discrete state space stochastic process with motives which have their roots in actual applications in manpower systems see for example [1,2,11], and also the review papers [12,13]. Since then, a large literature on theoretical developments on many aspects of a NHMS were published which also included the developments in [14,15] of NHMS's in a general state space. In [16] there appeared the link between the theory of NHMS and martingale theory. Lately, also another area of large interest has been the Law of Large Numbers in NHMS ([17]) which has its roots as a motive the study of Laws of Large numbers on homogeneous Markov chains by Markov himself. Also, many applications in areas with great diversity have also appeared in the literature. For example we could selectively refer to some of them. Let as start with [18–20] which are applications in the evolution of the HIV virus on the T-cells of the human body; population consisting with patients with asthma was studied in [21]; reliability studies were presented in [22]; applications in biomedical research appeared in [23,24]; various applications for human populations [25–29]; interesting application to consumption credit [30] infections of populations [31]; a very interesting application in DNA and web navigation [32]; interesting ecological applications [33]; results in Physical Chemistry [34]. Finally, there are a large number of publications by the research school of Prof McClean in hospital systems which are large manpower systems [35–39].

3. Non-Homogeneous Markov Set System

In Section 2 we defined the NHMS process and we will now define for the first time ever the non-homogeneous Markov set system. So far in the well developed theory of NHMS's the various perimeters are assumed to be exact while in practice they are naturally estimated by the data. Therefore, it is more realistic to be viewed as being contained in intervals with the desired probability confidence. In summary as we will see bellow a NHMSS is a NHMS for which its parameters are defined in intervals. In addition the study of NHMSS provides a new area of theoretical research with different mathematical tools in many instances than the corresponding theory of NHMS and a potential to be applied in other stochastic processes.

The practical advantages of NHMSS's are rather apparent since the assumptions on the parameters are less restrictive. The assumption of the parameters being in appropriate intervals absorbs in a way the errors of point estimates which increase their variability. In addition it provides the tools to study NHMS's whose parameters will be in "desired" intervals which increases considerably the control of the system since we could choose policies of the systems in intervals with desired outcomes for the expected relative population structures or to avoid trouble some situations.

We will start with the definition of an interval for a stochastic vector following [9] who first defined Markov set chains. Denote by $M_n(\mathbb{R})$ or simply M_n the set of all $n \times n$ matrices with elements from the field $\mathbb{R}$.

Definition 1. *Let SM_{1n} the set of all $1 \times n$ stochastic vectors. Also let λ and μ be non-negative $1 \times n$ vectors with $\lambda \leq \mu$ componentwise. Then define the corresponding interval in SM_{1n} by*

$$[\lambda, \mu] = \{\boldsymbol{p} : \boldsymbol{p} \in SM_{1,n} \text{ with } \lambda \leq \boldsymbol{p} \leq \mu\},$$

where λ,μ are chosen such that $[\lambda, \mu] \neq \varnothing$.

Example 1. *It is sometimes helpful to view mathematics geometrically. Let $SM_{1,3}$ the set of all 1×3 stochastic vectors, then it is easy to see that this is the convex hull of the vectors $\boldsymbol{e}_1 = \begin{pmatrix} 1 & 0 & 0 \end{pmatrix}$, $\boldsymbol{e}_2 = \begin{pmatrix} 0 & 1 & 0 \end{pmatrix}$ and $\boldsymbol{e}_3 = \begin{pmatrix} 0 & 0 & 1 \end{pmatrix}$ in $\mathbb{R}^3$. Now, all the non-negative vectors $\boldsymbol{x}$ such that $\lambda \leq \boldsymbol{x} \leq \mu$ are within and the surface of a rectangle the coordinates of which are determined by λ, μ. The interval $[\lambda, \mu]$ will be the intersection of the two above described spaces. We can visualize this more easily if we consider the triangle $\boldsymbol{e}_1\boldsymbol{e}_2\boldsymbol{e}_3$ in $\mathbb{R}^2$. Let $\lambda = \begin{pmatrix} 0.1 & 0.2 & 0.3 \end{pmatrix}$ and $\mu = \begin{pmatrix} 0.5 & 0.7 & 0.8 \end{pmatrix}$ then the interval $[\lambda, \mu]$ could be easily designed in the following way. Draw two parallel lines to the line $\boldsymbol{e}_2\boldsymbol{e}_3$ at the points $\lambda_1 = 0.1$ and $\mu_1 = 0.5$; also draw two parallel lines to the line $\boldsymbol{e}_1\boldsymbol{e}_3$ at the points $\lambda_2 = 0.2$ and $\mu_2 = 0.7$; finally draw two parallel lines to the line $\boldsymbol{e}_1\boldsymbol{e}_2$ at the points $\lambda_3 = 0.3$ and $\mu_3 = 0.8$. Then the interval $[\lambda, \mu]$ is the common area between these lines.*

Tight intervals are important in what follows:

Definition 2. *Let $[\lambda, \mu]$ be an interval, then if*

$$\lambda_i = \min_{x \in [\lambda, \mu]} x_i \quad \text{and} \quad \mu_i = \max_{x \in [\lambda, \mu]} x_i,$$

then λ_i, μ_i are called tight, respectively. If λ_i, μ_i are tight for all i, then the interval $[\lambda, \mu]$ is called tight.

Intervals can be tested for tightness using the following Lemma ([9]). Also, with the use of this Lemma an interval which is not tight, we can tighten it up using an algorithm without actually changing it. That is the new interval, the tightened one will contain the same stochastic vectors.

Lemma 1. *([9]). Let $[\lambda, \mu]$ be an interval. Then for each coordinate i*

$$(i)\ \lambda_i \text{ is tight if and only if } \lambda_i + \sum_{k \neq i} \mu_k \geq 1.$$

$$(ii)\ \mu_i \text{ is tight if and only if } \mu_i + \sum_{k \neq i} \lambda_k \geq 1.$$

We now need the following definition of when in a tight interval a vector is called *free*.

Definition 3. *Let $[\lambda, \mu]$ be a tight interval and $\boldsymbol{p} \in [\lambda, \mu]$. Then if $\lambda_i < p_i < \mu_i$ for some coordinate i, then the coordinate p_i in $\boldsymbol{p}$ is called free.*

Tight intervals and convex sets are well linked and play an important role in the preservation of many properties. In this respect, the following Lemma is very useful.

Lemma 2. *([9]). Let $[\lambda, \mu]$ be a tight an interval . Then $[\lambda, \mu]$ is a convex polytope. A vector $\boldsymbol{p} \in [\lambda, \mu]$ is a vertex of $[\lambda, \mu]$ if and only if $\boldsymbol{p}$ has at most one free component.*

We will now extend the definition of an interval of a vector to an interval of a matrix and to a tight interval of a matrix.

Definition 4. *Let SM_n the set of all $1 \times n$ stochastic matrices. Also let $\mathbf{\Lambda}$ and $\mathbf{M}$ be non-negative $n \times n$ matrices with $\mathbf{\Lambda} \leq \mathbf{M}$ componentwise. Then define the corresponding interval in SM_n by*

$$[\mathbf{\Lambda}, \mathbf{M}] = \{\mathbf{P} : \mathbf{P} \in SM_n \textit{ with } \mathbf{\Lambda} \leq \mathbf{P} \leq \mathbf{M}\},$$

where $\mathbf{\Lambda}, \mathbf{M}$ are chosen such that $[\mathbf{\Lambda}, \mathbf{M}] \neq \varnothing$.

We now proceed to define a tight interval of matrices:

Definition 5. *Let $[\mathbf{\Lambda}, \mathbf{M}]$ be an interval of matrices. If*

$$\lambda_{ij} = \min_{\mathbf{P} \in [\mathbf{\Lambda}, \mathbf{M}]} p_{ij} \textit{ and } \mu_{ij} = \max_{\mathbf{P} \in [\mathbf{\Lambda}, \mathbf{M}]} p_{ij},$$

for all i and j, then $[\mathbf{\Lambda}, \mathbf{M}]$ is called tight.

The interval $[\mathbf{\Lambda}, \mathbf{M}]$ can be constructed also by rows, i.e.,

$$[\mathbf{\Lambda}, \mathbf{M}] = \left\{ \begin{array}{c} \mathbf{P} : \mathbf{p}_i \in [\lambda_i, \mu_i] \text{ for all } i, \text{ with } \mathbf{p}_i, \lambda_i, \mu_i \\ \text{being the rows of the respective matrices } \mathbf{P}, \mathbf{\Lambda}, \mathbf{M} \end{array} \right\}.$$

In what follows we will define a non-homogeneous Markov set system. We will keep the entire notation introduced in Section 2 for a NHMS and we will build on that.

Let $[\mathbb{M}] = [\check{\mathbf{Q}}, \hat{\mathbf{Q}}]$ be an interval of $k \times k$ stochastic matrices with $\check{\mathbf{Q}} \leq \mathbf{Q}(t) \leq \hat{\mathbf{Q}}$ for every $t \in \mathbb{N}$ which is tight, i.e.,

$$[\mathbb{M}] = \{\mathbf{Q}(t) : \text{is an } k \times k \text{ stochastic matrix with } \check{\mathbf{Q}} \leq \mathbf{Q}(t) \leq \hat{\mathbf{Q}}\}$$

with

$$\check{q}_{ij} = \min_{\mathbf{Q}(t) \in [\check{\mathbf{Q}}, \hat{\mathbf{Q}}]} q_{ij}(t) \text{ for every } t \in \mathbb{N},$$

$$\hat{q}_{ij} = \max_{\mathbf{Q}(t) \in [\check{\mathbf{Q}}, \hat{\mathbf{Q}}]} q_{ij}(t) \text{ for every } t \in \mathbb{N},$$

and the notation $[\check{\mathbf{Q}}, \hat{\mathbf{Q}}]$ will be taken to imply that $[\check{\mathbf{Q}}, \hat{\mathbf{Q}}] \neq \varnothing$.

We will make now the following basic assumptions:

Assumption 1. *Let that the imbedded non-homogeneous Markov chain of the NHMS has all its probability matrices in $[\mathbb{M}]$.*

We call $[\mathbb{M}]$ the *probability transition matrix set (PTMS) of the imbedded non-homogeneous Markov chain.*

Now define by

$$[\mathbb{M}^2] = \{\mathbf{Q}(0)\mathbf{Q}(1) : \mathbf{Q}(0), \mathbf{Q}(1) \in [\mathbb{M}]\},$$

$$\cdots$$

$$\cdots$$

$$\cdots$$

$$[\mathbb{M}^n] = \{\mathbf{Q}(0)\mathbf{Q}(1)...\mathbf{Q}(n-1) : \mathbf{Q}(0), \ldots, \mathbf{Q}(n-1) \in [\mathbb{M}]\}.$$

We call the sequence $\{[\mathbb{M}^n]\}_{n=1}^{\infty}$ the *inherent or imbedded non-homogeneous Markov set chain.*

Assumption 2. *Let* $[\mathbb{S}_0]$ *be the set of* $1 \times k$ *stochastic vectors from which the initial distribution* $\boldsymbol{q}(0)$ *is chosen.*

$$[\mathbb{S}_0] = [\check{\boldsymbol{q}}_0, \hat{\boldsymbol{q}}_0] = \{\boldsymbol{q}(0) : \text{is a stochastic vector with } \boldsymbol{q}(0) \in [\mathbb{S}_0]\}.$$

Assumption 3. *Let* $[\mathbb{R}_0]$ *be the set of* $1 \times k$ *stochastic vectors from which the allocation probabilities* $\boldsymbol{p}_0(t)$ *are being selected. That is*

$$[\mathbb{R}_0] = [\check{\boldsymbol{p}}_0, \hat{\boldsymbol{p}}_0] = \{\boldsymbol{p}_0(t) : \text{ is a stochastic vector with } \boldsymbol{p}_0(t) \in [\mathbb{R}_0] \text{ for every } t\}.$$

We call a NHMS whose parameters are assumed to be in intervals as in Assumptions 1–3 a Non-homogeneous Markov Set System (NHMSS).

Note that it is apparent by now that Markov set chains that were initiated by [9,40–43] are special cases of a NHMSS.

4. The Set of the Expected Relative Population Structures of a NHMSS

In geometry the Minkowski sum (also known as dilation) of two sets of position vectors $\mathbb{A}$ and $\mathbb{B}$ in Euclidean space is formed by adding each vector in $\mathbb{A}$ to each vector in $\mathbb{B}$. That is

$$\mathbb{A} + \mathbb{B} = \{a + b : \ a \in \mathbb{A}, b \in \mathbb{B}\}.$$

Example 2. *If we have two sets* $\mathbb{A}$ *and* $\mathbb{B}$ *consisting of three position vectors (informally, three points) representing the vertices of two triangles in* $\mathbb{R}^2$ *with coordinates*

$$\mathbb{A} = \{(1,0), (0,1), (0,-1)\} \text{ and } \mathbb{B} = \{(0,0), (1,1), (1,-1)\},$$

then their Minkowski sum is

$$\mathbb{A} + \mathbb{B} = \{(1,0), (2,1), (2,-1), (0,1), (1,2), (0,-1), (1,-2)\},$$

which comprises the vertices of a hexagon.

For Minkowski addition, the zero set containing only the zero vector **0**, is an identity element for every subset $\mathbb{V}$ of a vector space, i.e., $\mathbb{V} + \{0\} = \mathbb{V}$.

The empty set is important in Minkowski addition because the empty set annihilates every other subset for every subset $\mathbb{V}$ of a vector space, its sum with the empty set is empty, i.e., $\mathbb{V} + \varnothing = \varnothing$.

We are now in a position to state the following Lemma ([44])

Lemma 3. *If* $\mathbb{V}$ *is a convex set then* $\mu\mathbb{V} + \ngeqq\mathbb{V}$ *is also a convex set and furthermore* $\mu\mathbb{V} + \ngeqq\mathbb{V} = (\mu + \lambda)\mathbb{V}$ *for every* $\lambda, \mu > 0$. *Conversely, if this "distributive property" holds for all non-negative real numbers* $\lambda, \mu > 0$ *then the set is convex.*

Remark 1. *For two convex polygons* $\mathbb{V}_1$ *and* $\mathbb{V}_2$ *in the plane with* m *and* n *vertices, their Minkowski sum is a convex polygon with at most* $m + n$ *vertices and may be computed in time* $O(m + n)$ *by a very simple procedure.*

We need the following sets for the Lemma that follows

$$Rng(\boldsymbol{q}) = \{\boldsymbol{y} : \boldsymbol{y} = \boldsymbol{q}\boldsymbol{Q} \text{ for some } \boldsymbol{Q} \in [\mathbb{M}] \text{ and any } \boldsymbol{q} \in SM_{1,n}\}.$$

Also

$$Rng(\mathbb{S}) = \cup_{\boldsymbol{q} \in \mathbb{S}} Rng(\boldsymbol{q}) = \{\boldsymbol{y} : \boldsymbol{y} = \boldsymbol{q}\boldsymbol{Q} \text{ for some } \boldsymbol{Q} \in [\mathbb{M}] \text{ and some } \boldsymbol{q} \in \mathbb{S}\}.$$

Lemma 4. *Let an NHMSS with $T(t) \geq 0$ and finite and which is expanding ($\Delta T(t) \geq 0$). Also let $[\mathbb{S}_0]$ the set from which the initial distribution of memberships is drawn, $[\mathbb{R}_0]$ the set from which the allocation probabilities in the various states are chosen at every time step and finally let $[\mathbb{M}]$ the set to which all the transition probability matrices of the inherent Markov chain of memberships belong. Then the set $Rng_t(\mathbb{S}_0, \mathbb{R}_0)$ of all possible expected relative population structures at time t is given by*

$$Rng_t(\mathbb{S}_0, \mathbb{R}_0) = \{\mathbb{E}[\boldsymbol{q}(0,t)] : \mathbb{E}[\boldsymbol{q}(0,t)] \in \frac{T(0)}{T(t)} \cup_{\boldsymbol{q}(0)\in[\mathbb{S}_0]} \boldsymbol{q}(0)\left[\mathbb{M}^t\right] \tag{5}$$

$$+\frac{1}{T(t)} \sum_{\tau=1}^{t} \Delta T(\tau) \cup_{\boldsymbol{p}_0(\tau)\in[\mathbb{R}_0]} \boldsymbol{p}_0(\tau)\left[\mathbb{M}^{t-\tau}\right]\},$$

with $\left[\mathbb{M}^0\right] = [\mathbb{M}]$ *and* $\left[\mathbb{M}^{-1}\right] = \{\boldsymbol{I}\}$.

Proof. Following the relevant proofs that lead to Equations (2.1), (2.2) and (3.4) in [10] or Equations (2) and (4) in the present, we could easily prove (5). □

We will need the following Lemma from ([9] p. 39):

Lemma 5. *In a non-homogeneous Markov set chain let the set of initial distributions $[\mathbb{S}_0]$ be convex and let $[\mathbb{M}]$ be a tight interval from which the transition probability sequence of matrices is being selected. Then the set $[\mathbb{S}_t]$ of all possible probability distributions at the various states at time t is a convex set.*

In the next theorem we show under which conditions the set of all possible expected relative population structures in a NHMSS is a convex set.

Theorem 1. *Let an NHMSS with $T(t)$ finite and which is expanding ($\Delta T(t) \geq 0$). If $[\mathbb{S}_0]$, $[\mathbb{R}_0]$ are convex sets and $[\mathbb{M}]$ a tight interval then the set of all possible expected relative population structures is a convex set.*

Proof. Define $\{\mathbb{E}[\mathring{\boldsymbol{q}}(0,t)]\}$ be the set of all possible expected relative structures of the initial memberships then from Lemma 4 we have

$$\{\mathbb{E}[\mathring{\boldsymbol{q}}(0,t)]\} = \{\mathbb{E}[\mathring{\boldsymbol{q}}(0,t)] : \mathbb{E}[\mathring{\boldsymbol{q}}(0,t)] \in \frac{T(0)}{T(t)} \cup_{\boldsymbol{q}(0)\in[\mathbb{S}_0]} \boldsymbol{q}(0)\left[\mathbb{M}^t\right]\}, \tag{6}$$

since $\frac{T(0)}{T(t)} \geq 0$, $[\mathbb{S}_0]$ is a convex set and $[\mathbb{M}]$ a tight interval, from Lemma 4 and 5 we get that $\{\mathbb{E}[\mathring{q}(0,t)]\}$ is a convex set. Also, the set

$$\{\mathbb{E}[\boldsymbol{r}_\tau]\} = \{\mathbb{E}[\boldsymbol{r}_\tau] : \mathbb{E}[\boldsymbol{r}_\tau] \in \Delta T(\tau) \cup_{\boldsymbol{p}(\tau)\in[\mathbb{R}_0]} \boldsymbol{p}_0(\tau)\left[\mathbb{M}^{t-\tau}\right]\},$$

is the set of all possible expected structures of new memberships at time t which entered in the system at time τ. Now, since the system is expanding, i.e., $\Delta T(\tau) \geq 0$, $[\mathbb{R}_0]$ is a convex set and $[\mathbb{M}]$ a tight interval, then with the same reasoning as in (6), we get that $\{\mathbb{E}[\boldsymbol{r}_\tau]\}$ is a convex set. Also, from Remark 1 we get that the Minkowski sum of sets

$$\frac{1}{T(t)} \sum_{\tau=1}^{t} \{\mathbb{E}[\boldsymbol{r}_\tau]\},$$

is a convex set. Hence, since the two sets in the right-hand side of Equation (5) are convex and then according to Remark 1 their Minkowski sum $Rng_t(\mathbb{S}_0, \mathbb{R}_0)$ is a convex set. □

We will now borrow the following Theorem from ([9] p. 40).

Theorem 2. *In a non-homogeneous Markov set chain let the set of initial distributions* $[\mathbb{S}_0]$ *be a convex polytope and let* $[\mathbb{M}]$ *be a tight interval from which the transition probability sequence of matrices is being selected. Then the set* $[\mathbb{S}_t]$ *of all possible probability distributions at the various states at time t is a convex polytope with vertices of the form* $\mathcal{E}_i E_{i_1} ... E_{i_k}$ *for some vertices* $\mathcal{E}_i$ *of* $[\mathbb{S}_0]$ *and some vertices* E_{i_j} *of* $[\mathbb{M}]$.

In the next theorem we show under which conditions the set of all possible expected relative population structures in a NHMSS is a convex polytope.

Theorem 3. *Let an NHMSS with* $T(t)$ *finite and which is expanding* $(\Delta T(t) \geq 0)$*. If* $[\mathbb{S}_0], [\mathbb{R}_0]$ *are convex polytopes and* $[\mathbb{M}]$ *a tight interval then the set of all possible expected relative population structures is a convex polytope.*

Proof. The proof follows the steps of the proof of Theorem 1 using Theorem 2. □

5. Asymptotic Behavior of NHMSS

The problem of asymptotic behavior has been one of central importance for, homogeneous Markov chains, non-homogeneous Markov chains, NHMS, and non-homogeneous Markov set chains. In the present section we will prove a series of theorems with which we establish the asymptotic behavior of NHMSS.

Since Markov himself and his student Dobrushin the coefficient of ergodicity $\mathcal{T}(Q)$ of a $k \times k$ stochastic matrix $Q = \{q_{ij}\}_{i,j=1}^{k}$, has been a fundamental tool in the study of Markov chains. We have

$$\mathcal{T}(Q) = \frac{1}{2} \max_{ij} \sum_{l=1}^{k} \left| q_{il} - q_{jl} \right|, \tag{7}$$

thus $0 \leq \mathcal{T}(Q) \leq 1$. We clarify in here that if $\mathcal{T}(Q) < 1$ the stochastic matrix Q is called *scrambling*. Scrambling matrices are regular, but not all regular matrices are scrambling. Yet if Q is a regular stochastic matrix then some power of Q, say Q^n is scrambling. We define by

$$\bar{\mathcal{T}}([\mathbb{M}]) = \max_{Q \in [\check{Q}, \hat{Q}]} \mathcal{T}(Q), \tag{8}$$

if $\bar{\mathcal{T}}([\mathbb{M}]) < 1$ we say that $[\mathbb{M}]$ is *uniformly scrambling*. More on uniform scrambling and the interpretation of the coefficient of ergodicity of a matrix $A \in \mathbb{R}^{n \times n}$ as a matrix norm when the norm is restricted to a specified subspace could be found, in [45]. For explicit forms for ergodicity coefficients and properties see also [46,47]. In what follows we will use the following norm for a matrix $A \in M_n(\mathbb{R})$

$$\|A\| = \max_{i} \sum_{j=1}^{k} |a_{ij}|$$

We will use the concept of $\bar{\mathcal{T}}([\mathbb{M}]) < 1$ to study asymptotic behavior in a NHMSS. It is important to note that if we consider as $\left([\mathbb{M}^C], \|.\|\right)$ the space of non-empty compact subsets of $([\mathbb{M}], \|.\|)$ then $\left([\mathbb{M}^C], \|.\|\right)$ is a metric space ([48]). This space can be topologized using the Hausdorff metric $d(.,.)$ defined by

$$d(\mathbb{S}_1, \mathbb{S}_2) = \max\{\delta(\mathbb{S}_1, \mathbb{S}_2), \delta(\mathbb{S}_2, \mathbb{S}_1)\}, \tag{9}$$

where

$$\delta(\mathbb{S}_1, \mathbb{S}_2) = \max_{Q_1 \in S_1} \min_{Q_2 \in S_2} \|Q_1 - Q_2\| \text{ and } \mathbb{S}_1, \mathbb{S}_2 \in \left[\mathbb{M}^C\right], \tag{10}$$

From [48] we get also that $\left([\mathbb{M}^C], d(.,.)\right)$ is a metric space.

We will need the following Lemmas

Lemma 6 ([49]). *The following statements are equivalent:*

(i) *Sequence* $\{\Delta T(t)\}_{t=0}^{\infty}$ *converges to zero with geometrical rate;*
(ii) *Sequence* $\{T(t)\}_{t=0}^{\infty}$ *converges to* T *with geometrical rate,*

Lemma 7 ([50] p. 541). *Suppose that*

$$\left\{\frac{\Delta T(t)}{T(t)}\right\}_{t=0}^{\infty}$$

converges to zero as $t \to \infty$ *geometrically fast with* $T(t) \geq T(t-1)$. *Then* $\{T(t)\}_{t=0}^{\infty}$ *converges geometrically fast.*

Remark 2. *The restriction*

$$\lim_{t\to\infty}\frac{\Delta T(t)}{T(t)} = 0,$$

is a general assumption with the physical interpretation that the proportional growth rate vanishes in the limit. This assumption allows $lim_{t\to\infty}T(t) = \infty$.

We will now study what happens asymptotically to the two sets of the expected relative population structures when the initial structures belong to two different sets as well as the allocation probabilities in the respective cases.

Theorem 4. *Let an NHMSS for which* $[\mathbb{M}]$ *is a tight interval which is uniformly scrambling. Assume that*

$$\left\{\frac{\Delta T(t)}{T(t)}\right\}_{t=0}^{\infty} \text{ converges to zero geometrically fast.}$$

Also, assume that $[\mathbb{S}], [\mathbb{R}_0]$ *are compact and convex. Let* $\mathbb{S}_1, \mathbb{S}_2 \subseteq [\mathbb{S}^C]$ *and* $\mathbb{R}_1, \mathbb{R}_2 \subseteq [\mathbb{R}_0^C]$ *then*

$$d(Rng_t(\mathbb{S}_1,\mathbb{R}_1), Rng_t(\mathbb{S}_2,\mathbb{R}_2)) \to_{t\to\infty} 0 \text{ geometrically fast.}$$

Proof. From (5) we get that

$$\begin{aligned} Rng_t(\mathbb{S}_1,\mathbb{R}_1) &= \{\mathbb{E}[\boldsymbol{q}(0,t)] : \mathbb{E}[\boldsymbol{q}(0,t)] \in \frac{T(0)}{T(t)} \cup_{\boldsymbol{q}(0)\in[\mathbb{S}_1]} \boldsymbol{q}(0)\left[\mathbb{M}^t\right] + \\ & \frac{1}{T(t)}\sum_{\tau=1}^{t}\Delta T(\tau) \cup_{\boldsymbol{p}_0(\tau)\in[\mathbb{R}_1]} \boldsymbol{p}_0(\tau)\left[\mathbb{M}^{t-\tau}\right]\}, \end{aligned} \tag{11}$$

and from (11) if we replace $\boldsymbol{q}(0)$ with $\bar{\boldsymbol{q}}(0)$, $[\mathbb{S}_1]$ with $[\mathbb{S}_2]$; $\boldsymbol{p}_0(\tau)$ with $\bar{\boldsymbol{p}}_0(\tau)$; $[\mathbb{R}_1]$ with $[\mathbb{R}_2]$ and $\mathbb{E}[\boldsymbol{q}(0,t)]$ with $\mathbb{E}[\bar{\boldsymbol{q}}(0,t)]$ we get $Rng_t(\mathbb{S}_2,\mathbb{R}_2)$. Since$[\mathbb{S}_1], [\mathbb{R}_1]$ are compact and $[\mathbb{M}]$ is a tight interval, it is not difficult to show that $Rng_t(\mathbb{S}_1,\mathbb{R}_1)$ is compact. The same applies for $Rng_t(\mathbb{S}_2,\mathbb{R}_2)$ since $[\mathbb{S}_2], [\mathbb{R}_2]$ are also compact. Hence, we may take their Hausdorff metric. Now, we have that

$$\begin{aligned} & \delta(Rng_t(\mathbb{S}_1,\mathbb{R}_1), Rng_t(\mathbb{S}_2,\mathbb{R}_2)) \\ = & \max_{\substack{\boldsymbol{q}(0)\in[\mathbb{S}_1],\boldsymbol{p}_0(\tau)\in[\mathbb{R}_1], \\ \boldsymbol{Q}(t)\in[\mathbb{M}] \text{ for all } \tau,t}} \ \min_{\substack{\bar{\boldsymbol{q}}(0)\in[\mathbb{S}_2],\bar{\boldsymbol{p}}_0(\tau)\in[\mathbb{R}_2] \\ \bar{\boldsymbol{Q}}(t)\in[\mathbb{M}] \text{ for all } \tau,t}} \|\mathbb{E}[\boldsymbol{q}(0,t)] - \mathbb{E}[\bar{\boldsymbol{q}}(0,t)]\|. \end{aligned} \tag{12}$$

by continuity of

$$f(\mathbb{E}[\bar{q}(0,t)]) = \min\|\mathbb{E}[q(0,t)] - \mathbb{E}[\bar{q}(0,t)]\|$$

then for some $\boldsymbol{q}^*(0) \in [\mathbb{S}_1]$, $\boldsymbol{p}_0^*(\tau) \in [\mathbb{R}_1]$ for every $\tau \in [1,t]$, $\boldsymbol{Q}^*(t) \in [\mathbb{M}]$ for every $t \in [0,t]$ and by denoting with

$$\mathbb{E}[\boldsymbol{q}^*(0,t)] = \frac{T(0)}{T(t)}\boldsymbol{q}^*(0)\boldsymbol{Q}^*(0,t) + \frac{1}{T(t)}\sum_{\tau=1}^{t} \Delta T(\tau)\boldsymbol{p}_0^*(\tau)\boldsymbol{Q}^*(\tau,t), \tag{13}$$

we get

$$\delta(Rng_t(\mathbb{S}_1,\mathbb{R}_1), Rng_t(\mathbb{S}_2,\mathbb{R}_2)) = \tag{14}$$

$$\min_{\substack{\bar{\boldsymbol{q}}(0)\in[\mathbb{S}_2],\bar{\boldsymbol{p}}_0(\tau)\in[\mathbb{R}_2]\\ \bar{\boldsymbol{Q}}(t)\in[\mathbb{M}] \text{ for all } \tau,t}} \|\mathbb{E}[\boldsymbol{q}^*(0,t)] - \mathbb{E}[\bar{\boldsymbol{q}}(0,t)]\|$$

$$\leq \|\mathbb{E}[\boldsymbol{q}^*(0,t)] - \mathbb{E}[\bar{\boldsymbol{q}}^*(0,t)]\|,$$

where $\mathbb{E}[\bar{\boldsymbol{q}}^*(0,t)]$ is given by (13) if we replace $\boldsymbol{q}^*(0)$ with any $\bar{\boldsymbol{q}}^*(0) \in [\mathbb{S}_2]$, $\boldsymbol{p}_0^*(\tau)$ with any $\bar{\boldsymbol{p}}_0^*(\tau) \in [\mathbb{R}_2]$ for every $\tau \in [1,t]$. Now, from (13) and (14) we get that

$$\delta(Rng_t(\mathbb{S}_1,\mathbb{R}_1), Rng_t(\mathbb{S}_2,\mathbb{R}_2)) \leq \frac{T(0)}{T(t)}\|[\boldsymbol{q}^*(0) - \bar{\boldsymbol{q}}^*(0)]\boldsymbol{Q}^*(0,t)\|$$

$$+\frac{1}{T(t)}\sum_{\tau=1}^{t} \Delta T(\tau)\|\boldsymbol{p}_0^*(\tau)\boldsymbol{Q}^*(\tau,t) - \bar{\boldsymbol{p}}_0^*(\tau)\boldsymbol{Q}^*(\tau,t)\|$$

$$\leq \frac{T(0)}{T(t)}\mathcal{T}(\boldsymbol{Q}^*(0,t))\|[\boldsymbol{q}^*(0) - \bar{\boldsymbol{q}}^*(0)]\|$$

$$+\frac{1}{T(t)}\sum_{\tau=1}^{t} \Delta T(\tau)\mathcal{T}(\boldsymbol{Q}^*(\tau,t))\|\boldsymbol{p}_0^*(\tau) - \bar{\boldsymbol{p}}_0^*(\tau)\|$$

$$\left(\text{since } [\mathbb{M}] \text{ is uniformly scrambling } \tilde{\mathcal{T}} = \max_{\boldsymbol{Q}^*(t)\in[\mathbb{M}]} \mathcal{T}(\boldsymbol{Q}^*(t)) < 1\right)$$

$$\leq \frac{T(0)}{T(t)}\tilde{\mathcal{T}}^t\|[\boldsymbol{q}^*(0) - \bar{\boldsymbol{q}}^*(0)]\|$$

$$+\frac{1}{T(t)}\sum_{\tau=1}^{t} \Delta T(\tau)\tilde{\mathcal{T}}^{t-\tau}\|\boldsymbol{p}_0^*(\tau) - \bar{\boldsymbol{p}}_0^*(\tau)\|$$

$$\leq \frac{T(0)}{T(t)}\tilde{\mathcal{T}}^t\delta(\mathbb{S}_1,\mathbb{S}_2) + \frac{1}{T(t)}\sum_{\tau=1}^{t} \Delta T(\tau)\tilde{\mathcal{T}}^{t-\tau}\delta(\mathbb{R}_1,\mathbb{R}_2). \tag{15}$$

Since $\left\{\frac{\Delta T(t)}{T(t)}\right\}_{t=0}^{\infty}$ converges to zero geometrically fast, from Lemmas 6 and 7 we get that there are $c_1, c_2 > 0$ and $0 < b_1, b_2 < 1$ such that

$$\left|\frac{1}{T(t)} - \frac{1}{T}\right| \leq c_1 b_1^t \quad \text{and} \quad \left|\frac{\Delta T(t)}{T(t)}\right| \leq c_2 b_2^t. \tag{16}$$

From (15), (16) we get that

$$\delta(Rng_t(\mathbb{S}_1,\mathbb{R}_1), Rng_t(\mathbb{S}_2,\mathbb{R}_2)) \leq T(0)\left(\frac{1}{T} + c_1 b_1^t\right)\tilde{\mathcal{T}}^t\delta(\mathbb{S}_1,\mathbb{S}_2)$$

$$+\delta(\mathbb{R}_1,\mathbb{R}_2)\left(\frac{1}{T} + c_1 b_1^t\right)\sum_{\tau=1}^{t} c_2 b_2^\tau \tilde{\mathcal{T}}^{t-\tau},$$

$$(\text{assuming } \tilde{\mathcal{T}} < b_2)$$

$$\leq T(0)\left(\frac{1}{T} + c_1 b_1^t\right)\tilde{\mathcal{T}}^t\delta(\mathbb{S}_1,\mathbb{S}_2) + \delta(\mathbb{R}_1,\mathbb{R}_2)\left(\frac{1}{T} + c_1 b_1^t\right)c_2 b_2^{t-1}\sum_{\tau=1}^{t}\left(\frac{\tilde{\mathcal{T}}}{b_2}\right)^{t-\tau}$$

$$\leq T(0)\left(\frac{1}{T}+c_1b_1^t\right)\tilde{\mathcal{T}}^t\delta(\mathbb{S}_1,\mathbb{S}_2)+\delta(\mathbb{R}_1,\mathbb{R}_2)\left(\frac{1}{T}+c_1b_1^t\right)c_2b_2^{t-1}\left(1-\frac{\tilde{\mathcal{T}}}{b_2}\right)^{-1}. \quad (17)$$

From (17) we conclude that $\delta(Rng_t(\mathbb{S}_1,\mathbb{R}_1), Rng_t(\mathbb{S}_2,\mathbb{R}_2))$ converges to zero geometrically fast. Similarly, we get the same conclusion if $\tilde{\mathcal{T}} > b_2$.

Now, in a similar way we may prove that $\delta(Rng_t(\mathbb{S}_2,\mathbb{R}_2), Rng_t(\mathbb{S}_1,\mathbb{R}_1))$ converges to zero geometrically fast. Thus, we arrive at the desired conclusion. □

We will now establish under what conditions the convergence to the limiting set in a NHMSS is geometrically fast.

Theorem 5. *Let an NHMSS for which $[\mathbb{M}]$ is a tight interval which is uniformly scrambling. Assume that*

$$\left\{\frac{\Delta T(t)}{T(t)}\right\}_{t=0}^{\infty} \text{ converges to zero geometrically fast.}$$

Also, let that $[\mathbb{S}]$, $[\mathbb{R}_0]$ are compact and convex. Then we have that

$$d(Rng_t(\mathbb{S},\mathbb{R}_0), Rng_\infty(\mathbb{S},\mathbb{R}_0) \to 0, \text{ geometrically fast.} \quad (18)$$

Proof. We will first show that

$$Rng(Rng_t(\mathbb{S},\mathbb{R}_0)) = Rng_{t+1}(\mathbb{S},\mathbb{R}_0). \quad (19)$$

By the evaluation of the function $Rng(.,.)$ given in Lemma 4 we get that

$$Rng(Rng_t(\mathbb{S},\mathbb{R}_0)) = \frac{T(t)}{T(t+1)} \cup_{\mathbb{E}[\boldsymbol{q}(0,t)]\in Rng_t(\mathbb{S},\mathbb{R}_0)} \mathbb{E}[\boldsymbol{q}(0,t)][\mathbb{M}]$$

$$+\frac{1}{T(t+1)}\Delta T(t+1) \cup_{\boldsymbol{p}_0(t)\in[\mathbb{R}_0]} \boldsymbol{p}_0(t) =$$

$$\frac{T(t)}{T(t+1)}\frac{T(0)}{T(t)} \cup_{\boldsymbol{q}(0)\in[\mathbb{S}]} \boldsymbol{q}(0)\left[\mathbb{M}^t\right][\mathbb{M}]+$$

$$\frac{T(t)}{T(t+1)}\frac{1}{T(t)}\sum_{\tau=1}^{t}\Delta T(\tau) \cup_{\boldsymbol{p}_0(\tau)\in[\mathbb{R}_0]} \boldsymbol{p}_0(\tau)\left[\mathbb{M}^{t-\tau}\right][\mathbb{M}]$$

$$+\frac{1}{T(t+1)}\Delta T(t+1) \cup_{\boldsymbol{p}_0(t)\in[\mathbb{R}_0]} \boldsymbol{p}_0(t)$$

$$=\frac{T(0)}{T(t+1)} \cup_{\boldsymbol{q}(0)\in[\mathbb{S}]} \boldsymbol{q}(0)\left[\mathbb{M}^{t+1}\right]+$$

$$\frac{1}{T(t+1)}\sum_{\tau=1}^{t+1}\Delta T(\tau) \cup_{\boldsymbol{p}_0(\tau)\in[\mathbb{R}_0]} \boldsymbol{p}_0(\tau)\left[\mathbb{M}^{t-\tau+1}\right] = Rng_{t+1}(\mathbb{S},\mathbb{R}_0).$$

Now, we will show that

$$Rng_t(Rng_\infty(\mathbb{S},\mathbb{R}_0)) = Rng_\infty(\mathbb{S},\mathbb{R}_0). \quad (20)$$

Assume that $\mathbb{E}[\boldsymbol{q}(\infty)]$ is an element of $Rng_\infty(\mathbb{S},\mathbb{R}_0)$ then from Lemma 4 we get that

$$Rng(Rng_\infty(\mathbb{S},\mathbb{R}_0)) = \frac{T}{T} \cup_{\mathbb{E}[\boldsymbol{q}(\infty)]\in Rng_\infty(\mathbb{S},\mathbb{R}_0)} \mathbb{E}[\boldsymbol{q}(\infty)][\mathbb{M}]+$$

$$\frac{1}{T}(T-T) \cup_{\boldsymbol{p}_0(\infty)\in[\mathbb{R}_0]} \boldsymbol{p}_0(\infty) = Rng_\infty(\mathbb{S},\mathbb{R}_0)[\mathbb{M}] = Rng_\infty(\mathbb{S},\mathbb{R}_0), \quad (21)$$

where the last equality will be proved in Theorem 10. From (21) we recursively get (20).

Since the conditions of Theorem 4 hold in the present theorem we get that

$$d(Rng_t(\mathbb{S}_1,\mathbb{R}_1),Rng_t(\mathbb{S}_2,\mathbb{R}_2)) \to 0 \text{ geometrically fast.} \tag{22}$$

Now, in (22) replace $Rng_t(\mathbb{S}_1,\mathbb{R}_1)$ with $Rng_t(Rng_\infty(\mathbb{S},\mathbb{R}_0))$ and $Rng_t(\mathbb{S}_2,\mathbb{R}_2)$ with $Rng_t(\mathbb{S},\mathbb{R}_0)$, then we get that

$$d(Rng_t(Rng_\infty(\mathbb{S},\mathbb{R}_0)),Rng_t(\mathbb{S},\mathbb{R}_0)) \to 0, \text{ geometrically fast.} \tag{23}$$

From (20), (23) we arrive at

$$d(Rng_t(\mathbb{S},\mathbb{R}_0),Rng_\infty(\mathbb{S},\mathbb{R}_0) \to 0, \text{ geometrically fast.}$$

□

The above Theorems 4 and 5 have an important consequence since it provides a generalization, by relaxing important assumptions, of the basic asymptotic theory for an NHMS. The theorem is the following and could be proved by just following the analogous steps as in the proofs of Theorems 4 and 5 and Theorem 3.3 in [11]:

Theorem 6. *Consider an NHMS and assume that*

$$\left\{\frac{\Delta T(t)}{T(t)}\right\}_{t=0}^{\infty} \text{ converges geometrically fast.}$$

Let $\{Q(t)\}_{t=0}^{\infty}$ *be the sequence of transition matrices and* $\{p_0(t)\}_{t=0}^{\infty}$ *be the sequence of allocation probabilities. If* $\sup_t \tilde{\mathcal{T}}(Q(t)) < 1$ *then*

$$\lim_{t\to\infty} \mathbb{E}(q(0,t)) = q(\infty) \text{ geometrically fast,}$$

where $q(\infty)$ *is the row of the stable matrix* $Q(\infty) = \lim_{t\to\infty} Q(0,t)$.

The basic asymptotic theorem for an NHMS and which has been used in many papers to provide further results is:

Theorem 7. *Let an NHMS and let that* (a) $\lim_{t\to\infty}\|Q(t)-Q\| = 0$ *and* Q *a regular stochastic matrix;* (b) $lim_{t\to\infty}\|p_0(t)-p_0\| = 0$;*and* (c) $\lim_{t\to\infty}[\Delta T(t)/T(t)] = 0$. *Then*

$$\lim_{t\to\infty}\|\mathbb{E}(q(0,t)) - q(\infty)\| = 0$$

where $q(\infty)$ *is the row of the stable matrix* $Q(\infty) = \lim_{t\to\infty} Q^t$.

Theorem 7 has been used extensively in the theory of NHMS to produce further results see for example [10–12,17,51], and the relaxation of the necessary conditions in Theorem 6 is apparent.

Another consequence is that Theorem 4 provides conditions under which two different NHMS, with the same number of states and population, but different initial states and different allocation probabilities of memberships if they have the same transition probabilities sequence of memberships (PTMS of the embedded Markov chains), they converge in the same expected relative population structure geometrically fast. This is stated in detail in the following theorem which could be proved by just following the analogous steps as in the proof of Theorems 4 and 5.

Theorem 8. *Let two NHMS* $\boldsymbol{a}$ *and* $\boldsymbol{b}$ *which have the same number of states, a common sequence of transition probability matrices of memberships* $\{\boldsymbol{Q}(t)\}_{t=0}^{\infty}$ *and the same total population of memberships. Assume that*

$$\left\{\frac{\Delta T(t)}{T(t)}\right\}_{t=0}^{\infty} \text{ converges geometrically fast.}$$

Let $\boldsymbol{q}_a(0)$ *and* $\boldsymbol{q}_b(0)$ *be their initial relative population structures respectively, and* $\{\boldsymbol{p}_a(t)\}_{t=0}^{\infty}$ *and* $\{\boldsymbol{p}_b(t)\}_{t=0}^{\infty}$ *their sequences of allocation probabilities. If* $\sup_t \tilde{\mathcal{T}}(\boldsymbol{Q}(t)) < 1$ *then*

$$\lim_{t\to\infty} \mathbb{E}(\boldsymbol{q}_a(0,t)) = \lim_{t\to\infty} \mathbb{E}(\boldsymbol{q}_b(0,t)) = \boldsymbol{q}(\infty) \text{ geometrically fast,}$$

where $\boldsymbol{q}(\infty)$ *is the row of the stable matrix* $\boldsymbol{Q}(\infty) = \lim_{t\to\infty} \boldsymbol{Q}(0,t)$.

6. Properties of the Limit Set

In the present section we establish some important properties of the limit set $Rng_\infty(\mathbb{S}, \mathbb{R}_0)$. We start with the following definition:

Definition 6. *Let* n *be an integer such that* $\mathcal{T}(Q_1 Q_2 ... Q_n) < 1$ *for all* $Q_1, Q_2, ..., Q_n \in [\mathbb{M}]$. *Then* $[\mathbb{M}]$ *is said to be product scrambling and* n *its scrambling integer.*

We will make use of the following Theorem 3.3 in [9]:

Theorem 9. *([9]). Let* $\boldsymbol{x}, \boldsymbol{y}$ *be non-compact subsets of* $SM_{1,n}$. *Then using the Hausdorff metric we have*

$$d(\boldsymbol{x}\mathbb{M}, \boldsymbol{y}\mathbb{M}) \le \mathcal{T}(\mathbb{M}) d(\boldsymbol{x}, \boldsymbol{y}).$$

We will now establish with a Theorem some important properties of the limit set.

Theorem 10. *Consider an NHMSS for which* $[\mathbb{M}]$ *is an interval which is uniformly scrambling. Assume that* $T(t) \ge T(t-1) > 0$ *for all* $t = 1, 2, ...$ *and*

$$\left\{\frac{\Delta T(t)}{T(t)}\right\}_{t=0}^{\infty} \text{ converges to zero geometrically fast.}$$

Also, assume that $[\mathbb{S}_0], [\mathbb{R}_0]$ *are compact and convex. Then if we define by*

$$Rng_\infty(\mathbb{S}_0, \mathbb{R}_0) = \lim_{t\to\infty} Rng_t(\mathbb{S}_0, \mathbb{R}_0) = \cap_{t=1}^{\infty} Rng_t(\mathbb{S}_0, \mathbb{R}_0),$$

the limit set $Rng_\infty(\mathbb{S}_0, \mathbb{R}_0)$ *satisfies*

$$Rng_\infty(\mathbb{S}_0, \mathbb{R}_0) = Rng_\infty(\mathbb{S}_0, \mathbb{R}_0)[\mathbb{M}].$$

If in addition $[\mathbb{M}]$ *is product scrambling with integer* n *then it is the unique set that has this property.*

Proof. For the first part of the Theorem 10 since $Rng_\infty(\mathbb{S}_0, \mathbb{R}_0)$ is compact, it is sufficient to show that

$$d(Rng_\infty(\mathbb{S}_0, \mathbb{R}_0), Rng_\infty(\mathbb{S}_0, \mathbb{R}_0)[\mathbb{M}]) = 0.$$

In this respect

$$\delta(Rng_\infty(\mathbb{S}_0, \mathbb{R}_0), Rng_\infty(\mathbb{S}_0, \mathbb{R}_0)[\mathbb{M}]) =$$

$$\max_{\substack{\boldsymbol{q}(0)\in[\mathbb{S}_0],\\ \boldsymbol{p}_0(\tau)\in[\mathbb{R}_0]\\ Q(t)\in[\mathbb{M}] \text{ for all } \tau,t}} \quad \min_{\substack{\boldsymbol{q}(0)\in[\mathbb{S}_0],\\ \boldsymbol{p}_0(\tau)\in[\mathbb{R}_0]\\ Q(t)\in[\mathbb{M}] \text{ for all } \tau,t}} \left\| \lim_{t\to\infty}\{\mathbb{E}[\boldsymbol{q}(0,t+1)]\} - \lim_{t\to\infty}\{\mathbb{E}[\boldsymbol{q}(0,t)]\}[\mathbb{M}] \right\|,$$

where

$$\{\mathbb{E}[\boldsymbol{q}(0,t+1)]\} = \cup_{\boldsymbol{q}(0)\in[\mathbb{S}_0]} \frac{T(0)}{T(t+1)} \boldsymbol{q}(0)\left[\mathbb{M}^{t+1}\right]$$

$$+\frac{1}{T(t+1)} \sum_{\tau=1}^{t+1} \Delta T(\tau) \cup_{\boldsymbol{p}_0(\tau)\in[\mathbb{R}_0]} \boldsymbol{p}_0(\tau)\left[\mathbb{M}^{t-\tau+1}\right].$$

Now there exists a $\boldsymbol{q}^*(0) \in [\mathbb{S}_0]$, $\boldsymbol{Q}^*(t) \in [\mathbb{M}]$ for $t \in [0,t]$, $\boldsymbol{p}_0(\tau) \in [\mathbb{R}_0]$ for $\tau \in [1,t]$, i.e.,

$$\mathbb{E}[\boldsymbol{q}(0,t+1)] = \frac{T(0)}{T(t+1)} \boldsymbol{q}^*(0)\boldsymbol{Q}^*(0,t+1)+$$

$$\frac{1}{T(t+1)} \sum_{\tau=1}^{t+1} \Delta T(\tau)\boldsymbol{p}_0^*(\tau)\boldsymbol{Q}^*(\tau,t+1),$$

such that

$$\delta(Rng_\infty(\mathbb{S}_0,\mathbb{R}_0), Rng_\infty(\mathbb{S}_0,\mathbb{R}_0)[\mathbb{M}]) =$$

$$\min_{\substack{\boldsymbol{q}(0)\in[\mathbb{S}_0],\boldsymbol{p}_0(\tau)\in[\mathbb{R}_0],\\ \boldsymbol{Q}(t)\in[\mathbb{M}] \text{ for all } \tau,t}} \left\| \lim_{t\to\infty} \mathbb{E}[\boldsymbol{q}^*(0,t+1)] - \lim_{t\to\infty} \{\mathbb{E}[\boldsymbol{q}(0,t)]\}[\mathbb{M}] \right\|. \tag{24}$$

Now, for any values of the parameters of $\mathbb{E}[\boldsymbol{q}(0,t)]$ that does not maximize it, the difference above is still greater or equal than the ones which minimize $\mathbb{E}[\boldsymbol{q}(0,t)]$. Thus, we are free to choose the parameters $\boldsymbol{q}^*(0) \in [\mathbb{S}_0]$, $\boldsymbol{Q}^*(0,t) \in [\mathbb{M}^t]$ and $\boldsymbol{p}_0^*(\tau) \in [\mathbb{R}_0]$ for every $\tau \in [0,t]$. With the same reasoning we could choose $\boldsymbol{Q}^*(t)$ in the place of $[\mathbb{M}]$. Thus, we get that

$$\delta(Rng_\infty(\mathbb{S}_0,\mathbb{R}_0), Rng_\infty(\mathbb{S}_0,\mathbb{R}_0)[\mathbb{M}]) =$$

$$\leq \lim_{t\to\infty} \left\| \frac{T(0)}{T(t+1)} \boldsymbol{q}^*(0)\boldsymbol{Q}^*(0,t+1) - \frac{T(0)}{T(t)} \boldsymbol{q}^*(0)\boldsymbol{Q}^*(0,t+1) \right\|$$

$$+ \lim_{t\to\infty} \| \frac{1}{T(t+1)} \sum_{\tau=1}^{t+1} \Delta T(\tau)\boldsymbol{p}_0^*(\tau)\boldsymbol{Q}^*(\tau,t+1) -$$

$$\frac{1}{T(t)} \sum_{\tau=1}^{t} \Delta T(\tau)\boldsymbol{p}_0^*(\tau)\boldsymbol{Q}^*(\tau,t+1) \|$$

$$\leq \lim_{t\to\infty} \left| \frac{T(0)}{T(t+1)} - \frac{T(0)}{T(t)} \right| \|\boldsymbol{q}^*(0)\|1\|\boldsymbol{Q}^*(0,t+1)\| +$$

$$\lim_{t\to\infty} \left| \frac{1}{T(t+1)} - \frac{1}{T(t)} \right| \left\| \sum_{\tau=1}^{t} \Delta T(\tau)\boldsymbol{p}_0^*(\tau)\boldsymbol{Q}^*(\tau,t+1) \right\|$$

$$+ \lim_{t\to\infty} \left| \frac{\Delta T(t)}{T(t+1)} \right| \|\boldsymbol{p}_0^*(t)\|. \tag{25}$$

Since $[\mathbb{M}]$ is uniformly scrambling we have that $\bar{\mathcal{T}} = max_{t\in\mathbb{N}} \mathcal{T}(\boldsymbol{Q}^*(t)) < 1$ and thus $\|\boldsymbol{Q}^*(\tau,t)\| < \bar{\mathcal{T}}^{t-\tau}$ and

$$\left\| \sum_{\tau=1}^{t} \Delta T(\tau)\boldsymbol{p}_0^*(\tau)\boldsymbol{Q}^*(\tau,t+1) \right\| \leq \sum_{\tau=1}^{t} \Delta T(\tau)\bar{\mathcal{T}}^{t-\tau+1},$$

which goes to zero as $t \to \infty$. Hence

$$\delta(Rng_\infty(\mathbb{S}_0,\mathbb{R}_0), Rng_\infty(\mathbb{S}_0,\mathbb{R}_0)[\mathbb{M}]) = 0.$$

In a similar way we could prove that

$$\delta(Rng_\infty(\mathbb{S}_0,\mathbb{R}_0)[\mathbb{M}], Rng_\infty(\mathbb{S}_0,\mathbb{R}_0)) = 0,$$

which lead us to the conclusion

$$d(Rng_{\infty}(\mathbb{S}_0,\mathbb{R}_0)[\mathbb{M}],Rng_{\infty}(\mathbb{S}_0,\mathbb{R}_0))=0.$$

For the second part of the theorem in addition we have that $[\mathbb{M}]$ is product scrambling with index n. Assume that there is a second $Rng^{*}_{\infty}(\mathbb{S}_0,\mathbb{R}_0)$ which is compact for which

$$Rng^{*}_{\infty}(\mathbb{S}_0,\mathbb{R}_0)[\mathbb{M}]=Rng^{*}_{\infty}(\mathbb{S}_0,\mathbb{R}_0).$$

Then we get that

$$\begin{aligned}d(Rng^{*}_{\infty}(\mathbb{S}_0,\mathbb{R}_0),Rng_{\infty}(\mathbb{S}_0,\mathbb{R}_0))&=d(Rng^{*}_{\infty}(\mathbb{S}_0,\mathbb{R}_0)[\mathbb{M}^n],Rng_{\infty}(\mathbb{S}_0,\mathbb{R}_0)[\mathbb{M}^n])\\&\leq(\text{by Theorem 9 })\\&\leq\mathcal{T}[\mathbb{M}^n]d(Rng^{*}_{\infty}(\mathbb{S}_0,\mathbb{R}_0),Rng_{\infty}(\mathbb{S}_0,\mathbb{R}_0)),\end{aligned}$$

and since $\mathcal{T}[\mathbb{M}^n]<1$ we get

$$d(Rng^{*}_{\infty}(\mathbb{S}_0,\mathbb{R}_0),Rng_{\infty}(\mathbb{S}_0,\mathbb{R}_0))=0$$

from which we conclude that $Rng^{*}_{\infty}(\mathbb{S}_0,\mathbb{R}_0)$ and $Rng_{\infty}(\mathbb{S}_0,\mathbb{R}_0)$ are the same set. □

We will now establish an interesting result, that under certain conditions if we have two different inherent Markov set chains for two NHMSS, then the Hausdorff metric of the two sets of all possible expected relative structures of the NHMSS is less than the multiplication of a function of the common bound of the two uniform coefficients of ergodicity of the two intervals and the Hausdorff metric of the two intervals. We will need the following Lemma from ([9] p. 70), or [52].

Lemma 8. *Let $q(0),\tilde{q}(0)$ be stochastic vectors and $Q(t)\in[\mathbb{M}];\tilde{Q}(t)\in[\tilde{\mathbb{M}}]$ for $t=1,2,...$ for $\mathcal{T}(\mathbb{M})\leq\mathcal{T}<1$ and $\mathcal{T}(\tilde{\mathbb{M}})\leq\mathcal{T}<1$. Then*

$$\|q(0)Q(1,t)-\tilde{q}(0)\tilde{Q}(1,t)\|\leq\mathcal{T}^t\|q(0)-\tilde{q}(0)\|1+\left(\mathcal{T}^{t-1}+...+1\right)D$$

where $D=\max_n\|Q(n)-\tilde{Q}(n)\|$.

Theorem 11. *Let two NHMSS with inherent Markov set chains with two different tight intervals $[\mathbb{M}]$ and $[\tilde{\mathbb{M}}]$ which have a common bound, i.e., $\mathcal{T}(\mathbb{M})\leq\mathcal{T}<1$ and $\mathcal{T}(\tilde{\mathbb{M}})\leq\mathcal{T}<1$. Let that for the first NHMSS we have $q(0)\in[\mathbb{S}_1]$ and $p_0(\tau)\in[\mathbb{R}_1]$ for $\tau=1,\dots,t$ where $[\mathbb{S}_1]$, $[\mathbb{R}_1]$ are convex and compact. For the second NHMSS we assume that $\tilde{q}(0)\in[\mathbb{S}_2]$ and $\tilde{p}_0(\tau)\in[\mathbb{R}_2]$ for $\tau=1,\dots,t$ where $[\mathbb{S}_2]$, $[\mathbb{R}_2]$ are convex and compact. Let that the two NHMSS have common total population of memberships $\{T(t)\}_{t=0}^{\infty}$which is known. Assume that $T(t)\geq T(t-1)>0$ and*

$$\left\{\frac{\Delta T(t)}{T(t)}\right\}_{t=0}^{\infty}\ \text{converges to zero geometrically fast.}$$

Denote by $Rng_t(\mathbb{S}_1,\mathbb{R}_1,\mathbb{M})$ the set of all possible expected relative population structures for the first NHMSS and $Rng_t(\mathbb{S}_2,\mathbb{R}_2,\tilde{\mathbb{M}})$ for the second one, respectively. Then the Hausdorff metric of the two limit sets of expected relative population structures is bounded by

$$d(Rng_{\infty}(\mathbb{S}_1,\mathbb{R}_1,\mathbb{M}),Rng_{\infty}(\mathbb{S}_2,\mathbb{R}_2,\tilde{\mathbb{M}}))\leq(1-\mathcal{T})^{-1}d(\mathbb{M},\tilde{\mathbb{M}}).$$

Proof. From Lemma 4 we get that

$$Rng_t(\mathbb{S}_1,\mathbb{R}_1,\mathbb{M})=\{\mathbb{E}[q(0,t)]:\mathbb{E}[q(0,t)]\in\frac{T(0)}{T(t)}\cup_{q(0)\in[\mathbb{S}_1]}q(0)\left[\mathbb{M}^t\right]+$$

$$+\frac{1}{T(t)}\sum_{\tau=1}^{t}\Delta T(\tau)\cup_{p_0(\tau)\in[\mathbb{R}_1]}p_0(\tau)\left[\mathbb{M}^{t-\tau}\right]\}, \tag{26}$$

and an analogous description is valid for the set $Rng_t(\mathbb{S}_2,\mathbb{R}_2,\tilde{\mathbb{M}})$. Since $[\mathbb{S}_1]$, $[\mathbb{R}_1]$, $[\mathbb{S}_2]$, $[\mathbb{R}_2]$ are convex and compact and $[\mathbb{M}]$ and $[\tilde{\mathbb{M}}]$ are tight intervals from Theorem 1 we get that the sets $Rng_t(\mathbb{S}_1,\mathbb{R}_1,\mathbb{M})$ and $Rng_t(\mathbb{S}_2,\mathbb{R}_2,\tilde{\mathbb{M}})$ are convex and compact, hence there is a meaning to get their Hausdorff metric

$$\begin{gathered}d\left(Rng_t(\mathbb{S}_1,\mathbb{R}_1,\mathbb{M}),Rng_t(\mathbb{S}_2,\mathbb{R}_2,\tilde{\mathbb{M}})\right)=\\ \max\{\delta\left(Rng_t(\mathbb{S}_1,\mathbb{R}_1,\mathbb{M}),Rng_t(\mathbb{S}_2,\mathbb{R}_2,\tilde{\mathbb{M}})\right),\\ \delta\left(Rng_t(\mathbb{S}_2,\mathbb{R}_2,\tilde{\mathbb{M}}),Rng_t(\mathbb{S}_1,\mathbb{R}_1,\mathbb{M})\right)\}.\end{gathered} \tag{27}$$

Now, we have that

$$\begin{gathered}\delta\left(Rng_t(\mathbb{S}_1,\mathbb{R}_1,\mathbb{M}),Rng_t(\mathbb{S}_2,\mathbb{R}_2,\tilde{\mathbb{M}})\right)=\\ \max_{\substack{q(0)\in[\mathbb{S}_1],p_0(\tau)\in[\mathbb{R}_1],\\ Q(t)\in[\mathbb{M}]\text{ for all }\tau,t}}\ \min_{\substack{\tilde{q}(0)\in[\mathbb{S}_2],\tilde{p}_0(\tau)\in[\mathbb{R}_2],\\ \tilde{Q}(t)\in[\tilde{\mathbb{M}}]\text{ for all }\tau,t}}\|\mathbb{E}[q(0,t)]-\mathbb{E}[\tilde{q}(0,t)]\|.\end{gathered} \tag{28}$$

There exists some $q^*(0)\in[\mathbb{S}_1], p_0^*(\tau)$ for every $\tau\in[1,t]$, $Q^*(t)\in[\mathbb{M}]$ for every $t\in[0,t]$ which determines a $\mathbb{E}[q^*(0,t)]$; also for any $\tilde{q}^*(0)\in[\mathbb{S}_2]$, any $\tilde{p}_0^*(\tau)$ for every $\tau\in[1,t]$, and finally any $\tilde{Q}^*(t)\in[\tilde{\mathbb{M}}]$ for every $t\in[0,t]$ which determines a $\mathbb{E}[\tilde{q}^*(0,t)]$ for which we have

$$\begin{aligned}\delta\left(Rng_t(\mathbb{S}_1,\mathbb{R}_1,\mathbb{M}),Rng_t(\mathbb{S}_2,\mathbb{R}_2,\tilde{\mathbb{M}})\right)&\le\|\mathbb{E}[q^*(0,t)]-\mathbb{E}[\tilde{q}^*(0,t)]\|\\ &\le\frac{T(0)}{T(t)}\|q^*(0)Q^*(0,t)-\tilde{q}^*(0)\tilde{Q}^*(0,t)\|\\ &+\frac{1}{T(t)}\sum_{\tau=1}^{t}\Delta T(\tau)\left\|p_0^*(\tau)Q^*(\tau,t)-\tilde{p}_0^*(\tau)\tilde{Q}^*(\tau,t)\right\|.\end{aligned} \tag{29}$$

Using Lemma 8 we get that

$$\begin{aligned}&\frac{T(0)}{T(t)}\left\|q^*(0)Q^*(0,t)-\tilde{q}^*(0)\tilde{Q}^*(0,t)\right\|\le\\ &\frac{T(0)}{T(t)}\mathcal{T}^t\|q^*(0)-\tilde{q}^*(0)\|+\frac{T(0)}{T(t)}\left\|\mathcal{T}^{t-1}+...+1\right\|\max_t\left\|Q^*(t)-\tilde{Q}^*(t)\right\|\\ &\le\frac{T(0)}{T(t)}\mathcal{T}^t\delta(\mathbb{S}_1,\mathbb{S}_2)+\frac{T(0)}{T(t)}\left\|\mathcal{T}^{t-1}+...+1\right\|\delta(\mathbb{M},\tilde{\mathbb{M}}).\end{aligned} \tag{30}$$

Similarly, we have

$$\begin{aligned}&\frac{1}{T(t)}\sum_{\tau=1}^{t}\Delta T(\tau)\left\|p_0^*(\tau)Q^*(\tau,t)-\tilde{p}_0^*(\tau)\tilde{Q}^*(\tau,t)\right\|\\ &\le\frac{1}{T(t)}\sum_{\tau=1}^{t}\Delta T(\tau)\delta(\mathbb{R}_1,\mathbb{R}_2)\mathcal{T}^{t-\tau}+\\ &\frac{1}{T(t)}\sum_{\tau=1}^{t}\Delta T(\tau)\left(\mathcal{T}^{t-1}+...+1\right)\delta(\mathbb{M},\tilde{\mathbb{M}}).\end{aligned} \tag{31}$$

From (29), (30) and (31) as $t\to\infty$ we get that

$$\delta\left(Rng_\infty(\mathbb{S}_1,\mathbb{R}_1,\mathbb{M}),Rng_\infty(\mathbb{S}_2,\mathbb{R}_2,\tilde{\mathbb{M}})\right)\le(1-\mathcal{T})^{-1}\delta(\mathbb{M},\tilde{\mathbb{M}}), \tag{32}$$

since

$$\lim_{t\to\infty} \frac{1}{T(t)} \sum_{\tau=1}^{t} \Delta T(\tau)\Big(\mathcal{T}^{t-1} + ... + 1\Big)\delta(\mathbb{M}, \tilde{\mathbb{M}})$$

$$= \lim_{t\to\infty} \frac{1}{T(t)} \delta(\mathbb{M}, \tilde{\mathbb{M}}) \sum_{\tau=1}^{t} \Delta T(\tau) \sum_{k=\tau}^{t-1} \mathcal{T}^{t-k} \leq$$

$$\lim_{t\to\infty} \frac{1}{T(t)} (1-\mathcal{T})^{-1}\delta(\mathbb{M}, \tilde{\mathbb{M}}) \sum_{\tau=1}^{t} \Delta T(\tau) = \left[1 - \frac{T(0)}{T}\right](1-\mathcal{T})^{-1}\delta(\mathbb{M}, \tilde{\mathbb{M}}),$$

also

$$\lim_{t\to\infty} \frac{T(0)}{T(t)} \mathcal{T}^t \delta(\mathbb{S}_1, \mathbb{S}_2) + \lim_{t\to\infty} \frac{T(0)}{T(t)} \left\|\mathcal{T}^{t-1} + ... + 1\right\| 1\delta(\mathbb{M}, \tilde{\mathbb{M}})$$

$$= \frac{T(0)}{T}(1-\mathcal{T})^{-1}\delta(\mathbb{M}, \tilde{\mathbb{M}})$$

and finally

$$\lim_{t\to\infty} \frac{1}{T(t)} \sum_{\tau=1}^{t} \Delta T(\tau)\delta(\mathbb{R}_1, \mathbb{R}_2)\mathcal{T}^{t-\tau+1} = 0,$$

as we have seen in the proof of Theorem 8. Similarly, as we proved (32) we could prove that

$$\delta(Rng_\infty(\mathbb{S}_2, \mathbb{R}_2, \tilde{\mathbb{M}}), Rng_\infty(\mathbb{S}_1, \mathbb{R}_1, \mathbb{M})) \leq (1-\mathcal{T})^{-1}\delta(\tilde{\mathbb{M}}, \mathbb{M}). \quad (33)$$

From (32) and (33) we arrive at the conclusion of the Theorem. □

7. An Illustrative Representative Example

In the present section we present an illustrative representative example to a Geriatric and Stroke Patients system and through it we will present the methodology in terms of computational geometry algorithms needed for an application to any population system. The NHMS model used is a general Coxian phase type model, special forms of which has been used by the school of research by McClean and her co-authors [38,53–59]. We distinguish three states which are called hospital pathways. For the system of Geriatric and Stroke Patients these stages are labeled as "Acute Care", the "Rehabilitative" and the "Long Stay". From each stay we have movements outside the hospital due to discharge or death. Also, geriatric patients may be thought of as progressing through stages of acute care, rehabilitation and long-stay care, where most patients are eventually rehabilitated and discharged. Geriatric medical services are an important asset in the care of elderly and their quality is certainly an indication of the level of civilization in a society. At the same time their funding could be easily reduced due to political pressure on savings in health care expenditure.

It is apparent that the number of pathways could be increased as much as it is needed to accommodate any important characteristics of any patients systems. However, there is no need to consider in here a larger number of states due to the restriction of space. Also, the internal movements in a population of patients could be of any number to accommodate any important characteristics.

Consider a hospital which starts with $T(0) = 400$ patients and in a very short time reaches its full capacity of 435 patients, i.e., $T(1) = 420$, $T(2) = 430$, $T(3) = 435$. Assume three hospital pathways and let that the initial relative population structure be any stochastic vector which lies in the set

$$\mathbb{S}_0 = \{[0 \;\; 0 \;\; 0], \;\; [1 \;\; 1 \;\; 1]\}.$$

The physical meaning of selecting $\mathbb{S}_0$ as above is that the initial relative structure could be any stochastic vector, i.e., $\mathbb{S}_0$ contains all possible initial structures. For example $q(0) = \begin{bmatrix}0.2 & 0.3 & 0.5\end{bmatrix}$ means that 20% of the patients are in pathway 1, 30% are in pathway 2, and 50% are in pathway 3. Now, there are some initial structures which might not be acceptable for the management of the hospital, such as for example let say $q(0) = \begin{bmatrix} 0 & 0 & 1.0 \end{bmatrix}$. In this case in the initial design of the hospital measures should be taken that such a situation will be avoided in cooperation with other nearby hospitals. Then $\mathbb{S}_0$ could be chosen to be

$$\mathbb{S}_0 = \{[0\ 0\ 0],\ [0.1\ 1\ 0.9]\}.$$

This is a convex set which excludes the initial relative structure $q(0) = \begin{bmatrix} 0 & 0 & 1.0 \end{bmatrix}$. However, it also needs to be tight . How to make it tight is explained below with the use of the Algorithm 1. Naturally, we could exclude more than one relative structure from the chosen initial relative structure but the procedure will be the same.

Most new patients enter the system in hospital pathway one, either by taking an empty place or as a virtual replacement of a discharged patient. Hence, let $\mathbb{R}_0$ be the convex set from which allocation probabilities are drawn and let it be the interval

$$\mathbb{R}_0 = \{[0.5\ 0.2\ \ 0.1],\ [0.7\ 0.4\ \ 0.1]\}.$$

The set of stochastic vectors of allocation probabilities that are in the above interval is a convex set with vertices $r_1 = [0.7\ 0.2\ 0.1]$ and $r_2 = [0.5\ 0.4\ 0.1]$. How to find the vertices will be explained below. The physical meaning of the above interval is that any stochastic vector that belongs in the interval $\mathbb{R}_0$ is a candidate to represent a registration policy for the hospital. Naturally, we can restrict our interval $\mathbb{R}_0$ to an interval which will contain the desired recruitment policies of the hospital management and find in this way using the results of the present paper the consequences of these policies. The recruitment vectors are the best control variables for human populations ([2]), as are the hospitals in this case. Methods of control by recruitment could be found in [50,60–62]. However, the interval $\mathbb{R}_0$ needs to be tight and how to make it tight is explained below with the use of the Algorithm 1.

Now, by observing past data it is not difficult to determine an interval of matrices $[\mathbb{M}] = [\check{Q}, \hat{Q}]$ where all stochastic transition matrices of the movements of memberships lie. Please note that the fact that the matrices $\check{Q}, \hat{Q}$ are not necessarily stochastic matrices makes this task easy. We need that $[\mathbb{M}] = [\check{Q}, \hat{Q}]$ should be tight. In order to test that interval $[\mathbb{M}]$ is tight we use Lemma 1 and Definition 5

We chose $[\mathbb{M}]$ to be

$$\check{Q} = \begin{pmatrix} 0.5 & 0.2 & 0.1 \\ 0.2 & 0.6 & 0.2 \\ 0.3 & 0.2 & 0.3 \end{pmatrix} \text{ and } \hat{Q} = \begin{pmatrix} 0.7 & 0.4 & 0.1 \\ 0.2 & 0.6 & 0.2 \\ 0.5 & 0.5 & 0.3 \end{pmatrix},$$

in order that $[\mathbb{M}]$ is tight every row should be tight hence using Lemma 1 we could see that the interval is tight. Hence, the stochastic matrices that belong to $[\mathbb{M}]$ is a convex polytope and we need to find its vertices. The same applies for $\mathbb{S}_0$ and $\mathbb{R}_0$ which are tight, and we need to find the vertices of the convex polytope on which all stochastic vectors of the interval lie.

Applying Lemma 1 we find that the set $\mathbb{S}_0$ is tight and applying Algorithm 1 we get that $\mathbb{S}_0$ is convex with vertices $v_1 = (1\ 0\ 0)$, $v_2 = (0\ 1\ 0)$ and $v_3 = (0\ 0\ 1)$. The physical meaning of the set $\mathbb{S}_0$ in here is that at the start of our study at time 0 we allow that the hospital could have any relative population structure. Applying Lemma 1 we find that the set $\mathbb{R}_0$ is tight and applying Algorithm 1 we get that $\mathbb{R}_0$ is convex with vertices $r_1 = (0.7\ 0.2\ 0.1)$, $r_2 = (0.5\ 0.4\ 0.1)$.

Algorithm 1 ([9]) **Finding the vertices of a tight interval** $[\boldsymbol{p}, \boldsymbol{q}]$.

For each $i = 1, 2, 3$ construct the vectors:

For $i = 1$ and the vector $\boldsymbol{p}$

$$[p_1\ p_2\ p_3]\,,\ [p_1\ \ p_2\ \ q_3]\,,\ [p_1\ q_2\ \ p_3]\,,\ [p_1\ q_2\ \ q_3],$$

replace p_1 with $\tilde{p}_1$ such that the above vectors will become stochastic.
If any of the four resulting stochastic vectors

$$[\tilde{p}_1\ p_2\ p_3]\,,\ [\tilde{p}_1\ \ p_2\ \ q_3]\,,\ [\tilde{p}_1\ q_2\ \ p_3]\,,\ [\tilde{p}_1\ q_2\ \ q_3],$$

belongs in the interval $[\boldsymbol{p}, \boldsymbol{q}]$ then it is a vertex.

Do the same for the vector $\boldsymbol{q}$.

END.

To find the vertices of the convex set of stochastic matrices that belong to the tight interval $[\mathbb{M}]$ we apply Algorithm 1 for each row vector in $[\check{\boldsymbol{Q}}, \hat{\boldsymbol{Q}}]$ and we find that the vertices are

$$\boldsymbol{V}_1 = \begin{pmatrix} 0.7 & 0.2 & 0.1 \\ 0.2 & 0.6 & 0.2 \\ 0.5 & 0.2 & 0.3 \end{pmatrix}, \quad \boldsymbol{V}_2 = \begin{pmatrix} 0.7 & 0.2 & 0.1 \\ 0.2 & 0.6 & 0.2 \\ 0.3 & 0.4 & 0.3 \end{pmatrix},$$

$$\boldsymbol{V}_3 = \begin{pmatrix} 0.5 & 0.4 & 0.1 \\ 0.2 & 0.6 & 0.2 \\ 0.5 & 0.2 & 0.3 \end{pmatrix}, \quad \boldsymbol{V}_4 = \begin{pmatrix} 0.5 & 0.4 & 0.1 \\ 0.2 & 0.6 & 0.2 \\ 0.3 & 0.4 & 0.3 \end{pmatrix}.$$

Now we compute all the row vectors $\boldsymbol{v}_i \boldsymbol{V}_j$ for $i = 1, 2, 3$ and $j = 1, 2, 3, 4$:

$$\begin{pmatrix}0.7 & 0.2 & 0.1\end{pmatrix}, \begin{pmatrix}0.7 & 0.2 & 0.1\end{pmatrix}, \begin{pmatrix}0.5 & 0.4 & 0.1\end{pmatrix}, \begin{pmatrix}0.5 & 0.4 & 0.1\end{pmatrix},$$

$$\begin{pmatrix}0.2 & 0.6 & 0.2\end{pmatrix}, \begin{pmatrix}0.2 & 0.6 & 0.2\end{pmatrix}, \begin{pmatrix}0.2 & 0.6 & 0.2\end{pmatrix}, \begin{pmatrix}0.2 & 0.6 & 0.2\end{pmatrix}$$

$$\begin{pmatrix}0.5 & 0.2 & 0.3\end{pmatrix}, \begin{pmatrix}0.3 & 0.4 & 0.5\end{pmatrix}, \begin{pmatrix}0.5 & 0.2 & 0.3\end{pmatrix}, \begin{pmatrix}0.3 & 0.4 & 0.3\end{pmatrix}.$$

These 12 stochastic vectors belong to a convex set, hence using any of the computational geometry methods in [61] we find that the vertices of the convex hull of these vectors are

$$\omega_1 = \begin{pmatrix}0.7 & 0.2 & 0.1\end{pmatrix}, \omega_2 = \begin{pmatrix}0.5 & 0.4 & 0.1\end{pmatrix}, \omega_1 = \begin{pmatrix}0.2 & 0.6 & 0.2\end{pmatrix},$$

$$\omega_4 = \begin{pmatrix}0.5 & 0.2 & 0.3\end{pmatrix}, \omega_5 = \begin{pmatrix}0.3 & 0.4 & 0.3\end{pmatrix}.$$

Hence

$$Rng_1(\mathbb{S}_0, \mathbb{R}_0) = \frac{T(0)}{T(1)} conv\{\omega_1, \omega_2, \omega_3, \omega_4, \omega_5\} +$$

$$\frac{1}{T(1)}[T(1) - T(0)] conv\{r_1, r_2\},$$

from which we get that $Rng_1(\mathbb{S}_0, \mathbb{R}_0)$ is the convex set with vertices given by the above Minkowski sum thus

$$\begin{pmatrix}0.7 & 0.2 & 0.1\end{pmatrix}, \begin{pmatrix}0.51 & 0.39 & 0.1\end{pmatrix}, \begin{pmatrix}0.22 & 0.58 & 0.2\end{pmatrix}, \begin{pmatrix}0.51 & 0.2 & 0.29\end{pmatrix},$$

$$\begin{pmatrix}0.32 & 0.39 & 0.29\end{pmatrix}, \begin{pmatrix}0.69 & 0.21 & 0.1\end{pmatrix}, \begin{pmatrix}0.5 & 0.4 & 0.1\end{pmatrix},$$

$$\begin{pmatrix}0.21 & 0.59 & 0.2\end{pmatrix}, \begin{pmatrix}0.5 & 0.21 & 0.29\end{pmatrix}, \begin{pmatrix}0.31 & 0.4 & 0.29\end{pmatrix},$$

$$\begin{pmatrix}0.51 & 0.2 & 0.29\end{pmatrix}.$$

Taking into account the rounding errors done with all the multiplications and additions we compute the vertices with one decimal point of accuracy and we get that $Rng_1(\mathbb{S}_0, \mathbb{R}_0)$ is the convex hull of the vertices

$$(0.7 \quad 0.2 \quad 0.1), \ (0.5 \quad 0.4 \quad 0.1), \ (0.2 \quad 0.6 \quad 0.2), \ (0.5 \quad 0.2 \quad 0.3),$$
$$\text{and} \ (0.3 \quad 0.4 \quad 0.3).$$

This result verifies Theorem 1. With now the vertices of the convex set $Rng_1(\mathbb{S}_0, \mathbb{R}_0)$ we repeat the previous process to find $Rng_2(\mathbb{S}_0, \mathbb{R}_0)$.

Now, in this way at every point in time we have the convex compact space $Rng_t(\mathbb{S}_0, \mathbb{R}_0)$ of all possible expected population structures. If any of these are problematic in some way then apparently the hospital has a lead time to adapt new policies and an instrument to visualize their consequences.

To verify that a tight interval of transition probability matrices is uniformly scrambling we need a sufficient condition as a criterion. This is given in the following easily proved Lemma.

Lemma 9. *Let* $[\mathbb{M}] = [\check{Q}, \hat{Q}]$ *then* $[\mathbb{M}]$ *is uniformly scrambling if the following holds* $\mathcal{T}(\hat{Q} - \check{Q}) < 1$.

Let $[\mathbb{M}]$ be the interval of the application we are working so far, then it is easy to check that $\mathcal{T}(\hat{Q} - \check{Q}) < 1$ and hence any stochastic matrix selected from $[\mathbb{M}]$ will be scrambling. We select as $Q(1), Q(2), Q(3), Q(4)$ the four vertices of the convex set of stochastic matrices in $[\mathbb{M}]$ and as $Q(6), Q(7)$ any convex combination of them. We also select as $q(0) = (0.5 \quad 0.25 \quad 0.25)$ and as vectors of allocation probabilities we select $p_0(1) = (0.6 \quad 0.3 \quad 0.1)$ and $p_0(2) = (0.5 \quad 0.4 \quad 0.1)$ which both belong to $\mathbb{R}_0$. Then we compute

$$\mathbb{E}[q(0,1)] = (0.5 \quad 0.3 \quad 0.2)$$
$$\mathbb{E}[q(0,2)] = (0.48 \quad 0.36 \quad 0.16)$$
$$\mathbb{E}[q(0,3)] = (0.4 \quad 0.43 \quad 0.17)$$
$$\mathbb{E}[q(0,4)] = (0.34 \quad 0.48 \quad 0.18)$$
$$\mathbb{E}[q(0,5)] = (0.35 \quad 0.46 \quad 0.19)$$
$$\mathbb{E}[q(\infty)] = (0.4 \quad 0.4 \quad 0.2).$$

Hence the expected relative population structure converges in six steps, that is geometrically fast which verifies Theorem 6. Also, it is easy to see that

$$\mathbb{E}[q(\infty)]V_1 = (0.4 \quad 0.4 \quad 0.2)\begin{pmatrix} 0.7 & 0.2 & 0.1 \\ 0.2 & 0.6 & 0.2 \\ 0.5 & 0.2 & 0.3 \end{pmatrix} = (0.4 \quad 0.4 \quad 0.2),$$

and the same happens with all the vertices of $[\mathbb{M}]$ which was proved in Theorem 10.

The hospital now has beforehand knowledge with a good lead time where its policies and tendencies of the hospital system will converge in terms of relative expected population structure. Hence it is able to decide if this is a desirable situation; to find out if it can cope with the resources available in doctors, nurses and medical material; if its medical facilities are adequate; it can also have an estimate of the cost of the system see [62,63].

Consider now that the previous NHMS is system a and let b be a second NHMS with initial population structure $q(0) = (0.6 \quad 0.2 \quad 0.2)$; allocation probabilities $p_0(1) = (0.5 \quad 0.4 \quad 0.1)$ and $p_0(2) = (0.7 \quad 0.2 \quad 0.1)$ and the remaining parameters the same. Then the asymptotically relative population structure is again $\mathbb{E}[q(\infty)] = (0.4 \quad 0.4 \quad 0.2)$ which verifies Theorem 8. The physical meaning of the previous result is that when the hospital is at full capacity for some time, then with different initial structures and allocation probabilities the expected relative population structure remains unchanged under the condition that the maximum ergodicity coefficient of the transition probability matrices is less than one.

To be able to use for the benefit of the hospital the last theorem, that is Theorem 11, we need a way to find a numerical value that will replace $d(\mathbb{M}, \tilde{\mathbb{M}})$. The following Lemma, which is not difficult to be proved, provides a solution to the problem.

Lemma 10. *Let $\mathbb{M}, \tilde{\mathbb{M}}$ be two tight intervals of transition probability matrices for the memberships. Let $V_1, V_2, \ldots, V_m$ be the vertices of the convex set $\mathbb{M}$, and $U_1, U_2, \ldots, U_n$ be the vertices of the convex set $\tilde{\mathbb{M}}$. Then*

$$\delta(\mathbb{M}, \tilde{\mathbb{M}}) = \|a_1^* V_1 + a_2^* V_2 + \ldots + a_m^* V_m - [b_1^* U_1 + b_2^* U_2 + \ldots + b_n^* U_n]\|,$$

where $a_1^*, a_2^*, \ldots, a_m^*$ is the solution of the optimization problem

$$\max[a_1 V_1 + a_2 V_2 + \ldots + a_m V_m] \text{ with}$$

$$a_1 + a_2 + \ldots + a_m = 1, a_1 \geq 0, a_2 \geq 0, \ldots, a_m \geq 0.$$

and b_1^*,b_2^*,..., b_n^* is the solution of the optimization problem

$$\min[b_1 U_1 + b_2 U_2 + \ldots + b_n U_n] \text{ with}$$

$$b_1 + b_2 + \ldots + b_n = 1, b_1 \geq 0, b_2 \geq 0, \ldots, b_m \geq 0.$$

In what follows we summarize in Algorithm 2 for convenience of the interest readers the previous steps which are necessary for using the present results in any population system.

Algorithm 2

Use Lemma 1 to check that $[\mathbb{S}_0]$ is tight.
Apply **Algorithm 1** to find the vertices of the convex set $[\mathbb{S}_0]$:

$$v_1, v_2, \ldots, v_s.$$

Use Lemma 1 to check that $[\mathbb{R}_0]$ is tight.
Apply **Algorithm 1** to find the vertices of the convex set $[\mathbb{R}_0]$:

$$r_1, r_2, \ldots, r_\pi.$$

Use Lemma 1 to check that each row in $[\check{Q}, \hat{Q}]$ is tight. If yes then $[\mathbb{M}]$ is tight.
Apply **Algorithm 1** for each row vector in $[\check{Q}, \hat{Q}]$ to find the vertices of $[\mathbb{M}]$:

$$V_1, V_2, \ldots, V_m.$$

Compute all the raw vectors

$$v_i V_j \text{ for all } i = 1, 2, \ldots, s; j = 1, 2, \ldots, m.$$

Use any of the computational geometry methods in [64].
to find the vertices of the convex hull of the vectors $v_i V_j$ for all $i = 1, 2, \ldots, s; j = 1, 2, \ldots, m$.
Let that

$$\omega_1, \omega_2, \ldots, \omega_v,$$

the vertices found. Compute using properties of the Minkowski sum of vectors

$$Rng_1(\mathbb{S}_0, \mathbb{R}_0) = \frac{T(0)}{T(1)} conv\{\omega_1, \omega_2, \omega_3, \omega_4, \omega_5\} +$$

$$\frac{1}{T(1)}[T(1) - T(0)] conv\{r_1, r_2\}.$$

Repeat the process until $Rng_\infty(\mathbb{S}_0, \mathbb{R}_0)$ geometrically fast, i.e., in 6 to 8 steps.
Use Lemma 9 to verify that $[\mathbb{M}] = [\check{Q}, \hat{Q}]$ is uniformly scrambling, i.e., $\mathcal{T}(\hat{Q} - \check{Q}) < 1$.
Use Lemma 10 to find bounds for

$$\delta(\mathbb{M}, \tilde{\mathbb{M}}) < \mu \text{ and } \delta(\tilde{\mathbb{M}}, \mathbb{M}) < \tilde{\mu}.$$

Set

$$d(\mathbb{M}, \tilde{\mathbb{M}}) < \max(\mu, \tilde{\mu}).$$

8. Conclusions

The concept of the non-homogeneous Markov set system was introduced which is a NHMS with its parameters in an interval. It was established under which conditions in a NHMSS the set of all possible expected relative population structures at a certain point in time is a convex set and a convex polygon. Then it was founded that if in an NHMSS the sets of initial structures are different but compact and convex; also, the sets of allocation probabilities of the memberships are different but convex and compact; the inherent non-homogeneous Markov set chain is common; then the Hausdorff metric of the two different sets of all possible expected relative structures asymptotically goes to zero geometrically fast, i.e., asymptotically they coincide geometrically fast. Then we established that in an NHMSS if the total population of memberships converge in a finite number geometrically fast, and the sets of initial structures and allocation probabilities of memberships are compact and convex, then the set of all possible expected relative population structure converge to a limit set geometrically fast. Then it was proved that these results generalize certain well-known results for NHMS's. Then it was proved that under some mild conditions the limit set of the expected relative population structures of an NHMSS remains invariant if any selected transition probability matrix of the inherent non-homogeneous Markov chain from the respective interval is multiplied by it from the right. It was also proved that the limit set is the only set with this property if the interval of selection of transition probabilities of the inherent non-homogeneous Markov chain is product scrambling. Finally it was assumed that two different NHMSS in the sense that they have different sets of selecting initial distributions, different sets of selecting allocation probabilities and different intervals of selecting the transition probabilities of the inherent non-homogeneous Markov chains, while they have in common that their respective intervals are uniformly scrambling with a common bound and they have the same total population of memberships. Then it was proved that the Hausdorff metric of the limit sets of the expected relative population structures of the two NHMSS is bounded by the multiplication of a function of the Hausdorff metric of the two tight intervals of selection of the stochastic matrices of the inherent non-homogeneous Markov set chains and the bound of their uniform coefficients of ergodicity.

Funding: This research received no external funding.

Institutional Review Board Statement: Not applicable.

Informed Consent Statement: Not applicable.

Conflicts of Interest: The authors declare no conflict of interest.

References

1. Vassiliou, P.-C.G. Asymptotic behavior of Markov systems. *J. Appl. Probab.* **1982**, *19*, 851–857. [CrossRef]
2. Bartolomew, D.J. *Stochastic Models for Social Processes*; Wiley: New York, NY, USA, 1982.
3. Young, A.; Vassiliou, P.-C.G. A non-linear model on the promotion of staff. *J. R. Stat. Soc. A* **1974**, *138*, 584–595. [CrossRef]
4. Vassiliou, P.-C.G. A markov model for wastage in manpower systems. *Oper. Res. Quart.* **1976**, *27*, 57–70. [CrossRef]
5. Vassiliou, P.-C.G. A high order non-linear Markovian model for prediction in manpower systems. *J. R. Stat. Soc. A* **1978**, *141*, 86–94. [CrossRef]
6. Neumaier, A. *Interval Methods for Systems of Equations*; Cambridge University Press: Cambridge, UK, 1990.
7. Alfred, G.; Herzberger, J. *Introduction to Interval Computations*; Academic Press: New York, NY, USA, 1983.
8. Ben-Haim, Y.; Elishakoff, I. *Convex models of uncertainty in Applied Mathematics*; Elsevier: New York, NY, USA, 1990.
9. Hartfiel, D.J. *Markov Set-Chains*; Springer: Heidelberg, Germany, 1998.
10. Georgiou, A.C.; Vassiliou, P.-C.G. Periodicity of asymptotically attainable structures in non-homogeneous Markov systems. *Linear Algebra Its Appl.* **1992**, *176*, 137–174. [CrossRef]
11. Vassiliou, P.-C.G. On the limiting behaviour of a non-homogeneous Markov chain model in manpower systems. *Biometrika* **1981**, *68*, 557–561.
12. Vassiliou, P.-C.G. The evolution of the theory of non-homogeneous Markov systems. *Appl. Stoch. Models Data Anal.* **1997**, *13*, 159–176. [CrossRef]
13. Ugwuogo, F.I.; Mc Clean, S.I. Modelling heterogeneity in manpower systems. A review. *Appl. Stoch. Models Bus. Ind.* **2000,** *2*, 99–110. [CrossRef]

14. Vassiliou, P.-C.G. Markov systems in a General State Space. *Commun. Stat.-Theory Methods* **2014**, *43*, 1322–1339. [CrossRef]
15. Vassiliou, P.-C.G. Rate of Convergence and Periodicity of the Expected Population Structure of Markov Systems that Live in General State Space. *Mathematics* **2020**, *8*, 1021. [CrossRef]
16. Vassiliou, P.-C.G. Exotic properties of non-homogeneous Markov and semi-Markov systems. *Commun. Stat. Theory Methods* **2013**, *42*, 2971–2990. [CrossRef]
17. Vassiliou, P.-C.G. Laws of Large numbers for non-homogeneous Markov systems. *Methodol. Comput. Appl. Prob.* **2020**, 1631–1658. [CrossRef]
18. Mathew, E.; Foucher, Y.; Dellamonika, P.; Daures, J.-P. *Parametric and Non-Homogeneous Semi-Markov Process for HIV Control;* Working Paper; Archer Hospital: Nice, France, 2006.
19. Foucher, Y.; Mathew, E.; Saint Pierre, P.; Durant, J.-F.; Daures, J.-P. A semi-Markov model based on generalized Weibull distribution with an illustration for HIV disease. *Biom. J.* **2005**, *47*, 825–833. [CrossRef]
20. Dessie, Z.G. Modeling of HIV/AIDS dynamic evolution using non-homogeneous semi-Markov processes. *Springerplus* **2014**, *3*, 537. [CrossRef] [PubMed]
21. Saint Pierre, P. Modelles Multi-Etas de Type Markovien et Application a la Astme. Ph.D. Thesis, University of Monpellier I, Montpellier, France, 2005.
22. Barbu, V.; Boussement, M.; Limnios, N. Discrete-time semi-Markov model for reliability and survival analysis. *Commun. Stat. Theory Methods* **2004**, *33*, 2833–2886. [CrossRef]
23. Perez-Ocon, R.; Castro, J.E.R. A semi-Markov model in biomedical studies. *Commun. Stat. Theory Methods* **2004**, *33*, 437–455.
24. Ocana-Riola, R. Non-homogeneous Markov process for biomedical data analysis. *Biom. J.* **2005**, *47*, 369–376. [CrossRef]
25. De Freyter, T. Modelling heterogenity in Manpower Planning. Dividing the personel system in more homogeneous subgroups. *Appl. Stoch. Model Bus. Ind.* **2006**, *22*, 321–334. [CrossRef]
26. Niakantan, K.; Raghavendra, B.G. Control aspects in proportionality Markov manpower systems. *Appl. Stoch. Models Data Anal.* **2005**, *7*, 27–41.
27. Yadavalli, V.S.S.; Natarajan, R.; Udayabhaskaran, S. Optimal training policy for promotion-stochastic models of manpower systems. *Electron. Publ.* **2002**, *13*, 13–23. [CrossRef]
28. De Freyter, T.; Guerry, M. Markov manpower models: A review. In *Handbook of Optimization Theory: Decision Analysis and Applications*; Varela, J., Acuidja, S., Eds.; Nova Science Publishers: Hauppauge, NY, USA, 2011; pp. 67–88.
29. Guerry, M.A. Some results on the Embeddable problem for discrete time Markov models. *Commun. Stat. Theory Methods* **2014**, *43*, 1575–1584. [CrossRef]
30. Esquivel, M.L.; Fernandez, J.M.; Guerreiro, G.R. On the evaluation and asymptotic analysis of open Markov populations: Application to consumption credit. *Stoch. Model.* **2014**, *30*, 365–389. [CrossRef]
31. Esquivel, M.L.; Patricio, P.; Guerreiro, G.R. From ODE to open Markov chains, via SDE: An application to models of infections in individuals and populations. *Comput. Math.* **2020**, *8*, 180–197.
32. Papadopoulou, A.A. Some results on modeling biological sequences and web navigation with a semi-Markov chain. *Commun. Stat. Theory Methods* **2013**, *41*, 2853–2871. [CrossRef]
33. Patoucheas, P.D.; Stamou, G. Non-homogeneous Markovian models in ecological modelling: Astudy of the zoobenthos dynamics in Thermaikos Gulf, Greece. *Ecol. Model.* **1993**, *66*, 197–215. [CrossRef]
34. Crooks, G.E. Path ensemble averages in system driven far from equilibrium. *Phys. Rev. E* **2000**, *61*, 2361–2366. [CrossRef]
35. Faddy, M.J.; McClean, S.J. Markov chain for for geriatricpatient care. *Methods Inf. Med.* **2005**, *44*, 369–373. [CrossRef]
36. Gurnescu, F.; McClean, S.J.; Millard, P.H. A queueing model for bed-occupancy management and planning of hospitals. *J. Oper. Res. Soc.* **2002**, *5*, 307–312. [CrossRef]
37. Gurnescu, F.; McClean, S.J.; Millard, P.H. Using a queueing model to help plan bed allocation in a department of geriatric medicine. *Health Care Manag. Sci.* **2004**, *7*, 285–289.
38. Marshall, A.H.; McClean, S.I. Using Coxian phase type distributions to identify patient characteristics for duration of stay in hospital. *Health Care Manag. Sci.* **2004**, *7*, 285–289. [CrossRef] [PubMed]
39. McClean, S.I.; Millard, P. Where to treat the older patient? Can Markov models help us better understand the relationship between hospital and community care. *J. Oper. Res. Soc.* **2007**, *58*, 255–261. [CrossRef]
40. Hartfiel, D.J.; Seneta, E. On the theory of Markov set chains. *Adv. Appl. Probab.* **1994**, *26*, 947–964. [CrossRef]
41. Hartfiel, D.J. Sequential limits in Markov set chains. *J. Appl. Probab.* **1991**, *28*, 910–913. [CrossRef]
42. Hartfiel, D.J. Homogeneous Markov chains with bounded transition matrix. *Stoch. Proc. Appl.* **1994**, *50*, 275–279. [CrossRef]
43. Hartfiel, D.J. Intermediate Markov systems. *Appl. Math. Comput.* **1995**, *72*, 51–59.
44. Eggleston, H.G. *Convexity*; Cambridge University Press: Cambridge, UK, 2008.
45. Hartfiel, D.J.; Rothblum, U. Convergence of inhomogeneous products of matrices and coefficients of ergodicity. *Linear Algebra Its Appl.* **1998**, *277*, 1–9. [CrossRef]
46. Rhodius, A. On explicit forms for ergodicity coefficients. *Linear Algebra Its Appl.* **1993**, *194*, 71–83. [CrossRef]
47. Seneta, E. Sensitivity of finite Markov chains under perturbation. *Stat. Probab. Lett.* **1993**, *17*, 163–168. [CrossRef]
48. Berge, C. *Topological Spaces*; Oliver Boyd: Edinburgh, UK, 1963.
49. Vassiliou, P.-C.G.; Tsaklidis, G. The rate of convergence of the vector of variances and covariances in non-homogeneous Markov systems. *J. Appl. Probab.* **1989**, *26*, 776–783. [CrossRef]

50. Vassiliou, P.-C.G.; Georgiou, A.C. Asymptotically attainable structures in nonhomogeneous Markov systems. *Oper. Res.* **1990**, *38*, 537–545 [CrossRef]
51. Vassiliou, P.-C.G. On the periodicity of non-homogeneous Markov chains and systems. *Linear Algebra Its Appl.* **2015**, *471*, 654–684. [CrossRef]
52. Hartfiel, D.J. Results on limiting sets of Markov set chains. *Linear Algebra Its Appl.* **1993**, *105*, 155–163. [CrossRef]
53. McClean, S.I.; Millard, P. A three compartment model of the patient flows in a geriatric department: A decision support approach. *Health Care Manag. Sci.* **1998**, *7*, 285–289.
54. McClean, S.I.; McAlea, B.; Millard, P. Using a Markov reward model to estimate spend-down costs for geriatric department. *J. Oper. Res. Soc.* **1998**, *49*, 1021–1025. [CrossRef]
55. Taylor, G.J.; McClean, S.I.; Millard, P. Stochastic models of geriatric patient bed occupancy behavior. *J. R. Stat. Soc. A* **2000**, *163*, 39–48. [CrossRef]
56. Marshall, A.H.; McClean, S.I.; Shapcott, C.M.; Millard, P. Modelling patient duration of stay to facilitate resource management of geriatric hospitals. *Health Care Manag. Sci.* **2002**, *5*, 313–319. [CrossRef]
57. Marshall, A.H.; McClean, S.I. Conditional phase type distributionsfor modelling patient length of stay in hospital. *Int. Trans. Oper. Res.* **2003**, *10*, 565–576. [CrossRef]
58. McClean, S.I.; Gillespie, J.; Garg, L.; Barton, M.; Scotney, B.; Fullerton, K. Using phase-type models to cost stroke patient care across health, social and community services. *Eur. J. Oper. Res.* **2014**, *236*, 190–199. [CrossRef]
59. McClean, S.I.; Garg, L.; Fullerton, K. Costing mixed Coxian phase type systems with Poisson arrivals. *Commun. Stat. Methods* **2014**, *43*, 1437–1452. [CrossRef]
60. Vassiliou, P.-C.G.; Tsantas, N. Maintainability of structures in non-homogeneous Markov systems under cyclic behavior and input control. *SIAM J. Appl. Math.* **1984**, *44*, 1014–1022. [CrossRef]
61. Vassiliou, P.-C.G.; Tsantas, N. Stochastic control in non-homogeneous Markov systems. *Int. J. Comput.* **1984**, *16*, 139–155.
62. Georgiou, A.C.; Vassiliou, P.-C.G. Cost models in non-homogeneous Markov systems. *Eur. J. Oper. Res.* **1997**, *100*, 81–96. [CrossRef]
63. De Feyter, T.; Guerry, M.-A.; Komarudin. Optimizing cost-effectiveness in stochastic Markov manpwer planning system under control by recruitment. *Ann. Oper. Res.* **2017**, *253*, 117–131. [CrossRef]
64. Preparata, F.P.; Shamos, M.I. *Computational Geometry. An Introduction*; Springer: New York, NY, USA, 1985

mathematics

Article

Period-Life of a Branching Process with Migration and Continuous Time

Khrystyna Prysyazhnyk [1,*], Iryna Bazylevych [2], Ludmila Mitkova [3] and Iryna Ivanochko [4]

1 Artificial Intelligence Department, Institute of Computer Sciences and Information Technologies, Lviv Polytechnic National University, 79013 Lviv, Ukraine
2 Faculty of Mechanics and Mathematics, Ivan Franko National University of Lviv, 79000 Lviv, Ukraine; i_bazylevych@yahoo.com
3 Department of Economics and Finance, Comenius University, 82005 Bratislava, Slovakia; ludmila.mitkova@fm.uniba.sk
4 Department of Management and International Business, Lviv Polytechnic National University, 79000 Lviv, Ukraine; irene.ivanochko@gmail.com
* Correspondence: yakymyshyn_hrystyna@ukr.net or khrystyna.m.yakymyshyn@lpnu.ua

Abstract: The homogeneous branching process with migration and continuous time is considered. We investigated the distribution of the period-life τ, i.e., the length of the time interval between the moment when the process is initiated by a positive number of particles and the moment when there are no individuals in the population for the first time. The probability generating function of the random process, which describes the behavior of the process within the period-life, was obtained. The boundary theorem for the period-life of the subcritical or critical branching process with migration was found.

Keywords: branching process; migration; continuous time; generating function; period-life

MSC: 60J80; 60J85

Citation: Prysyazhnyk, K.; Bazylevych, I.; Mitkova, L.; Ivanochko, I. Period-Life of a Branching Process with Migration and Continuous Time. *Mathematics* **2021**, *9*, 868. https://doi.org/10.3390/math9080868

Academic Editors: Panagiotis-Christos Vassiliou and Andreas C. Georgiou

Received: 22 March 2021
Accepted: 12 April 2021
Published: 15 April 2021

Publisher's Note: MDPI stays neutral with regard to jurisdictional claims in published maps and institutional affiliations.

Copyright: © 2021 by the authors. Licensee MDPI, Basel, Switzerland. This article is an open access article distributed under the terms and conditions of the Creative Commons Attribution (CC BY) license (https://creativecommons.org/licenses/by/4.0/).

1. Introduction

Branching processes (BPs) are often used as mathematical models of different real processes, in particular, chemical [1], biological [2], genetic [3], demographic [4], technical [5] and others. In addition, BPs can describe the population dynamics of particles of different natures, in particular, they can be photons, electrons, neutrons, protons, atoms, molecules, cells, microorganisms, plants, animals, individuals, prices, information, etc. This list can be continued. Thus, a BP is quite widely used in various sciences. Since third party factors often exist, there is a need to study different modifications of this process. Among them are BPs with immigration, emigration, or a combination of two processes, namely processes with migration for the case of discrete or continuous time.

The theory of Non-Homogeneous Markov systems first introduced in [6]. The case of the Non-Homogeneous Markov systems in continuous time in its latest results exist in Dimitriou and Georgiou [7].

For the first time, the term period-life (PL) τ for the Galton–Watson BP and the Markov BP with immigration was considered by Zubkov A.M. in [8]. He obtained asymptotic formulas for the distribution tails as a function of the PL and found the necessary and sufficient conditions for the process to obtain zero for the corresponding Markov chain. Vatutin V.A. [9] continued the study of PL for the critical case and obtained a limit theorem on the behavior of the process at $\tau > t$ and provided that the beginning of the PL $T = 0$. PL for a critical BP with random migration and discrete time were studied by Yanev N.M. and Mitov K.V. [10]. The distribution PL of the BP with immigration in a limited environment and its behavior in the PL was investigated by Boyko R.V. [11]. Formanov Sh. K., Yasin M.T. [12] obtained boundary theorems for the PL of critical BP Galton–Watson with migration. The case of the Bellman–Harris BP with immigration was studied in [13,14]. The distribution of the PL for the subcritical and critical BP with immigration in a random environment was studied in the works [15,16]. However, the asymptotic properties of the PL of the branching process with migration and continuous time (BPMCT) have not been considered to date.

The case of a BPMCT is considered in [17–21]. Chen A. Y. and Renshaw E. [17] have considered a case of the process in which large immigration, i.e., the sum of immigration rates, is infinite; excessively high population levels are avoided by allowing the carrying capacity of the system to be controlled by mass emigration. Rahimov I. and Al-Sabah W.S. [18] have investigated a family of independent, equally distributed with a continuous time Markov BP. The migration was determined as follows: the particles first immigrate and stay in the population for some time, and then emigrate. Srivastava O. P. and Gupta S. [19] have considered a branching process in which the migration and the emigration of the particle occur independently of each other and with the same probability. Pakes A. G [20] has studied the process of when a batch of immigrants arrive in a region at event times of a renewal process and the individuals grow according to a Bellman–Harris branching process. Tribal emigration allows the possibility that all descendants of a group of immigrants collectively leave the region at some instant. Balabaev I. S. [21] considered the Khan–Nagaev process [22] as a nested BP.

In this article, we consider a more general model of the BPMCT [23]. Immigration, emigration, and evolution occur at random moments in time are determined by the intensity of the transition probabilities.

The purpose of this work is to study the PL of the BPMCT and find the distribution of the PL and the boundary distribution in the case of the subcritical or critical process. This model can be used to model real processes, including biological and demographic, which allow migration processes with continuous time.

The structure of this article is as follows. The first part contains a brief overview of PL that studies different types of BP. Then comes a model of the process with migration and continuous time. In the next section, we find the form of the differential equation and the generating function for the random process, which describes the behavior of the process within the PL. The boundary theorem for PL of the subcritical and critical processes is given below. The section of the conclusion emphasizes the obtained results.

2. Description of a Branching Process Model with Migration and Continuous Time

Consider a Markov BP with one type of particles and migration $\mu(t)$, $t \in [0, \infty)$. Let $\mu(t)$ denote the number of particles at time $t \in [0, \infty)$.

We suppose that at time t = 0, the process starts with one particle in the system:

$$\mu(0) = 1. \tag{1}$$

The process $\mu(t)$, $t \in [0, \infty)$ then $\Delta t \to 0$ is given by the transition probabilities:

$$P\{\mu(t+\Delta t)=j|\mu(t)=i\}=$$

$$=\begin{cases} 1+q_0\Delta t+o(\Delta t), & i=j=0; \\ q_j\Delta t+o(\Delta t), & i=0, j=1,2,...; \\ (p_0+\sum\limits_{l=1}^{m} r_l)\Delta t+o(\Delta t), & i=1, j=0; \\ 1+(q_0+r_0+p_1)\Delta t+o(\Delta t), & i=1, j=1; \\ (p_j+q_{j-1})\Delta t+o(\Delta t), & i=1, j=2,...; \\ \sum\limits_{l=i}^{m} r_l\Delta t+o(\Delta t) & 1<i\le m, j=0; \\ (p_0+r_1)\Delta t+o(\Delta t), & i=2,3,..., j=i-1; \\ r_{i-j}\Delta t+o(\Delta t), & i=2,3,..., j<i-1; \\ 1+(q_0+r_0+ip_1)\Delta t+o(\Delta t), & i=2,3,..., i=j; \\ (ip_{j-i+1}+q_{j-i})\Delta t+o(\Delta t), & i=2,3,..., i<j; \\ o(\Delta t), & \text{in other cases}, \end{cases} \quad (2)$$

where m is a fixed integer, and p_k, q_k, r_n satisfy the conditions:

$$p_k\ge 0,\ k\ne 1,\ p_1<0,\ \sum_{k=0}^{\infty} p_k=0,$$

$$q_k\ge 0,\ k\ne 0,\ q_0<0,\ \sum_{k=0}^{\infty} q_k=0,$$

$$r_n\ge 0,\ n=\overline{1,m},\ r_0<0,\ \sum_{k=0}^{m} r_k=0.$$

We note that p_k $(k=0,1,...)$ is the intensity of the reproduction particle, q_k $(k=0,1,...)$ is the intensity of immigration, and r_n $(n=\overline{0,m})$ is the intensity of emigration.

We introduce the following notation:

$$f(s)=\sum_{n=0}^{\infty} p_n s^n,\ |s|\le 1,\ s\in C,$$

$$g(s)=\sum_{n=0}^{\infty} q_n s^n,\ |s|\le 1,\ s\in C,$$

$$r(s)=\sum_{n=0}^{m} r_n s^{-n},\ 0<|s|\le 1.$$

We let $\widehat{F}(t,s)$ be the probability generating functions (PGFs) of a BP with continuous time (without migration) ([24], page 24).

3. Results

In this section, we find a differential equation for the PGF and PGF random process, which describes the behavior of the process within the PL of the BPMCT.

The method of PGF is widely used in the study of processes with continuous time, because in some cases, it can be found in the form of its generation, and then the corresponding probabilities of the process are calculated. The PGF of the process will uniquely determine the distribution of the process and the limiting behavior of the process.

3.1. PGF of the Random Process, Which Describes the Behavior of the Process within the PL

Definition 1. *[8] τ is the PL of a BP within which immigration begins at the moment T and has length τ if $P\{\mu(T+\tau)=0\}=P\{\mu(T-\Delta t)=0\}=0$ and $P\{\mu(t)=0\}>0$ for all $t\in[T,T+\tau)$ (the trajectories of the process are assumed to be continuous from the right).*

Let:

$$u(t) = P\{\tau > t\},$$

and define a random process $\nu(t)$, which describes the behavior of the process within the PL:

$$\nu(t) = \begin{cases} \mu(t), & t \le \tau, \\ 0, & t > \tau, \end{cases}$$

obviously that:

$$\nu(0) = \mu(0).$$

We define a PGF for $\nu(t)$:

$$N(t,s) = \sum_{k=0}^{\infty} P\{\nu(t) = k\}s^k.$$

Theorem 1. *Let τ be the PL of the BPMCT, then:*
1. The PGF $N(t,s)$ satisfies the differential equation:

$$\frac{\partial N(t,s)}{\partial t} = \frac{\partial N(t,s)}{\partial s} f(s) + N(t,s)(g(s) + r(s))$$

$$-(g(s) + r(s))P\{\nu(t) = 0\} + \sum_{n=1}^{m} P\{\nu(t) = n\} \sum_{k=n}^{m} r_k(1 - s^{n-k}), \tag{3}$$

with the initial condition:

$$N(0,s) = \frac{q_0 - g(s)}{q_0}. \tag{4}$$

2. The PGF for $\nu(t)$ has the form:

$$N(t,s) = V\left(t + \int_0^s \frac{du}{f(u)}\right) e^{\int_0^t (g(\widehat{F}(u,s)) + r(\widehat{F}(u,s)))du}$$

$$-\int_0^t P\{\nu(t) = 0\}\left(g(\widehat{F}(t-x,s)) + r(\widehat{F}(t-x,s))\right) e^{\int_0^{t-x} (g(\widehat{F}(u,s)) + r(\widehat{F}(u,s)))du} dx$$

$$+\int_0^t \sum_{n=1}^{m} P\{\nu(t) = n\} \sum_{k=n}^{m} r_k(1 - \widehat{F}^{n-k}(t-x,s)) e^{\int_0^{t-x} (g(\widehat{F}(u,s)) + r(\widehat{F}(u,s)))du} dx, \tag{5}$$

where $V(\cdot)$ is some continuous-differentiated function that satisfies:

$$\begin{cases} V\left(\int_0^s \frac{du}{f(u)}\right) = \frac{q_0 - g(s)}{q_0}, \\ V(\infty) = 1. \end{cases} \tag{6}$$

Proof of Theorem 1. We prove the first part of the theorem.
Consider:

$$\sum_{k=0}^{\infty} P\{\nu(t + \Delta t) = k | \nu(t) = n\}s^k.$$

For $n = 0$:

$$\sum_{k=0}^{\infty} P\{\nu(t + \Delta t) = k | \nu(t) = 0\}s^k = 1s^0 = 1.$$

If $n = 1$ then:

$$\sum_{k=0}^{\infty} P\{\nu(t+\Delta t) = k|\nu(t) = 1\}s^k = \sum_{k=0}^{\infty} P\{\mu(t+\Delta t) = k|\mu(t) = 1\}s^k$$

$$= s + (f(s) + g(s)s + \sum_{k=1}^{m} r_k + r_0 s)\Delta t + o(\Delta t).$$

In case $1 < n \leq m$:

$$\sum_{k=0}^{\infty} P\{\nu(t+\Delta t) = k|\nu(t) = n\}s^k = \sum_{k=0}^{\infty} P\{\mu(t+\Delta t) = k|\mu(t) = n\}s^k$$

$$= s^n + (ns^{n-1}f(s) + s^n g(s) + s^n \sum_{k=0}^{n-1} r_k s^{-k} + \sum_{k=n}^{m} r_k)\Delta t + o(\Delta t).$$

For $n > m$:

$$\sum_{k=0}^{\infty} P\{\nu(t+\Delta t) = k|\nu(t) = n\}s^k = \sum_{k=0}^{\infty} P\{\mu(t+\Delta t) = k|\mu(t) = n\}s^k$$

$$= s^n + (ns^{n-1}f(s)t + s^n g(s) + s^n r(s))\Delta t + o(\Delta t).$$

Hence, we obtain the following:

$$\sum_{k=0}^{\infty} P\{\nu(t+\Delta t) = k|\nu(t) = n\}s^k$$

$$= \begin{cases} 1, & n = 0; \\ s + (f(s) + g(s)s + \sum\limits_{k=1}^{m} r_k + r_0 s)\Delta t + o(\Delta t), & n = 1; \\ s^n + (ns^{n-1}f(s) + s^n(g(s) + \sum\limits_{k=0}^{n-1} r_k s^{-k}) + \sum\limits_{k=n}^{m} r_k)\Delta t + o(\Delta t), & 1 < n \leq m; \\ s^n + (ns^{n-1}f(s) + s^n(g(s) + r(s)))\Delta t + o(\Delta t), & n > m. \end{cases}$$

Consider the PGF $N(t,s)$.
Let T be the beginning of the PL of the process $\mu(t)$, then:

$$N(0,s) = \sum_{k=0}^{\infty} P\{\nu(0) = k\}s^k$$

$$= \sum_{k=0}^{\infty} P\{\mu(T) = k|\mu(T-\Delta t) = 0,\ \mu(0) > 0\}s^k$$

$$= \lim_{\Delta t \to 0} \sum_{k=1}^{\infty} \frac{(q_k \Delta t + o(\Delta t))s^k}{\sum\limits_{k=1}^{\infty} q_k \Delta t + o(\Delta t)} = \sum_{k=1}^{\infty} \frac{q_k s^k}{-q_0} = \frac{q_0 - g(s)}{q_0}.$$

Since it is the initial condition, we obtain (4).
Thus, we derive:

$$N(t+\Delta t, s) = \sum_{k=0}^{\infty} \sum_{n=0}^{\infty} P\{\nu(t+\Delta t) = k|\nu(t) = n\}P\{\nu(t) = n\}s^k = P\{\nu(t) = 0\}$$

$$+ \sum_{n=1}^{m} P\{\nu(t) = n\}\left(s^n + (ns^{n-1}f(s) + s^n g(s) + s^n \sum_{k=0}^{n-1} r_k s^{-k} + \sum_{k=n}^{m} r_k)\Delta t + o(\Delta t)\right)$$

$$+\sum_{n=m+1}^{\infty} P\{\nu(t)=n\}\left(s^n+ns^{n-1}f(s)\Delta t+s^n g(s)\Delta t+s^n r(s)\Delta t+o(\Delta t)\right)$$

$$=\sum_{n=0}^{\infty} P\{\nu(t)=n\}s^n+\sum_{n=1}^{\infty} P\{\nu(t)=n\}ns^{n-1}f(s)\Delta t+\sum_{n=1}^{\infty} P\{\nu(t)=n\}g(s)s^n\Delta t$$

$$+\sum_{n=1}^{m} P\{\nu(t)=n\}(r(s)s^n+\sum_{k=n}^{m} r_k(1-s^{n-k}))\Delta t+o(\Delta t)$$

$$=N(t,s)+\frac{\partial N(t,s)}{\partial s}f(s)\Delta t+N(t,s)g(s)\Delta t-g(s)P\{\nu(t)=0\}\Delta t$$

$$+N(t,s)r(s)\Delta t-r(s)P\{\nu(t)=0\}\Delta t+\sum_{n=1}^{m} P\{\nu(t)=n\}\sum_{k=n}^{m} r_k(1-s^{n-k})\Delta t+o(\Delta t)$$

$$=N(t,s)+\left(\frac{\partial N(t,s)}{\partial s}f(s)+N(t,s)(g(s)+r(s))-(g(s)+r(s))P\{\nu(t)=0\}\right.$$

$$\left.+\sum_{n=1}^{m} P\{\nu(t)=n\}\sum_{k=n}^{m} r_k(1-s^{n-k})\right)\Delta t+o(\Delta t).$$

Consider $\frac{N(t+\Delta t,s)-N(t,s)}{\Delta t}$ and directing $\Delta t\to 0$, we obtain (5).

We turn to the second part of the theorem. In proving the second part, we will use the notation $g(s)+r(s)=\gamma(s)$.

Consider the equation of the characteristics:

$$dt=-\frac{ds}{f(s)}=\frac{dN(t,s)}{(N(t,s)-P\{\nu(t)=0\})\gamma(s)+\sum\limits_{n=1}^{m} P\{\nu(t)=n\}\sum\limits_{k=n}^{m} r_k(1-s^{n-k})}.$$

The first integral of this equation:

$$C_1=t+\int_0^s \frac{du}{f(u)}.$$

We obtain the second integral of this equation:

$$dt=\frac{dN(t,s)}{(N(t,s)-P\{\nu(t)=0\})\gamma(s)+\sum\limits_{n=1}^{m} P\{\nu(t)=n\}\sum\limits_{k=n}^{m} r_k(1-s^{n-k})},$$

and rewrite it in the form:

$$\frac{\partial N(t,s)}{\partial t}=(N(t,s)-P\{\nu(t)=0\})\gamma(s)+\sum_{n=1}^{m} P\{\nu(t)=n\}\sum_{k=n}^{m} r_k(1-s^{n-k}).$$

We find the solution of the corresponding homogeneous equation:

$$N(t,s)=C_2 e^{\int_0^t \gamma(\widehat{F}(u,s))du}.$$

Using the method of the variation of constants, we obtain a partial solution of the corresponding inhomogeneous equation:

$$N(t,s)=-\int_0^t P\{\nu(t)=0\}\gamma(\widehat{F}(t-x,s))e^{\int_0^{t-x}\gamma(\widehat{F}(u,s))du}dx$$

$$+\int_0^t \sum_{n=1}^m P\{v(t)=n\} \sum_{k=n}^m r_k(1-\widehat{F}^{n-k}(t-x,s)) e^{\int_0^{t-x} \gamma(\widehat{F}(u,s))du} dx.$$

Thus, the general solution of the inhomogeneous equation will take the form:

$$N(t,s) = C_2 e^{\int_0^t \gamma(\widehat{F}(u,s))du} - \int_0^t P\{v(t)=0\}\gamma(\widehat{F}(t-x,s)) e^{\int_0^{t-x} \gamma(\widehat{F}(u,s))du} dx$$

$$+\int_0^t \sum_{n=1}^m P\{v(t)=n\} \sum_{k=n}^m r_k(1-\widehat{F}^{n-k}(t-x,s)) e^{\int_0^{t-x} \gamma(\widehat{F}(u,s))du} dx.$$

Hence, the second integral of the equation:

$$C_2 = N(t,s) e^{-\int_0^t \gamma(\widehat{F}(u,s))du} + \int_0^t P\{v(t)=0\}\gamma(\widehat{F}(t-x,s)) e^{\int_0^{t-x} \gamma(\widehat{F}(u,s))du} dx e^{-\int_0^t \gamma(\widehat{F}(u,s))du}$$

$$-\int_0^t \sum_{n=1}^m P\{v(t)=n\} \sum_{k=n}^m r_k(1-\widehat{F}^{n-k}(t-x,s)) e^{\int_0^{t-x} \gamma(\widehat{F}(u,s))du} dx e^{-\int_0^t \gamma(\widehat{F}(u,s))du}.$$

According to ([25] page 97), we obtain:

$$V\left(t+\int_0^s \frac{du}{f(u)}\right) = N(t,s) e^{-\int_0^t \gamma(\widehat{F}(u,s))du}$$

$$+\int_0^t P\{v(t)=0\}\gamma(\widehat{F}(t-x,s)) e^{\int_0^{t-x} \gamma(\widehat{F}(u,s))du} dx e^{-\int_0^t \gamma(\widehat{F}(u,s))du}$$

$$-\int_0^t \sum_{n=1}^m P\{v(t)=n\} \sum_{k=n}^m r_k(1-\widehat{F}^{n-k}(t-x,s)) e^{\int_0^{t-x} \gamma(\widehat{F}(u,s))du} dx e^{-\int_0^t \gamma(\widehat{F}(u,s))du},$$

where $V(\cdot)$ is some continuous-differentiated function.

Thus, we obtain a PGF:

$$N(t,s) = V\left(t+\int_0^s \frac{du}{f(u)}\right) e^{\int_0^t \gamma(\widehat{F}(u,s))du}$$

$$-\int_0^t P\{v(t)=0\}\gamma(\widehat{F}(t-x,s)) e^{\int_0^{t-x} \gamma(\widehat{F}(u,s))du} dx$$

$$+\int_0^t \sum_{n=1}^m P\{v(t)=n\} \sum_{k=n}^m r_k(1-\widehat{F}^{n-k}(t-x,s)) e^{\int_0^{t-x} \gamma(\widehat{F}(u,s))du} dx.$$

From the initial condition, we obtain:

$$V\left(\int_0^s \frac{du}{f(u)}\right) = \frac{q_0 - g(s)}{q_0}.$$

When $s = 1$, then:

$$N(t,1) = V\left(t + \int_0^1 \frac{du}{f(u)}\right) e^{\int_0^t \gamma(\widehat{F}(u,1))du}$$

$$-\int_0^t P\{v(t) = 0\}\gamma(\widehat{F}(t-x,1)) e^{\int_0^{t-x} \gamma(\widehat{F}(u,1))du} dx$$

$$+\int_0^t \sum_{n=1}^m P\{v(t) = n\} \sum_{k=n}^m r_k (1 - \widehat{F}^{n-k}(t-x,1)) e^{\int_0^{t-x} \gamma(\widehat{F}(u,1))du} dx.$$

From ([24], p. 69), it is known that:

$$\int_0^1 \frac{du}{f(u)} = \infty,$$

then $V(\infty) = 1$. Thus, we obtain (6).

Considering [9], and $g(s) + r(s) = \gamma(s)$ we obtain (5), where $V(\cdot)$ is some continuous-differentiated function that satisfies (6). □

3.2. The Limit Theorem for PL of the Subcritical and Critical BPMCT

Let $\xi(t)$ be BP (without migration) with continuous time ([24], page 24), then we obtain the following result:

Theorem 2. *Let $M\xi(t) \leq 0$, then:*

$$\lim_{t\to\infty} u(t) = 0. \tag{7}$$

Proof of Theorem 2. Consider:

$$u(t) = P\{\tau > t\} = P\{v(t) > 0\} = 1 - N(t,0)$$

$$= 1 - V\left(t + \int_0^0 \frac{du}{f(u)}\right) e^{\int_0^t (g(\widehat{F}(u,0)) + r(\widehat{F}(u,0)))du}$$

$$+\int_0^t P\{v(t) = 0\}\Big(g(\widehat{F}(t-x,0)) + r(\widehat{F}(t-x,0))\Big) e^{\int_0^{t-x} (g(\widehat{F}(u,0)) + r(\widehat{F}(u,0)))du} dx$$

$$-\int_0^t \sum_{n=1}^m P\{v(t) = n\} \sum_{k=n}^m r_k (1 - \widehat{F}^{n-k}(t-x,0)) e^{\int_0^{t-x} (g(\widehat{F}(u,0)) + r(\widehat{F}(u,0)))du} dx$$

$$= 1 - V\Big(t\Big) e^{\int_0^t (g(\rho(t)) + r(\rho(t)))du}$$

$$+\int_0^t P\{v(t) = 0\}\Big(g(\rho(t-x)) + r(\rho(t-x))\Big) e^{\int_0^{t-x} (g(\rho(u)) + r(\rho(u)))du} dx$$

$$-\int_0^t \sum_{n=1}^m P\{v(t) = n\} \sum_{k=n}^m r_k (1 - \rho(t-x))^{n-k} e^{\int_0^{t-x} (g(\rho(u)) + r(\rho(u)))du} dx,$$

where $\rho(t) = P\{\xi(t) = 0\}$.

In the case of the subcritical and critical process $\xi(t)$, the probability of degeneration is 1 and $\lim_{t\to\infty} P\{\xi(t) = 0\} = q$. Hence, we obtain (7). □

4. Conclusions

This article investigates a more general model of the BPMCT than in [17–21]. The form of the differential equation and the PGF for the random process $\nu(t)$, which describes the behavior of the process within the PL, was determined. The boundary theorem for the PL of the subcritical and critical BPMCT has been proven.

This model of the development process can be used to describe the popularization of countries or species in the external territory or to predict epidemics. One of our next works will be to describe the development of the COVID-19 pandemic through an extensive migration process.

Author Contributions: Conceptualization, K.P. and I.B.; methodology, K.P.; software, I.I.; validation, K.P., I.B. and L.M.; formal analysis, I.B.; investigation, K.P.; resources, K.P. and I.I.; data curation, L.M.; writing—original draft preparation, K.P. and I.B.; writing—review and editing, K.P., I.B. and L.M.; supervision, I.B.; project administration, I.B.; funding acquisition, L.M. and I.I. All authors have read and agreed to the published version of the manuscript.

Funding: This research received no external funding.

Institutional Review Board Statement: Not applicable.

Informed Consent Statement: Not applicable.

Data Availability Statement: Not applicable.

Conflicts of Interest: The authors declare no conflict of interest.

Abbreviations

The following abbreviations are used in this manuscript:

PL	period-life
BP	branching process
BPMCT	branching process with migration and continuous time
PGF	probability generating functions

References

1. Demetrius, L.; Schuster, P.; Sigmund, K. Polynucleotide evolution and branching processes. *Bull. Math. Biol.* **1985**, *47*, 239–262. [CrossRef]
2. Kimmel, M.; Axelrod, D. *Branching Processes in Biology*; Springer: New York, NY, USA, 2002.
3. Sawyer, S. Branching Diffusion Processes in Population Genetics. *Adv. Appl. Probab.* **1976**, *8*, 659–689. [CrossRef]
4. Caron-Lormier, G.; Masson, J.P.; Mènard, N.; Pierre, J.S. A branching process, its application in biology: Influence of demographic parameters on the social structure in mammal groups. *J. Theor. Biol.* **2005**, *238*, 564–574. [CrossRef]
5. Gleeson, J.P.; Onaga, T.; Fennell, P.; Cotter, J.; Burke, R.; O'Sullivan, D. Branching process descriptions of information cascades on Twitter. *J. Complex Netw.* **2021**, *8*, 1–29.
6. Vassiliou, P.-C.G. Asymptotic behaviour of Markov systems. *J. Appl. Prob.* **1982**, *19*, 851–857. [CrossRef]
7. Dimitriou, V.A.; Georgiou, A.C. Introduction, analysis and asymptotic behavior of a multi-level manpower planning model in a continuous time setting under potential department contraction. *Commun. Stat. Theory Methods* **2021**, *50*, 1173-1199. [CrossRef]
8. Zubkov, A.M. Life-Periods of a Branching Process with Immigration. *Theory Probab. Appl.* **1972**, *17*, 174–183. [CrossRef]
9. Vatutin, V.A. A conditional limit theorem for a critical Branching process with immigration. *Math. Notes Acad. Sci. USSR* **1977**, *21*, 405–411. [CrossRef]
10. Yanev, N.M.; Mitov, K.V. Lifetimes of Critical Branching Processes with Random Migration. *Theory Probab. Appl.* **1984**, *28*, 481–491. [CrossRef]
11. Boiko, R.V. Lifetime of branching process with immigration in limiting environment. *Ukr. Math. J.* **1983**, *35*, 242–247. [CrossRef]
12. Formanov, S.K.; Yasin, M.T. Limit theorems for life periods for critical Galton-Watson branching processes with migration. *Izv. Akad. Nauk UzSSR Ser. Fiz-Mat. Nauk* **1989**, *1*, 40–44.

13. Badalbaev, I.S.; Ganikhodjaev, A.N. Limit theorem for the branching Bellman-Harris process with immigration under the condition of non-zero. *Izv. Akad. Nauk UzSSR Ser. Fiz-Mat. Nauk* **1989**, *3*, 8–14.
14. Badalbaev, I.S.; Mashrabbaev, A. Life spans of a Bellman-Harris branching process with immigration. *J. Sov. Math.* **1987**, *38*, 2198–2210. [CrossRef]
15. Li, D.; Vatutin, V.; Zhang, M. Subcritical branching processes in random environment with immigration stopped at zero. *J. Theor. Probab.* **2020**, 1–23. [CrossRef]
16. Dyakonova, E.; Li, D.; Vatutin, V.; Zhang, M. Branching processes in random environment with immigration stopped at zero. *J. Appl. Probab.* **2020**, *57*, 237–249. [CrossRef]
17. Chen, A.Y.; Renshaw, E. Markov branching processes regulated by emigration and large immigration. *Stoch. Process. Appl.* **1995**, *57*, 339–359. [CrossRef]
18. Rahimov, I.; Al-Sabah, W.S. Branching processes with decreasing immigration and tribal emigration. *Arab. J. Math. Sci.* **2000**, *6*, 81–97 .
19. Srivastava, O.P.; Gupta, S.C. On a countinuous-time branching process with migration. *Statistica* **1989**, *49*, 547–552.
20. Pakes, A.G. Some properties of a branching process whith group immigration and emigration. *Adv. Appl. Prob.* **1986**, *18*, 628–645. [CrossRef]
21. Balabaev, I.S. Limit theorems for a critical Markov branching process with continuous time and migration. *Uzbek Math. J.* **1994**, *2*, 12–15.
22. Nagaev, S.V.; Khan, L.V. Limit theorems for Galton-Watson branching processes with migration. *Theory Probab. Appl.* **1980**, *25*, 523–534.
23. Yakymyshyn, K. Equation for generation function for branching processes with migration. *Visnyk Lviv Univ. Ser. Mech. Math.* **2017**, *84*, 119–125.
24. Sevastyanov, B.A. *Branching Processes*; Nauka: Moscow, Russia, 1971; p. 436.
25. Zaitsev, V.F.; Polyanin, A.D. *Handbook of First Order Partial Differential Equations*; Fizmatlit: Moscow, Russia, 2003; p. 416.

mathematics

Article

Optimizing a Multi-State Cold-Standby System with Multiple Vacations in the Repair and Loss of Units

Juan Eloy Ruiz-Castro

Department of Statistics and O.R., Math Institute (IMAG), University of Granada, 18071 Granada, Spain; jeloy@ugr.es

Abstract: A complex multi-state redundant system with preventive maintenance subject to multiple events is considered. The online unit can undergo several types of failure: both internal and those provoked by external shocks. Multiple degradation levels are assumed as both internal and external. Degradation levels are observed by random inspections and, if they are major, the unit goes to a repair facility where preventive maintenance is carried out. This repair facility is composed of a single repairperson governed by a multiple vacation policy. This policy is set up according to the operational number of units. Two types of task can be performed by the repairperson, corrective repair and preventive maintenance. The times embedded in the system are phase type distributed and the model is built by using Markovian Arrival Processes with marked arrivals. Multiple performance measures besides the transient and stationary distribution are worked out through matrix-analytic methods. This methodology enables us to express the main results and the global development in a matrix-algorithmic form. To optimize the model, costs and rewards are included. A numerical example shows the versatility of the model.

Keywords: reliability; redundant systems; preventive maintenance; multiple vacations

Citation: Ruiz-Castro, J.E. Optimizing a Multi-State Cold-Standby System with Multiple Vacations in the Repair and Loss of Units. *Mathematics* **2021**, *9*, 913. https://doi.org/10.3390/math9080913

Academic Editor: Panagiotis-Christos Vassiliou

Received: 12 March 2021
Accepted: 16 April 2021
Published: 20 April 2021

Publisher's Note: MDPI stays neutral with regard to jurisdictional claims in published maps and institutional affiliations.

Copyright: © 2021 by the author. Licensee MDPI, Basel, Switzerland. This article is an open access article distributed under the terms and conditions of the Creative Commons Attribution (CC BY) license (https://creativecommons.org/licenses/by/4.0/).

1. Introduction

Redundant systems and preventive maintenance are of fundamental importance in ensuring reliability, preventing system failures and reducing costs. These questions, therefore, are of considerable research interest.

The occurrence of total, unexpected system failure can provoke severe damage and major financial loss. To avoid such an outcome, various reliability-enhancing methods can be applied, chief among which are redundancy and preventive maintenance. In this respect, cold, hot and warm redundant standby and k-out-of-n systems have been proposed. Among researchers who have addressed these questions, Levitin et al. [1] considered an optimal standby element sequencing problem (SESP) for 1-out-of-N: G heterogeneous warm-standby systems, while Zhai et al. [2] constructed a multi-value decision diagram with which to analyse a demand-based warm standby system. In related papers, Cha et al. [3] considered preventive maintenance for items operating in a random environment subjected to a shock Poisson process, Levitin et al. [4] evaluated the probability of mission success given an arbitrary redundancy level, and Osaki et al. [5] analysed the behaviour of a two-unit standby redundant system.

Preventive maintenance enhances system reliability and performance, reduces costs, for both repairable and non-repairable systems, and decreases the probability of sudden equipment failure. Various maintenance systems were studied by [6,7] who developed a new model for the hybrid preventive maintenance of systems with partially observable degradation. Levitin et al. (2021) [8] modelled the (time-consuming) procedure of task transfer, in an event transition-based reliability analysis of standby systems in which preventive replacements are performed according to a predetermined schedule. The aim of this approach is to optimise preventive replacement scheduling and hence to maximise reliability. In another approach to this situation, Yang et al. [9] discussed a preventive maintenance policy for a single-unit system subject to failure by internal deterioration and/or sudden shock, according to a non-homogeneous Poisson process whereby the process of internal failure is partitioned into two stages.

Complex systems that have a finite number of performance levels and various failure modes, each producing different effects on system performance, are termed multi-state systems (MSS). This concept was first discussed by [10] and has since been developed extensively. For example, Levitin et al. [11] described various MSS measures and considered problems of MSS optimisation, and Lisnianski et al. [12] conducted a comprehensive analysis of the question.

One of the main problems encountered with multi-state complex models is the existence of intractable expressions for their modelling and/or of difficulties in their interpretation. In this respect, matrix-analytic methods are a valuable means of analysing complex systems, preserving the Markovian structure and obtaining manageable results. This approach is usually based on two elements—phase-type distributions (PHD) and Markovian arrival processes (MAP)—which enable the results to be expressed and complex systems modelled in an algorithmic, computational form. PHD were first introduced and detailed by [13]. MAP is a counting process in which PH distributions play an important role. This method was described by [14] and comprehensively reviewed by [15,16]. A special case is that of the MAP with marked arrivals (MMAP), which enables us to count different types of arrival. Moreover, the arrival probabilities of events, for the discrete case, can be customised for different situations. MAP and MMAP theory were further developed by [16].

Many multi-state reliability systems, over time, are subject to events such as repairable or non-repairable failure, inspections or external shocks. These systems can be modelled using appropriate Markov processes, i.e., PHD and MAP ([17,18]). In parallel, unitary complex systems subject to multiple events have been discussed by [19,20]. Matrix algorithmic methods have also been used to model multi-state complex redundant systems. Ruiz-Castro (2020) [21] developed a k-out-of-n: G system, in which the units are subject to repairable and/or non-repairable failure and receive random inspections. In this system, the potential loss of units is included; thus, when a non-repairable failure occurs, the unit is removed and the system continues to be operational. In the context of complex models, a repair facility with a single repairperson is usually assumed. Thus, Ruiz-Castro et al. [22,23] analyse redundant complex systems with a general number of repairpersons and the potential loss of units, determining the optimum number of repairpersons in each case.

In brief, redundancy and preventive maintenance are incorporated into complex systems in order to enhance their reliability, and must also be included in the modelling of such systems. In theory, a unit is repaired either immediately after failure if the system is unitary or when the element in next in line in the repair facility queue. However, this might not be the case in a real scenario. For example, a failed unit might not be repaired immediately in a small or medium-sized firm that cannot afford to employ a full-time repairperson. Furthermore, when there is no failed unit to be attended to in the repair facility, what should a repairperson do? Instead of remaining idle during this period, the repairperson may take a 'vacation' and/or use the time to do other work, thus optimising resources and reducing costs. A repairperson is on vacation when absent from the repair facility, whether or not it is empty. The economic implications of this situation should be considered, taking into account that the vacation policy applied might impact both on performance and also on economic rewards/costs. In studies of this question, two time points are of particular importance: the start and end times of the vacation. Moreover, the services provided may be exhaustive or non-exhaustive. In the first case, the repairperson cannot be on vacation when the repair facility is not empty, but in the second, even if an item has been sent to the repair facility, the repairperson may be on vacation. Another possibility that must be considered is that of interruption, i.e., the repairperson may take a vacation while a unit is being repaired. The vacation end time determines when the repairperson resumes work. Finally, depending on the maintenance system adopted, the vacation may occupy a single period of time or multiple periods.

Vacation policies have been considered in queuing theory and in reliability analysis, among other areas. Thus, Doshi [24] provided a wide-ranging analysis of vacation system models and Ke et al. [25] examined the application of two vacation policies (one single and the other multiple) in a repairable system. Zaiming et al. [26] developed a reliability system with multiple, but finite, vacation periods and Wu et al. [27] analysed the reliability of a two-unit cold standby system with a single repairperson, entitled to take a vacation.

Vacation periods have also been considered for systems governed by a Markov model. In this respect, Shrivastava et al. [28] presented the case of an exhaustive vacation policy, whereby the repairperson could only take a vacation when the repair facility was empty. Under the Markovian modelling described by [29], the repairperson could take a vacation if there were no failed units in need of repair, but had to return as soon as any unit failed. In another approach, Zhang et al. [30] modelled a k-out-of-n system with a single repairperson, assuming a phase-type distribution for the vacation time and an exponential distribution for the lifetime of the units. In this system, the repairperson could take a vacation whenever there was no failed component in the system. On return, the repairperson might or might not encounter failed components waiting for repair. In the second case, the repairperson would remain within the repair facility, idle, until a failed component arrived. Finally, Ruiz-Castro et al. [31] modelled a multi-state complex system subject to multiple events and where preventive maintenance was applied. In this case, the repairperson had various duties and, moreover, was entitled to take a vacation.

In the present study, we model a cold standby system with the potential loss of units. The system evolves in discrete time; the online unit is multi-state and subject to internal failure, repairable or otherwise, to external shocks with diverse consequences, and to random inspection. When a non-repairable failure occurs, the faulty unit is removed and the system continues working with one unit less. An external shock may provoke any of the following consequences: degraded system performance, a repairable failure of the online unit or its total (non-repairable) failure. Damage to the internal performance of the online unit may be minor or major. During system inspection, the internal status of the online unit is observed. If major damage is present, the faulty unit is sent to the repair facility for preventive maintenance. According to the case presented, the repairperson may perform corrective repair or preventive maintenance. The complexity of the system is determined as follows. In modelling the system, the vacation policy employed in the repair facility is determined by the number of operational units included. A general number R of operational units is considered. If the repairperson returns from a vacation period and there are fewer than R operational units, the repairperson must then remain in the facility. Otherwise, a new vacation period begins. As the system is subject to a loss of units, when there are fewer than R units in the system, the repairperson must remain in the facility while this situation persists. The times embedded are PH distributed and a MMAP is constructed to model the system. In modelling this system, the following measures are calculated: availability, reliability and expected times (in both transient and stationary regimes). Rewards and costs are incorporated, and a numerical optimisation is performed to determine the optimum threshold R and to decide whether preventive maintenance is profitable or not.

The rest of this paper is organised as follows. In Section 2, we describe the system to be modelled, after which we present the corresponding MMAP in Section 3. In Section 4, we detail the measures applied to the transient and stationary regimes, and calculate the transient and stationary distributions. The latter is obtained both algorithmically and computationally. The system costs, rewards and associated measures are then derived in Section 5. Taking advantage of the favourable properties of PHD and MMAP, the study findings are obtained in a matrix algorithmic form. Section 6 presents a numerical example, including an optimisation exercise. Finally, the main conclusions drawn are summarised in Section 7.

2. Assumptions of the System: The State Space

A cold standby system composed of n units is initially assumed. One unit is online and the others are waiting on standby without degrading. The online unit is multi-state, where the internal performance is partitioned into major and minor states. It is subject to multiple events. This can suffer internal failures, repairable or not, and external shocks. Each external shock can provoke three different consequences: total failure (non-repairable), modification in the internal behaviour or an internal repairable or non-repairable failure. When a repairable failure occurs, the unit goes to the repair facility for corrective repair. The corrective repair time distribution is PH. The repair facility is composed of one repairperson who can take vacations. As it has been mentioned above, the internal performance of the online unit is partitioned into major and minor states. A major state is a state from where the online unit has a greater risk of suffering a failure. To avoid serious damage and major financial losses random inspections are carried out. The inspector observes the online unit and if this one is operational in major damage, the unit goes to the repair facility for preventive maintenance. Preventive maintenance time is also PH distributed. When the online unit undergoes a failure, one cold standby occupies the online place, if any. The new online unit will start executing from the initial distribution of the internal performance, because after repairing or preventive maintenance the unit is as good as new. The system is also subject to loss of units. After a non-repairable failure the unit is removed and the system continues working until there are no units in the system. If only one unit is in the system and a non-repairable failure occurs, the system is restarted.

One repairperson can be in the repair facility who can develop two different tasks: corrective repair and preventive maintenance. To optimise the system, the repairperson is allowed to take vacations, for a random duration, according to certain criteria.

Initially, all units are operational and the repairperson is on vacation. After returning, a new vacation begins if there are R or more operational units in the system. Equivalently, if there are $\mathrm{k} - \mathrm{R} + 1 = N$ or more failed units needing to be repaired, where k is the number of units in the system, $k = 1,\ldots, n$, the repairperson must remain in the repair facility.

After finishing a repair, the repairperson begins a new period of vacation if R units are then operational. As the system can lose units, the repairperson must always remain in the facility (or interrupt the vacation to return) when fewer than R units are in the system.

The following Section *"The Assumptions"* specifies the assumptions of the system.

The Assumptions

The system follows the following assumptions.

Assumption 1. The internal performance time is PH distributed with representation $(\boldsymbol{\alpha}, \mathbf{T})$, and with order m (number of internal stages). The internal failure probability depends on the states. The column vectors $\mathbf{T}_r^0$ and $\mathbf{T}_{nr}^0$ contains the probabilities of repairable and non-repairable failures, respectively.

Assumption 2. The internal performance of the online unit is multi-state where the n_1 first units are minor and the rest are major according to damage.

Assumption 3. The external events occur according to a PH-renewal process where the time between two consecutive shocks is a PH distribution with representation $(\boldsymbol{\gamma}, \mathbf{L})$, with order t.

Assumption 4. An external shock can provoke a total non-repairable failure of the online unit with a probability equal to ω^0.

Assumption 5. After an external shock the internal performance state can undergo a modification. This modification between any two internal states occurs according to the transition probability matrix $\mathbf{W}$. The column vectors $\mathbf{W}_r^0$ and $\mathbf{W}_{nr}^0$ contains the probabilities of repairable and non-repairable failures respectively provoked by an external shock.

Assumption 6. The time between two consecutive random inspections is PH distributed with representation $(\boldsymbol{\eta}, \mathbf{M})$, with order ε.

Assumption 7. The vacation time is distributed following a PH distribution with representation $(\mathbf{v}, \mathbf{V})$, with order v.

Assumption 8. The corrective repair time is PH distributed with representation $(\boldsymbol{\beta}_1, \mathbf{S}_1)$, with order z_1.

Assumption 9. The preventive maintenance time is PH distributed with representation $(\boldsymbol{\beta}_2, \mathbf{S}_2)$, with order z_2.

The behaviour of the system is shown in Figure 1, for inspection and repairable failure, Figure 2 for non-repairable failure, and Figure 3 for the vacation policy.

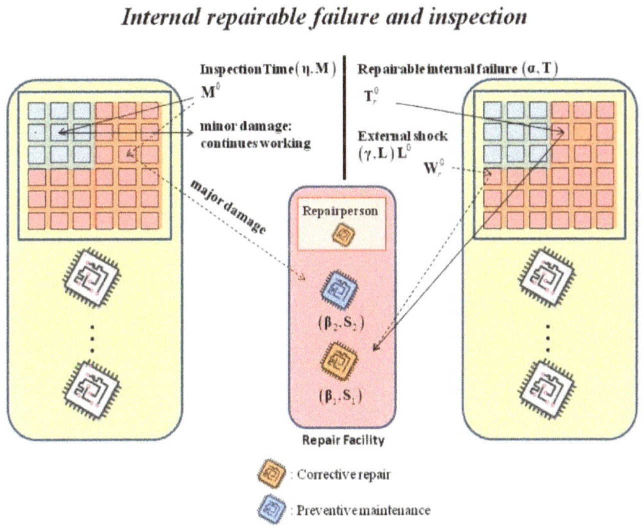

Figure 1. Internal repairable failure and inspection in the system.

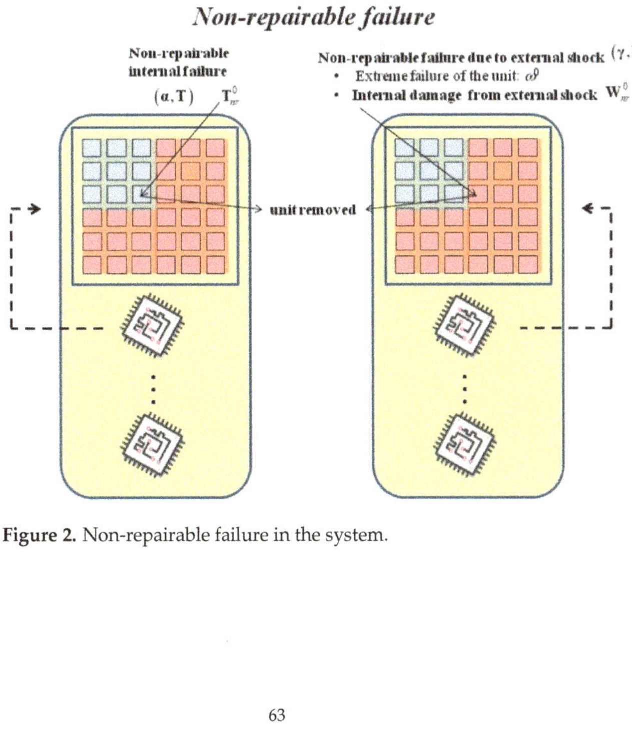

Figure 2. Non-repairable failure in the system.

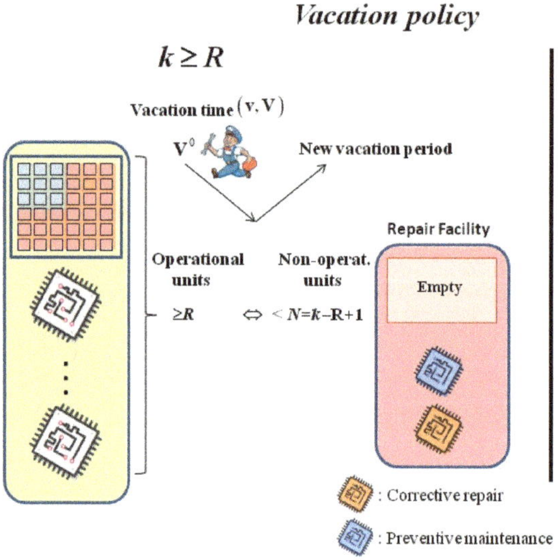

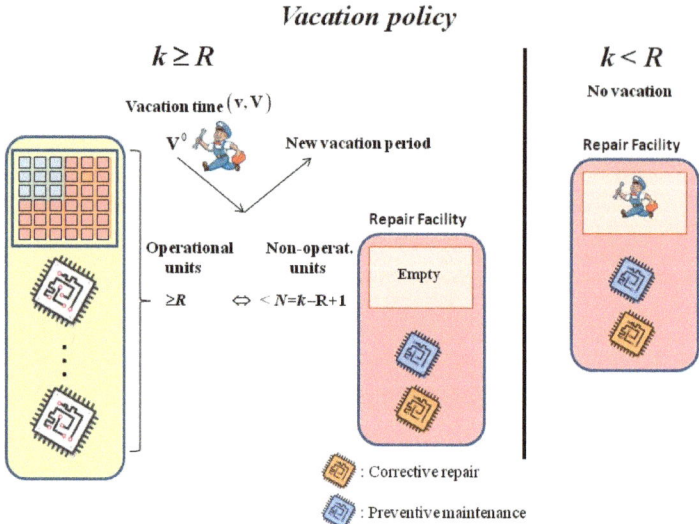

Figure 3. Vacation policy in the system.

3. Modelling the System. The Markovian Arrival Process with Marked Arrivals

The system is governed by a Markov process vector in discrete time. In this section the state space is described and, to model the proposed complex system, the behaviour of the online unit and of the repair facility is developed separately.

3.1. *The State-Space*

The state-space is composed of macro-states and it is denoted by $S = \{\mathbf{U}^n, \mathbf{U}^{n-1}, \ldots, \mathbf{U}^1\}$, where $\mathbf{U}^k$ contains the phases when there are k units in the system. In turn, these macro-states are partitioned as follows

$$\mathbf{U}^k = \left\{\mathbf{E}_0^{k,v}, \mathbf{E}_1^{k,v}, \ldots, \mathbf{E}_{N-1}^{k,v}, \mathbf{E}_N^{k,v}, \mathbf{E}_{N+1}^{k,v}, \ldots, \mathbf{E}_k^{k,v}, \mathbf{E}_N^{k,nv}, \mathbf{E}_{N+1}^{k,nv}, \ldots, \mathbf{E}_k^{k,nv}\right\};\ k \geq R$$

$$\mathbf{U}^k = \left\{\mathbf{E}_0^{k,nv}, \mathbf{E}_1^{k,nv}, \ldots, \mathbf{E}_k^{k,nv}\right\};\ k \geq R$$

where $\mathbf{E}_s^{k,x}$ contains the phases when there are k units in the system and s of them are in the repair facility and the superscript x indicates if the repairperson in on vacation (v) or not (nv). Initially the repairperson begins to operate the first time that he comes back from vacation and the system has at least $N = k - R + 1$ units in the repair facility. He remains working until $N - 1$ units are in the repair facility. At this moment the repairperson goes on vacation. In any case, the order of the units in the repair facility has to be saved in memory, and there are two types of repair, corrective and preventive maintenance. For this reason, the macro-state $\mathbf{E}_s^{k,x}$ is composed of the first level of macro-states $\mathbf{E}_{i_1,\ldots,i_s}^{k,x}$.

These macro-states contain the phases when there are k units in the system, with s of them in the repair facility, and the type of repair is given by the ordered sequence i_1, …, i_s. The values of i_l are equal to 1 or 2 if the unit is in corrective repair or preventive maintenance, respectively.

When the number of units in the system is $R - 1$ units, then the repairperson occupies his place work immediately. The inspection time is restarted each time that one unit occupies the online place.

- For $k = 1, \ldots, R - 1$
  $\mathbf{E}_0^{k,nv} = \{(k, 0; i, j, u); i = 1, \ldots, m,\ j = 1, \ldots, t,\ u = 1, \ldots, \varepsilon\}$

$\mathbf{E}_s^{k,nv} = \left\{ \mathbf{E}_{i_1,\ldots,i_s}^{k,nv}; i_l = 1,2; l = 1,\ldots,s \right\}$ for s = 1, ... , k where

$\mathbf{E}_{i_1,\ldots,i_s}^{k,nv} = \left\{ (k,s;i,j,u,r); i = 1,\ldots,m,\ j = 1,\ldots,t,\ u = 1,\ldots,\varepsilon, r = 1,\ldots,z_{i_1} \right\}$ for $s < k$

$\mathbf{E}_{i_1,\ldots,i_k}^{k,nv} = \left\{ (k,k;j,r);\ j = 1,\ldots,t,\ u = 1,\ldots,\varepsilon, r = 1,\ldots,z_{i_1} \right\}$

- For $k = N, \ldots, n$

$\mathbf{E}_0^{k,v} = \{ (k,0;i,j,u,v); i = 1,\ldots,m,\ j = 1,\ldots,t,\ u = 1,\ldots,\varepsilon, v = 1,\ldots,v \}$

$\mathbf{E}_s^{k,v} = \left\{ \mathbf{E}_{i_1,\ldots,i_s}^{k,v}; i_l = 1,2; l = 1,\ldots,s \right\}$ for s = 1, ... , k where

$\mathbf{E}_{i_1,\ldots,i_s}^{k,v} = \{ (k,s;i,j,u,v); i = 1,\ldots,m,\ j = 1,\ldots,t,\ u = 1,\ldots,\varepsilon, v = 1,\ldots,v \}$ for $s < k$

$\mathbf{E}_{i_1,\ldots,i_k}^{k,v} = \{ (k,k;j,u,v); j = 1,\ldots,t,\ u = 1,\ldots,\varepsilon, v = 1,\ldots,v \}$

$\mathbf{E}_s^{k,nv} = \left\{ \mathbf{E}_{i_1,\ldots,i_s}^{k,nv}; i_l = 1,2; l = 1,\ldots,s \right\}$ for s = N, ... , k where

$\mathbf{E}_{i_1,\ldots,i_s}^{k,nv} = \left\{ (k,s;i,j,u,r); i = 1,\ldots,m,\ j = 1,\ldots,t, u = 1,\ldots,\varepsilon, r = 1,\ldots,z_{i_1} \right\}$ for $s < k$

$\mathbf{E}_{i_1,\ldots,i_k}^{k,nv} = \left\{ (k,k;j,u,r); j = 1,\ldots,t, u = 1,\ldots,\varepsilon, r = 1,\ldots,z_{i_1} \right\}$

The phase $(k,s;i,j,u,m,r)$ indicates that there are k units in the system, with s in the repair facility; the internal performance of the online unit is in state i, the external shock time is in state j, the cumulative damage caused by external shocks is given by u, m is the current phase of the inspection time and r is the corrective repair/preventive maintenance phase for the unit currently being attended to in the repair facility. If the repairperson is taking a vacation, the phase is indicated by v.

The order of these macro-states is as follows:

$$o_{\mathbf{E}_0^{k,nv}} = m \cdot t \cdot \varepsilon;\ s < k,\ o_{\mathbf{E}_s^{k,nv}} = m \cdot t \cdot \varepsilon \cdot (z_1 + z_2) 2^{s-1};\ s = k,\ o_{\mathbf{E}_k^{k,nv}} = t \cdot (z_1 + z_2) 2^{k-1}$$

$$o_{\mathbf{E}_0^{k,v}} = m \cdot t \cdot \varepsilon;\ s < k,\ o_{\mathbf{E}_s^{k,v}} = m \cdot t \cdot \varepsilon \cdot 2^s;\ s = k,\ o_{\mathbf{E}_k^{k,v}} = t \cdot 2^{k-1}$$

3.2. Modelling the Online Unit

The online unit can undergo different types of event at any time. These are noted and defined as:

A: Internal repairable failure

B: Major revision

C: Non-repairable failure

O: No events

Two of them are described below, and the rest are given in Appendix A.

The elements of auxiliary matrices $\mathbf{U}_1$ and $\mathbf{U}_2$ are defined as

$$\mathbf{U}_1(i,j) = \begin{cases} 1 & ; \quad i = j; i = 1,\ldots,n_1 \\ 0 & ; \quad \text{otherwise} \end{cases} ; \mathbf{U}_2(i,j) = \begin{cases} 1 & ; \quad i = j; i = n_1 + 1,\ldots,n \\ 0 & ; \quad \text{otherwise} \end{cases}$$

Throughout this work the symbol $\otimes$ denotes the Kronecker product and, given a matrix $\mathbf{A}$, we denote this as $\mathbf{A}^0$ to the column vector $\mathbf{A}^0 = \mathbf{e} - \mathbf{A}\mathbf{e}$, $\mathbf{e}$ being a column vector of units with appropriate order.

3.3. No Events at a Certain Time (O)

We assume that the online unit is operational and at this time it continues working. This occurs because of different situations:

- The internal performance continues in the same phase or changes to another, equally operational state. There is no external shock ($\mathbf{T} \otimes \mathbf{L}$), and no inspection takes place ($\mathbf{M}$). The matrix that governs this transition for the online unit is given by $\mathbf{T} \otimes \mathbf{L} \otimes \mathbf{M}$.

- The online undergoes an external shock but total failure does not occur ($\mathbf{L}^0\boldsymbol{\gamma}(1-\omega^0)$). This external shock might modify the internal performance but does not produce internal failure (**TW**). No inspection takes place (**M**). The matrix is $\left(\mathbf{TW}\otimes\mathbf{L}^0\boldsymbol{\gamma}(1-\omega^0)\right)\otimes\mathbf{M}$.
- An inspection takes place and the time preceding the next one begins ($\mathbf{M}^0\boldsymbol{\eta}$). The inspector observes that the online unit does not need preventive maintenance and no external shock occurs ($\mathbf{U}_1\mathbf{T}\otimes\mathbf{L}$). The matrix is $\mathbf{U}_1\mathbf{T}\otimes\mathbf{L}\otimes\mathbf{M}^0\boldsymbol{\eta}$.
- An inspection takes place and the time preceding the next one begins ($\mathbf{M}^0\boldsymbol{\eta}$). One external shock also takes place without total failure ($\mathbf{L}^0\boldsymbol{\gamma}(1-\omega^0)$). This shock provokes a change in the internal performance without failure and the inspection observes minor damage ($\mathbf{U}_1\mathbf{TW}$). This matrix is $\left(\mathbf{U}_1\mathbf{TW}\otimes\mathbf{L}^0\boldsymbol{\gamma}(1-\omega^0)\right)\otimes\mathbf{M}^0\boldsymbol{\eta}$.

Therefore, the matrix that governs this transition for the online unit is given by

$$\mathbf{H}_O=\left(\mathbf{T}\otimes\mathbf{L}+\mathbf{TW}\otimes\mathbf{L}^0\boldsymbol{\gamma}\left(1-\omega^0\right)\right)\otimes\mathbf{M}+\left(\mathbf{U}_1\mathbf{T}\otimes\mathbf{L}+\mathbf{U}_1\mathbf{TW}\otimes\mathbf{L}^0\boldsymbol{\gamma}\left(1-\omega^0\right)\right)\otimes\mathbf{M}^0\boldsymbol{\eta}$$

3.4. Non-Repairable Failure (C)

The online unit is assumed to be operational and at the next time point a non-repairable failure occurs, because:

- An internal non-repairable failure occurs with no external shock, $\mathbf{T}^0_{nr}\boldsymbol{\alpha}\otimes\mathbf{L}$.
- An external shock occurs, but does not provoke total failure. This shock provokes a non-repairable internal failure or, irrespective of the shock, the online unit may experience a non-repairable internal failure. The matrix is $\left(\mathbf{T}^0_{nr}+\mathbf{TW}^0_{nr}\right)\boldsymbol{\alpha}\otimes\mathbf{L}^0\boldsymbol{\gamma}(1-\omega^0)$.
- An external shock provokes total failure. In this case the internal behaviour is irrelevant. The matrix is $\mathbf{e}\boldsymbol{\alpha}\otimes\mathbf{L}^0\boldsymbol{\gamma}\omega^0$.

This transition is independent of the inspection time. After the online unit experiences a non-repairable failure, the online place is occupied by a substitute, identical unit. Then, the matrix is given by

$$\mathbf{H}_C=\left[\mathbf{T}^0_{nr}\boldsymbol{\alpha}\otimes\mathbf{L}+\left(\mathbf{T}^0_{nr}+\mathbf{TW}^0_{nr}\right)\boldsymbol{\alpha}\otimes\mathbf{L}^0\boldsymbol{\gamma}\left(1-\omega^0\right)+\mathbf{e}\boldsymbol{\alpha}\otimes\mathbf{L}^0\boldsymbol{\gamma}\omega^0\right]\otimes\mathbf{e}\boldsymbol{\eta}.$$

If only one unit is operational and online (i.e., all others are under repair), this unit experiences a non-repairable failure and no repair occurs, no immediate substitution can be made and therefore the system does not restart. The matrix is given by

$$\mathbf{H}'_C=\left[\mathbf{T}^0_{nr}\otimes\mathbf{L}+\left(\mathbf{T}^0_{nr}+\mathbf{TW}^0_{nr}\right)\otimes\mathbf{L}^0\boldsymbol{\gamma}\left(1-\omega^0\right)+\mathbf{e}\otimes\mathbf{L}^0\boldsymbol{\gamma}\omega^0\right]\otimes\mathbf{e}$$

3.5. The Markovian Arrival Process with Marked Arrivals (MMAP)

The behaviour of the system is governed by a MMAP. The representation of this MMAP is given from the types of event shown below:

A: Internal repairable failure (default without D)
B: Major revision (default without D)
C: Non-repairable failure (default without D)
D: The repairperson resumes to work (default without A, B, C)
AD: Internal repairable failure and the repairperson resumes work
BD: Major revision and the repairperson resumes work
CD: Non-repairable failure and the repairperson resumes work
NS: New system
O: No events

The representation of the MMAP is $\left(\mathbf{D}^{O}, \mathbf{D}^{A}, \mathbf{D}^{B}, \mathbf{D}^{C}, \mathbf{D}^{D}, \mathbf{D}^{AD}, \mathbf{D}^{BD}, \mathbf{D}^{CD}, \mathbf{D}^{NS}\right)$.

The transition probability matrix associated to the embedded Markov chain from the MMAP is given by $\mathbf{D} = \sum_{Y} \mathbf{D}^{Y}$.

Two matrices $\mathbf{D}^{Y}$ are described in the next section. The rest are given in Appendix B.

The Matrices $\mathbf{D}^{A}$ and $\mathbf{D}^{B}$

The matrices $\mathbf{D}^{A}$ and $\mathbf{D}^{B}$ govern the transition when a repairable failure or a major inspection takes place, respectively. These matrices are composed of matrix blocks that contain the transitions between macro-states $\mathbf{U}^{k}$. This is a diagonal matrix block given that the number of units in the system does not change in this transition. The matrix $\mathbf{D}_{k}^{Y}$ contains the transition probabilities when there are k units in the system and the event Y occurs for $Y = A$ or B and $k = 1, \ldots, n$. Then,

$$\mathbf{D}^{Y} = \begin{pmatrix} \mathbf{D}_{n}^{Y} & & & & \\ & \mathbf{D}_{n-1}^{Y} & & & \\ & & \mathbf{D}_{n-2}^{Y} & & \\ & & & \ddots & \\ & & & & \mathbf{D}_{1}^{Y} \end{pmatrix} \quad \text{for } Y = A, B.$$

These blocks are composed of further blocks.

- If the number of units is less than $R-1$, the repairperson is always in his workplace. Then, for $k = 1, \ldots, R-1$

$$\mathbf{D}_{k}^{Y} = \begin{array}{c} \\ E_{0}^{k,nv} \\ E_{1}^{k,nv} \\ \vdots \\ E_{k-1}^{k,vn} \\ E_{k}^{k,nv} \end{array} \begin{array}{c} \begin{array}{ccccc} E_{0}^{k,nv} & E_{1}^{k,nv} & E_{2}^{k,nv} & E_{k-1}^{k,nv} & E_{k}^{k,nv} \end{array} \\ \begin{pmatrix} \mathbf{0} & \mathbf{D}_{01}^{Y,k,nv} & & & \\ & \mathbf{D}_{11}^{Y,k,nv} & \mathbf{D}_{12}^{Y,k,nv} & & \\ & & \ddots & \ddots & \\ & & & \mathbf{D}_{k-1,k-1}^{Y,k,nv} & \mathbf{D}_{k-1,k}^{Y,k,nv} \\ & & & & \mathbf{0} \end{pmatrix} \end{array}.$$

The block $\mathbf{D}_{i,j}^{Y,k,nv}$ contains the transition, when there are k units in the system, from i units in the repair facility to j (a type event Y occurs) and the repairperson is in his workplace. For instance, the cases $\mathbf{D}_{01}^{A,k,nv}$ and $\mathbf{D}_{01}^{B,k,nv}$(transition $E_{0}^{k,nv} \to E_{1}^{k,nv}$ for type A and B respectively) are analyzed.

In both cases, there are k units in the system and none of these is in the repair facility (all operational). The online unit goes to the repair facility if it undergoes an internal repairable failure ($\mathbf{H}_{A}$) or a major inspection ($\mathbf{H}_{B}$). In both cases a new unit will occupy the online place if the number of units in the system is greater than one. If the event is a repairable failure, then the unit will begin the repair given that the repairperson is not on vacation ($\boldsymbol{\beta}_{1}$). If the event is a major inspection, the initial distribution for the preventive maintenance would be $\boldsymbol{\beta}_{2}$.

- If the number of units is greater or equal than R, the repairperson can be on vacation or not. If the repairperson returns and there are less than R operational units then he remains at his workplace. Given that these events A and B occur when a repairable or major inspection occurs (without returning to work) then, for $k = R$, $\ldots, n$ ($N = k - R + 1$, the limit of the number of units in the repair facility for the repairperson to remain):

$$
\mathbf{D}_k^Y = \begin{array}{c} E_0^{k,v} \\ E_1^{k,v} \\ \vdots \\ E_{N-1}^{k,v} \\ E_N^{k,v} \\ E_{N+1}^{k,v} \\ \vdots \\ E_{k-1}^{k,v} \\ E_k^{k,v} \\ E_N^{k,nv} \\ E_{N+1}^{k,nv} \\ \vdots \\ E_{k-1}^{k,nv} \\ E_k^{k,nv} \end{array}
\left(\begin{array}{ccccccccc|ccccc}
\mathbf{0} & \mathbf{D}_{0,1}^{Y,k,v} & \cdots & & & & & & & & & & & \\
 & \mathbf{0} & \ddots & & & & & & & & & & & \\
 & & \ddots & \ddots & & & & & & & & & & \\
 & & & \mathbf{0} & \mathbf{D}_{N-1,N}^{Y,k,v} & & & & & & & & & \\
 & & & & \mathbf{0} & \mathbf{D}_{N,N+1}^{Y,k,v} & & & & & & & & \\
 & & & & & \mathbf{0} & \ddots & & & & & & & \\
 & & & & & & \ddots & \ddots & & & & & & \\
 & & & & & & & \mathbf{0} & \mathbf{D}_{k-1,k}^{Y,k,v} & & & & & \\
 & & & & & & & & \mathbf{0} & & & & & \\
\hline
 & & & & & & & & & \mathbf{D}_{N,N}^{Y,k,nv} & \mathbf{D}_{N,N+1}^{Y,k,nv} & & & \\
 & & & & & & & & & & \mathbf{D}_{N+1,N+1}^{Y,k,nv} & \mathbf{D}_{N+1,N+2}^{Y,k,nv} & & \\
 & & & & & & & & & & & \ddots & \ddots & \\
 & & & & & & & & & & & & \mathbf{D}_{k-1,k-1}^{Y,k,nv} & \mathbf{D}_{k-1,k}^{Y,k,nv} \\
 & & & & & & & & & & & & & \mathbf{0}
\end{array}\right)
$$

with columns $E_0^{k,v}\ E_1^{k,v}\ \ldots\ E_{N-1}^{k,v}\ E_N^{k,v}\ E_{N+1}^{k,v}\ \ldots\ E_{k-1}^{k,v}\ E_k^{k,v}\ E_N^{k,nv}\ E_{N+1}^{k,nv}\ E_{N+2}^{k,nv}\ E_{k-1}^{k,nv}\ E_k^{k,nv}$.

This matrix is partitioned into two great matrix blocks depending on the transition between macro states; continues on vacation and continues in the repair facility.

The block $\mathbf{D}_{i,j}^{Y,k,v}$ contains the transition, when there are k units in the system, from i units in the repair facility to j (type Y) and the repairperson continues on vacation. For instance, the cases $\mathbf{D}_{01}^{A,k,v}$ and $\mathbf{D}_{01}^{B,k,v}$ correspond to the transition $E_0^{k,v} \to E_1^{k,v}$ for type A and B, respectively.

These matrices are for $k = 1, \ldots, n$ and $R > 1$

$\mathbf{D}_{01}^{A,k,nv} = \left(\left(I_{\{k>1\}}\mathbf{H}_A + I_{\{k=1\}}\mathbf{H}'_A\right) \otimes \boldsymbol{\beta}_1, \mathbf{0}\right); \mathbf{D}_{01}^{B,k,nv} = \left(\mathbf{0}, \left(I_{\{k>1\}}\mathbf{H}_B + I_{\{k=1\}}\mathbf{H}'_B\right) \otimes \boldsymbol{\beta}_2\right).$

The rest of matrices for this matrix block are as follows.

$$\mathbf{D}_{1,1}^{A,k,nv} = \begin{pmatrix} \mathbf{H}_A \otimes \mathbf{S}_1^0 \otimes \boldsymbol{\beta}_1 & \mathbf{0} \\ \mathbf{H}_A \otimes \mathbf{S}_2^0 \otimes \boldsymbol{\beta}_1 & \mathbf{0} \end{pmatrix}; \mathbf{D}_{1,1}^{B,k,nv} = \begin{pmatrix} \mathbf{0} & \mathbf{H}_B \otimes \mathbf{S}_1^0 \otimes \boldsymbol{\beta}_2 \\ \mathbf{0} & \mathbf{H}_B \otimes \mathbf{S}_2^0 \otimes \boldsymbol{\beta}_2 \end{pmatrix}$$

For $r = 2, \ldots, k-1$

$$\mathbf{D}_{r,r}^{A,k,nv} = \begin{pmatrix} I_{2^{r-2}} \otimes \left(\mathbf{H}_A \otimes \mathbf{S}_1^0 \otimes \boldsymbol{\beta}_1, 0\right) & \mathbf{0} \\ \mathbf{0} & I_{2^{r-2}} \otimes \left(\mathbf{H}_A \otimes \mathbf{S}_1^0 \otimes \boldsymbol{\beta}_2, \mathbf{0}\right) \\ I_{2^{r-2}} \otimes \left(\mathbf{H}_A \otimes \mathbf{S}_2^0 \otimes \boldsymbol{\beta}_1, \mathbf{0}\right) & \mathbf{0} \\ \mathbf{0} & I_{2^{r-2}} \otimes \left(\mathbf{H}_A \otimes \mathbf{S}_2^0 \otimes \boldsymbol{\beta}_2, \mathbf{0}\right) \end{pmatrix}$$

$$\mathbf{D}_{r,r}^{B,k,nv} = \begin{pmatrix} I_{2^{r-2}} \otimes \left(\mathbf{0}, \mathbf{H}_B \otimes \mathbf{S}_1^0 \otimes \boldsymbol{\beta}_1\right) & \mathbf{0} \\ \mathbf{0} & I_{2^{r-2}} \otimes \left(\mathbf{0}, \mathbf{H}_B \otimes \mathbf{S}_1^0 \otimes \boldsymbol{\beta}_2\right) \\ I_{2^{r-2}} \otimes \left(\mathbf{0}, \mathbf{H}_B \otimes \mathbf{S}_2^0 \otimes \boldsymbol{\beta}_1\right) & \mathbf{0} \\ \mathbf{0} & I_{2^{r-2}} \otimes \left(\mathbf{0}, \mathbf{H}_B \otimes \mathbf{S}_2^0 \otimes \boldsymbol{\beta}_2\right) \end{pmatrix}$$

For $r = \max\{1, k - R + 1\}, \ldots, k - 1$

$$\mathbf{D}_{r,r+1}^{A,k,nv} = \begin{pmatrix} \mathbf{I}_{2^{r-1}} \otimes \left(\left(I_{\{r<k-1\}}\mathbf{H}_A + I_{\{r=k-1\}}\mathbf{H}'_A\right) \otimes \mathbf{S}_1, \mathbf{0}\right) & \mathbf{0} \\ \mathbf{0} & \mathbf{I}_{2^{r-1}} \otimes \left(\left(I_{\{r<k-1\}}\mathbf{H}_A + I_{\{r=k-1\}}\mathbf{H}'_A\right) \otimes \mathbf{S}_2, \mathbf{0}\right) \end{pmatrix}$$

$$\mathbf{D}_{r,r+1}^{B,k,nv} = \begin{pmatrix} \mathbf{I}_{2^{r-1}} \otimes \left(\mathbf{0}, \left(I_{\{r<k-1\}}\mathbf{H}_B + I_{\{r=k-1\}}\mathbf{H}'_B\right) \otimes \mathbf{S}_1\right) & \mathbf{0} \\ \mathbf{0} & \mathbf{I}_{2^{r-1}} \otimes \left(\mathbf{0}, \left(I_{\{r<k-1\}}\mathbf{H}_A + I_{\{r=k-1\}}\mathbf{H}'_A\right) \otimes \mathbf{S}_2\right) \end{pmatrix}$$

For $r = 1, \ldots, k-1$ and $k \geq R$

$\mathbf{D}_{0,1}^{A,k,v} = \left(\mathbf{H}_A \otimes \left(\mathbf{V} + I_{\{k \geq R+1\}}\mathbf{V}^0\boldsymbol{\nu}\right), \mathbf{0}\right); \mathbf{D}_{0,1}^{B,k,v} = \left(\mathbf{0}, \mathbf{H}_B \otimes \left(\mathbf{V} + I_{\{k \geq R+1\}}\mathbf{V}^0\boldsymbol{\nu}\right)\right)$

$\mathbf{D}_{r,r+1}^{A,k,v} = \mathbf{I}_{2^r} \otimes \left(\left(I_{\{r<k-1\}}\mathbf{H}_A + I_{\{r=k-1\}}\mathbf{H}'_A\right) \otimes \left(\mathbf{V} + I_{\{r<N-1\}}\mathbf{V}^0\boldsymbol{\nu}\right), \mathbf{0}\right)$

$\mathbf{D}_{r,r+1}^{B,k,v} = \mathbf{I}_{2^r} \otimes \left(\mathbf{0}, \left(I_{\{r<k-1\}}\mathbf{H}_B + I_{\{r=k-1\}}\mathbf{H}'_B\right) \otimes \left(\mathbf{V} + I_{\{r<N-1\}}\mathbf{V}^0\boldsymbol{\nu}\right)\right).$

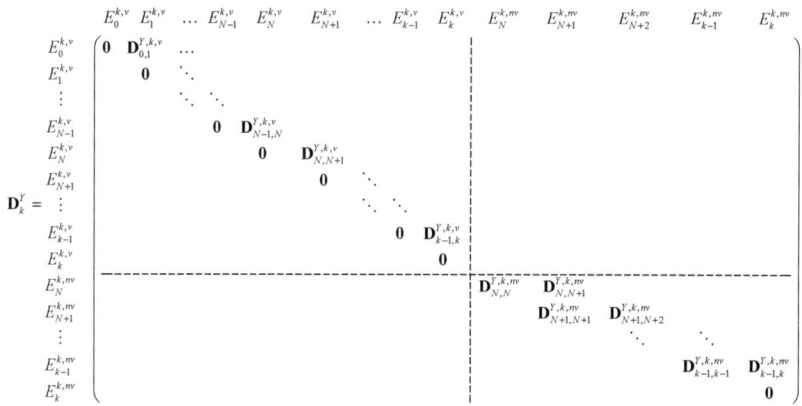

4. Measures

Multiple interesting measures in transient and stationary regime can be worked out and are described in this section.

4.1. The Transient and the Stationary Distribution

The transient distribution is determined by the initial distribution and the transition probability matrix of the vector Markov process given in Section 3.3.

Initially the online unit is new and the inspection time begins. Then, the initial distribution of the Markov process is $\phi = [\boldsymbol{\alpha} \otimes \boldsymbol{\gamma}_{st} \otimes \boldsymbol{\eta}, 0]$ where $\boldsymbol{\gamma}_{st}$ is the stationary distribution of the phase-type renewal process with transition probability matrix $\mathbf{L} + \mathbf{L}^0\boldsymbol{\gamma}$. Therefore, $\boldsymbol{\gamma}_{st} = [1,0]\left[\mathbf{e}\middle|\left(\mathbf{L} + \mathbf{L}^0\boldsymbol{\gamma} - \mathbf{I}\right)^*\right]^{-1}$.

The probability of occupying the macro-state $E_s^{k,a}$ at time ν is worked out by matrix blocks as $\mathbf{p}^{\nu}_{E_s^{k,a}} = (\phi\mathbf{D}^{\nu})_{I_s^{k,a}}$ where $I_s^{k,a}$ indicates the range for the corresponding states. Evidently, $\mathbf{p}^{\nu}$ is the transient distribution at time ν.

To calculate the stationary distribution in a matrix-algorithmic form, we have partitioned the matrix $\mathbf{D}$ for the transitions between the macro-states $\mathbf{U}^j$ into the following blocks,

$$\mathbf{D} = \begin{pmatrix} \mathbf{D}_{n,n} & \mathbf{D}_{n,n-1} & \mathbf{0} & \cdots & \mathbf{0} & \mathbf{0} \\ \mathbf{0} & \mathbf{D}_{n-1,n-1} & \mathbf{D}_{n-1,n-2} & \cdots & \mathbf{0} & \mathbf{0} \\ \vdots & \vdots & \ddots & \ddots & \vdots & \vdots \\ \mathbf{0} & \mathbf{0} & \cdots & \cdots & \mathbf{D}_{22} & \mathbf{D}_{21} \\ \mathbf{D}_{1n} & \mathbf{0} & \cdots & \cdots & \cdots & \mathbf{D}_{11} \end{pmatrix}$$

where

$\mathbf{D}_{ii} = \mathbf{D}_i^O + \mathbf{D}_i^A + \mathbf{D}_i^B + \mathbf{D}_i^D + \mathbf{D}_i^{AD} + \mathbf{D}_i^{BD}; i = 1, \ldots, n$

$\mathbf{D}_{i,i-1} = \mathbf{D}_i^C + \mathbf{D}_i^{CD}; i = 2, \ldots, n$

$\mathbf{D}_{1,n} = \mathbf{D}_1^{NS}$.

The stationary distribution $\boldsymbol{\pi}$ verifies the balance equations $\boldsymbol{\pi}\mathbf{D} = \boldsymbol{\pi}$ and the normalization equation $\boldsymbol{\pi}\mathbf{e} = 1$. This vector is partitioned into the macro-states $\mathbf{U}^j$, j units in the system, then, $\boldsymbol{\pi} = \{\boldsymbol{\pi}_n, \boldsymbol{\pi}_{n-1}, \ldots, \boldsymbol{\pi}_1\}$ for the macro-states $\mathbf{U}^n, \ldots, \mathbf{U}^1$, respectively.

The solution of this matrix system is $\boldsymbol{\pi}_j = \boldsymbol{\pi}_1\mathbf{R}_j; j = 2, \ldots, n$, being $\mathbf{R}_j = \mathbf{R}_{j+1}\mathbf{G}_{j+1,j} = \mathbf{G}_{1n}\mathbf{G}_{n,n-1}\cdots\mathbf{G}_{j+1,j}$; $j = 2, \ldots, n-1$, $\mathbf{R}_n = \mathbf{G}_{1,n}$ and $\mathbf{G}_{ij} = \mathbf{D}_{ij}\left(\mathbf{I} - \mathbf{D}_{jj}\right)^{-1}$ for $(i,j) \in \{(1,n), (n,n-1), (n-1,n-2), \ldots, (3,2)\}$

The transition probability vector for the macro-state $\mathbf{U}^1$ can be worked out from the normalization condition and one balanced equation as

$$\boldsymbol{\pi}_1 = (1,0)\left(\mathbf{e} + \sum_{j=2}^{n}\mathbf{R}_j\mathbf{e}\middle|\left(\mathbf{I} - \mathbf{D}_{11} - \mathbf{R}_2\mathbf{D}_{21}\right)^*\right)^{-1},$$

where * is the corresponding matrix without the first column.

From the stationary distribution and considering the macro-states, multiple proportional time measures can be defined:

- Proportional time that the system has k units: $\boldsymbol{\pi}_{\mathbf{U}^k}$.
- Proportional time that the repairperson is in the workplace:
  $\Upsilon_{nv} = \sum_{k=1}^{R-1}\sum_{s=0}^{k}\boldsymbol{\pi}_{E_s^{k,nv}}\mathbf{e} + \sum_{k=R}^{n}\sum_{s=k-R+1}^{k}\boldsymbol{\pi}_{E_s^{k,nv}}\mathbf{e}$.
- Proportional time that the repairperson is on vacation:
  $\Upsilon_v = 1 - \Upsilon_{nv}$
- Proportional time that the repairperson is working:
  $\Upsilon_w = \sum_{k=1}^{R-1}\sum_{s=1}^{k}\boldsymbol{\pi}_{E_s^{k,nv}}\mathbf{e} + \sum_{k=R}^{n}\sum_{s=k-R+1}^{k}\boldsymbol{\pi}_{E_s^{k,nv}}\mathbf{e}$

- Proportional time that the repairperson is idle:
  $\Upsilon_i = \Upsilon_{nv} - \Upsilon_w$.

4.2. Availability and Mean Times

It is interesting to calculate the availability of the system, the mean time in each macro-state and the mean operational time. This has been summed up in Table 1 in both regimes, transient and stationary.

Table 1. Availability and mean times in transient and stationary regime.

	Transient Regime (up to Time ν)	Stationary Regime
Availability	$A(\nu) = 1 - \sum_{k=R}^{n}\left(\mathbf{p}^{\nu}_{E_k^{k,v}} \cdot \mathbf{e} + \mathbf{p}^{\nu}_{E_k^{k,nv}} \cdot \mathbf{e}\right) - \sum_{k=1}^{R-1} \mathbf{p}^{\nu}_{E_k^{k,nv}} \cdot \mathbf{e}$	$A = 1 - \sum_{k=R}^{n}\left(\boldsymbol{\pi}_{E_k^{k,v}} \cdot \mathbf{e} + \boldsymbol{\pi}_{E_k^{k,nv}} \cdot \mathbf{e}\right) - \sum_{k=1}^{R-1} \boldsymbol{\pi}_{E_k^{k,nv}} \cdot \mathbf{e}$
Mean time in $\mathbf{E}_s^{k,v}$; $\mathbf{E}_s^{k,nv}$	$\psi_{k,s}(\nu) = \sum_{m=0}^{\nu}\left(\mathbf{p}^{m}_{E_s^{k,v}} \cdot \mathbf{e} + \mathbf{p}^{m}_{E_s^{k,nv}} \cdot \mathbf{e}\right)$	$\psi_{k,s} = \boldsymbol{\pi}_{E_s^{k,v}} \cdot \mathbf{e} + \boldsymbol{\pi}_{E_s^{k,nv}} \cdot \mathbf{e}$
Mean time in $\mathbf{E}^k$	$\psi_k(\nu) = \sum_{s=0}^{k} \psi_{k,s}(\nu)$	$\psi_k = \sum_{s=0}^{k} \psi_{k,s}$
Mean operational time	$\mu_{op}(\nu) = \sum_{k=1}^{K}\sum_{s=0}^{k-1} \psi_{k,s}(\nu)$	$\mu_{op} = \sum_{k=1}^{K}\sum_{s=0}^{k-1} \psi_{k,s}$

4.3. Time up to First Time That the System Is Replaced

A system composed of n units is replaced by a new and identical one when all units undergo a non-repairable failure. The time up to this event is phase-type distributed with representation $(\phi, \mathbf{D}')$ where $\mathbf{D}' = \mathbf{D}^O + \mathbf{D}^A + \mathbf{D}^B + \mathbf{D}^C + \mathbf{D}^D + \mathbf{D}^{AD} + \mathbf{D}^{BD} + \mathbf{D}^{CD}$.

4.4. Expected Number of Events

The expected number of events up to time ν is determined using the Markovian Arrival Process with Marked arrivals developed in Section 3.3. If the event considered is denoted by Y then the corresponding expected number of events is given by

$$\Lambda^Y(\nu) = \sum_{u=1}^{\nu} \mathbf{p}^{u-1}\mathbf{D}^Y\mathbf{e},$$

For $Y = A, B, C, D, AD, BD, CD, NS$. This value in stationary regime is $\Lambda^Y = \boldsymbol{\pi}\mathbf{D}^Y\mathbf{e}$. Another mean number of events can be calculated as follows.

4.5. Mean Number of Repairable Failures

A repairable failure can occur when the repairperson resumes work or not at the same time. Then, the mean number up to time ν is $\Lambda^{rep}(\nu) = \sum_{u=1}^{\nu} \mathbf{p}^{u-1}(\mathbf{D}^A + \mathbf{D}^{AD})\mathbf{e}$ and in stationary regime it is $\Lambda^{rep} = \boldsymbol{\pi}(\mathbf{D}^A + \mathbf{D}^{AD})\mathbf{e}$.

4.6. Mean Number of Major Inspections

Analogously to the repairable case, a major inspection can occur when the repairperson occupies the workplace or not at the same time. Then, it is in transient regime $\Lambda^{mi}(\nu) = \sum_{u=1}^{\nu} \mathbf{p}^{u-1}(\mathbf{D}^B + \mathbf{D}^{BD})\mathbf{e}$ and in the stationary case it is $\Lambda^{mi} = \boldsymbol{\pi}(\mathbf{D}^B + \mathbf{D}^{BD})\mathbf{e}$.

4.7. Mean Number of Non-Repairable Failures (No Provoking System Failure)

The mean number of non-repairable failures up to time ν is

$$\Lambda^{nr}(\nu) = \sum_{u=1}^{\nu} \mathbf{p}^{u-1}\left(\mathbf{D}^{C} + \mathbf{D}^{CD}\right)\mathbf{e}.$$

This value in the stationary case is $\Lambda^{nr} = \boldsymbol{\pi}(\mathbf{D}^{C} + \mathbf{D}^{CD})\mathbf{e}$.

4.8. Mean Number of Times That the Repairperson Resumes to Work

The mean number that the repairperson resumes and remains in his workplace up to a certain time is given by

$$\Lambda^{rejoined}(\nu) = \sum_{u=1}^{\nu} \mathbf{p}^{u-1}\left(\mathbf{D}^{D} + \mathbf{D}^{AD} + \mathbf{D}^{BD} + \mathbf{D}^{CD}\right)\mathbf{e}$$

In the stationary case this is $\Lambda^{rejoined} = \boldsymbol{\pi}(\mathbf{D}^{D} + \mathbf{D}^{AD} + \mathbf{D}^{BD} + \mathbf{D}^{CD})\mathbf{e}$.

4.9. Mean Number of Times That the Repairperson Resumes and Begins a New Period of Vacation

The mean number that the repairperson resumes and begins a new period of vacation up to a certain time is given by

$$\Lambda^{r-b}(\nu) = \sum_{u=1}^{\nu} \mathbf{p}^{u-1}\mathbf{Q}\mathbf{e}.$$

where $\mathbf{Q}$ is a matrix described in Appendix C. In the stationary case it is $\Lambda^{r-b} = \boldsymbol{\pi}\mathbf{Q}\mathbf{e}$.

4.10. Mean Number of New Systems

When the system is composed of only one unit and a non-repairable failure occurs, the system is restarted with n new units. The mean number of new systems up to time ν is

$$\Lambda^{NS}(\nu) = \sum_{u=1}^{\nu} \mathbf{p}^{u-1}\mathbf{D}^{NS}\mathbf{e}.$$

This measure in stationary case is $\Lambda^{NS} = \boldsymbol{\pi}\mathbf{D}^{NS}\mathbf{e}$.

5. Rewards and Costs

To analyze the effectiveness of the model from an economic point of view, costs and rewards have been taken into account. A net profit vector associated to the state-space is built. Previously, multiple values are introduced:

B: Gross profit per unit of time if the system is operational.

$\mathbf{c_0}$: expected cost per unit of time depending on the operational phase while the system is operational.

$\mathbf{cr_1}$: expected corrective repair cost per unit of time depending on the repair phase.

$\mathbf{cr_2}$: expected preventive maintenance cost per unit of time for a unit that was observed with major damage depending on the preventive maintenance phase.

H: repairperson cost per unit of time while the repairperson in idle.

C: loss per unit of time while the system is not operational

G: fixed cost associated to each return of the repairperson (independently of if he stays or not).

fcr: fixed cost each time that the online unit undergoes a repairable failure from the online unit.

fmi: fixed cost each time that the online unit undergoes a major inspection.

fnu: cost for a new unit ($n \cdot fnu$ cost of a new system).

5.1. Net Profit Vector

When the system occupies a determined state, a net profit value is produced. Costs and rewards from the online unit and the cost provoked by the repairperson have been taken into account to build the net profit vector.

5.1.1. Online Unit

If only the online unit is considered when the system visits the macro-state $\mathbf{E}_s^{k,nv}$, a net reward for the phases of this macro-state is worked out. The profit net vector for the online unit if the repairperson is in his workplace ($\mathbf{E}_s^{k,nv}$) is for $k = 1, \ldots, n$,

$$\mathbf{nr}_s^{k,nv} = \begin{cases} B\mathbf{e}_{mt\varepsilon} - \mathbf{c}_0 \otimes \mathbf{e}_{t\varepsilon} & ; \quad s = 0 \\ B\mathbf{e}_{mt\varepsilon 2^{s-1}(z_1+z_2)} - \mathbf{c}_0 \otimes \mathbf{e}_{t\varepsilon 2^{s-1}(z_1+z_2)} & ; \quad s = 1, \ldots, k-1 \\ -C \cdot \mathbf{e}_{t2^{s-1}(z_1+z_2)} & ; \quad s = k. \end{cases}$$

This can be expressed for any number of units in the repair facility as the following column vector $\mathbf{nr}_{Total}^{k,nv} = \left(\mathbf{nr}_0^{k,nv\prime}; \ldots; \mathbf{nr}_k^{k,nv\prime}\right)'$.

If the number of units in the repair facility is N or more, then the repairperson remains at his workplace without vacation. In this case we define $\mathbf{nr}_{fromN}^{k,nv} = \left(\mathbf{nr}_N^{k,nv\prime}; \ldots; \mathbf{nr}_k^{k,nv\prime}\right)'$.

For cased when the repairperson is on vacation, the profit net vector for the online unit for the macro-state $\mathbf{E}_s^{k,v}$ is

$$\mathbf{nr}_s^{k,v} = \begin{cases} B\mathbf{e}_{mt\varepsilon v} - \mathbf{c}_0 \otimes \mathbf{e}_{t\varepsilon v} & ; \quad s = 0 \\ B\mathbf{e}_{mt\varepsilon v 2^s} - \mathbf{c}_0 \otimes \mathbf{e}_{t\varepsilon v 2^s} & ; \quad s = 1, \ldots, k-1 \\ -C \cdot \mathbf{e}_{tv2^s} & ; \quad s = k. \end{cases}$$

For any number of units in the repair facility the column vector $\mathbf{nr}_{Total}^{k,v} = \left(\mathbf{nr}_0^{k,v\prime}; \ldots; \mathbf{nr}_k^{k,v\prime}\right)'$ is defined.

Then, if the total state space is considered then the net reward, according to the state visited, for the online unit is

$$\mathbf{nr} = \left(\mathbf{nr}_{Total}^{n,v\;\prime}; \mathbf{nr}_{fromN}^{n,nv\;\prime}; \mathbf{nr}_{Total}^{n-1,v\prime}; \mathbf{nr}_{fromN}^{n-1,nv\prime}; \ldots; \mathbf{nr}_{Total}^{N,v\;\prime}; \mathbf{nr}_{fromN}^{N,nv\;\prime}; \mathbf{nr}_{Total}^{N-1,nv\prime}; \mathbf{nr}_{Total}^{N-1,nv\prime}; \ldots; \mathbf{nr}_{Total}^{1,nv\;\prime}\right)'$$

5.1.2. Repair Facility

If only the repair facility is considered, when the system visits the macro-states $\mathbf{E}_s^{k,nv}$, a cost vector for the phases of the corresponding macro-state, for $k = 1, \ldots, n$ is

$$\mathbf{nc}_s^{k,nv} = \begin{cases} H \cdot \mathbf{e}_{mt\varepsilon} & ; \quad s = 0 \\ \mathbf{e}_{t(m\varepsilon)^{I_{\{s<k\}}}} \otimes \begin{pmatrix} \mathbf{e}_{2^{s-1}} \otimes \mathbf{cr}_1 \\ \mathbf{e}_{2^{s-1}} \otimes \mathbf{cr}_2 \end{pmatrix} & ; \quad s = 1, \ldots, k. \end{cases}$$

For any number of units in the repair facility, the following column vectors are defined,

$$\mathbf{nc}_{Total}^{k,nv} = \left(\mathbf{nc}_0^{k,nv\prime}; \ldots; \mathbf{nc}_k^{k,nv\prime}\right)', \quad \mathbf{nc}_{fromN}^{k,nv} = \left(\mathbf{nc}_N^{k,nv\prime}; \ldots; \mathbf{nc}_k^{k,nv\prime}\right)'$$

For any k and s while the repairperson is on vacation the cost of the repair facility is zero, then the following column vector is defined for this case as $\mathbf{nc}_s^{k,v} = 0_{(m\varepsilon)^{I_{\{s<k\}}} tv2^s}$. For any number of units in the repair facility it is defined as $\mathbf{nc}_{Total}^{k,v} = \left(\mathbf{nc}_0^{k,v\prime}; \ldots; \mathbf{nc}_k^{k,v\prime}\right)'$.

Then, the cost vector associated to the state space due to repair is given by

$$\mathbf{nc} = \left(\mathbf{nc}^{n,v\prime}; \mathbf{nc}_{fromN}^{n,nv\;\prime}; \mathbf{nc}^{n-1,v\prime}; \mathbf{nc}_{fromN}^{n-1,nv\prime}; \ldots; \mathbf{nc}^{N,v\prime}; \mathbf{nc}_{fromN}^{N,nv\;\prime}; \mathbf{nc}_{Total}^{N-1,nv\prime}; \mathbf{nc}_{Total}^{N-1,nv\prime}; \ldots; \mathbf{nc}_{Total}^{1,nv\;\prime}\right)'$$

Therefore, the net profit vector corresponding to the online unit and the repair facility for the global state space is given by

$$\mathbf{c} = \mathbf{nr} - \mathbf{nc} = \begin{pmatrix} \mathbf{c}^{n} \\ \mathbf{c}^{n-1} \\ \vdots \\ \mathbf{c}^{1} \end{pmatrix},$$

where
$\mathbf{c}^{k} = \left(\mathbf{nr}^{k,nv\ \prime}_{Total} - \mathbf{nc}^{k,nv\ \prime}_{Total}\right)'$ for $k = 1, \ldots, R-1$,
$\mathbf{c}^{k} = \left(\mathbf{nr}^{k,v\prime} - \mathbf{nc}^{k,v\prime}; \mathbf{nc}^{k,nv\ \prime}_{fromN} - \mathbf{nc}^{k,nv\ \prime}_{fromN}\right)'$ for $k = R, \ldots, n$.

5.2. *Expected Net Profits and Total Net Profit*

Net reward measures are worked out, in transient and stationary regimes, to analyze the effectiveness of the system from an economic point of view.

5.2.1. Expected Net Profit from the Online Unit Up to Time ν

The expected net profit up to time ν by considering only the online unit is

$$\Phi_{w}^{\nu} = \sum_{m=0}^{\nu} \mathbf{p}^{m} \cdot \mathbf{nr}.$$

In stationary regime this is given by $\Phi_{w\_s} = \boldsymbol{\pi} \cdot \mathbf{nr}$.

5.2.2. Expected Cost from Corrective Repair and Preventive Maintenance

The expected cost because of corrective repair and preventive maintenance up to time ν is calculated. This is respectively

$$\Phi_{cr}^{\nu} = \sum_{m=0}^{\nu} \mathbf{p}^{m} \cdot \mathbf{mc}^{cr} \text{ and } \Phi_{pm}^{\nu} = \sum_{m=0}^{\nu} \mathbf{p}^{m} \cdot \mathbf{mc}^{pm}$$

where $\mathbf{mc}^{cr}$ is the vector $\mathbf{nc}$ with $\mathbf{cr}_2 = 0_{z_2}$ and $\mathbf{mc}^{pm}$ is the vector $\mathbf{nc}$ with $\mathbf{cr}_1 = 0_{z_1}$, being $\mathbf{0}_a$ a column vector of 0s with order a.

If the stationary regime is considered, then

$$\Phi_{cr\_s} = \boldsymbol{\pi} \cdot \mathbf{mc}^{cr} \text{ and } \Phi_{pm\_s} = \boldsymbol{\pi} \cdot \mathbf{mc}^{pm}$$

5.2.3. Total Net Profit

If costs, fixed costs and profits are considered, the total net profit up to time ν is

$$\Phi^{\nu} = \Phi_{w}^{\nu} - \Phi_{cr}^{\nu} - \Phi_{pm}^{\nu} - \left(1 + \Lambda^{NS}(\nu)\right) \cdot n \cdot fnu - \Lambda^{rep}(\nu) \cdot fcr - \Lambda^{mi}(\nu) \cdot fmi - \Lambda^{r-b}(\nu) \cdot G$$

In the stationary case this is

$$\Phi = \Phi_{w} - \Phi_{cr} - \Phi_{pm} - \left(1 + \Lambda^{NS}\right) \cdot n \cdot fnu - \Lambda^{rep} \cdot fcr - \Lambda^{mi} \cdot fmi - \Lambda^{r-b} \cdot G.$$

6. A Numerical Example

The system modelled in this paper can be applied to real-world engineering problems. It would be interesting to examine whether or not preventive maintenance is profitable and to determine the optimum distribution for vacation time and hence the corresponding value of R.

6.1. The System

We assume a standby system composed of four units initially as described in this work. Each unit is composed of four performance internal states where the first two are considered minor damage and the last two as major damage. The transition probability matrix for wearing out time is given by

$$\mathbf{T} = \begin{pmatrix} 0.96 & 0.03 & 0 & 0 \\ 0 & 0.97 & 0.01 & 0 \\ 0 & 0 & 0.85 & 0.06 \\ 0 & 0 & 0 & 0.6 \end{pmatrix},$$

Beginning in the initial state (α = (1, 0, 0, 0)). From each state, only a transition to failure or to next performance level state can occur. The transition probability to repairable and non-repairable failure depending on the performance state are given by the column vectors $\mathbf{T}_r^0 = \begin{pmatrix} 0.008 \\ 0.016 \\ 0.072 \\ 0.32 \end{pmatrix}$ and $\mathbf{T}_{nr}^0 = \begin{pmatrix} 0.002 \\ 0.004 \\ 0.018 \\ 0.080 \end{pmatrix}$ respectively.

The online unit is subject to external shocks. The time between two consecutive external shocks follows a phase-type distribution with representation $(\boldsymbol{\gamma}, \mathbf{L})$ being $\boldsymbol{\gamma} = (1,0)$ and $\mathbf{L} = \begin{pmatrix} 0.9 & 0.05 \\ 0 & 0.5 \end{pmatrix}$.

The mean time between two consecutive accidental external failures is equal to 11 units of time.

Each time that the system suffers an external shock the internal performance can be modified by producing a repairable or non-repairable failure. The matrix that governs the changes into the operational states is

$$\mathbf{W} = \begin{pmatrix} 0.2 & 0.1 & 0.3 & 0.1 \\ 0 & 0.1 & 0.3 & 0.1 \\ 0 & 0 & 0.3 & 0.1 \\ 0 & 0 & 0 & 0.1 \end{pmatrix}$$

and the change to a repairable and non-repairable is $\mathbf{W}_r^0 = \begin{pmatrix} 0.3 \\ 0.4 \\ 0.5 \\ 0.6 \end{pmatrix}$ and $\mathbf{W}_{nr}^0 = \begin{pmatrix} 0 \\ 0.1 \\ 0.1 \\ 0.3 \end{pmatrix}$ respectively.

When an external shock occurs, a total failure can also be produced with a probability equal to ω^0 = 0.2.

Inspections occur randomly where the inter-inspection time is phase-type distributed with representation $(\boldsymbol{\eta}, \mathbf{M})$ being

$$\boldsymbol{\eta} = (1,0),\ \mathbf{M} = \begin{pmatrix} 0.85 & 0.1 \\ 0.45 & 0.4 \end{pmatrix}.$$

When a unit undergoes a repairable failure or inspection observes major damage, this goes to the repair facility. Therefore, two types of tasks can be developed by the repairperson, corrective repair and preventive maintenance. Both are phase-type distributed with representation for the corrective repair time,

$$\boldsymbol{\beta}_1 = (1,0,0) \text{ and } \mathbf{S}_1 = \begin{pmatrix} 0.2 & 0.4 & 0.3 \\ 0.2 & 0.2 & 0.5 \\ 0.3 & 0.2 & 0.3 \end{pmatrix}$$

and for the preventive maintenance time,

$$\boldsymbol{\beta}_2 = (1,0,0) \text{ and } \mathbf{S}_2 = \begin{pmatrix} 0.2 & 0.3 & 0.1 \\ 0.1 & 0.1 & 0.4 \\ 0.2 & 0.2 & 0.2 \end{pmatrix}.$$

The mean corrective repair time is 7.3810 units of time and for the preventive maintenance case this is equal to 2.5 units of time.

6.2. *Costs and Rewards*

Different costs and rewards have been considered as described in Section 5. We assume a gross profit while the system is operational, equal to $B = 60$. This is also the loss per unit of time while the system is not operational, $C = 60$. The online unit has a cost while it is operational depending on the operational phase. This vector is $\mathbf{c}_0 = (5, 12, 30, 40)'$. The repairperson can be on vacation or in his workplace. Each time that the repairperson returns on his vacation a cost equal to $G = 20$ is produced. While the repairperson is idle, a cost equal to $H = 15$ is produced.

The online unit can undergo a repairable failure. In this case, the unit goes to the repair facility for corrective repair. A fixed cost is considered for each failure equal to $fcr = 10$. Once in corrective repair, a cost depending on the state is given by $\mathbf{cr}_1 = (18,18,18)'$.

When inspection observes major damage, the unit also goes to the repair facility for preventive maintenance. A fixed cost is produced, $fmi = 5$. Once in the repair facility the cost will depend on the preventive maintenance state. This is given by the vector $\mathbf{cr}_2 = (15.5, 15.5, 15.5)'$. Finally, when all units undergo a non-repairable failure the system is re-started. It has a cost per unit equal to $fnu = 100$.

6.3. *Optimization Analysis*

The repairperson can take a vacation, for a random duration, and inspections may take place at random intervals. This circumstance raises two interesting questions. Firstly, if a distribution class is assumed for the duration of the vacation, from an economic standpoint what is the optimum distribution and the optimum value of R (i.e., the limit value of the number of operational units needed to require the repairperson to remain in the facility on returning from vacation) from an economic standpoint? Secondly, is it profitable to perform preventive maintenance?

To answer these questions, we consider two classes of distributions, the geometric distribution and the Erlang distribution, from which optimum values for R and the other parameters can be determined.

6.3.1. The Geometric Distribution Case

We assume that the vacation time of the repairperson is distributed geometrically with parameter p. Then, the p.m.f. is $P\{X = n\} = p^{n-1}(1 - p)$; $n = 0, 1, 2, \ldots$

The stationary net profit depending on p for the system with and without preventive maintenance is shown in Figure 4. This has been worked out from Section 5.2. We can see that, when the geometric distribution is considered, the optimum value is reached for the preventive maintenance case with $p = 0.8$ and $R = 3$. In this case, and in the stationary case, the net profit per unit of time would be equal to 22.0571.

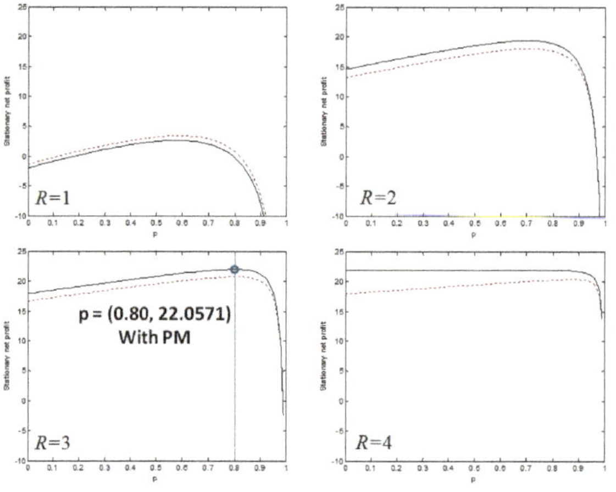

Figure 4. Stationary net profit depending on p and R (with preventive maintenance, continuous line; without preventive maintenance, dashed line).

6.3.2. The Generalized Erlang Distribution Case

Analogously to the geometric case, we assume now that the vacation time is distributed as a Generalized Erlang distribution with parameter shape equal to 2. This distribution can be expressed as a phase-type with representation $(\mathbf{v}, \mathbf{V})$ being

$$\mathbf{v} = (1,0);\ \mathbf{V} = \begin{pmatrix} p_1 & 1-p_1 \\ 0 & p_2 \end{pmatrix}.$$

Figures 5 and 6 show the stationary net profit depending on the parameters p_1 and p_2 and R for the case without preventive maintenance and with preventive maintenance, respectively.

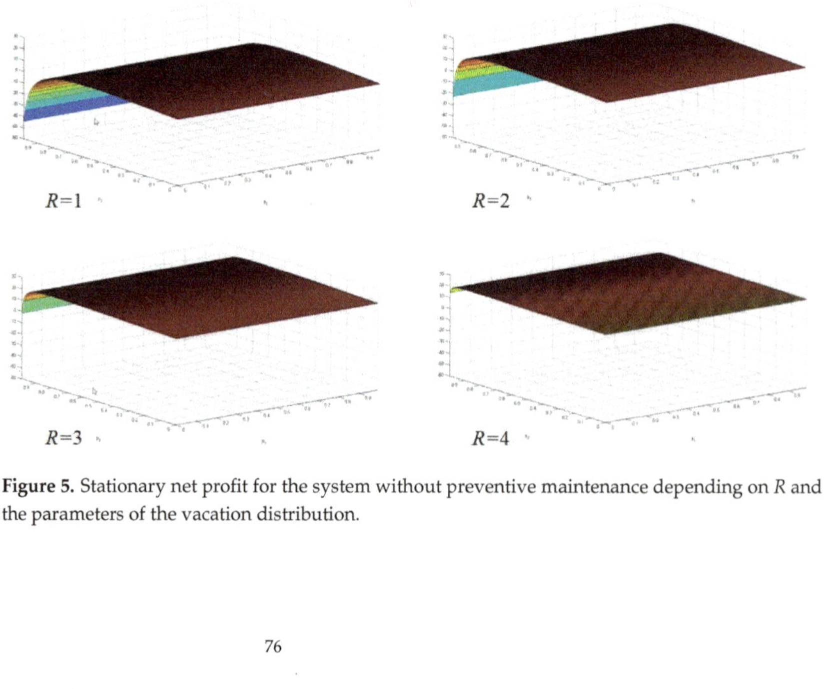

Figure 5. Stationary net profit for the system without preventive maintenance depending on R and the parameters of the vacation distribution.

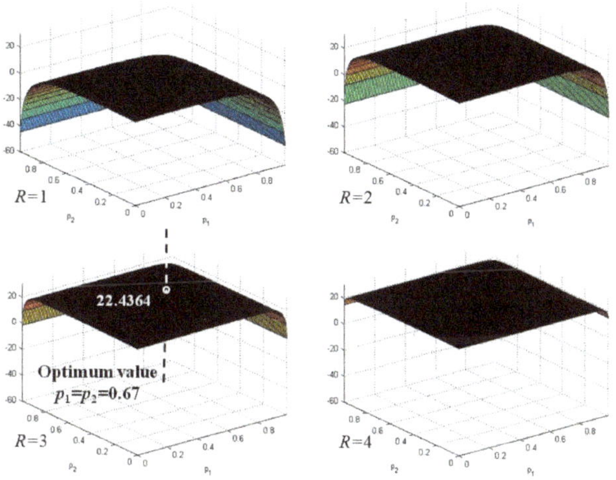

Figure 6. Stationary net profit for the system with preventive maintenance depending on R and the parameters of the vacation distribution.

We can see that, when the generalized Erlang distribution is considered for the vacation time, the optimum value is reached for the preventive maintenance case with $p_1 = p_2 = 0.67$ and $R = 3$. In this case, a stationary case, the net profit per unit of time would be equal to 22.4364.

6.4. The Optimum System with the Generalized Erlang Distribution

In section above we have worked out the optimum system. It is given when the generalized Erlang distribution is considered with parameters (2, 0.67, 0.67) and $R = 3$. In this section the performance measures of this system are analysed.

Firstly, the time up to first time that the system is replaced (all units undergo a non-repairable failure), described in Section 4.3, has been analysed. The reliability function is plotted in Figure 7. Two cases are shown, with and without inspection.

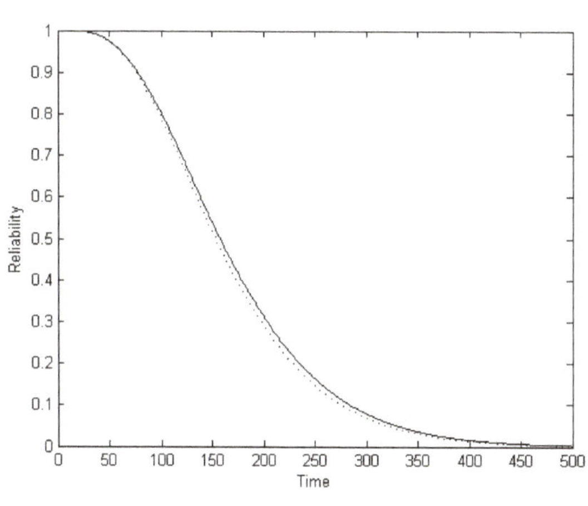

Figure 7. Reliability function of the time up to a new system (with inspection, continuous line; without inspection, dashed line).

From the corresponding phase-type distribution, the mean time up to a new system has been calculated in both cases. Thus, the mean time up to replacing the system for the case without inspection is 167.7631 u.t., and with inspection 172.5269 u.t.

Multiple measures have been achieved for this system with and without inspection. These measures are described in Section 4. Table 2 shows the stationary distribution for macro-states $\mathbf{U}^k$, k units in the system. They can be interpreted as the proportional time that the system is in these macro-states.

Table 2. Proportional time in macro-state $\mathbf{U}^k$.

	$\pi_{\mathbf{U}^1}$	$\pi_{\mathbf{U}^2}$	$\pi_{\mathbf{U}^3}$	$\pi_{\mathbf{U}^4}$
Without inspection	0.3043	0.2411	0.2306	0.2240
With inspection	0.3057	0.2410	0.2299	0.2234

Performance measures are developed for the optimum system with and without inspection following Section 4. Table 3 shows the results.

Table 3. Performance measures for the optimum system (without inspection between parentheses).

Υ_{nv}	Υ_v	Υ_w	Υ_i	Λ^{rep}	Λ^{mi}	Λ^{NS}	Φ	A
0.6806 (0.6826)	0.3194 (0.3174)	0.3139 (0.3187)	0.3667 (0.3639)	0.0409 (0.0432)	0.0049	0.0058 (0.0059)	22.4364 (21.2077)	0.8772 (0.8752)

The proportional time that the repairperson is on vacation is 0.3194. This fact is of interest for the total cost. Therefore, the repairperson is in his workplace for 0.6806 proportion of time and working for 0.3139 proportion of time. Then, the 46.12% of the time that the repairperson is in his workplace, he is working. The remaining time he is idle.

Regarding the mean number of events per unit of time we can observe that this is 0.0409 for repairable failures, 0.0049 for major inspection and 0.0058 for new systems. Thus, for each 10,000 units of time 58 new systems are expected to be re-started. The availability is also worked out. For 87.72% of the time the system is operational, a 0.23% increase than the without inspection case. Really this is low but the difference between both net profits is important, 5.79% maximum for the case with preventive maintenance.

7. Conclusions

Matrix analysis methods can be used to model a complex discrete cold standby system subject to multiple events. This method facilitates the algorithmic and computational development of multi-state complex systems. In the case in question, the online unit within the system is subject to wear and external shocks and may undergo periodic or random inspection. The repair facility is composed of a single repairperson, who may take a vacation (absence) from the repair facility. This repairperson may perform corrective repair and/or preventive maintenance.

The system described is not the standard one in which units are replaced when they undergo a non-repairable failure. In the present study, the analysis takes account of the loss of units following the occurrence of a non-repairable failure. When such a failure occurs, the system continues working with one less unit. This outcome often occurs in practice, and is reflected in the study method presented.

The (indeterminate) number of units within the repair facility and the vacation policy applied determine the behaviour of the repairperson. The vacation time begins when the number of operational units exceeds a given value, and the repairperson will remain in place, without taking a vacation, if the number of operational units in the system is below a pre-determined value.

The system is modelled in an algorithmic and computational form by means of a Markovian Arrival Process with marked arrivals. Matrix-analytic methods are used to obtain the stationary distributions, and multiple measures are derived using a matrix. These measures are related to system performance and financial results.

The method presented in this paper enables us to analyse optimization problems in multi-state complex systems. A numerical example of such an optimization is presented. The results obtained show whether preventive maintenance is profitable and reveal the optimum number of operational units, hence determining the appropriate policy for the repairperson's vacation times.

Funding: This paper is partially supported by the project FQM-307 of the Government of Andalusia (Spain) and by the project MTM2017-88708-P of the Spanish Ministry of Science, Innovation and Universities (also supported by the European Regional Development Fund program, ERDF).

Institutional Review Board Statement: Not applicable.

Informed Consent Statement: Not applicable.

Data Availability Statement: Not applicable.

Conflicts of Interest: The author declares no conflict of interest.

Appendix A. Transition Probability Matrix Blocks for the Online Unit Depending on Type of Event

$$
\begin{aligned}
\mathbf{H}_O &= (\mathbf{T}\otimes\mathbf{L}+\mathbf{TW}\otimes\mathbf{L}^0\gamma(1-\omega^0))\otimes\mathbf{M}+(\mathbf{U}_1\mathbf{T}\otimes\mathbf{L}+\mathbf{U}_1\mathbf{TW}\otimes\mathbf{L}^0\gamma(1-\omega^0))\otimes\mathbf{M}^0\eta \\
\mathbf{H}_A &= \mathbf{T}_r^0\alpha\otimes\mathbf{L}\otimes\mathbf{e}\eta+\left(\mathbf{T}_r^0+\mathbf{TW}_r^0\right)\alpha\otimes\mathbf{L}^0\gamma(1-\omega^0)\otimes\mathbf{e}\eta \\
\mathbf{H}'_A &= \mathbf{T}_r^0\otimes\mathbf{L}\otimes\mathbf{e}+\left(\mathbf{T}_r^0+\mathbf{TW}_r^0\right)\otimes\mathbf{L}^0\gamma(1-\omega^0)\otimes\mathbf{e}. \\
\mathbf{H}_B &= \left[\mathbf{U}_2(\mathbf{e}-\mathbf{T}^0)\alpha\otimes\mathbf{L}+\mathbf{U}_2\mathbf{T}\left(\mathbf{e}-\mathbf{W}^0\right)\alpha\otimes\mathbf{L}^0\gamma(1-\omega^0)\right]\otimes\mathbf{M}^0\eta \\
\mathbf{H}'_B &= \left[\mathbf{U}_2(\mathbf{e}-\mathbf{T}^0)\otimes\mathbf{L}+\mathbf{U}_2\mathbf{T}\left(\mathbf{e}-\mathbf{W}^0\right)\otimes\mathbf{L}^0\gamma(1-\omega^0)\right]\otimes\mathbf{M}^0 \\
\mathbf{H}_C &= \left[\mathbf{T}_{nr}^0\alpha\otimes\mathbf{L}+\left(\mathbf{T}_{nr}^0+\mathbf{TW}_{nr}^0\right)\alpha\otimes\mathbf{L}^0\gamma(1-\omega^0)+\mathbf{e}\alpha\otimes\mathbf{L}^0\gamma\omega^0\right]\otimes\mathbf{e}\eta. \\
\mathbf{H}'_C &= \left[\mathbf{T}_{nr}^0\otimes\mathbf{L}+\left(\mathbf{T}_{nr}^0+\mathbf{TW}_{nr}^0\right)\otimes\mathbf{L}^0\gamma(1-\omega^0)+\mathbf{e}\otimes\mathbf{L}^0\gamma\omega^0\right]\otimes\mathbf{e}
\end{aligned}
$$

Appendix B

Appendix B.1. Matrices for the Markovian Arrival Process Depending on the Type of Event

The matrices $\mathbf{D}^A$ and $\mathbf{D}^B$ are developed in the text. The rest are given below.

Appendix B.2. The Matrix $\mathbf{D}^O$

The matrix D^O contains the transitions when a none-event occurs. This matrix is composed of blocks according to the transitions between the macro-states $\mathbf{U}^k$ for k = 1, … ,n. It is given by

$$
\mathbf{D}^O = \begin{pmatrix} \mathbf{D}_n^O & & & & \\ & \mathbf{D}_{n-1}^O & & & \\ & & \mathbf{D}_{n-2}^O & & \\ & & & \ddots & \\ & & & & \mathbf{D}_1^O \end{pmatrix}.
$$

Therefore, for the different macro-states, this is given by:

- For $k = 1, \ldots, R-1$

$$
\mathbf{D}_k^O = \begin{array}{c} E_0^{k,nv} \\ E_1^{k,nv} \\ \vdots \\ E_{k-1}^{k,vn} \\ E_k^{k,nv} \end{array}
\overset{\begin{array}{ccccc} E_0^{k,nv} & E_1^{k,nv} & E_2^{k,nv} & E_{k-1}^{k,nv} & E_k^{k,nv} \end{array}}{\begin{pmatrix} \mathbf{D}_{00}^{O,k,nv} & & & & \\ \mathbf{D}_{10}^{O,k,nv} & \mathbf{D}_{11}^{O,k,nv} & & & \\ & \ddots & \ddots & & \\ & & \mathbf{D}_{k-1,k-2}^{O,k,nv} & \mathbf{D}_{k-1,k-1}^{O,k,nv} & \\ & & & \mathbf{D}_{k,k-1}^{O,k,nv} & \mathbf{D}_{k,k}^{O,k,nv} \end{pmatrix}}
$$

- For $k = R, \ldots, n$

$$
\mathbf{D}_k^O = \begin{array}{c|cccccccccc|ccccc}
 & E_0^{k,v} & E_1^{k,v} & \ldots & E_{N-1}^{k,v} & E_N^{k,v} & E_{N+1}^{k,v} & \ldots & E_{k-1}^{k,v} & E_k^{k,v} & & E_N^{k,nv} & E_{N+1}^{k,nv} & E_{N+2}^{k,nv} & E_{k-1}^{k,nv} & E_k^{k,nv} \\
\hline
E_0^{k,v} & \mathbf{D}_{00}^{O,k,v} & & \ldots & & & & & & & & & & & & \\
E_1^{k,v} & & \mathbf{D}_{11}^{O,k,v} & & & & & & & & & & & & & \\
\vdots & & & \ddots & & & & & & & & & & & & \\
E_{N-1}^{k,v} & & & & \mathbf{D}_{N-1,N-1}^{O,k,v} & & & & & & & & & & & \\
E_N^{k,v} & & & & & \mathbf{D}_{N,N}^{O,k,v} & & & & & & & & & & \\
E_{N+1}^{k,v} & & & & & & \mathbf{D}_{N+1,N+1}^{O,k,v} & & & & & & & & & \\
\vdots & & & & & & & \ddots & & & & & & & & \\
E_{k-1}^{k,v} & & & & & & & & \mathbf{D}_{k-1,k-1}^{O,k,v} & & & & & & & \\
E_k^{k,v} & & & & & & & & & \mathbf{D}_{k,k}^{O,k,v} & & & & & & \\
\hline
E_N^{k,nv} & & & & \mathbf{D}_{N,N-1}^{O,k,v} & & & & & & & \mathbf{D}_{N,N}^{O,k,nv} & & & & \\
E_{N+1}^{k,nv} & & & & & & & & & & & \mathbf{D}_{N+1,N}^{O,k,nv} & \mathbf{D}_{N+1,N+1}^{O,k,nv} & & & \\
\vdots & & & & & & & & & & & & \ddots & \ddots & & \\
E_{k-1}^{k,nv} & & & & & & & & & & & & & \mathbf{D}_{k-1,k-2}^{O,k,nv} & \mathbf{D}_{k-1,k-1}^{O,k,nv} & \\
E_k^{k,nv} & & & & & & & & & & & & & & \mathbf{D}_{k,k-1}^{O,k,nv} & \mathbf{D}_{k,k}^{O,k,nv}
\end{array}
$$

with

$$\boldsymbol{\theta} = \boldsymbol{\alpha} \otimes \left(\mathbf{L} + \mathbf{L}^0\boldsymbol{\gamma}\right) \otimes \boldsymbol{\eta},$$

$$\mathbf{D}_{N,N-1}^{O,k,v} = \begin{pmatrix} \mathbf{I}_{2^{N-1}} \otimes \left(I_{\{k=N\}}\boldsymbol{\theta} + I_{\{k\neq N\}}\mathbf{H}_O\right) \otimes \mathbf{S}_1^0 \otimes \boldsymbol{\upsilon} \\ \mathbf{I}_{2^{N-1}} \otimes \left(I_{\{k=N\}}\boldsymbol{\theta} + I_{\{k\neq N\}}\mathbf{H}_O\right) \otimes \mathbf{S}_2^0 \otimes \boldsymbol{\upsilon} \end{pmatrix}$$

$$\mathbf{D}_{r,r}^{O,k,v} = \mathbf{I}_{2^r} \otimes \left(I_{\{r<k\}}\mathbf{H}_O + I_{\{r=k\}}\left(\mathbf{L} + \mathbf{L}^0\boldsymbol{\gamma}\right)\right) \otimes \left(\mathbf{V} + I_{\{r<N\}}\mathbf{V}^0\upsilon\right), r = 0, \ldots, k$$

$$\mathbf{D}_{00}^{O,k,nv} = \mathbf{H}_O$$

For $r = 1, \ldots, k$

$$\mathbf{D}_{r,r}^{O,k,nv} = \begin{pmatrix} \mathbf{I}_{2^{r-1}} \otimes \left(I_{\{r<k\}}\mathbf{H}_O + I_{\{r=k\}}\left(\mathbf{L} + \mathbf{L}^0\boldsymbol{\gamma}\right)\right) \otimes \mathbf{S}_1 & \mathbf{0} \\ \mathbf{0} & \mathbf{I}_{2^{r-1}} \otimes \left(I_{\{r<k\}}\mathbf{H}_O + I_{\{r=k\}}\left(\mathbf{L} + \mathbf{L}^0\boldsymbol{\gamma}\right)\right) \otimes \mathbf{S}_2 \end{pmatrix}$$

$$\mathbf{D}_{10}^{O,k,nv} = \begin{pmatrix} \left(I_{\{k>1\}}\mathbf{H}_O + I_{\{k=1\}}\boldsymbol{\theta}\right) \otimes \mathbf{S}_1^0 \\ \left(I_{\{k>1\}}\mathbf{H}_O + I_{\{k=1\}}\boldsymbol{\theta}\right) \otimes \mathbf{S}_2^0 \end{pmatrix}$$

For $r = 2, \ldots, k$

$$\mathbf{D}_{r,r-1}^{O,k,nv} = \begin{pmatrix} I_{2^{r-2}} \otimes \left(I_{\{r<k\}}\mathbf{H}_O + I_{\{r=k\}}\boldsymbol{\theta}\right) \otimes \mathbf{S}_1^0 \otimes \boldsymbol{\beta}_1 & \mathbf{0} \\ \mathbf{0} & I_{2^{r-2}} \otimes \left(I_{\{r<k\}}\mathbf{H}_O + I_{\{r=k\}}\boldsymbol{\theta}\right) \otimes \mathbf{S}_1^0 \otimes \boldsymbol{\beta}_2 \\ I_{2^{r-2}} \otimes \left(I_{\{r<k\}}\mathbf{H}_O + I_{\{r=k\}}\boldsymbol{\theta}\right) \otimes \mathbf{S}_2^0 \otimes \boldsymbol{\beta}_1 & \mathbf{0} \\ \mathbf{0} & I_{2^{r-2}} \otimes \left(I_{\{r<k\}}\mathbf{H}_O + I_{\{r=k\}}\boldsymbol{\theta}\right) \otimes \mathbf{S}_2^0 \otimes \boldsymbol{\beta}_2 \end{pmatrix}$$

Appendix B.3. The Matrix $\mathbf{D}^D$

The matrix $\mathbf{D}^D$ contains the transitions when the repairperson resumes work without any other event. The structure of this matrix is

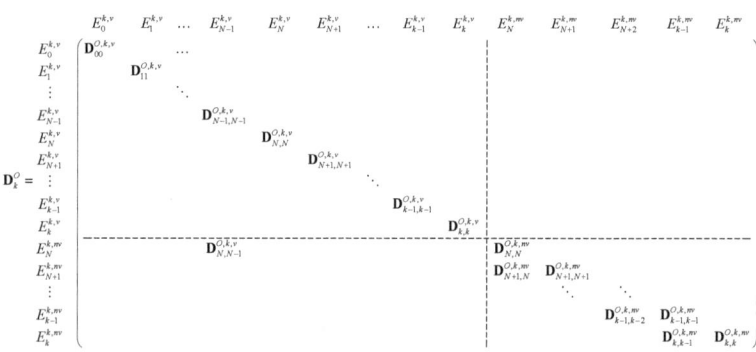

$$\mathbf{D}^{D} = \begin{pmatrix} \mathbf{D}^{D}_{n} & & & & & & \\ & \mathbf{D}^{D}_{n-1} & & & & & \\ & & \ddots & & & & \\ & & & \mathbf{D}^{D}_{R} & & & \\ & & & & \mathbf{0} & & \\ & & & & & \ddots & \\ & & & & & & \mathbf{0} \end{pmatrix}$$

- For $k = R, \ldots, n$

$$\mathbf{D}^{Y}_{k} = \begin{array}{c} \\ E_0^{k,v} \\ E_1^{k,v} \\ \vdots \\ E_{N-1}^{k,v} \\ E_N^{k,v} \\ E_{N+1}^{k,v} \\ \vdots \\ E_{k-1}^{k,v} \\ E_k^{k,v} \\ E_N^{k,nv} \\ E_{N+1}^{k,nv} \\ \vdots \\ E_{k-1}^{k,nv} \\ E_k^{k,nv} \end{array} \begin{array}{c} \begin{array}{cccccccccc|ccccc} E_0^{k,v} & E_1^{k,v} & \ldots & E_{N-1}^{k,v} & E_N^{k,v} & E_{N+1}^{k,v} & \ldots & E_{k-1}^{k,v} & E_k^{k,v} & & E_N^{k,nv} & E_{N+1}^{k,nv} & \ldots & E_{k-1}^{k,nv} & E_k^{k,nv} \end{array} \\ \left(\begin{array}{c|ccccc} & & & & & \\ & & & & & \\ & & & & & \\ & & & & & \\ & \mathbf{D}^{Y,k,nv}_{N,N} & & & & \\ & & \mathbf{D}^{Y,k,nv}_{N+1,N+1} & & & \\ & & & \ddots & & \\ & & & & \mathbf{D}^{Y,k,nv}_{k-1,k-1} & \\ & & & & & \mathbf{D}^{Y,k,nv}_{k,k} \\ \hline & & & & & \\ & & & & & \\ & & & & & \\ & & & & & \\ & & & & & \end{array}\right) \end{array}$$

For $r = N, \ldots, k$

$$\mathbf{D}^{D,k,nv}_{r,r} = \begin{pmatrix} \mathbf{I}_{2^{r-1}} \otimes \left(I_{\{r=k\}} \left(\mathbf{L} + \mathbf{L}^{0}\boldsymbol{\gamma} \right) + I_{\{r<k\}} \mathbf{H}_{O} \right) \otimes \mathbf{V}^{0} \otimes \boldsymbol{\beta}_1 & \mathbf{0} \\ \mathbf{0} & \mathbf{I}_{2^{r-1}} \otimes \left(I_{\{r=k\}} \left(\mathbf{L} + \mathbf{L}^{0}\boldsymbol{\gamma} \right) + I_{\{r<k\}} \mathbf{H}_{O} \right) \otimes \mathbf{V}^{0} \otimes \boldsymbol{\beta}_2 \end{pmatrix}$$

Appendix B.4. The Matrix $\mathbf{D}^{AD}$ and $\mathbf{D}^{BD}$

The matrices $\mathbf{D}^{AD}$ and $\mathbf{D}^{BD}$ contain the transitions when the repairperson resumes work and at same time a repairable failure or major inspection occur. In this case, for Y = *AD*, *BD* we have

$$\mathbf{D}^{Y} = \begin{pmatrix} \mathbf{D}^{Y}_{n} & & & & & & \\ & \mathbf{D}^{Y}_{n-1} & & & & & \\ & & \ddots & & & & \\ & & & \mathbf{D}^{Y}_{R} & & & \\ & & & & 0 & & \\ & & & & & \ddots & \\ & & & & & & 0 \end{pmatrix}.$$

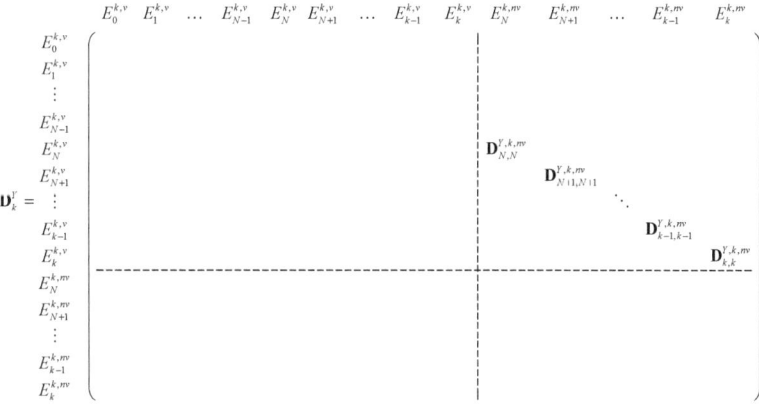

- For $k = R, \ldots, n$

$$
\mathbf{D}_k^{Y} = \begin{array}{c} \\ E_0^{k,v} \\ E_1^{k,v} \\ \vdots \\ E_{N-1}^{k,v} \\ E_N^{k,v} \\ E_{N+1}^{k,v} \\ \vdots \\ E_{k-1}^{k,v} \\ E_k^{k,v} \\ E_N^{k,nv} \\ E_{N+1}^{k,nv} \\ \vdots \\ E_{k-1}^{k,nv} \\ E_k^{k,nv} \end{array}
\begin{array}{c}
\begin{array}{ccccccccccccc} E_0^{k,v} & E_1^{k,v} & \ldots & E_{N-1}^{k,v} & E_N^{k,v} & E_{N+1}^{k,v} & \ldots & E_{k-1}^{k,v} & E_k^{k,v} & E_N^{k,nv} & E_{N+1}^{k,nv} & \ldots & E_{k-1}^{k,nv} \; E_k^{k,nv} \end{array} \\
\left(\begin{array}{c|cccc}
 & & & & \\
 & & & & \\
 & & & & \\
 & \mathbf{D}_{N-1,N}^{Y,k,nv} & & & \\
 & & \mathbf{D}_{N,N+1}^{Y,k,nv} & & \\
 & & & \ddots & \\
 & & & & \ddots \\
 & & & & \mathbf{D}_{k-1,k}^{Y,k,nv} \\
 & & & & \mathbf{0} \\
\hline
 & & & & \\
 & & & & \\
 & & & & \\
 & & & & \\
 & & & &
\end{array}\right)
\end{array}
$$

- For $r = N-1, \ldots, k-1$

$$
\mathbf{F}_{r,r+1}^{AD,k,nv} = \begin{pmatrix} \mathbf{I}_{2^{r-1}} \otimes \left(\left(I_{\{r=k-1\}} \mathbf{H}'_A + I_{\{r<k-1\}} \mathbf{H}_A \right) \otimes \mathbf{V}^0 \otimes \boldsymbol{\beta}_1, 0 \right) & \mathbf{0} \\ 0 & \mathbf{I}_{2^{r-1}} \otimes \left(\left(I_{\{r=k-1\}} \mathbf{H}'_A + I_{\{r<k-1\}} \mathbf{H}_A \right) \otimes \mathbf{V}^0 \otimes \boldsymbol{\beta}_2, 0 \right) \end{pmatrix}
$$

$$
\mathbf{F}_{r,r+1}^{BD,k,nv} = \begin{pmatrix} \mathbf{I}_{2^{r-1}} \otimes \left(\mathbf{0}, \left(I_{\{r=k-1\}} \mathbf{H}'_B + I_{\{r<k-1\}} \mathbf{H}_B \right) \otimes \mathbf{V}^0 \otimes \boldsymbol{\beta}_1 \right) & \mathbf{0} \\ 0 & \mathbf{I}_{2^{r-1}} \otimes \left(\mathbf{0}, \left(I_{\{r=k-1\}} \mathbf{H}'_B + I_{\{r<k-1\}} \mathbf{H}_B \right) \otimes \mathbf{V}^0 \otimes \boldsymbol{\beta}_2 \right) \end{pmatrix}
$$

Appendix B.5. The Matrix $\mathbf{D}^C$

The matrix $\mathbf{D}^C$ contains the transitions when only a non-repairable failure occurs. In this case the matrix is

$$
\mathbf{D}^C = \begin{pmatrix} \mathbf{0} & \mathbf{D}_n^C & & & \\ & \mathbf{0} & \mathbf{D}_{n-1}^C & & \\ & & \mathbf{0} & \ddots & \\ & & & \ddots & \mathbf{D}_2^C \\ \mathbf{0} & & & & \mathbf{0} \end{pmatrix}.
$$

- For $k = 2, \ldots, R-1$ and $k \neq R \geq 3$

$$
\mathbf{D}_k^C = \begin{array}{c} \\ E_0^{k,nv} \\ E_1^{k,nv} \\ \vdots \\ E_{k-1}^{k,vn} \\ E_k^{k,nv} \end{array}
\begin{pmatrix}
E_0^{k-1,nv} & E_1^{k-1,nv} & \ldots & E_{k-2}^{k-1,nv} & E_{k-1}^{k-1,nv} \\
\mathbf{D}_{00}^{C,k,nv} & & & & \\
\mathbf{D}_{10}^{C,k,nv} & \mathbf{D}_{11}^{C,k,nv} & & & \\
 & \ddots & & \ddots & \\
 & & & \mathbf{D}_{k-1,k-2}^{C,k,nv} & \mathbf{D}_{k-1,k-1}^{C,k,nv} \\
 & & & & 0
\end{pmatrix}
$$

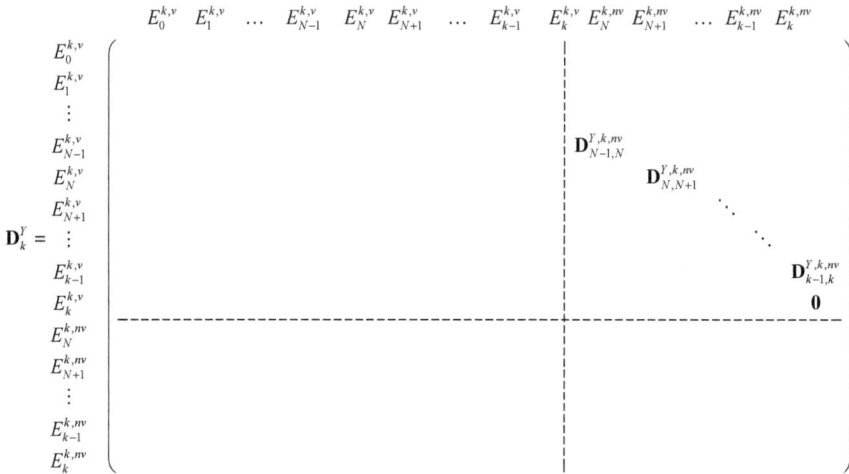

- For $k = R \geq 2$

$$
\mathbf{D}_k^C = \begin{array}{c} \\ E_0^{k,v} \\ E_1^{k,v} \\ \vdots \\ E_{k-1}^{k,v} \\ E_k^{k,v} \\ E_{N=1}^{k,nv} \\ E_2^{k,nv} \\ \vdots \\ E_{k-1}^{k,nv} \\ E_k^{k,nv} \end{array}
\begin{array}{c} \begin{array}{cccc} E_0^{k-1,nv} & E_1^{k-1,nv} & \dots & E_{k-2}^{k-1,nv} \quad E_{k-1}^{k-1,nv} \end{array} \\
\left(\begin{array}{cccc}
\mathbf{0} & & & \mathbf{0} \\
 & & & \\
 & & \ddots & \\
 & & & \\
\mathbf{0} & & & \mathbf{0} \\ \hdashline
\mathbf{D}_{10}^{C,k,nv} & \mathbf{D}_{11}^{C,k,nv} & & \\
 & \mathbf{D}_{21}^{C,k,nv} & \ddots & \\
 & & \ddots & \ddots \\
 & & \mathbf{D}_{k-1,k-2}^{C,k,nv} & \mathbf{D}_{k-1,k-1}^{C,k,nv} \\
\mathbf{0} & \dots & \dots & \mathbf{0}
\end{array} \right) \end{array}
$$

- For $k = R+1, \dots, n$ with $R \leq n-1$

$$
\mathbf{D}_k^C = \begin{array}{c} \\ E_0^{k,v} \\ E_1^{k,v} \\ \vdots \\ E_{N-1}^{k,v} \\ E_N^{k,v} \\ E_{N+1}^{k,v} \\ \vdots \\ E_{k-1}^{k,v} \\ E_k^{k,v} \\ E_N^{k,nv} \\ E_{N+1}^{k,nv} \\ \vdots \\ E_{k-1}^{k,nv} \\ E_k^{k,nv} \end{array}
\left(\begin{array}{ccccccccc:ccccc}
E_0^{k-1,v} & E_1^{k-1,v} & \dots & E_{N-1}^{k-1,v} & E_N^{k-1,v} & E_{N+1}^{k-1,v} & \dots & E_{k-2}^{k-1,v} & E_{k-1}^{k-1,v} & E_{N-1}^{k-1,nv} & E_N^{k-1,nv} & \dots & E_{k-2}^{k-1,nv} & E_{k-1}^{k-1,nv} \\
\mathbf{D}_{00}^{C,k,v} & & & & & & & & & & & & & \\
 & \mathbf{D}_{11}^{C,k,v} & & & & & & & & & & & & \\
 & & \ddots & & & & & & & & & & & \\
 & & & \mathbf{D}_{N-1,N-1}^{C,k,v} & & & & & & & & & & \\
 & & & & \mathbf{D}_{N,N}^{C,k,v} & & & & & & & & & \\
 & & & & & \mathbf{D}_{N+1,N+1}^{C,k,v} & & & & & & & & \\
 & & & & & & \ddots & & & & & & & \\
 & & & & & & & & \mathbf{D}_{k-1,k-1}^{C,k,v} & & & & & \\
 & & & & & & & & \mathbf{0} & & & & & \\ \hdashline
 & & & & & & & & & \mathbf{D}_{N,N-1}^{C,k,nv} & \mathbf{D}_{N,N}^{C,k,nv} & & & \\
 & & & & & & & & & & \mathbf{D}_{N+1,N}^{C,k,nv} & \mathbf{D}_{N+1,N+1}^{C,k,nv} & & \\
 & & & & & & & & & & \ddots & \ddots & & \\
 & & & & & & & & & & & \ddots & \mathbf{D}_{k-1,k-2}^{C,k,nv} & \mathbf{D}_{k-1,k-1}^{C,k,nv} \\
 & & & & & & & & & & & & & \mathbf{0}
\end{array} \right)
$$

For $r = 0, \dots, k-1$, $\mathbf{D}_{r,r}^{C,k,v} = \mathbf{I}_{2^r} \otimes \left(I_{\{r=k-1\}} \mathbf{H}'_C + I_{\{r<k-1\}} \mathbf{H}_C \right) \otimes \left(\mathbf{V} + I_{\{r<N-1\}} \mathbf{V}^0 v \right)$,

$\mathbf{D}_{00}^{C,k,nv} = \mathbf{H}_C$;

For $r = 1, \dots, k-1$;

$$
\mathbf{D}_{r,r}^{C,k,nv} = \begin{pmatrix} \mathbf{I}_{2^{r-1}} \otimes \left(I_{\{r=k-1\}} \mathbf{H}'_C + I_{\{r<k-1\}} \mathbf{H}_C \right) \otimes \mathbf{S}_1 & \mathbf{0} \\ \mathbf{0} & \mathbf{I}_{2^{r-1}} \otimes \left(I_{\{r=k-1\}} \mathbf{H}'_C + I_{\{r<k-1\}} \mathbf{H}_C \right) \otimes \mathbf{S}_2 \end{pmatrix}
$$

$$
\mathbf{D}_{10}^{C,k,nv} = \begin{pmatrix} \mathbf{H}_C \otimes \mathbf{S}_1^0 \\ \mathbf{H}_C \otimes \mathbf{S}_2^0 \end{pmatrix}
$$

For $r = 2, \dots, k-1$, $\mathbf{D}_{r,r-1}^{C,k,nv} = \begin{pmatrix} I_{2^{r-2}} \otimes \mathbf{H}_C \otimes \mathbf{S}_1^0 \otimes \boldsymbol{\beta}_1 & \mathbf{0} \\ \mathbf{0} & I_{2^{r-2}} \otimes \mathbf{H}_C \otimes \mathbf{S}_1^0 \otimes \boldsymbol{\beta}_2 \\ I_{2^{r-2}} \otimes \mathbf{H}_C \otimes \mathbf{S}_2^0 \otimes \boldsymbol{\beta}_1 & \mathbf{0} \\ \mathbf{0} & I_{2^{r-2}} \otimes \mathbf{H}_C \otimes \mathbf{S}_2^0 \otimes \boldsymbol{\beta}_2 \end{pmatrix}$

Appendix B.6. The Matrix $\mathbf{D}^{CD}$

The matrix $\mathbf{D}^{CD}$ contains the transitions when a non-repairable failure occurs and the repairperson resumes his work. In this case the matrix is

$$\mathbf{D}^{CD} = \begin{pmatrix} \mathbf{0} & \mathbf{D}_n^{CD} & & & & & \\ & \ddots & \ddots & & & & \\ & & \mathbf{0} & \mathbf{D}_R^{CD} & & & \\ & & & \mathbf{0} & \mathbf{0} & & \\ & & & & \ddots & \ddots & \\ & & & & & \mathbf{0} & \mathbf{0} \\ & & & & & & \mathbf{0} \end{pmatrix}$$

- For $k = R$

$$\mathbf{D}_k^{CD} = \begin{array}{c} \\ E_0^{k,v} \\ E_1^{k,v} \\ \vdots \\ E_{k-1}^{k,v} \\ E_k^{k,v} \\ E_{N=1}^{k,nv} \\ E_2^{k,nv} \\ \vdots \\ E_{k-1}^{k,nv} \\ E_k^{k,nv} \end{array} \begin{array}{c} \begin{array}{ccccc} E_0^{k-1,nv} & E_1^{k-1,nv} & \dots & E_{k-2}^{k-1,nv} & E_{k-1}^{k-1,nv} \end{array} \\ \left(\begin{array}{cccc} \mathbf{D}_{00}^{CD,k,nv} & & & \mathbf{0} \\ & \mathbf{D}_{11}^{CD,k,nv} & & \\ & & \ddots & \\ & & & \mathbf{D}_{k-1,k-1}^{CD,k,nv} \\ \mathbf{0} & & & \mathbf{0} \\ \hline \mathbf{0} & \dots & \dots & \mathbf{0} \\ & & & \\ & & & \\ & & & \\ \mathbf{0} & \dots & \dots & \mathbf{0} \end{array} \right) \end{array}$$

The matrix blocks for the case $k = R$ are $\mathbf{D}_{00}^{CD,k,nv} = \mathbf{H}_C \otimes \mathbf{e}$

For $r = 1, \dots, k-1$

$$\mathbf{D}_{r,r}^{CD,k,nv} = \begin{pmatrix} \mathbf{I}_{2^{r-1}} \otimes \left(I_{\{r=k-1\}} \mathbf{H}'_C + I_{\{r<k-1\}} \mathbf{H}_C \right) \otimes \mathbf{e} \otimes \boldsymbol{\beta}_1 & \mathbf{0} \\ \mathbf{0} & \mathbf{I}_{2^{r-1}} \otimes \left(I_{\{r=k-1\}} \mathbf{H}'_C + I_{\{r<k-1\}} \mathbf{H}_C \right) \otimes \mathbf{e} \otimes \boldsymbol{\beta}_2 \end{pmatrix}$$

- For $k = R+1, \dots, n$ and $R \leq n-1$

$$\mathbf{D}_k^{CD} = \begin{array}{c} \\ E_0^{k,v} \\ E_1^{k,v} \\ \vdots \\ E_{N-1}^{k,v} \\ E_N^{k,v} \\ E_{N+1}^{k,v} \\ \vdots \\ E_{k-1}^{k,v} \\ E_k^{k,v} \\ E_N^{k,nv} \\ E_{N+1}^{k,nv} \\ \vdots \\ E_{k-1}^{k,nv} \\ E_k^{k,nv} \end{array} \begin{array}{c} \begin{array}{cccccccccccccc} E_0^{k-1,v} & E_1^{k-1,v} & \dots & E_{N-1}^{k-1,v} & E_N^{k-1,v} & E_{N+1}^{k-1,v} & \dots & E_{k-2}^{k-1,v} & E_{k-1}^{k-1,nv} & E_{N-1}^{k-1,nv} & E_N^{k-1,nv} & \dots & E_{k-2}^{k-1,nv} & E_{k-1}^{k-1,nv} \end{array} \\ \left(\begin{array}{c|cccc} & & & & \\ & & & & \\ & & & & \\ & \mathbf{D}_{N-1,N-1}^{CD,k,nv} & & & \\ & & \mathbf{D}_{N,N}^{CD,k,nv} & & \\ & & & & \\ & & & \ddots & \\ & & & & \mathbf{D}_{k-1,k-1}^{CD,k,nv} \\ & & & & \mathbf{0} \\ \hline & & & & \\ & & & & \\ & & & & \\ & & & & \\ & & & & \end{array} \right) \end{array}$$

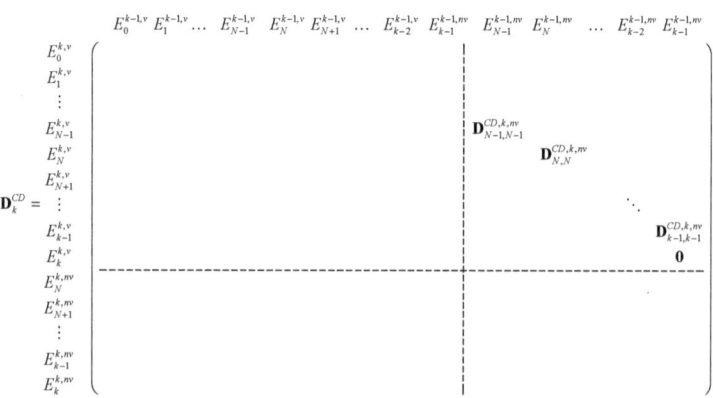

- The matrix blocks for the case $k = R+1, \ldots, n$ are
  For $r = N-1, \ldots, k-1$

$$\mathbf{D}_{r,r}^{CD,k,nv} = \begin{pmatrix} \mathbf{I}_{2^{r-1}} \otimes \left(I_{\{r<k-1\}}\mathbf{H}_C + I_{\{r=k-1\}}\mathbf{H}'_C \right) \otimes \mathbf{V}^0 \otimes \boldsymbol{\beta}_1 & \mathbf{0} \\ \mathbf{0} & \mathbf{I}_{2^{r-1}} \otimes \left(I_{\{r<k-1\}}\mathbf{H}_C + I_{\{r=k-1\}}\mathbf{H}'_C \right) \otimes \mathbf{V}^0 \otimes \boldsymbol{\beta}_2 \end{pmatrix}$$

Appendix B.7. The Matrix $\mathbf{D}^{NS}$

The matrix $\mathbf{D}^{NS}$ contains the transitions when a failure provokes the system to be restarted. Obviously, in this case the system is composed of only one unit. When this one is broken, a new system with n units re-starts. When this occurs, the vacation time begins again. The structure of the matrix is

$$\mathbf{D}^{NS} = \begin{pmatrix} \mathbf{0} & & & & \\ & \mathbf{0} & & & \\ & & \mathbf{0} & & \\ & & & \ddots & \\ \mathbf{D}_1^{NS} & & & & \mathbf{0} \end{pmatrix}$$

- If $R = 1$

$$\mathbf{D}_1^{NS} = \begin{matrix} \\ E_0^{1,v} \\ E_1^{1,v} \\ E_1^{1,nv} \end{matrix} \begin{matrix} E_0^{n,v} & E_1^{n,v} & \ldots & E_{N-1}^{n,v} & E_N^{n,v} & E_{N+1}^{n,v} & \ldots & E_{k-2}^{n,v} & E_{k-1}^{n,nv} & E_N^{n,nv} & E_{N+1}^{n,nv} & \ldots & E_{k-2}^{n,nv} & E_{k-1}^{n,nv} \\ \mathbf{D}_{00}^{NS,1,v} & \mathbf{0} & \ldots & & & & & & & & & & \ldots & \mathbf{0} \\ \mathbf{0} & \mathbf{0} & \ldots & & & & & & & & & & \ldots & \mathbf{0} \\ \mathbf{0} & \mathbf{0} & \ldots & & & & & & & & & & & \mathbf{0} \end{matrix}$$

with $\mathbf{D}_{00}^{NS,1,v} = \mathbf{H}_C \otimes \mathbf{e}_v \mathbf{v}$.

- If $R > 1$

$$\mathbf{D}_1^{NS} = \begin{matrix} \\ E_0^{1,nv} \\ E_1^{1,nv} \end{matrix} \begin{matrix} E_0^{n,v} & E_1^{n,v} & \ldots & E_{N-1}^{n,v} & E_N^{n,v} & E_{N+1}^{n,v} & \ldots & E_k^{n,v} & E_1^{n,nv} & E_2^{n,nv} & E_3^{n,nv} & \ldots & E_{n-1}^{n,nv} & E_n^{n,nv} \\ \mathbf{D}_{00}^{NS,1,v} & \mathbf{0} & \ldots & & & & & & & & & & \ldots & \mathbf{0} \\ \mathbf{0} & \mathbf{0} & \ldots & & & & & & & & & & \ldots & \mathbf{0} \end{matrix}$$

with $\mathbf{D}_{00}^{NS,1,v} = \mathbf{H}_C \otimes \mathbf{v}$.

Appendix C

To calculate the expected times that the repairperson returns to the workplace, independently of whether he remains or begins another period of vacation, the following matrix $\mathbf{Q}$ is defined. This matrix is built analogously to the matrix $\mathbf{D}$, but any return is considered. Therefore, the matrix $\mathbf{Q}$ is the addition of the following matrices

$$\mathbf{Q} = \mathbf{D}_{r-b}^{O} + \mathbf{D}_{r-b}^{A} + \mathbf{D}_{r-b}^{B} + \mathbf{D}_{r-b}^{C} + \mathbf{D}^{D} + \mathbf{D}^{AD} + \mathbf{D}^{BD} + \mathbf{D}^{CD} + \mathbf{D}_{r-b}^{NS}.$$

The matrices $\mathbf{D}^D, \mathbf{D}^{AD}, \mathbf{D}^{BD}, \mathbf{D}^{CD}$ are described in Appendix B. The other matrices have the same structure for the corresponding event given in Appendix B. These matrices are of zeros, excepting the following blocks.

- For $r = 0, \ldots, k-R$ and $k \geq R$
  $\mathbf{D}_{r,r}^{O,k,v} = \mathbf{I}_{2^r} \otimes \mathbf{H}_O \otimes \mathbf{V}^0 v$
- For $r = 1, \ldots, k-R-1$ and $k \geq R+2$
  $\mathbf{D}_{0,1}^{A,k,v} = (\mathbf{H}_A \otimes \mathbf{V}^0 \mathbf{v}, 0)$; $\mathbf{D}_{0,1}^{B,k,v} = (0, \mathbf{H}_B \otimes \mathbf{V}^0 \mathbf{v})$

$$\mathbf{D}_{r,r+1}^{A,k,v} = \mathbf{I}_{2^r} \otimes \left(\mathbf{H}_A \otimes \mathbf{V}^0\mathbf{v}, 0\right)$$
$$\mathbf{D}_{r,r+1}^{B,k,v} = \mathbf{I}_{2^r} \otimes \left(0, \mathbf{H}_B \otimes \mathbf{V}^0\mathbf{v}\right).$$

- For $r = 0, \ldots, k-R-1$ and $k \geq R+1$
$$\mathbf{D}_{r,r}^{C,k,v} = \mathbf{I}_{2^r} \otimes \mathbf{H}_C \otimes \mathbf{V}^0 v$$
- If $R = 1$,
$$\mathbf{D}_{00}^{NS,1,v} = \mathbf{H}_C \otimes \mathbf{V}^0\mathbf{v}$$

References

1. Levitin, G.; Xing, L.; Dai, Y. Optimal sequencing of warm standby elements. *Comput. Ind. Eng.* **2013**, *65*, 570–576. [CrossRef]
2. Zhai, Q.; Peng, R.; Xing, L.; Yang, J. Reliability of demand-based warm standby systems subject to fault level coverage. *Appl. Stoch. Model. Bus. Ind.* **2015**, *31*, 380–393. [CrossRef]
3. Cha, J.H.; Finkelstein, M.; Levitin, G. On preventive maintenance of systems with lifetimes dependent on a random shock process. *Reliab. Eng. Syst. Saf.* **2017**, *168*, 90–97. [CrossRef]
4. Levitin, G.; Finkelstein, M.; Dai, Y. Redundancy optimization for series-parallel phased mission systems exposed to random shocks. *Reliab. Eng. Syst. Saf.* **2017**, *167*, 554–560. [CrossRef]
5. Osaki, S.; Asakura, T. A two-unit standby redundant system with repair and preventive maintenance. *J. Appl. Probab.* **1970**, *7*, 641–648. [CrossRef]
6. Nakagawa, T. *Maintenance Theory of Reliability*; Springer Series in Reliability Engineering; Springer: London, UK, 2005. [CrossRef]
7. Finkelstein, M.; Cha, J.H.; Levitin, G. A hybrid preventive maintenance model for systems with partially observable degradation. *IMA J. Manag. Math.* **2020**, *31*, 345–365. [CrossRef]
8. Levitin, G.; Xing, L.; Xiang, Y. Optimizing preventive replacement schedule in standby systems with time consuming task transfers. *Reliab. Eng. Syst. Saf.* **2021**, *205*, 107227. [CrossRef]
9. Yang, L.; Ma, X.; Peng, R.; Zhai, Q.; Zhao, Y. A preventive maintenance policy based on dependent two-stage deterioration and external shocks. *Reliab. Eng. Syst. Saf.* **2017**, *160*, 201–211. [CrossRef]
10. Murchland, J.D. Fundamental Concepts and Relations for Reliability Analysis of Multi-State Systems. 1975. Available online: https://inis.iaea.org/search/search.aspx?orig_q=RN:8291134 (accessed on 19 April 2021).
11. Levitin, G.; Lisnianski, A. Multi-state system reliability analysis and optimization (universal generating function and genetic algorithm approach). In *Handbook of Reliability Engineering*; Springer: London, UK, 2003; pp. 61–90. [CrossRef]
12. Lisnianski, A.; Frenkel, I.; Ding, Y. *Multi-State System Reliability Analysis and Optimization for Engineers and Industrial Managers*; Springer: London, UK, 2010. [CrossRef]
13. Neuts, M.F. *Matrix-Geometric Solutions in Stochastic Models: An Algorithmic Approach*; Dover Publications: Mineola, NY, USA, 1981; p. 332.
14. Neuts, M.F. A versatile Markovian point process. *J. Appl. Probab.* **1979**, *16*, 764–779. [CrossRef]
15. Artalejo, J.R.; Antonio, G.-C. Markovian arrivals in stochastic modelling: A survey and some new results (invited article with discussion: Rafael Pérez-Ocón, Miklos Telek and Yiqiang Q. Zhao). *SORT-Statistics Oper. Res. Trans.* **2011**, *34*. Available online: https://www.raco.cat/index.php/SORT/article/view/217210 (accessed on 19 April 2021).
16. He, Q.-M. *Fundamentals of Matrix-Analytic Methods*; Springer: New York, NY, USA, 2014. [CrossRef]
17. Peng, R.; Xiao, H.; Liu, H. Reliability of multi-state systems with a performance sharing group of limited size. *Reliab. Eng. Syst. Saf.* **2017**, *166*, 164–170. [CrossRef]
18. Yu, J.; Zheng, S.; Pham, H.; Chen, T. Reliability modeling of multi-state degraded repairable systems and its applications to automotive systems. *Qual. Reliab. Eng. Int.* **2018**, *34*, 459–474. [CrossRef]
19. Ruiz-Castro, J.E.; Dawabsha, M. A discrete MMAP for analysing the behaviour of a multi-state complex dynamic system subject to multiple events. *Discret. Event Dyn. Syst. Theory Appl.* **2019**, *29*, 1–29. [CrossRef]
20. Ruiz-Castro, J.E. Markov counting and reward processes for analysing the performance of a complex system subject to random inspections. *Reliab. Eng. Syst. Saf.* **2016**, *145*, 155–168. [CrossRef]
21. Ruiz-Castro, J.E. A complex multi-state k-out-of-n: G system with preventive maintenance and loss of units. *Reliab. Eng. Syst. Saf.* **2020**, *197*, 106797. [CrossRef]
22. Ruiz-Castro, J.E.; Dawabsha, M.; Alonso, F.J. Discrete-time Markovian arrival processes to model multi-state complex systems with loss of units and an indeterminate variable number of repairpersons. *Reliab. Eng. Syst. Saf.* **2018**, *174*, 114–127. [CrossRef]
23. Ruiz-Castro, J.E.; Dawabsha, M. A multi-state warm standby system with preventive maintenance, loss of units and an indeterminate multiple number of repairpersons. *Comput. Ind. Eng.* **2020**, *142*, 106348. [CrossRef]
24. Doshi, B.T. Queueing systems with vacations? A survey. *Queueing Syst.* **1986**, *1*, 29–66. [CrossRef]
25. Ke, J.-C.; Wang, K.-H. Vacation policies for machine repair problem with two type spares. *Appl. Math. Model.* **2007**, *31*, 880–894. [CrossRef]

26. Zaiming, L.; Renbin, L. Reliability Analysis of the Repair Facility for an n-Unit Series Repairable System with an Unreliable Repair Facility and Finite Vacations. In Proceedings of the 2010 3rd International Conference on Information Management, Innovation Management and Industrial Engineering, Kunming, China, 26–28 November 2010; IEEE: New York, NY, USA, 2010; pp. 443–448. [CrossRef]
27. Wu, Q.; Wu, S. Reliability analysis of two-unit cold standby repairable systems under Poisson shocks. *Appl. Math. Comput.* **2011**, *218*, 171–182. [CrossRef]
28. Shrivastava, R.; Kumar Mishra, A. Analysis of Queuing Model for Machine Repairing System with Bernoulli Vacation Schedule. *Int. J. Math. Trends Technol.* **2014**, *10*, 85–92. [CrossRef]
29. Jain, M.; Meena, R.K. Fault tolerant system with imperfect coverage, reboot and server vacation. *J. Ind. Eng. Int.* **2017**, *13*, 171–180. [CrossRef]
30. Zhang, Y.; Wu, W.; Tang, Y. Analysis of an k-out-of-n:G system with repairman's single vacation and shut off rule. *Oper. Res. Perspect.* **2017**, *4*, 29–38. [CrossRef]
31. Ruiz-Castro, J.E. A Complex Multi-State System with Vacations in the Repair. *J. Math. Stat.* **2019**, *15*, 225–232. [CrossRef]

mathematics

MDPI

Article

Using Markov Models to Characterize and Predict Process Target Compliance

Sally McClean

School of Computing, Ulster University, Belfast BT37 0QB, Northern Ireland, UK; si.mcclean@ulster.ac.uk

Abstract: Processes are everywhere, covering disparate fields such as business, industry, telecommunications, and healthcare. They have previously been analyzed and modelled with the aim of improving understanding and efficiency as well as predicting future events and outcomes. In recent years, process mining has appeared with the aim of uncovering, observing, and improving processes, often based on data obtained from logs. This typically requires task identification, predicting future pathways, or identifying anomalies. We here concentrate on using Markov processes to assess compliance with completion targets or, inversely, we can determine appropriate targets for satisfactory performance. Previous work is extended to processes where there are a number of possible exit options, with potentially different target completion times. In particular, we look at distributions of the number of patients failing to meet targets, through time. The formulae are illustrated using data from a stroke patient unit, where there are multiple discharge destinations for patients, namely death, private nursing home, or the patient's own home, where different discharge destinations may require disparate targets. Key performance indicators (KPIs) of this sort are commonplace in healthcare, business, and industrial processes. Markov models, or their extensions, have an important role to play in this work where the approach can be extended to include more expressive assumptions, with the aim of assessing compliance in complex scenarios.

Keywords: process mining; process modelling; phase-type models; process target compliance

Citation: McClean, S. Using Markov Models to Characterize and Predict Process Target Compliance. *Mathematics* **2021**, *9*, 1187. https://doi.org/10.3390/math9111187

Academic Editors: Andreas C. Georgiou and Panagiotis-Christos Vassiliou

Received: 7 May 2021
Accepted: 21 May 2021
Published: 24 May 2021

Publisher's Note: MDPI stays neutral with regard to jurisdictional claims in published maps and institutional affiliations.

Copyright: © 2021 by the author. Licensee MDPI, Basel, Switzerland. This article is an open access article distributed under the terms and conditions of the Creative Commons Attribution (CC BY) license (https://creativecommons.org/licenses/by/4.0/).

1. Introduction

Processes are widespread, encompassing disparate areas such as business, production, telecommunications, and healthcare. They have previously been analyzed and modelled with the aim of improving understanding and efficiency as well as predicting future events and outcomes. With the burgeoning capability of IT systems to collect, process, store, and exchange data, and the upsurge of suitable technologies for Big Data, recently, process mining has appeared, providing a bridge between data mining and process modelling [1]. Process mining provides an opportunity and framework for service design and improvement, as well as a scientific rationale for decision-making. In general, we consider processes comprising several tasks each with start and end times and associated durations. A process instance completes these tasks according to the logic and rules prevailing in the real-world setting. The process data features mainly consist of data such as duration, customer id, etc., and are held in log files. Hence, such log files provide an automated time-stamped record of tasks performed during the execution of a given process.

Consequently, process mining may include discovering the tasks and trajectories that comprise the process, predicting trajectories, or identifying anomalies. Such activities can employ traditional methods for data mining such as classification, clustering, regression, association rules, sequence mining, or deep learning. However, model-based approaches can also provide opportunities for incorporating structural process knowledge into the analysis, thereby facilitating improved understanding and prediction. As such, process mining can be employed in diverse areas, such as manufacturing [2], telecommunications [3], financial processing, and healthcare [4].

A mathematical model is often used as a simplified version of a process, where simulation uses the model to imitate the behaviour of the process, without interfering with the actual process [5]. Correctness, conformance, and performance are some of the most important problems for complex processes, where models have often been used to resolve such issues. Performance analysis typically focuses on the dynamic behaviour of the process, based on metrics such as response time, uptime, or output. Our emphasis here is on measuring if a process meets its targets. For example, a business process might have order completion targets to meet, and an accident and emergency department could have discharge targets, while service-level agreements (SLAs) are commonly used to characterize cloud performance targets.

(Stochastic) process algebras have been implemented in formal languages to describe a system model. For example, Petri nets [6] were introduced by Carl Adam Petri in 1962 to characterize and analyze concurrent systems. They are based on mathematical specification alongside a mathematical theory for interpretation and analysis. For example, Petri nets have been used for workflow modelling [7]. In addition, stochastic Petri nets [8], including queueing Petri nets, have been developed.

A Markov model is a type of probabilistic process model that can describe such systems where it is assumed that the Markov property is followed, i.e., future states only depend on present states, and not additionally on previous ones. This enables both individual probabilistic predictive modelling [9] and group forecasting for individuals moving through a process [10]. Higher-order Markov models may alternatively be employed if the Markov assumption is not appropriate. In addition, continuous-time Markov chains (CTMCs) are commonly used where the Markov property translates into exponentially distributed durations. Such models can be used to find "interesting" (in)frequent pathways [11].

In this paper, we extend our previous initial work on using Markov models to predict process target compliance [12]. Several formulae are obtained and used for a process concerning stroke patient pathways achieving targets for the duration of hospitalization and subsequent discharge to different types of community care. In what follows, we formulate the problem for a general phase-type Markov model, where previously we focus on Coxian models. We also extend the work to situations where there are multiple absorbing (discharge) states and also for groups of individuals (e.g., patients) moving through the system towards discharge targets.

2. Background

Markov models have been used to represent various types of process applications, including call centres [13], sensor networks [14], telecommunications [15], production modelling [16] and healthcare [10]. Phase-type models are a special case of a Markov model where there are transient states (or phases) and, typically, a single absorbing state where generally the interest is in duration of stay in the set of transient states. In healthcare, we typically have some hospital states followed by one, or more, absorbing states in the community. These models can be used to predict individual patient movements or to predict future resource requirements or costs for groups of patients [17]. They facilitate conceptualization of flows, e.g., for hospital patients, through testing, diagnosis, treatment, and rehabilitation. Such phase-type distributions (PHDs) can be utilized to describe duration in a group of states where the PHD represents the time from admission to the transient states until absorption into the absorbing state. In particular, Coxian phase-type distributions (C-PHDs) are a useful special case where a process always starts in the first transient state and can never return to a state once it has left it. Transition from a transient state to the absorbing state is also allowed (Figure 1). Such PHDs provide a simple model for a key performance indicator (KPI) such as length of stay in the transient states, e.g., duration of a particular activity, or from order placement to delivery. Parameter estimation for PHDs is also typically straightforward [9]. In general, phase-type models (PHDs) are well suited to a range of situations, including healthcare [18–21], community care [22], accident and emergency [23], and activity recognition [24]. They are also understandable as we can conceptualize a patient or customer as moving through the phases.

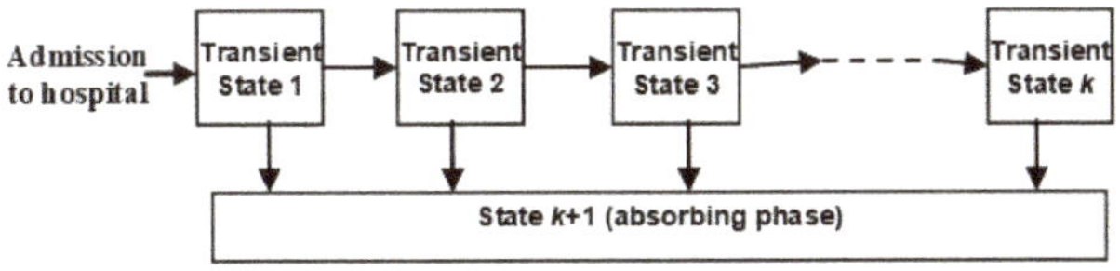

Figure 1. A k state patient Coxian phase-type distribution.

In addition, PHDs have the following advantages: (i) their mathematical simplicity; (ii) parsimonious parameterization; (iii) flexibility in terms of fitting different shapes of distribution; and (iv) ease of migration to more complex settings, either using a mathematical or simulation approach.

In the current paper, as in our previous work, e.g., [25,26], we include covariates, or additional features, into the model by allowing the initial and transition probabilities to depend explicitly on these covariates. The specific functional form of these covariate models will be described in the next section.

3. Phase-Type Models

3.1. The Basic Phase-Type Model

As in [5,12], we use here a phase-type Markov process model. This representation of a process by a Markov, or more specifically, a phase-type model allows us to incorporate variability into process tasks, thus facilitating implementation and adaptation. As discussed, the phase-type model provides a useful way of describing process duration and also has other advantages, such as computational efficiency.

We begin by defining k transient phases $S_1, \ldots, S_k$, with phase S_{k+1} being the only absorbing state. Writing the initial vector as: $\boldsymbol{\alpha} = (\alpha_1, \ldots \alpha_k)$, where α_i is the probability of entry to phase S_i for $i = 1, \ldots, k$, we obtain the probability density function (p.d.f.) of the distribution of duration until transition to the absorbing state as:

$$f(x) = \boldsymbol{\alpha} \exp(\mathbf{T}x)\boldsymbol{t}_A, \tag{1}$$

where $\mathbf{T} = \{t_{ij}\}$ is the $k \times k$ generator matrix for the transition rates between the transient states and $i = 1, \ldots, k$, $j = 1, \ldots, k$. Here, $\boldsymbol{t}_A$ is the column vector of transition rates from the transient states to the absorbing state and $\boldsymbol{t}_A = -\mathbf{T}\mathbf{1}$ where $\mathbf{1}$ is a column vector of 1's, pointing to the fact that the row sums of the generator matrix are zero.

Integrating the p.d.f., we obtain the cumulative distribution function (c.d.f.) as

$$F_X(y;\, \boldsymbol{\alpha},\, \mathbf{T}) = 1 - \boldsymbol{\alpha} \exp(\mathbf{T}y)\mathbf{1}; y \geq 0. \tag{2}$$

which describes the probability of meeting a given duration target y for length of stay in the transient states. Similarly, the probability of missing a duration target y is given by

$$\overline{F}_x(y;\, \boldsymbol{\alpha},\, \mathbf{T}) = \boldsymbol{\alpha} \exp(\mathbf{T}y)\mathbf{1}; y \geq 0. \tag{3}$$

We note that the inverse problem of ascertaining an appropriate duration target, given a required percentage compliance, can be obtained from Equation (3) by solving to find y for a given F. Here F can be a service-level agreement in a management or industrial context. So, for example, we may require that 95% of tasks are completed within a given target duration in the transient states. Although we cannot solve Equation (3) explicitly for y, we can a use a numerical solution, such as Newton–Raphson, where the estimate of y is given at the $(n + 1)$th iteration by

$$y_{n+1} = y_n - F(y_n)/F'(y_n), \tag{4}$$

where $F'(y_n) = \boldsymbol{\alpha}\ \exp(\mathbf{T}y_n)\mathbf{T1}$.

In this way, we can not only characterize the relative likelihoods of compliance and non-compliance with a target but also consider the most likely state trajectories. Using the approach of [26], we determine the conditional probability of meeting (or otherwise) a target of duration y given that an amount of time d has already passed. This probability is given by

$$F_{X|X>d}(y;\ \boldsymbol{\alpha},\ \mathbf{T}) = 1 - \frac{\boldsymbol{\alpha}\ \exp\ (\mathbf{T}y)\mathbf{1}}{\boldsymbol{\alpha}\ \exp\ (\mathbf{T}d)\mathbf{1}}; y \geq d, \tag{5}$$

which represents the probability of meeting a given target y. Also, the probability of missing a target y is given by

$$\overline{F}_{X|X>d}(y;\ \boldsymbol{\alpha},\ \mathbf{T}) = \frac{\boldsymbol{\alpha}\ \exp\ (\mathbf{T}y)\mathbf{1}}{\boldsymbol{\alpha}\ \exp\ (\mathbf{T}d)\mathbf{1}}; y \geq d. \tag{6}$$

In a similar manner, conditional means can be calculated by integrating the conditional densities, as previously discussed in McClean et al. [12].

3.2. Multiple Absorbing States with Different Targets

To date, we have assumed that the target for absorption will be the same, irrespective of the initial state. While this may be the case in many situations, it is clearly not always the case. For example, for stroke patients, as we will discuss in our case study, the three initial states are (1) haemorrhagic stroke, (2) cerebral infarction stroke, and (3) transitory ischaemic attack (TIA). However, in this example, the anticipated length of stay in hospital depends on the type of stroke, with haemorrhagic stroke being more severe than cerebral infarction and cerebral infarction being more severe than TIA. In addition, the expected length of stay will vary with the discharge destination, with more severe strokes leading to destinations which require community settings which provide more support for the patient. This observation underpins our model, where we assume that the patients progress from one transient state (phase) to another less severe one. It is therefore likely that, for such situations, the individual targets will differ across initial phases. So, for the stroke patient example, we might expect the target for haemorrhagic patients to be greater than that of cerebral infarction patients and the target for cerebral infarction patients should be larger than that for TIA patients, corresponding to greater stroke severity generally requiring longer hospitalization.

Previously, we extended this model to incorporate the occurrence of multiple absorbing states into the phase-type model [26], as follows.

The infinitesimal generator matrix $\mathbf{Q}$ is given by

$$\mathbf{Q}(\mathbf{x}) = \begin{pmatrix} \mathbf{T}(\mathbf{x}) & t_A(\mathbf{x}) \\ 0_{AT} & 0_{AA} \end{pmatrix}. \tag{7}$$

Here $\mathbf{T} = \{t_{ij}\}$ is a $k \times k$ matrix of transition rates between the k transient states, given by

$$\mathbf{T}(\mathbf{x}) = \begin{pmatrix} -\Lambda_1(\mathbf{x}) & \cdots & \lambda_{1k}(\mathbf{x}) \\ \vdots & \ddots & \vdots \\ 0 & \cdots & -\Lambda_k(\mathbf{x}) \end{pmatrix}, \tag{8}$$

where $\Lambda_i(\mathbf{x}) = \sum_{j=2}^{k} \lambda_{ij}(\mathbf{x}) + \sum_{j=1}^{m} \mu_{ij}(\mathbf{x})$.

Here we allow the transition rates to depend on covariates $\mathbf{x} = \{x_i\}$; for example, for stroke patients these could be age and gender, where the $\mu_{ij}(\mathbf{x})$ terms represent transition rates from transient state S_i to absorbing state S_j for $i = 1, \ldots, k$ and $j = 1, \ldots, m$, and m represents the number of absorbing states. The $k \times m$ matrix t_A is then given by $t_A = \{\mu_{ij}(\mathbf{x})\}$.

Finally, $\mathbf{0}_{AT}$ and $\mathbf{0}_{AA}$ are zero matrices of suitable dimensions and $\mathbf{0}$ is a zero column vector. These elements satisfy the conditions $t_{ii} < 0$ for $I = 1, \ldots, k$ and $t_{ij} \geq 0$ for $i = 1, \ldots, k$; for $j = 1, \ldots m$. Also, $\mathbf{T}$ and $\boldsymbol{t}_A$ satisfy $\boldsymbol{t}_A \mathbf{1}_m = -\mathbf{T}\mathbf{1}_k$ where $\mathbf{1}_m$ is an m-dimensional column vector of ones.

In a similar way to Equation (1) we obtain $f(\mathrm{t}) = \{f_i(\mathrm{t})\}$ where $f_i(\mathrm{t})$ is the unconditional (degenerate) p.d.f. of the time spent in the transient states prior to discharge to absorbing state S_{k+i} for $i = 1, \ldots, m$, and

$$f(t) = \boldsymbol{\alpha}\ \exp\left(\mathbf{T}t\right)\boldsymbol{t}_A \tag{9}$$

The probability of meeting target τ_i for absorbing state S_{k+i} is therefore given by

$$M_i(\tau_i;\ \boldsymbol{\alpha},\ \mathbf{T}) = \int_0^{\tau_i} \boldsymbol{\alpha}\ \exp(\mathbf{T}y)\boldsymbol{t}_A\ \mathbf{I}_i dy; y \geq 0,\ i = 1,\ \ldots,\ m, \tag{10}$$

where $\mathbf{I}_i$ is an m-dimensional column vector with 1 in the ith position and zeros elsewhere; $\boldsymbol{t}_A\ \mathbf{I}_i$ is therefore the ith column of $\boldsymbol{t}_A$.

Integrating this expression, we obtain

$$\begin{aligned} M_i(\tau_i;\ \boldsymbol{\alpha},\ \mathbf{T}) &= \left\{\boldsymbol{\alpha}\ \exp(\mathbf{T}\tau_i)\mathbf{T}^{-1}\boldsymbol{t}_A\ \mathbf{I}_i - \boldsymbol{\alpha}\ \mathbf{T}^{-1}\boldsymbol{t}_A\ \mathbf{I}_i\right\} \\ &= \boldsymbol{\alpha}\ (\mathbf{I} - \exp(\mathbf{T}\tau_i))\left(-\mathbf{T}^{-1}\right)\boldsymbol{t}_A\ \mathbf{I}_i\ \ i = 1,\ \ldots,\ m. \end{aligned} \tag{11}$$

Here, when the targets are equal across all absorbing states, i.e., $\tau_i = \tau\ \forall\ i$, the total probability of meeting the target is $\sum_{i=1}^{m} M_i(\tau;\ \boldsymbol{\alpha},\ \mathbf{T}) = \boldsymbol{\alpha}\ \exp\left(\mathbf{T}\tau\right)\mathbf{1}$, as for Equation (3).

We note that these formulae, for the probability of meeting targets when there are multiple "risks", are related to those used in epidemiology for cumulative incidence, e.g., [27].

We can also obtain the conditional probability of meeting the target τ_i for the absorbing state S_{k+i}, given eventual absorption is to this state, which is given by

$$L_i(\tau_i;\ \boldsymbol{\alpha},\ \mathbf{T}) = \left\{\boldsymbol{\alpha}\ (\mathbf{I} - \exp(\mathbf{T}\tau_i))\left(-\mathbf{T}^{-1}\right)\boldsymbol{t}_A\ \mathbf{I}_i\right\} / \left\{\boldsymbol{\alpha}\ \left(-\mathbf{T}^{-1}\right)\boldsymbol{t}_A\ \mathbf{I}_i\right\}\ \ i = 1,\ \ldots,\ m. \tag{12}$$

This expression is useful in terms of allowing us to determine the profile of different groups of patients characterized by their final destination and quantifying how likely they are to meet the given possible targets with regard to duration in the transient states. While our previous expressions are more geared towards making and meeting targets for individuals, Equation (12) allows us to move towards thinking about cohorts of individuals meeting overall targets for the system of transient states. For example, in the stroke patient situation we explore below, the performance of a stroke unit in terms of meeting hospital targets can be measured in terms of the different discharge destinations (absorbing states), namely death, private nursing home and own homes. Mathematically, this is achieved through the entry vector $\boldsymbol{\alpha}$, which here represents an overall probability distribution across the different types of stroke. We now focus further on such population models for setting targets.

3.3. Poisson Arrivals

So far, we have considered individual movements through the transient states, with eventual absorption into one of a number of possible exit states. Our focus here has thus been on providing expressions for target achievement. However, for such processes, there is often an interest in characterizing the movements of a number of individuals moving through the system in parallel where, for example, we may want to characterize and/or predict the numbers of individuals attaining a target in a given time interval. As such, our focus now shifts to a Markov system; for further details of such systems and a discussion of various possible extensions, see, for example, [28].

We consider a situation where new arrivals to the Markov process occur according to a Poisson process, rate ω where we have an initial probability vector $\boldsymbol{\alpha}$, k transient states, and one absorbing state, as before. We are interested in determining the probability distribution of the number of individuals arriving in time interval $(0, \infty)$ who fail to meet a fixed target d.

Let $M(t)$ be the number of individuals who arrive in $(0, t)$ according to a Poisson process, rate ω, and fail to comply. Each of these individuals fails to comply with probability Φ where $\Phi = \boldsymbol{\alpha} \exp\{\mathbf{T}d\}\ \mathbf{1}_k$, using Equation (3). Then, the distribution of $N(t)$, the total number of arrivals in $(0, d)$, is Poisson (ωd) and the distribution of $M(d)$ is a compound distribution, consisting of a binomial choice from a Poisson number of failures.

The probability generating function (p.g.f.) of a r.v. N~Poisson(ωt) is given by $E_N[z^N]$ = $G(z) = \exp(\omega d\ (z - 1))$, and the p.g.f. of a random variable (r.v.) M~Binomial (N, p) is $E_M[z^M] = G(z) = (q + pz)^M$, where $q = 1 - p$. The p.g.f. of the required compound distribution is therefore

$$H_M(z) = E_N\left[E_M\left[z^M \middle| N\right]\right] = G(F(z)) = \exp\{\ \omega d\ ((1-\Phi)+\Phi z)-1)\} = \exp\{\ \omega d\ \Phi(z-1)\}. \tag{13}$$

So, the number of failures who comply with target d from individuals arriving in $(0, t)$ is a Poisson with mean (and variance):

$$\omega t\ \boldsymbol{\alpha} \exp\{\mathbf{T}d\}\ \mathbf{1}_k. \tag{14}$$

Similarly to the situation considered previously, where we have m absorbing states, we again have a compound distribution of a Poisson (arrival) rate ω and a binomial (transition to absorbing state i after duration d_i). Then, integrating Equation (12) we obtain the result that the number of individuals arriving in $(0, t)$ who meet target d_i for absorbing state S_{k+i} is a Poisson with mean (and variance):

$$\omega t\left\{\boldsymbol{\alpha}\ (\mathbf{I} - \exp(\mathbf{T}\tau_i))\left(-\mathbf{T}^{-1}\right)\mathbf{t}_A\ \mathbf{I}_i\right\}, \tag{15}$$

where $\mathbf{I}_i$ is an m-dimensional column vector with 1 in the ith position, as before.

Based on this result, we can understand and predict the variability of numbers of individuals moving through the transient states in terms of their likelihood of meeting targets. The mathematical development in this section suggests that such variability is likely to be high and increase with time. This further highlights the importance of setting achievable targets.

4. Results

4.1. The Stroke Care Case Study

In practice, it is often the case that a number of absorbing states are possible, with possibly different targets. Previously, we have discussed phase-type models which contain multiple absorbing states [5,26]. We now apply our model to such a situation involving stroke patients using data spreading over 5 years. Here, we have described a phase-type model with four transient states corresponding to different types of stroke with contrasting severity and related admission probabilities for differing stroke severity. The data contain three types of stroke: haemorrhagic (the most severe, caused by bleeding in the brain), cerebral infarction (less severe, due to blood clots), and transient ischaemic attack or TIA (the least severe, a minor stroke caused by a small clot). Following hospitalization, there are three possible discharge destinations: (1) following the patient's death, (2) with a discharge to a private nursing home, and (3) with a discharge to the patient's own home. These different situations can be described by defining the exit matrix $\mathbf{t}_A$ as

$$\mathbf{t}_A = \begin{pmatrix} \mu_1 & \nu_1 & \rho_1 \\ \mu_2 & \nu_2 & \rho_2 \\ \mu_3 & \nu_3 & \rho_3 \\ \mu_4 & \nu_4 & \rho_4 \end{pmatrix} \tag{16}$$

For this special case of $\mathbf{t}_A$, each column relates to a different hospital discharge event, while the rows correspond to the transient phases of hospitalization [26].

In this study, data were collected over a 5-year period, on admission date, discharge date, diagnosis on admission, and discharge destination, alongside other covariates, such as age on admission and gender. The transition rates of the model may depend upon the age and stroke type of the patient, or may not depend on age [5]. We note in passing that the Poisson admission assumption was previously tested using chi-square and Kolmogorov–Smirnov tests and shown to be acceptable for our Belfast City Hospital stroke patient data [5].

So far, we have not discussed the possibility of covariates playing a significant role in the Markov model. However, as is often the case, for the stroke patient case we have additional covariates, namely age and gender. In our previous work, we determined that while gender does not have a significant effect, age does and has therefore been included in the model, as follows. Other covariates were not available for this dataset but, in general, the results of tests or diagnostics might be relevant covariates.

For $i = 1, 2$, let $\lambda_i(x)$ be the transition intensity from phase S_i to phase S_{i+1} for a patient of age x, where $\lambda_i(x) = exp(\gamma_i + \beta_i\, x)$. Also, $p(x)$ is the probability that a TIA patient aged x enters phase S_4 upon admission to hospital, representing the least severe type of stroke. Consequently, a more severe TIA patient starts in phase S_3 with probability $1 - p(x)$. We assume that $p(x) = \exp\{-\exp(\theta_0 + \theta_1\, x)\}$. The exponential functions here used in modelling $\lambda_i(\mathrm{x})$ and (x) are standard representations, which constrain the probability values to the required ranges. Such functions are found in the literature for log link and complementary log–log link functions for generalized linear models, e.g., [29]. As seen in Figure 2, it is assumed that $\mu_4 = \nu_4 = 0$, representing the fact that patients with a minor TIA (S_4) are always discharged to their own home. Similarly, for the other transitions from the transient phases (S_1, S_2, and S_3) to each absorbing state, we assume that $\nu_1 = p_1 = 0$. We note that transitions absent in Figure 2, and corresponding zero parameters, have been found by statistical testing based on likelihood ratio tests; for further details, see [26].

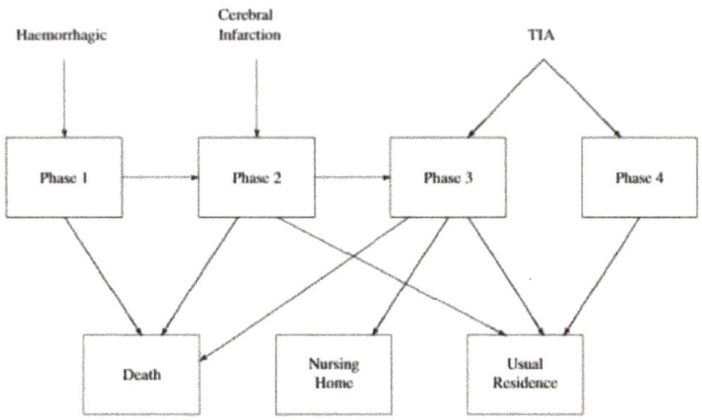

Figure 2. Stroke care transition diagram.

4.2. Findings

The following findings are based on model parameter values as described in [26]. These were estimated using a 5-year retrospective dataset consisting of 1985 patients. Figure 3 presents the cumulative probability of discharge from hospital by age for (a) haemorrhagic stroke, (b) cerebral infarction, and (c) TIA. The 95% compliance is also presented in these plots to make it easier to evaluate the compliance target, in days, for a commonly used compliance probability. In all three plots, we can see that the older the patient, the longer the stay in hospital and the less likely patients are to comply with a given target, as expected. Here, we see that, for a given compliance probability, the haemorrhagic patients typically spend much longer in hospital and, similarly, TIA patients spend much shorter periods in hospitals, so a lower target would be appropriate for them. This is as we would anticipate, with more serious, or more infirm, patients staying longer in hospital. Patients with cerebral infarction are intermediate, in this regard, as we would expect.

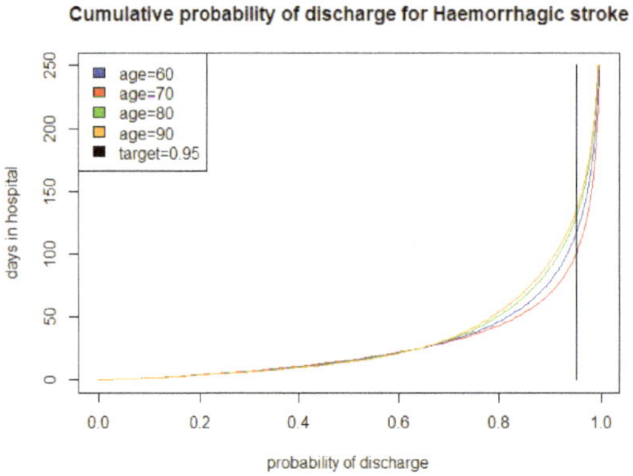

(**a**)

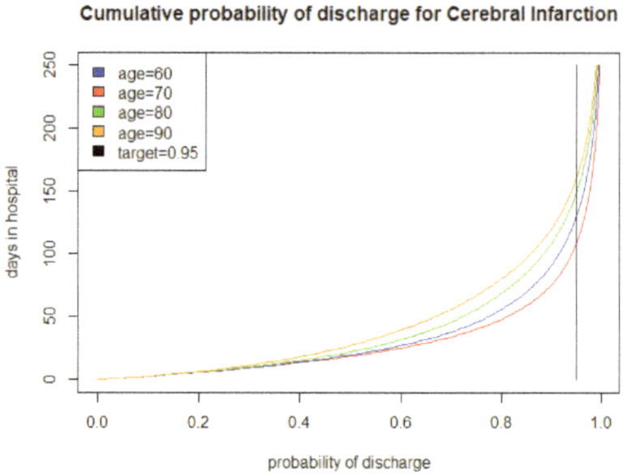

(**b**)

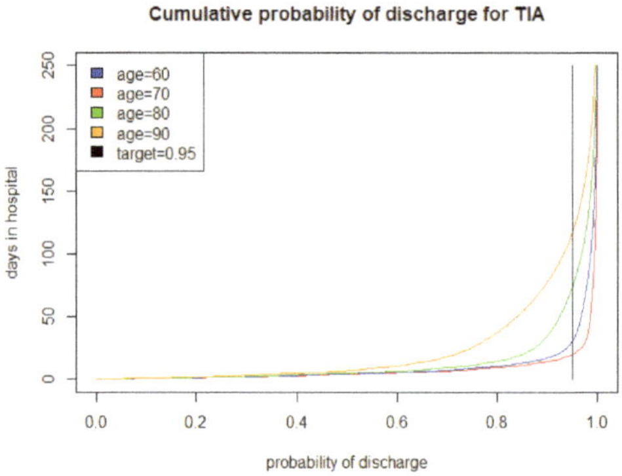

(**c**)

Figure 3. Cumulative probability of discharge from hospital by age for (**a**) haemorrhagic stroke, (**b**) cerebral infarction, and (**c**) TIA.

In Figure 4 we present the duration of stay in hospital by age for compliance with different targets for (**a**) haemorrhagic stroke, (**b**) cerebral infarction, and (**c**) TIA. We see from the plots that, as before, the more serious the stroke, the longer the patients need to be allocated to reach a given target, as prolonged rehabilitation is needed for such patients to move through the different treatment and recovery phases before discharge. Moreover, as the targets become more severe, they become increasingly harder to achieve, for all patients.

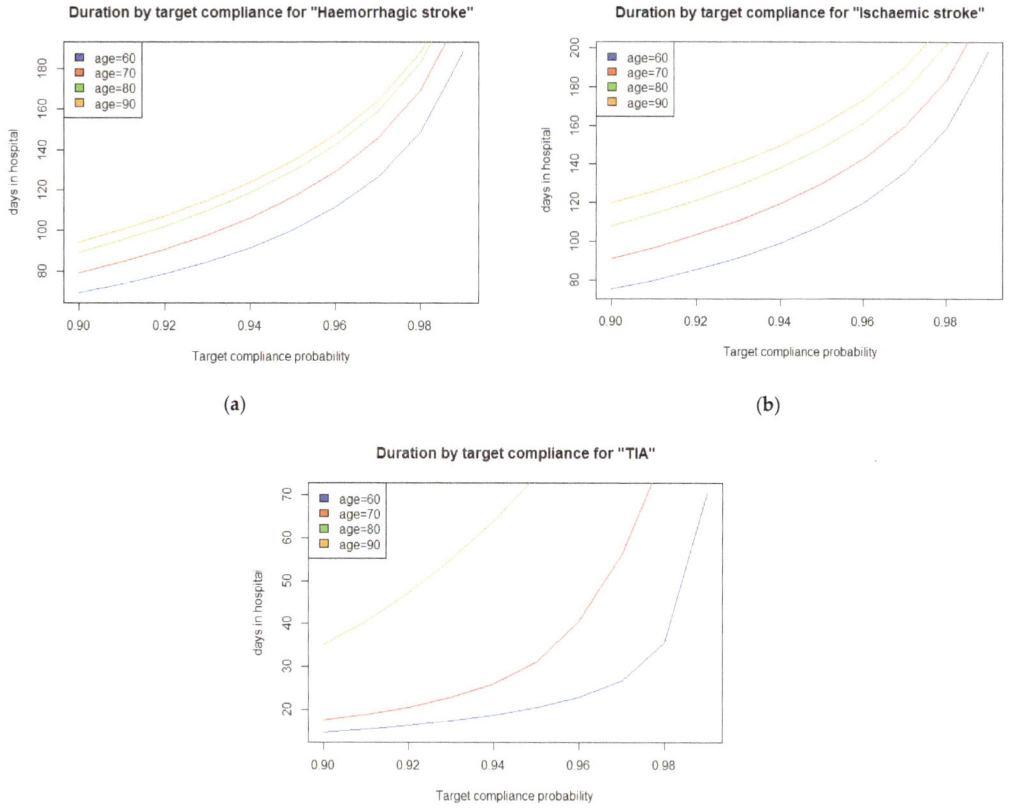

Figure 4. Duration of stay in hospital by age for compliance with different targets for (**a**) haemorrhagic stroke, (**b**) cerebral infarction, and (**c**) TIA.

In Figure 4 we present the duration of stay in hospital by age for compliance with different targets for (**a**) haemorrhagic stroke, (**b**) cerebral infarction, and (**c**) TIA. We see from the plots that, as before, the more serious the stroke, the longer the patients need to be given to reach a given target, as a longer period of rehabilitation is required for these patients to move through the treatment and recovery phases prior to discharge. Also, as the targets become more severe, they become increasingly harder to achieve, for all patients.

Figure 5 presents cumulative conditional probability of discharge from hospital by age conditional on eventual discharge to (a) death, (b) private nursing home, and (c) own home. We note that the admission vector here is across the population of stroke patients from all types of stroke, as we are thinking in terms of setting targets for the stroke unit rather than individual patients, as before. As we can see, these profiles are quite different across discharge distributions, highlighting the importance of different targets for private nursing homes and own homes. We have presented the graph for deaths in hospital as well for interest, although a target would be inappropriate here. Looking at the plots, we see that the longest durations are for patients who are discharged to their own home. The shortest are those who die in hospital, while those discharged to private nursing home are intermediate. This is reasonable as the patients who die in hospital are mainly very ill when they are admitted, while patients who are discharged to private nursing home are also quite ill and need a lot of rehabilitation before discharge. The patients who die do not display much variation between age groups, while older patients discharged to their home require longer periods in hospital than younger such patients, as they probably require more rehabilitation than younger patients. It is interesting that this age effect is reversed in patients discharged to private nursing homes, possibly because more time is spent trying unsuccessfully to rehabilitate them to a stage when they might manage at home, with a package.

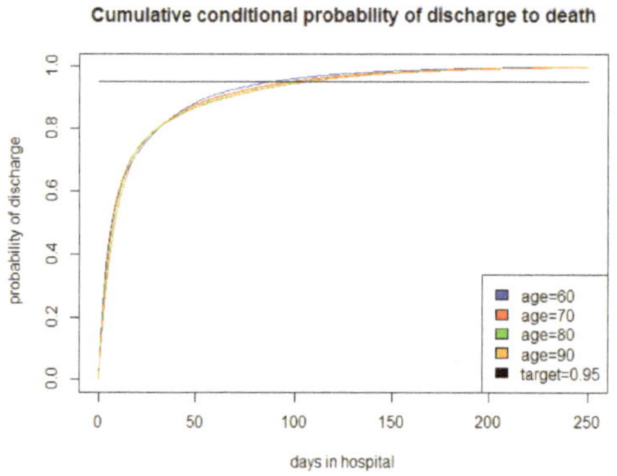

(**a**)

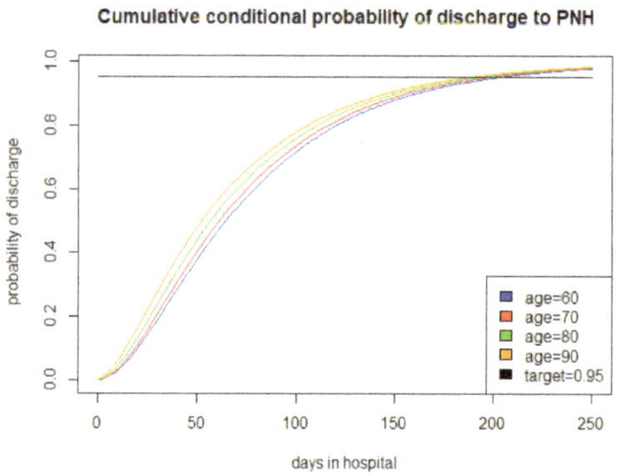

(**b**)

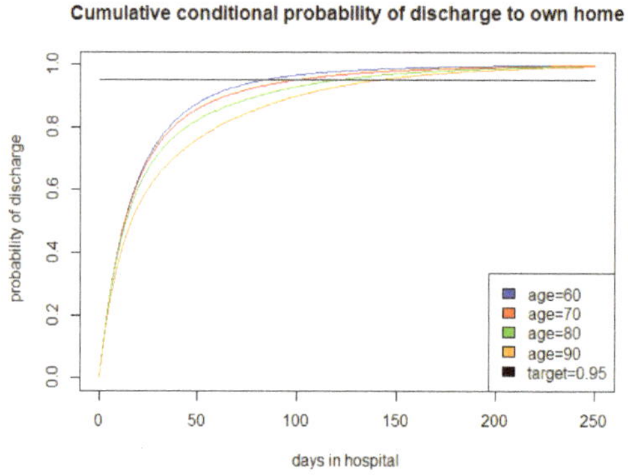

(**c**)

Figure 5. Cumulative conditional probability of discharge from hospital by age conditional on eventual discharge to (**a**) death, (**b**) private nursing home, and (**c**) own home.

5. Discussion

In our stroke patient example, Markov models can be used to describe the stroke patient care system using well-known clinical pathways, which integrate hospital and community services to provide ways of characterizing services, evaluating planned transformations, and predicting resourcing needs for future situations. Our previous paper [26] developed approaches to utilize routinely available discharge data to characterize patient admission patterns, movements through care, and release to suitable destinations. Such an approach can assist performance modelling, bed occupancy analysis, capacity planning, and patient destination prediction across different sectors of the patient care system. By using such an approach, we can compare different options and identify optimal policies. We note that stroke patient care provides an important paradigm example for healthcare processes generally, as there are numerous other specialties that encompass hospital and community services. Overall consumption of hospital resources and compliance with related targets are KPIs for healthcare services, and tools are thus needed to assess the effect of policies and their impact on patient hospitalization targets.

6. Conclusions

This paper described how process mining can provide suitable data from suitable datasets to populate phase-type models, which can then be used to quantify compliance with process targets or identify suitable targets given a required compliance percentage We described an example that uses phase-type models to describe stroke patient hospitalization and discharge, where there are multiple discharge destinations. Based on this use-case, various options have been investigated, with an emphasis on measuring target compliance; such performance indicators are frequently used in healthcare settings as well as in business and industrial environments. Multiple absorbing states quite commonly occur in such application domains. For example, there is an extensive literature on using Markov models for breast cancer patients where multiple absorbing states may come from different outcomes or using stratification to represent different characteristics of the patients [30].

Our current approach is part of initial efforts towards developing integrated process models, with the aim of supporting integrated management, planning, and resourcing. An important aspect of extending our framework, as described, is that it allows us to find the probability distribution of target compliance for multiple absorbing states and use Poisson processes to model arrivals; costs can also be associated with various parts of the system.

The approach is likely to be pertinent to business processes generally where phase-type models should have an important role to play.

Funding: This research was partly funded by the Invest Northern Ireland BTIIC project (BT Ireland Innovation Centre).

Institutional Review Board Statement: Not applicable.

Informed Consent Statement: Not applicable.

Data Availability Statement: Not applicable.

Conflicts of Interest: The author declares no conflict of interest. The funders had no role in the design of the study; in the collection, analyses, or interpretation of data; in the writing of the manuscript; or in the decision to publish the results.

References

1. Van der Aalst, W. Process mining: Overview and opportunities. *ACM TMIS* **2012**, *3*, 7. [CrossRef]
2. Van der Aalst, W.; Reijers, H.; Weijters, A.; van Dongen, B.; de Medeiros, A.A.; Song, M.; Verbeek, H. Business process mining: An industrial application. *Inf. Syst.* **2007**, *32*, 713–732. [CrossRef]
3. Taylor, P. Autonomic Business Processes. Ph.D. Thesis, University of York, York, UK, 2015.
4. Agrawal, R.; Imieliński, T.; Swami, A. Mining association rules between sets of items in large databases. *ACM SIGMOD Rec.* **1993**, *22*, 207–216. [CrossRef]
5. McClean, S.; Barton, M.; Garg, L.; Fullerton, K. A modeling framework that combines markov models and discrete-event simulation for stroke patient care. *ACM Trans. Model. Comput. Simul.* **2011**, *21*, 1–26. [CrossRef]
6. Peterson, J.L. *Petri Net Theory and the Modeling of Systems*; Prentice Hall PTR: Hoboken, FL, USA, 1981.
7. McChesney, I. Process support for continuous, distributed, multi-party healthcare processes—applying workflow modelling to an anticoagulation monitoring protocol. In Proceedings of the 10th International Conference on Ubiquitous Computing and Ambient Intelligence, Las Palmas de Gran Canaria, Spain, 29 November—2 December 2016; Springer: Berlin/Heidelberg, Germany, 2016; p. 255.
8. Haas, P.J. *Stochastic Petri Nets: Modelling, Stability, Simulation*; Springer Science & Business Media: Berlin/Heidelbaerg, Germany, 2016; p. 10.
9. Garg, L.; McClean, S.I.; Barton, M.; Meenan, B.J.; Fullerton, K. Intelligent Patient Management and Resource Planning for Complex, Heterogeneous, and Stochastic Healthcare Systems. *IEEE Trans. Syst. Man Cybern. Part A Syst. Humans* **2012**, *42*, 1332–1345. [CrossRef]
10. Gillespie, J.; McClean, S.; Garg, L.; Barton, M.; Scotney, B.; Fullerton, K. A multi-phase des modelling framework for pa-tient-centred care. *JORS* **2016**, *67*, 1239–1249.

11. Garg, L.; McClean, S.; Meenan, B.; Millard, P. Non-homogeneous Markov models for sequential pattern mining of healthcare data. *IMA J. Manag. Math.* **2008**, *20*, 327–344. [CrossRef]
12. McClean, S.; Stanford, D.; Garg, L.; Khan, N. Using Phase-type Models to Monitor and Predict Process Target Compliance. *Proc. Int. Conf. Oper. Res. Enterpr. Syst.* **2019**, *2019*, 82–90. [CrossRef]
13. Dudin, A.; Kim, C.; Dudina, O.; Dudin, S. Multi-server queueing system with a generalized phase-type service time dis-tribution as a model of call center with a call-back option. *Ann. Oper. Res.* **2016**, *239*, 401–428. [CrossRef]
14. Dudin, S.A.; Lee, M.H. Analysis of Single-Server Queue with Phase-Type Service and Energy Harvesting. *Math. Probl. Eng.* **2016**, *2016*, 1–16. [CrossRef]
15. Vishnevskii, V.M.; Dudin, A.N. Queueing systems with correlated arrival flows and their applications to modeling tele-communication networks. *Autom. Remote Control* **2017**, *78*, 1361–1403. [CrossRef]
16. Barron, Y.; Perry, D.; Stadje, W. A make-tostock production/inventory model with map arrivals and phase-type demands. *Ann. Oper. Res.* **2016**, *241*, 373–409. [CrossRef]
17. McClean, S.; Gillespie, J.; Garg, L.; Barton, M.; Scotney, B.; Kullerton, K. Using phase-type models to cost stroke patient care across health, social and community services. *Eur. J. Oper. Res.* **2014**, *236*, 190–199. [CrossRef]
18. Fackrell, M. Modelling healthcare systems with phase-type distributions. *Heal. Care Manag. Sci.* **2008**, *12*, 11–26. [CrossRef] [PubMed]
19. Tang, X.; Luo, Z.; Gardiner, J.C. Modeling hospital length of stay by Coxian phase-type regression with heterogeneity. *Stat. Med.* **2012**, *31*, 1502–1516. [CrossRef]
20. Marshall, A.H.; Zenga, M. Experimenting with the Coxian Phase-Type Distribution to Uncover Suitable Fits. *Methodol. Comput. Appl. Probab.* **2010**, *14*, 71–86. [CrossRef]
21. Griffiths, J.; Williams, J.; Wood, R. Modelling activities at a neurological rehabilitation unit. *Eur. J. Oper. Res.* **2013**, *226*, 301–312. [CrossRef]
22. Xie, H.; Chaussalet, T.; Millard, P. A Model-Based Approach to the Analysis of Patterns of Length of Stay in Institutional Long-Term Care. *IEEE Trans. Inf. Technol. Biomed.* **2006**, *10*, 512–518. [CrossRef]
23. Knight, V.A.; Harper, P.R. Modelling emergency medical services with phase-type distributions. *Heal. Syst.* **2012**, *1*, 58–68. [CrossRef]
24. Duong, T.; Phung, D.; Bui, H.H.; Venkatesh, S. Efficient duration and hierarchical modeling for human activity recogni-tion. *Artif. Intell.* **2009**, *173*, 830–856. [CrossRef]
25. Faddy, M.J.; McClean, S.I. Markov Chain Modelling for Geriatric Patient Care. *Methods Inf. Med.* **2005**, *44*, 369–373. [CrossRef] [PubMed]
26. Jones, B.; McClean, S.; Stanford, D. Modelling mortality and discharge of hospitalized stroke patients using a phase-type recovery model. *HCMS* **2018**, *22*, 1–19. [CrossRef] [PubMed]
27. McClean, S.; Gribbin, O. A non-parametric competing risks model for manpower planning. *Appl. Stoch. Model. Data Anal.* **1991**, *7*, 327–341. [CrossRef]
28. Vassiliou, P.-C.G. Rate of Convergence and Periodicity of the Expected Population Structure of Markov Systems that Live in a General State Space. *Mathematics* **2020**, *8*, 1021. [CrossRef]
29. Dobson, A.J.; Barnett, A.G. *An Introduction to Generalized Linear Models*; CRC Press: Boca Raton, FL, USA, 2008.
30. Otten, M.; Timmer, J.; Witteveen, A. Stratified breast cancer follow-up using a continuous state partially observable Markov decision process. *Eur. J. Oper. Res.* **2020**, *281*, 464–474. [CrossRef]

mathematics

MDPI

Article

Open Markov Type Population Models: From Discrete to Continuous Time

Manuel L. Esquível [1,*], Nadezhda P. Krasii [2] and Gracinda R. Guerreiro [1]

1 Department of Mathematics, FCT NOVA, and CMA New University of Lisbon, Campus de Caparica, 2829-516 Caparica, Portugal; grg@fct.unl.pt
2 Department of Higher Mathematics, Don State Technical University, 344000 Rostov-on-Don, Russia; krasnad@yandex.ru
* Correspondence: mle@fct.unl.pt

Abstract: We address the problem of finding a natural continuous time Markov type process—in open populations—that best captures the information provided by an open Markov chain in discrete time which is usually the sole possible observation from data. Given the open discrete time Markov chain, we single out two main approaches: In the first one, we consider a calibration procedure of a continuous time Markov process using a transition matrix of a discrete time Markov chain and we show that, when the discrete time transition matrix is *embeddable* in a continuous time one, the calibration problem has optimal solutions. In the second approach, we consider semi-Markov processes—and open Markov schemes—and we propose a direct extension from the discrete time theory to the continuous time one by using a known structure representation result for semi-Markov processes that decomposes the process as a sum of terms given by the products of the random variables of a discrete time Markov chain by time functions built from an adequate increasing sequence of stopping times.

Keywords: Markov chains; open population Markov chain models; Semi-Markov processes

Citation: Esquível, M.L.; Krasii, N.P.; Guerreiro, G.R. Open Markov Type Population Models: From Discrete to Continuous Time. *Mathematics* **2021**, *9*, 1496. https://doi.org/10.3390/math9131496

Academic Editors: Panagiotis-Christos Vassiliou and Andreas C. Georgiou

Received: 31 May 2021
Accepted: 23 June 2021
Published: 25 June 2021

Publisher's Note: MDPI stays neutral with regard to jurisdictional claims in published maps and institutional affiliations.

Copyright: © 2021 by the authors. Licensee MDPI, Basel, Switzerland. This article is an open access article distributed under the terms and conditions of the Creative Commons Attribution (CC BY) license (https://creativecommons.org/licenses/by/4.0/).

1. Introduction

After the first works introducing homogeneous open Markov population models in [1] followed by those in [2] and then in [3], further expanded by several authors and exposed in [4] and then in [5], the study of open populations in a finite state space in discrete time with a Markov chain structure became well established.

Following the pioneering work of Gani, introducing in [6] what now is known as *Cyclic Open Markov* population models, there were further extensions in [7], for non-homogeneous Markov chains and then, for cyclic non-homogeneous Markov systems or equivalently for non-homogeneous open Markov population processes, by the authors of [8,9]. Let us stress that continuous time non-homogeneous Markov systems have been studied lately in [10]. Furthermore, the recent work in [11] develops an approach to open Markov chains in discrete time—allowing a particle physics interpretation—for which there is a state space of the Markov chain—where distributions are studied by means of moment generating functions—there is an exit *reservoir*, which is tantamount to a cemetery state and, there is an incoming flow of particles, defined as a stochastic process in discrete time whose properties—e.g., stationarity—condition the distribution law of the particles in the state space.

Discrete time non-homogeneous semi-Markov systems or equivalently open semi-Markov population models were introduced and studied in [12,13]. The study of open populations in a finite state space in continuous time and governed by Markov laws, has already been carried in [14] and the references therein, and extensions to a general state space have been given in [15–17]. The continuous time framework has also been addressed, for instance, in [18–20], for the case of semi-Markov processes and for non-homogeneous semi-Markov systems [21]. We may also refer a framework of open Markov chains with finite state space—see in [22] and references therein—that has already seen applications in Actuarial or Financial problems—as, for instance, in [23,24]—but also in population dynamics (see [25]). The weaker formalism open Markov schemes, in discrete time—developed in [26]—allows for influxes of new elements in the population to be given as general time series models.

Another example was motivated by the study of a continuous time non homogeneous Markov chain model for Long Term Care, based on an estimated Markov chain transition matrix with a finite state space, in [27], by means of a method for calibrating the intensities on the continuous time Markov chain using the discrete time transition matrix in the context of usual existence theorems for ordinary differential equations (ODE); this method will be considered, in Section 3.2, in the more general context of Caratheodory existence theorems for ODE.

The main contribution of the present work is to extend results on open Markov chains in discrete time to some continuous time process of Markov type using different methods of associating a continuous process to an observed process in discrete time. One of these methods—presented in Sections 3.2 and 3.3—is by calibration of the transition intensities. Another method considered for open Markov schemes—in Section 4.2 and also, briefly, for some particular cases, in Section 4.3—is to exploit a natural representation of the continuous time Markov type process, in Formula (2) of Section 2.

2. From Discrete Time to Continuous Time via a Structural Approach

We present the main ideas on a structural representation for continuous time process of Markov type that are crucial to our approach. The structure of continuous time processes—for instance, Markov, semi-Markov, and Markov type schemes processes—allows us to consider a fairly general representation formula—Formula (2)—decoupling the continuous time process as a discrete time process and a sequence of time functions depending on the sequence of the jump stopping times.

Consider a complete probability space $(\Omega, \mathcal{F}, \mathbb{P})$, a continuous time stochastic process $(Y_t)_{t\geq 0}$ defined on this probability space and $\mathbb{F} = (\mathcal{F}_t)_{t\geq 0}$ the natural filtration associated to this process, that is, such that $\mathcal{F}_t := \sigma(Y_s : s \leq t)$ is the algebra-σ generated by the variables of the process until time t. Consider also a sequence of random variables $(Z_n)_{n\geq 0}$ taking values in a finite state space $\boldsymbol{\Theta} = \{\boldsymbol{\theta}_1, \boldsymbol{\theta}_2, \dots, \boldsymbol{\theta}_r\}$, the sequence being adapted to the filtration $\mathbb{F}$ and $0 \equiv \tau_0 < \tau_1 < \tau_2 < \cdots < \tau_n < \cdots$ an increasing sequence of $\mathbb{F}$-stopping times, denoted by $\mathcal{T}$, satisfying the following hypothesis:

Hypothesis 1. *Almost surely,* $\lim_{n\to+\infty} \tau_n = +\infty$ *and, for any* $T \in \mathbb{R}_+$ *and almost all* $\omega \in \Omega$*:*

$$\#\{k \geq 1 : \tau_k(\omega) \leq T\} < +\infty. \tag{1}$$

This hypothesis means that in every compact time interval $[0, T]$, for almost all $\omega \in \Omega$, there is only a finite number of stopping times realizations $\tau_k(\omega)$ in this interval.

Hypothesis 2. *The continuous time process* $(Y_t)_{t\geq 0}$ *admits a representation given, for* $t \geq 0$*, by*

$$Y_t = \sum_{n=0}^{+\infty} Z_n \mathbb{I}_{[\tau_n, \tau_{n+1}[}(t), \tag{2}$$

that is, a hypothesis on the structure of the continuous time process $(Y_t)_{t\geq 0}$.

It is well known—see in [28] (pp. 367–379) and in [29] (pp. 317–320)—that if $(Z_n)_{n\geq 0}$ is a Markov chain and the time intervals $(\tau_{n+1} - \tau_n)_{n\geq 0}$ are Exponentially distributed then $(Y_t)_{t\geq 0}$ can be taken to be a continuous time **homogeneous** Markov chain. If $(Z_n)_{n\geq 0}$ is a Markov chain and the time intervals $(\tau_{n+1} - \tau_n)_{n\geq 0}$ have a distribution that can depend on the present state as well as on the one visited next then $(Y_t)_{t\geq 0}$ can be taken to be a **semi-Markov** process (see in [30] (pp. 261–262) and in [31] (pp. 295–299), for brief references). In the case of a semi-Markov processes, a nice result of Ronald Pyke (see in [32] (p. 1236)), reproduced ahead in Theorem A7, guarantees that when the state space is finite the process is *regular* implying that almost all paths of such a semi-Markov process are step-functions over $[0, +\infty[$ and so, the paths satisfy Formula (1). In another important case (see Theorems A5 and A6 ahead, or [30] (pp. 262–266) and [31] (pp. 195–244)), adequate hypothesis on the distribution of the stopping times and on the sequence $(Z_n)_{n\geq 0}$ implies that $(Y_t)_{t\geq 0}$ will be a **non homogeneous** Markov chain process in continuous time, whose trajectories are step functions also satisfying Formula (1). The representation in Formula (2), thus covers the cases of homogeneous and non homogeneous Markov processes in continuous time as well as semi-Markov processes, providing a desired connection between a continuous time process and a discrete one that is a component of the former. We observe that there is a practical justification for Hypothesis 1, namely, the *identifiability* of the process; as can be read in [33] (p. 3): "*…Actually, in real systems the transition from one observable state into another takes some time.*" Being so, the existence of accumulation points in a compact interval would preclude estimation procedures for instance of the distribution of the sequence $(\tau_{n+1} - \tau_n)_{n\geq 1}$.

3. From Discrete to Continuous Time Markov Chains: A Calibration Approach

In this section, we consider a calibration approach in order to determine a set of probability densities that best approaches a sequence of discrete time transition matrices with respect to a quadratic loss function. We then show that *embeddable* stochastic matrices, according to Definition 1, are solutions of the calibration problem. For the reader's convenience, we recall in the first appendix the most important results on continuous time Markov chains with finite state space that are relevant for our study with emphasis on the crucial non-accumulation property of the jump times of a continuous time Markov chain (see Theorem A6 ahead). We will start by recalling the main information on embeddable chains. We then present one of the main contributions of this work, that is, a general result on the optimization problem of calibration and its relations with embeddable properties of discrete time Markov chains.

3.1. The Embedding of a Discrete Time Markov Chain in a Continuous One

The embedding of the discrete time Markov chain in a continuous one following the guidelines, for instance, in [34–40], can be considered as a method to connect a discrete time process with a continuous one. For notations on non-homogeneous continuous time Markov chains see Section 3.2.

Definition 1 (Embeddable stochastic matrix (see [38])). *A stochastic matrix $\boldsymbol{R}$ is said to be* ***embeddable*** *if there exists a time $t_{\boldsymbol{R}} > 0$ and a family of stochastic matrices $\boldsymbol{P}(s,t)$ continuously defined in the set of times $\{(s,t) \in \mathbb{R}^2 : 0 \leq s \leq t \leq t_{\boldsymbol{R}}\}$ such that*

$$\begin{cases} \boldsymbol{P}(s,t) = \boldsymbol{P}(s,u)\boldsymbol{P}(u,t) & 0 \leq s \leq u \leq t \leq t_{\boldsymbol{R}} \\ \boldsymbol{P}(s,s) = \boldsymbol{I} & 0 \leq s \leq t_{\boldsymbol{R}}. \\ \boldsymbol{P}(0,t_{\boldsymbol{R}}) = \boldsymbol{R}. & \end{cases} \tag{3}$$

We observe that by Theorem A2 ahead, the condition in Formulas (3) is tantamount to the definition of a continuous time Markov chain with transition probabilities given by $\boldsymbol{P}(s,t)$.

Remark 1 (Intrinsic time for embeddable chains). *Goodman in [41]—aiming at a more general result for the Kolmogorov differential equations—showed that with the change of time given by*

$\varphi(u) := -\log\det \mathbf{P}(0,u)$—which amounts to a change in the matrix coefficients of $\mathbf{P}(s,t)$—we have that

$$t_{\mathbf{R}} = -\log\det \mathbf{R}. \tag{4}$$

This remarkable representation for the embedding time $t_{\mathbf{R}}$ will be useful for a result in Section 3.2 devoted to the calibration approach. It has also been used for estimation in [42] (p. 330).

See the work in [35] for a definition similar to Definition 1 and for a summary of many important results on this subject. The characterization of an embeddable stochastic matrix in a form useful for practical purposes was recently achieved in [43]. More useful results were obtained in [44]. The connections between this kind of embedding and the other approaches, for the association of a discrete time Markov chain and a continuous time process, deserve further study.

3.2. Continuous Time Markov Chains Calibration with a Discrete Time Markov Transition Matrix

The calibration of transition intensities of a non homogeneous Markov chain, with a discrete time Markov chain transition matrix estimated from data, was proposed in [27]. In this section, we establish a general formulation of the existence a unicity result that subsumes the approach and we establish a connection with the *embedding* approach of Section 3.1. Notation and needed essential results on non-homogeneous Markov processes in continuous time were recalled in Appendix A.

The procedure for calibration of intensities consists in finding the intensities of a non homogeneous continuous time Markov chain using a probability transition matrix of a discrete time Markov chain and a given loss function—having as arguments the transition probabilities of the continuous time Markov chain and some function of the transition matrix of the discrete time Markov chain—in such a way that the loss function is minimized.

Previously to the consideration of the theorem on the calibration of intensities we discuss some motivation for this approach. It may happen that a phenomena that could be dealt—due to its characteristics—with a continuous time Markov chain model can only be observed at regularly spaced time intervals. This is the case of the periodic assessments of the healthcare status of patients that can change at any time but are only object of a comprehensive evaluation on, say, a weekly basis. With the data originated by these observations we can only determine transition probabilities—for a defined period, say, a week—and, most importantly we cannot determine the time stamps for the patient status change. The question naturally poses itself: is it possible to associate—in some canonical way—to an estimated discrete time Markov chain transition matrix a process in continuous time that encompasses the discrete time process? First steps in this direction are provided by Theorem 1 that we now present and the following Theorems 2 and 3.

We formulate Theorem 1 in the context of Caratheodory's general existence theory of solutions of ordinary differential equations that we briefly recall. One reason for this choice is that according to [41] (p. 169) and we quote: "*...This fact gives further evidence in support of the view that Caratheodory equations occupy a natural place in the theory of non-stationary Markov chains.*" Another reason is the fact that Caratheodory existence theory is particularly suited for regime switching models and these models are the object of Theorem 3 ahead. Following the work in [45] (pp. 41–44), we consider the definition of an **extended solution** for a Cauchy problem of a differential equation,

$$\mathbf{Y}'(t) = f(t, \mathbf{Y}(t)),\ \mathbf{Y}(0) = \xi, \tag{5}$$

or formulated in an equivalent form,

$$\mathbf{Y}(t) = \xi + \int_0^t f(s, \mathbf{Y}(s))ds, \tag{6}$$

for $f(t,y) : I \times \mathcal{D} \to \mathbb{R}^r$ a non-necessarily continuous function, with $I \subset [0,+\infty[$ and $\mathcal{D} \subset \mathbb{R}^r$, to be an **absolutely continuous** function $\mathbf{Y}(t)$ (see [46], pp. 144–150) such that

$f(t, \mathbf{Y}(t)) \in \mathcal{D}$ for $t \in I$ and Formula (5) is verified for all $t \in I$ possibly with the exception of a set of null Lebesgue measure. The well-known Caratheodory's existence theorem (see in [45], p. 43) ensures the existence of an extended solution with a given initial condition—given in a neighborhood of the initial time—under the conditions that $f(t, \boldsymbol{y})$ is measurable in the variable t, for fixed $\boldsymbol{y}$, and continuous in the variable $\boldsymbol{y}$, for fixed t, and moreover that there exists a Lebesgue integrable function $m(t)$, defined on a neighborhood of the initial time, let us say I, such that $|f(t, \boldsymbol{y})| \leq m(t)$ for $(t, \boldsymbol{y}) \in I \times \mathcal{D}$. The question of unicity of the solution is dealt, usually, either directly using Theorem 18.4.13 in [47] (p. 337) or using Osgood's uniqueness theorem—as exposed, for instance, in [48] (p. 58) or in [49] (pp. 149–151)—to conclude that the extended solution—that with Caratheodory's theorem we know to exist—is unique in the sense that two solutions may only differ on a set of Lebesgue measure equal to zero. For our purposes we need an existence and unicity theorem for ordinary differential equations with solutions depending continuously on a parameter such as the general result of Theorem 4.2 in [45] (p. 53) with an omitted proof that follows for a lengthy previous exposition of related matters. For completeness we now establish a result that is suited to our purposes as it deals with the particular type of Kolmogorov equations for continuous time Markov chains.

Theorem 1 (Calibration of intensities with Caratheodory's type ODE existence theorem hypothesis). *Let, for $1 \leq n \leq N$, $\mathbf{R}^{\tau_n} = \left[r_{ij}^{(\tau_n)}\right]_{i,j=1,\dots,r}$ be the generic element of a sequence of numerical transition matrices taken at sequence of increasing dates $(\tau_n)_{1 \leq n \leq N}$. Consider a set of intensities $\mathbf{Q}(t, \boldsymbol{\lambda}) = [q(u, i, j, \boldsymbol{\lambda})]_{i,j=1,\dots,r}$—with $\boldsymbol{\lambda} \in \boldsymbol{\Lambda} \subset \mathbb{R}^d$ being a parameter and $\boldsymbol{\Lambda}$ being a compact set—satisfying the following conditions:*

1. *For every fixed $\boldsymbol{\lambda}$ the functions $q(u, i, j, \boldsymbol{\lambda})$ are measurable as functions of u.*
2. *For every fixed u the functions $q(u, i, j, \boldsymbol{\lambda})$ are continuous as functions of λ.*
3. *There exists a locally integrable function $M : [0, +\infty[\mapsto [0, +\infty[$, such that for all $\lambda \in \boldsymbol{\Lambda}$, $i \in \mathcal{I}$, $u \in [0, +\infty[$ and $0 \leq s \leq t$, the following conditions are verified:*

$$- q(u, i, i, \boldsymbol{\lambda}) \leq M(u) \text{ and } \int_s^t M(u) du < +\infty. \tag{7}$$

Then, we have

1. *There exists $\mathbf{P}(s, t, \boldsymbol{\lambda}) = [p(s, i, t, j, \boldsymbol{\lambda})]_{i,j=1,\dots,r}$ a probability transition matrix, with entries absolutely continuous in s and t, such that conditions in Definition A2, the Chapman–Kolmogorov equations in Theorem A1 and Theorem A3 are verified.*
2. *For each fixed s_0, consider the loss function*

$$\mathcal{O}(s_0, \boldsymbol{\lambda}) := \sum_{i,j=1,\dots,r} \sum_{n=1}^{N} \left(p(s_0, i, \tau_n, j, \boldsymbol{\lambda}) - r_{ij}^{(\tau_n)} \right)^2. \tag{8}$$

Then, for the optimization problem $\inf_{\boldsymbol{\lambda} \in \boldsymbol{\Lambda}} \mathcal{O}(s_0, \boldsymbol{\lambda})$ there exists $\boldsymbol{\lambda}_0 \in \boldsymbol{\Lambda}$ such that

$$\mathcal{O}(s_0, \boldsymbol{\lambda}_0) = \min_{\boldsymbol{\lambda} \in \boldsymbol{\Lambda}} \mathcal{O}(s_0, \boldsymbol{\lambda}), \tag{9}$$

the unique minimum being attained at possibly several points $\boldsymbol{\lambda}_0 \in \boldsymbol{\Lambda}$.

Proof. We will prove, simultaneously, the existence of the probability transition matrix, the unicity in the extended solution sense and the continuous dependence of the parameter $\lambda \in \boldsymbol{\Lambda}$ following the lines of the proof of the result denominated Hostinsky's representation (see in [29], pp. 348–349). As we suppose that $\boldsymbol{\Lambda}$ is compact, the continuity of $\mathbf{P}(s_0, t, \boldsymbol{\lambda})$, as a function of $\lambda \in \boldsymbol{\Lambda}$ for every fixed t, will be enough to establish the second thesis.

We want to determine an extended solution of the Kolmogorov forward equation given in Formula (A11), that is an extended solution of

$$\begin{cases} \mathbf{P}'_t(s_0,t,\boldsymbol{\lambda}) = \mathbf{P}(s_0,t,\boldsymbol{\lambda})\mathbf{Q}(t,\boldsymbol{\lambda}) \\ \mathbf{P}(t,t) = \mathbf{I}, \end{cases} \tag{10}$$

an equation which, as seen in Formula (A12), can be read in integral form as,

$$\mathbf{P}(s_0,t,\boldsymbol{\lambda}) = \mathbf{I} + \int_{[s_0,t]} \mathbf{P}(s_0,s,\boldsymbol{\lambda})\mathbf{Q}(s,\boldsymbol{\lambda})ds. \tag{11}$$

As previously said, we will now follow the general idea of successive approximations in the proof of the Picard–Lindelöf theorem for proving existence and unicity of solutions of ordinary differential equations for the forward Kolmogorov equation. By replacing $\mathbf{P}(s_0,s,\boldsymbol{\lambda})$ in the right-hand member of Equation (11) by this right-hand member we get,

$$\mathbf{P}(s_0,t,\boldsymbol{\lambda}) = \mathbf{I} + \int_{[s_0,t]} \mathbf{Q}(s,\boldsymbol{\lambda})ds + \int_{[s_0,t]}\int_{[s_0,t_1]} \mathbf{P}(s_0,t_2,\boldsymbol{\lambda})\mathbf{Q}(t_1,\boldsymbol{\lambda})\mathbf{Q}(t_2,\boldsymbol{\lambda})dt_2 dt_1$$

and, by induction, we obtain

$$\begin{aligned} \mathbf{P}(s_0,t,\boldsymbol{\lambda}) &= \mathbf{I} + \int_{[s_0,t]} \mathbf{Q}(s,\boldsymbol{\lambda})ds + \\ &+ \sum_{n=2}^{k} \int_{[s_0,t]}\int_{[t_1,t]}\cdots\int_{[t_{n-1},t]} \mathbf{Q}(t_1,\boldsymbol{\lambda})\mathbf{Q}(t_2,\boldsymbol{\lambda})\cdots\mathbf{Q}(t_n,\boldsymbol{\lambda})dt_n\cdots dt_1 + \\ &+ \int_{[s_0,t]}\int_{[t_1,t]}\cdots\int_{[t_{k-1},t]} \mathbf{P}(s_0,t_k,\boldsymbol{\lambda})\mathbf{Q}(t_1,\boldsymbol{\lambda})\mathbf{Q}(t_2,\boldsymbol{\lambda})\cdots\mathbf{Q}(t_k,\boldsymbol{\lambda})dt_k\cdots dt_1. \end{aligned}$$

Now, considering the function $M(t)$ in the third hypothesis stated above about the intensity matrix, we have that, by Lemma A1 (see also Lemma 8.4.1 in [29], p. 348), since $M(t)$ is integrable over any compact set, considering the (i,j) component of the $r \times r$ matrix, we have that

$$\begin{aligned} &\left| \left[\int_{[s_0,t]}\int_{[t_1,t]}\cdots\int_{[t_{k-1},t]} \mathbf{P}(s_0,t_k,\boldsymbol{\lambda})\mathbf{Q}(t_1,\boldsymbol{\lambda})\mathbf{Q}(t_2,\boldsymbol{\lambda})\cdots\mathbf{Q}(t_k,\boldsymbol{\lambda})dt_k\cdots dt_1 \right]_{ij} \right| \leq \\ \leq\ & r^k \int_{[s_0,t]}\int_{[t_1,t]}\cdots\int_{[t_{k-1},t]} M(t_1)M(t_2)\cdots M(t_k)dt_k\cdots dt_1 = \\ =\ & \frac{\left(r\int_{[s_0,t]} M(s)ds\right)^k}{k!}. \end{aligned}$$

Finally, as

$$\lim_{k\to+\infty} \frac{\left(r\int_{[s_0,t]} M(s)ds\right)^k}{k!} = 0,$$

we have that the series for which the sum represents $\mathbf{P}(x,t,\boldsymbol{\lambda})$, that is,

$$\mathbf{P}(s_0,t,\boldsymbol{\lambda}) = \mathbf{I} + \sum_{n=1}^{+\infty}\left(\int_{[s_0,t]}\int_{[t_1,t]}\cdots\int_{[t_{n-1},t]} \mathbf{Q}(t_1,\boldsymbol{\lambda})\mathbf{Q}(t_2,\boldsymbol{\lambda})\cdots\mathbf{Q}(t_n,\boldsymbol{\lambda})dt_n\cdots dt_1 \right),$$

is a series—of absolutely continuous functions of the variable t which are also continuous as functions of the parameter $\boldsymbol{\lambda} \in \boldsymbol{\Lambda}$—converging normally and so the sum is an absolutely continuous function of the variable t and continuous function of the parameter $\boldsymbol{\lambda}$. With a similar reasoning applied to the backward Kolmogorov equation we also have that $\mathbf{P}(s,t_0,\boldsymbol{\lambda})$ is absolutely continuous in the variable s and, obviously, continuous as a function of the parameter $\boldsymbol{\lambda} \in \boldsymbol{\Lambda}$. We observe that it was stated in [41], pp. 166–167 (with a reference to a proof in [50] and proved also in [51]), that the separate absolute continuity of $\mathbf{P}(s,t,\boldsymbol{\lambda})$ in the variables s and t ensures the uniqueness of the solution. □

Remark 2 (An alternative path for the existence result). *We observe that, for every fixed value of the parameter $\boldsymbol{\lambda}$, by a direct application of Caratheodory's existence theorem to the forward and backward Kolmogorov equations in Theorem A3, we obtain a probability transition matrix $\mathbf{P}(s,t,\boldsymbol{\lambda}) = [p(s,i,t,j,\boldsymbol{\lambda})]_{i,j=1,\dots,r}$, such that conditions in Definition A2 and the Chapman–Kolmogorov equations in Theorem A1 are verified, that in addition has entries absolutely continuous in s and t and such that Kolmogorov's equations are satisfied almost everywhere. With this approach the continuous dependence of the probability transition matrix on the parameter $\boldsymbol{\lambda}$ requires further proof.*

Remark 3 (On the parametrized intensities and transition probabilities). *In a first application to Long-Term Care of a simpler version of Theorem 1 presented in [27], we chose as intensities a parametrized family—of Gompertz–Makeham type (see, for instance, in [52], p. 62)—with a three dimensional parameter. We observe that, in its actual formulation, Theorem 1 contemplates the case of a set of intensities—and of associated transition probabilities—not necessarily with the same functional form with varying parameters but merely with a finite set of different functional forms indexed by the parameters.*

Remark 4 (Only one transition matrix observation). *In the case where we only have one estimated transition matrix* $\mathbf{R}$, *we can consider the sequence of n step transition matrices given by the n fold product of the matrix* $\mathbf{R}$ *by itself. This situation will be addressed in Theorem 2 ahead, in the case of homogeneous Markov chains and in Theorem 3 for the non-homogeneous case.*

We also observe that in the case of a multidimensional parameter set $\boldsymbol{\Lambda}$—say r_1—and even in a reasonable state space of the discrete time Markov chain—say with r_2 states—the optimization problem of Formula (8) may require adequate algorithms to be solved as the number of variables is of the order of $r_1 \times r_2 \times (r_2 - 1)$. In [27] we opted for a modified grid search coupled with the numerical solutions of the Kolmogorov equations in order to recover the transition probabilities of the continuous time Markov chain.

Remark 5 (On the unicity of the solution of the calibration problem). *The unicity in law of the solution of the calibration problem deserves discussion. If there are several minimizers of the calibration problem, to each of these minimizers corresponds an intensity and to each intensity a, possible, different law for the stopping times of the continuous time Markov chain, as these laws are determined by the intensities (see Remark A2). The existence of criteria allowing to identify a distribution of inter-arrival times that stochastically dominates all other solutions is an open problem.*

We can establish a connection between the approach in Section 3.1 and Theorem 1 on calibration above, showing first—in Theorem 2—that, if a matrix is embeddable in a homogeneous continuous time Markov chain—with intensities depending continuously on a parameter—for a fixed value of the parameter, then this continuous time Markov chain solves the calibration problem in an optimum way. We recall that the continuous time Markov chain is homogeneous if, for all $0 \leq s,t$ the transition probabilities satisfy

$$\boldsymbol{P}(s,s+t) = \boldsymbol{P}(0,t),$$

and that the intensities matrix is constant as a function of time (see [41] (pp. 165–166) for definitions in this context).

Theorem 2 (Discrete chains embeddable in **homogeneous** continuous chains can be optimally calibrated). *Suppose that the matrix* $\mathbf{R}$ *is embeddable and let $t_{\mathbf{R}}$ and the transition probabilities $\mathbf{P}(s,t,\boldsymbol{\lambda}_1)$ satisfy Definition 1 in the case of a homogeneous continuous time Markov* *chain for some family of intensities* $\mathbf{Q}(\boldsymbol{\lambda}_1)$ *where* $\boldsymbol{\lambda}_1 \in \boldsymbol{\Lambda}$ *is a given parameter. Then, with* $\tau_n := nt_{\mathbf{R}}$ *for* $n \geq 1$ *and* $\mathbf{R}^{\tau_n} := \mathbf{R}^{(n)}$*—the n fold product of the matrix* $\mathbf{R}$ *by itself—we have that the optimization problem,* $\inf_{\boldsymbol{\lambda} \in \boldsymbol{\Lambda}} \mathcal{O}(\boldsymbol{\lambda})$ *with respect to the loss function given by Formula (8) has an optimal solution* $\mathbf{P}(s,t,\boldsymbol{\lambda}_1)$ *such that*

$$\mathcal{O}(\boldsymbol{\lambda}_1) = \min_{\boldsymbol{\lambda} \in \boldsymbol{\Lambda}} \mathcal{O}(\boldsymbol{\lambda}) = 0.$$

Proof. It is enough to observe that by Formulas (3) in Definition 1 we have, as $\tau_2 - \tau_1 = \tau_1$,

$$\begin{aligned} \boldsymbol{P}(0,\tau_2,\boldsymbol{\lambda}_1) &= \boldsymbol{P}(0,\tau_1,\boldsymbol{\lambda}_1)\boldsymbol{P}(\tau_1,\tau_2,\boldsymbol{\lambda}_1) = \boldsymbol{P}(0,\tau_1,\boldsymbol{\lambda}_1)\boldsymbol{P}(0,\tau_2-\tau_1,\boldsymbol{\lambda}_1) = \\ &= \boldsymbol{P}(0,\tau_1,\boldsymbol{\lambda}_1)\boldsymbol{P}(0,\tau_1,\boldsymbol{\lambda}_1) = \mathbf{R}^{(2)} = \mathbf{R}^{\tau_2}, \end{aligned}$$

and, by induction, that $\boldsymbol{P}(0,\tau_n,\boldsymbol{\lambda}_1) = \mathbf{R}^{\tau_n}$ and so in Formula (8) we have that $\mathcal{O}(\boldsymbol{\lambda}_1) = 0$. □

Remark 6 (On the skeletons of a homogeneous continuous time Markov chain)**.** *Another possible way to extend results from discrete time to continuous time is the approach of skeletons of Kingman and other authors (see [53,54], for instance). As we are more interested in non-homogeneous continuous time Markov chains we do not pursue this approach in the present work.*

We now address the case of non homogeneous Markov chain. In Theorem 3, we show that if every element of a sequence, with no gaps, of matrix powers of a discrete time Markov chain is embeddable then there is a regime switching process of Markov type that solves optimally the calibration problem.

Theorem 3 (Discrete *power-embeddable* discrete chains can be optimally calibrated)**.** *Suppose that all the powers* $\boldsymbol{R}^{(n)} = \left[r_{ij}^{(n)}\right]_{i,j=1,\dots,r}$*, for* $1 \leq n \leq N$*, of a discrete time Markov chain transition matrix* $\boldsymbol{R}$ *are embeddable and let* $\boldsymbol{P}_n(s,t,\boldsymbol{\lambda}_n)$ *be the transition probabilities of the embedding continuous time Markov chain for* $\boldsymbol{R}^{(n)}$ *given in their intrinsic time—defined in Remark 1—in such a way that the respective embedding times verifies* $t_{\boldsymbol{R}^{(n)}} = -n \log \det \boldsymbol{R}$ *(according to Formula (4)). We suppose that the intensities* $\mathbf{Q}_n(t,\boldsymbol{\lambda}_n)$ *for each of the transition probabilities* $\boldsymbol{P}_n(s,t,\boldsymbol{\lambda}_n)$ *depend on parameters* $\boldsymbol{\lambda}_n \in \boldsymbol{\Lambda}$*, possibly different but all in a common parameter set* $\boldsymbol{\Lambda}$*. With the convention* $t_{\boldsymbol{R}^{(0)}} = 0$*, and*

$$\boldsymbol{\lambda}(t) := \boldsymbol{\lambda}_n,\ t_{\boldsymbol{R}^{(n-1)}} \leq t \leq t_{\boldsymbol{R}^{(n)}},$$

let $\widetilde{\mathbf{P}}(s,t,\boldsymbol{\lambda}(t))$ *be defined by*

$$\widetilde{\mathbf{P}}(s,t,\boldsymbol{\lambda}(t)) := \boldsymbol{P}_n(s,t,\boldsymbol{\lambda}_n),\ 0 = t_{\boldsymbol{R}^{(0)}} \leq s \leq t_{\boldsymbol{R}^{(n)}},\ t_{\boldsymbol{R}^{(n-1)}} \leq t \leq t_{\boldsymbol{R}^{(n)}},\ s \leq t, \tag{12}$$

and thus satisfying $\widetilde{\mathbf{P}}(0,t_{\boldsymbol{R}^{(n)}},\boldsymbol{\lambda}(t)) = \boldsymbol{P}_n(0,t_{\boldsymbol{R}^{(n)}},\boldsymbol{\lambda}_n) = \boldsymbol{R}^{(n)}$*. Then, we have that the optimization problem,* $\inf_{\boldsymbol{\lambda} \in \boldsymbol{\Lambda}} \mathcal{O}(\boldsymbol{\lambda})$ *with respect to the loss function given by*

$$\mathcal{O}(\boldsymbol{\lambda}) := \sum_{i,j=1,\dots,r} \sum_{n=1}^{N} \left(\widetilde{\mathbf{P}}(0,t_{\boldsymbol{R}^{(n)}},\boldsymbol{\lambda}(t))_{ij} - r_{ij}^{(n)} \right)^2, \tag{13}$$

has an optimal solution $\widetilde{\mathbf{P}}(s,t,\boldsymbol{\lambda}(t))$ *such that*

$$\mathcal{O}(\boldsymbol{\lambda}(t)) = \min_{\boldsymbol{\lambda} \in \boldsymbol{\Lambda}} \mathcal{O}(\boldsymbol{\lambda}) = 0.$$

Proof. We observe that the definition in Formula (12) is coherent—see Figure 1—and then it is a simple verification with the definitions proposed. □

Remark 7 (An associated *regime switching* process). *The function $\widetilde{\mathbf{P}}(s,t,\boldsymbol{\lambda}(t))$ defined in Formula (12) was obtained by superimposing different transition probabilities for different Markov chains in continuous time. A natural question is to determine if there is—based on these different transitions probabilities—a regime switching Markov chain in continuous time that bears some connection with $\widetilde{\mathbf{P}}(s,t,\boldsymbol{\lambda}(t))$. From a brief analysis of Figure 1 we can guess the natural definition of a regime switching Markov chain based on the probabilities $\mathbf{P}_n(s,t,\boldsymbol{\lambda_n})$. Let*

$$\mathbf{P}(s,t,\boldsymbol{\lambda}(t)) := \mathbf{P}_n(s,t,\boldsymbol{\lambda_n}),\ t_{\mathbf{R}^{(n-1)}} \leq s \leq t \leq t_{\mathbf{R}^{(n)}}. \tag{14}$$

Formula (14) has the following interpretation. For each $1 \leq n \leq N$, consider continuous time Markov chain processes $(X_t^n)_{t\in[t_{\mathbf{R}^{(n-1)}},t_{\mathbf{R}^{(n)}}]}$ with transition probabilities $\mathbf{P}_n(s,t,\boldsymbol{\lambda_n})$ defined in the domains $\mathcal{R}_n := \{(s,t) \in \mathbb{R}^2 : t_{\mathbf{R}^{(n-1)}} \leq s \leq t \leq t_{\mathbf{R}^{(n)}}\}$ with the convention $t_{\mathbf{R}^{(0)}} = 0$. The regime switching process $(Y_t)_{t\in[0,t_{\mathbf{R}^{(n)}}]}$ is such that (compare with Formula (2)):

$$Y_t = X_t^n,\ t \in [t_{\mathbf{R}^{(n-1)}}, t_{\mathbf{R}^{(n)}}],$$

that is, the process $(Y_t)_{t\in[0,t_{\mathbf{R}^{(n)}}]}$ is obtained by gluing together $(X_t^n)_{t\in[t_{\mathbf{R}^{(n-1)}},t_{\mathbf{R}^{(n)}}]}$, the paths of the processes which are bona fide continuous time Markov processes in each of their—non-random—time intervals $[t_{\mathbf{R}^{(n-1)}}, t_{\mathbf{R}^{(n)}}]$. It is clear that $\mathbf{P}(s,t,\boldsymbol{\lambda}(t))$ can be interpreted as a transition probability only when restricted to some domain $\mathcal{R}_n$ and that, in general, it will not be a transition probability in the whole interval $[0, t_{\mathbf{R}^{(N)}}]$.

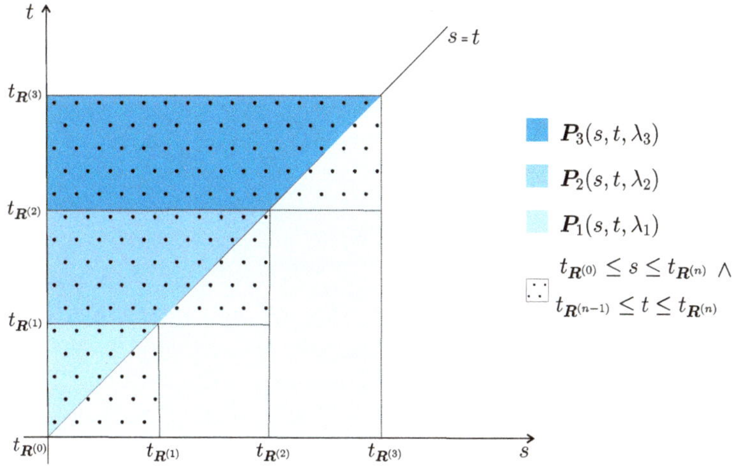

Figure 1. A representation of $\widetilde{\mathbf{P}}(s,t,\boldsymbol{\lambda}(t))$ in Formula (12) for the first three initial times.

Remark 8. *The regime switching process defined in Remark 7 deserves further study. We may, nevertheless, define transition probabilities $\widehat{\mathbf{P}}(s,t,\boldsymbol{\lambda}(t))$ for $t_{\mathbf{R}^{(k-1)}} \leq s \leq t_{\mathbf{R}^{(k)}} \leq t \leq t_{\mathbf{R}^{(k+1)}}$—with properties to be thoroughly investigated—by considering*

$$\widehat{\mathbf{P}}(s,t,\boldsymbol{\lambda}(t)) := \mathbf{P}_k(s,t_{\mathbf{R}^{(k)}},\boldsymbol{\lambda_k}) \cdot \mathbf{P}_{k+1}(t_{\mathbf{R}^{(k)}},t,\boldsymbol{\lambda_{k+1}}).$$

3.3. Conclusions on the Relations between Embeddable Matrices, Calibration, and Open Markov Chain Models

From Theorems 1–3, the following conclusions can be drawn. Given a discrete time Markov transition matrix,

- if the matrix is *embeddable*—according to Definition 1 of Section 3.1—there is an unique in law homogeneous Markov chain in continuous time that solves the calibration problem optimally; the unicity is a consequence of Remark A2 that shows that the laws of the stopping times $(\tau_n)_{n\geq 0}$ in the representation of Formula (A13) only depend on the intensities and these are uniquely determined whenever the discrete time Markov chain is embeddable.

- if the matrix is *power-embeddable*—that is, if all the matrices of a finite sequence with no gaps of powers of the matrix are embeddable—then there is an unique regime switching continuous time non-homogeneous Markov chain—in the sense of Remark 7—that solves the calibration problem optimally. In this case, the unicity has a justification similar to the previously referred case, that is, the laws of the stopping times only depends on the intensities and these are determined by the fact that the matrix is power-embeddable.

As a consequence, for our purposes, it appears of fundamental importance to determine if a discrete time Markov chain transition matrix is embeddable and to determine—if possible, explicitly—the embedding continuous time Markov chain. Regarding this problem the results in [43,55] deserve further consideration.

Remark 9 (Aplying Theorems 1–3). *Suppose that discrete time Markov chain transition matrix, of a Markov chain process* $(Z_n)_{n\geq 1}$ *is embeddable in a continuous time Markov chain* $(X_t)_{t\geq 0}$*. We have, for this continuous time process and for a determined sequence of stopping times* $(\tau_n)_{n\geq 1}$*, the representation given in Formula* (A13) *of Theorem A5, that is,*

$$X_t = \sum_{n=0}^{+\infty} X_{\tau_n} \mathbb{1}_{[\tau_n,\tau_{n+1}[}(t).$$

Now, as the Theorems referred to may consider that the process $(Z_n)_{n\geq 1}$ *is suitably approximated by* $(X_t)_{t\geq 0}$*, we can also consider that the continuous time process defined by*

$$\widetilde{X}_t := \sum_{n=0}^{+\infty} Z_{\tau_n} \mathbb{1}_{[\tau_n,\tau_{n+1}[}(t), \tag{15}$$

is an approximation of $(Z_n)_{n\geq 1}$ *in continuous time. For processes with a structural representation similar to the one of the process* $(\widetilde{X}_t)_{t\geq 0}$ *we propose in Section 4.3 a method to extend from discrete to continuous time the open populations methodology.*

4. More on Open Continuous Time Processes from Discrete Ones

In this section, we discuss an extension of the formalism of open Markov chains to the case of semi-Markov processes (sMp) and other continuous time processes, namely, the open Markov chain schemes introduced in [26]. For the reader's convenience we present in Appendix B a short summary on sMp and in the next Section 4.1 a review of the main results on the open Markov chain formalism for discrete time. Finally, we propose the second main contribution of this work, that is, an extension of the open Markov chain formalism in discrete time to continuous time in the case of sMp. We also briefly refer the case of open Markov schemes that, in some particular instances, can be dealt as the sMp case.

4.1. Open Markov Chain Modeling in Discrete Time: A Short Review

We now detail and comment the results that will be used in this paper on discrete time open Markov chains. The study of open Markov chain models we will present next relies on results and notations that were introduced in [56], further developed in [22] and that we reproduce next, for the readers convenience. We will suppose that, in general, the transition matrix of the Markov chain model may be written in the following form:

$$\mathbf{P} = \begin{bmatrix} \mathbf{K} & \mathbf{U}_1 \\ \mathbf{0} & \mathbf{V} \end{bmatrix} \tag{16}$$

where $\mathbf{K}$ is a $k \times k$ transition matrix between transient states, $\mathbf{U}_1$ a $k \times (r-k)$ matrix of transitions between the transient and the recurrent states, and $\mathbf{V}$ a $(r-k) \times (r-k)$ matrix of transitions between the recurrent states. A straightforward computation then shows that

$$\mathbf{P}^{(n)} = \begin{bmatrix} \mathbf{K}^{(n)} & \mathbf{U}_n \\ \mathbf{0} & \mathbf{V}^{(n)} \end{bmatrix}, \quad n \in \mathbb{N}$$

with $\mathbf{U}_n = \mathbf{U}_{n-1}\mathbf{V} + \mathbf{K}^{(n-1)}\mathbf{U}_1 = \sum_{i=0}^{n-1} \mathbf{K}^{(i)}\, \mathbf{U}_1\, \mathbf{V}^{(n-1-i)}$. We write the vector of the initial classification, for a time period i, as

$$\mathbf{c}_i^{\mathsf{T}} = \left[\mathbf{t}_i^{\mathsf{T}} | \mathbf{r}_i^{\mathsf{T}}\right], \quad i \in \mathbb{N} \tag{17}$$

with $\mathbf{t}_i$ the vector of the initial allocation probabilities for the transient states and $\mathbf{r}_i$ the vector of the initial allocation probabilities for the recurrent states. We suppose that at each epoch $i \geq 0$ there is an influx of new elements in the classes of the population—population that has its evolution governed by the Markov chain transition matrix—that is, a Poisson distributed with parameter λ_i. It is a consequence of the *randomized sampling* principle (see [57], pp. 216–217) that, if the incoming populations are distributed by the classes according with the multinomial distribution, then the sub-populations in the transient classes have independent Poisson distributions, with parameters given by the product of the Poisson parameter by the probability of the incoming new member being affected to the given class. With Formulas (16) and (17), we now notice that the vector of the Poisson parameters, for the population sizes in each state at an integer time N, may be written as

$$\boldsymbol{\lambda}_N^{++\mathsf{T}} = \left[\sum_{i=1}^{N} \lambda_i \mathbf{t}_i^{\mathsf{T}} \mathbf{K}^{(N-i)} \middle| \sum_{i=1}^{N} \lambda_i \left(\mathbf{t}_i^{\mathsf{T}} \mathbf{U}_{N-i} + \mathbf{r}_i^{\mathsf{T}} \mathbf{V}^{(N-i)} \right) \right]. \tag{18}$$

We observe that the first block corresponds to the transient states and the second block, the one in the right-hand side, corresponds to the recurrent states. From now on, as a first restricting hypothesis, we will also suppose that the transition matrix of the transient states, $\mathbf{K}$, is diagonalizable and so

$$\mathbf{K} = \sum_{j=1}^{k} \eta_j \boldsymbol{\alpha}_j \boldsymbol{\beta}_j^{\mathsf{T}},$$

with $(\eta_j)_{j \in \{1,\ldots,k\}}$ the eigenvalues, $(\boldsymbol{\alpha}_j)_{j \in \{1,\ldots,k\}}$ the left eigenvectors and $(\boldsymbol{\beta}_j)_{j \in \{1,\ldots,k\}}$ the right eigenvectors of matrix $\mathbf{K}$. We observe that $j \in \{1,\ldots,k\}$ corresponds to a transient state if and only if $\mid \eta_j \mid < 1$. We may write the powers of $\mathbf{K}$ as

$$\mathbf{K}^{(n)} = \sum_{j=1}^{k} \eta_j^n \boldsymbol{\alpha}_j \boldsymbol{\beta}_j^{\mathsf{T}}, \tag{19}$$

and so, as a consequence of (18), for the vector of the Poisson parameters corresponding only to the transient states, $\boldsymbol{\lambda}_N^{+\mathsf{T}}$, we have

$$\boldsymbol{\lambda}_N^{+\mathsf{T}} = \sum_{i=1}^{N} \lambda_i\, \mathbf{t}_i^{\mathsf{T}}\, \mathbf{K}^{(N-i)} = \sum_{j=1}^{k} \sum_{i=1}^{N} \lambda_i\, \eta_j^{N-i}\, \mathbf{t}_i^{\mathsf{T}}\, \boldsymbol{\alpha}_j \boldsymbol{\beta}_j^{\mathsf{T}}. \tag{20}$$

The main result describing the asymptotic behaviour, established in [22], is the following.

Theorem 4 (Asymptotic behavior of Poisson parameters of an open Markov chain with Poisson distributed influxes)**.** *Let a Markov chain driven system have a diagonalizable transition matrix between the transient states* $\mathbf{K} = \sum_{j=1}^{k} \eta_j \boldsymbol{\alpha}_j \boldsymbol{\beta}_j^{\mathsf{T}}$, *written in its spectral decomposition form. Suppose the system to be fed by Poisson inputs with intensities* $(\lambda_i)_{i \in \mathbb{N}}$ *and such that the* *vector of initial classification of the inputs in the transient states converges to a fixed value, that is,* $\lim_{i\to+\infty} \mathbf{t}_i^{\mathsf{T}} = \mathbf{t}_\infty^{\mathsf{T}} \neq \mathbf{0}$. *Then, with* $\boldsymbol{\lambda}_n^{+\mathsf{T}}$ *the vector of Poisson parameters of the transient sub-populations, at date* $n \in \mathbb{N}$, *we have the following:*

1. *If* $\lim_{n\to+\infty} \lambda_n = \lambda \in \mathbb{R}_+$, *then*

$$\boldsymbol{\lambda}_\infty^{+} = \lim_{n\to+\infty} \boldsymbol{\lambda}_n^{+\mathsf{T}} = \sum_{j=1}^{k} \frac{\lambda}{1-\eta_j} \mathbf{t}_\infty^{\mathsf{T}} \boldsymbol{\alpha}_j \boldsymbol{\beta}_j^{\mathsf{T}}. \tag{21}$$

2. *If* $\lim_{n\to+\infty} \lambda_n = +\infty$ *and there exists a constant* $C > 0$ *such that*

$$\max_{1\leq i\leq n} \left| \frac{\lambda_i - \lambda_{i+1}}{\lambda_n} \right| \leq C$$

then

$$\lim_{n\to+\infty} \frac{\boldsymbol{\lambda}_n^{+\mathsf{T}}}{\lambda_n} = \sum_{j=1}^{k} \frac{1}{1-\eta_j} \mathbf{t}_\infty^{\mathsf{T}} \boldsymbol{\alpha}_j \boldsymbol{\beta}_j^{\mathsf{T}}. \tag{22}$$

Remark 10. *We observe that proportions in the Markov chain transient classes, on both statements of the Theorem 4, only depend on the eigenvalues* $\eta_j, j = 1, \dots, k$. *In fact, whenever using Formula (21) to compute proportions these proportions do not depend on the value of* λ *as we have that*

$$\sum_{j=1}^{k} \frac{\lambda}{1-\eta_j} \mathbf{t}_\infty^{\mathsf{T}} \boldsymbol{\alpha}_j \boldsymbol{\beta}_j^{\mathsf{T}} = \lambda \left[\mathbf{t}_\infty^{\mathsf{T}} \cdot \left(\sum_{j=1}^{k} \frac{1}{1-\eta_j} \boldsymbol{\alpha}_j \boldsymbol{\beta}_j^{\mathsf{T}} \right) \right],$$

and the term in the right-hand side multiplying λ *is a vector with the dimension equal to the number of transient classes* k, *which is equal to the dimension of the square matrix* $\mathbf{K}$. *As so, when computing proportions, by normalizing this vector with the sum of its components,* $\lambda \neq 0$ *disappears.*

4.2. Open sMP from Discrete time Open Markov Chains

Let us suppose that the successive Poisson distributions of the influx of new members in the population are independent of the random time at which the influx of new members in the population occurs. For the notations used, see Appendix B. Consider a sMp given by the representation in Formula (A17), that is,

$$Y_t = \sum_{n=0}^{+\infty} Z_n \mathbb{1}_{[\tau_n, \tau_{n+1}[}(t),$$

in which $(Z_n)_{n\geq 0}$ is the embedded Markov chain and $(\tau_n)_{n\geq 0}$ are the jump times of the process. We now propose a method to extend the known method to study open Markov chains in discrete time to sMps.

(1) In applications we usually consider that we have the influx of new members in the population being modeled by Poisson random variables that at each time t has a parameter $\lambda(t)$. Being so, Formula (20) may be rewritten as

$$\boldsymbol{\lambda}_N^{+\mathsf{T}} = \sum_{i=1}^{i:t_i\leq N} \lambda(t_i)\, \mathbf{t}_i^{\mathsf{T}}\, \mathbf{K}^{(N-i)} = \sum_{j=1}^{k} \sum_{i=1}^{i:t_i\leq N} \lambda(t_i)\, \eta_j^{N-i}\, \mathbf{t}_i^{\mathsf{T}}\, \boldsymbol{\alpha}_j \boldsymbol{\beta}_j^{\mathsf{T}}, \tag{23}$$

where usually we can take $t_i = i$, as in a discrete time Markov chain, the actual time stamp is irrelevant as we only consider the sequence of epochs $i \geq 0$.

(2) In a sMp the only difference we have with respect to a discrete time Markov chain is that the dates τ_i corresponding to each epoch i are random; altogether, the structure of the changes in the sub-populations in the transient states is governed by the transition matrix of the Markov chain. In a sMp, the only possible observable changes are those that occur at the random times where it jumps; as so, we will suppose that **the**

influxes of the new members of the population only occur at these random times. As a consequence, we should have that the vector parameter of the Poisson parameters, in the transient classes, is random since it depends on the random times in each we consider influxes and so, Formula (23) becomes

$$\boldsymbol{\lambda}_N^{+\top}(\omega) = \sum_{i=1}^{i:\tau_i(\omega)\le N} \lambda(\tau_i(\omega))\, \mathbf{t}_i^\top\, \mathbf{K}^{(N-i)} = \sum_{j=1}^{k} \sum_{i=1}^{i:\tau_i(\omega)\le N} \lambda(\tau_i(\omega))\, \eta_j^{N-i}\, \mathbf{t}_i^\top\, \boldsymbol{\alpha}_j \boldsymbol{\beta}_j^\top . \quad (24)$$

(3) The parameters of interest will be the expected values of the random variables $\boldsymbol{\lambda}_N^{+\top}(\omega)$—with the correspondent asymptotic behavior of these expected values when N grows indefinitely—and these expected values can be computed whenever the joint laws of $(\tau_0, \tau_1, \dots, \tau_i)$ are known, for $i \ge 0$. In fact, we observe that by Formula (24) we have

$$\mathbb{E}\left[\boldsymbol{\lambda}_N^{+\top} \,|\, \tau_1, \dots \tau_i \dots\right] = \mathbb{E}\left[\sum_{j=1}^{k} \sum_{i=1}^{i:\tau_i\le N} \lambda(\tau_i)\, \eta_j^{N-i}\, \mathbf{t}_i^\top\, \boldsymbol{\alpha}_j \boldsymbol{\beta}_j^\top \,|\, \tau_1, \dots \tau_i \dots\right] =$$
$$= \sum_{j=1}^{k} \sum_{i=1}^{i:\tau_i\le N} \lambda(\tau_i)\, \eta_j^{N-i}\, \mathbf{t}_i^\top\, \boldsymbol{\alpha}_j \boldsymbol{\beta}_j^\top .$$

This formula has two consequences. The first one is that given an arbitrary strictly increasing sequence of dates $0 = t_0 < t_1 < \dots < t_i < \dots$ we have

$$\mathbb{E}\left[\boldsymbol{\lambda}_N^{+\top} \,|\, \tau_1 = t_1, \dots \tau_i = t_i \dots\right] = \sum_{j=1}^{k} \sum_{i=1}^{i:t_i\le N} \lambda(t_i)\, \eta_j^{N-i}\, \mathbf{t}_i^\top\, \boldsymbol{\alpha}_j \boldsymbol{\beta}_j^\top ,$$

thus justifying the assumption that given the strictly increasing of non accumulating stopping times dates $(\tau_1 = t_1, \dots \tau_i = t_i \dots)$ we can proceed as with the usual open Markov chain model in discrete time. The second consequence deserving mention is that in order to compute the expected value of the vector parameters of the transient classes sub-populations, while preserving the Poisson distribution of the influx new members, we compute

$$\mathbb{E}\left[\boldsymbol{\lambda}_N^{+\top}\right] = \mathbb{E}\left[\mathbb{E}\left[\boldsymbol{\lambda}_N^{+\top} \,|\, \tau_1, \dots \tau_i \dots\right]\right] = \mathbb{E}\left[\sum_{j=1}^{k} \sum_{i=1}^{i:\tau_i\le N} \lambda(\tau_i)\, \eta_j^{N-i}\, \mathbf{t}_i^\top\, \boldsymbol{\alpha}_j \boldsymbol{\beta}_j^\top\right],$$

using the joint laws of $(\tau_1, \dots, \tau_i)$ for $i \ge 0$, laws we will suppose to be given.

Theorem 6, in the following, is one possible extension of the open Markov chain formalism to the sMp case taking as a starting point a discrete time Markov chain. To prove this result we will need Theorem 5—a generalization of Lebesgue dominated convergence theorem with varying measures—that we quote from Theorem 3.5 in [58] (p. 390).

Theorem 5 (Lebesgue dominated convergence theorem with varying measures). *Consider $(\mathbf{X}, \mathcal{B}(\mathbf{X}))$ a locally compact, separable topological space endowed with its Borel σ-algebra. Suppose that the sequence of probability measures $(\mu_n)_{n\ge1}$—each one of them defined in $(\mathbf{X}, \mathcal{B}(\mathbf{X}))$—converges weakly to $\boldsymbol{\mu}$ on $(\mathbf{X}, \mathcal{B}(\mathbf{X}))$ and that the sequence of measurable functions $(f_n)_{n\ge1}$ converges continuously to f. Suppose additionally that, for some sequence of measurable functions $(f_n)_{n\ge1}$ defined on $\mathbf{X}$:*

1. *For all $\mathbf{t} \in \mathbf{X}$ and $n \ge 1$, we have that $|f_n(\mathbf{t})| \le g_n(\mathbf{t})$.*
2. *With the function g defined on $\mathbf{X}$ by*

$$g(\mathbf{t}) := \inf_{(t_n)_{n\ge1},\, \lim_{n\to+\infty} t_n = t} \left\{ \liminf_{n\to+\infty} g_n(t_n) \right\}$$

we have that

$$\limsup_{n\to+\infty}\int g_n(t)d\mu_n(t) \leq \int g(t)d\mu(t) < +\infty.$$

Then, we have

$$\lim_{n\to+\infty}\int f_n(t)d\mu_n(t) = \int f(t)d\mu(t) < +\infty.$$

As said, we will suppose that we only observe the influx of the new members of the population into the sMp classes at the random times where it jumps—but, of course, accounting the state before the jump and the state after the jump—which is a hypothesis that makes sense under the perspective that we usually observe trajectories of the process. We then have the following extension of Theorem 4 to the case of sMp.

Theorem 6 (On the stability of open sMp transient states). *Let a sMp given by the representation in Formula (A17), that is,*

$$Y_t = \sum_{n=0}^{+\infty} Z_n \mathbb{I}_{[\tau_n,\tau_{n+1}[}(t),$$

*in which $(Z_n)_{n\geq 0}$ is the embedded Markov chain and $(\tau_i)_{i\geq 0}$ are the jump times of the process. For the embedded Markov chain $(Z_n)_{n\geq 0}$, consider the notations of Section 4.2 and of Theorem 4 in this subsection. Suppose that the influx of new members in the population is modeled by Poisson random variables that at each time $t \in [0,+\infty[$ have a parameter $\lambda(t)$, with λ a **continuous** function. Suppose, furthermore, that the following hypothesis are verified.*

1. *The stopping times $(\tau_i)_{i\geq 0}$ are integrable, that is, $\mathbb{E}[\tau_i] < +\infty$ for all $i \geq 1$.*
2. *There exists $\lambda_\infty > 0$ such that, for every sequence of positive real numbers $(t_i)_{i\geq 1}$ such that $\lim_{i\to+\infty} t_i = +\infty$ we have*
$$\lim_{i\to+\infty} \lambda(t_i) = \lambda_\infty \tag{25}$$

Then, we have that the asymptotic behavior of the expected value vector of parameters of Poisson distributed sub-populations in the transient classes of an open sMp, submitted to a Poisson influx of new members at the jump times of the sMp, is given by

$$\lim_{N\to+\infty} \mathbb{E}\left[\boldsymbol{\lambda}_N^{+\mathsf{T}}\right] = \lim_{N\to+\infty} \mathbb{E}\left[\sum_{j=1}^{k}\sum_{i=1}^{i:\tau_i\leq N} \lambda(\tau_i)\,\eta_j^{N-i}\,\mathbf{t}_i^{\mathsf{T}}\,\boldsymbol{\alpha}_j\boldsymbol{\beta}_j^{\mathsf{T}}\right] = \sum_{j=1}^{k}\frac{\lambda_\infty}{1-\eta_j}\mathbf{t}_\infty^{\mathsf{T}}\boldsymbol{\alpha}_j\boldsymbol{\beta}_j^{\mathsf{T}}. \tag{26}$$

Proof. For each $n \geq 1$, let $F_{(\tau_1,\ldots,\tau_n)}$ be the joint distribution function of $(\tau_1,\ldots,\tau_n)$. We want to compute the following limit of expectations:

$$\begin{aligned}
\lim_{N\to+\infty} \mathbb{E}\left[\boldsymbol{\lambda}_N^{+\mathsf{T}}\right] &= \lim_{N\to+\infty} \mathbb{E}\left[\boldsymbol{\lambda}_N^{+\mathsf{T}},\ \tau_1<\cdots<\tau_i\leq N\right] = \\
&= \lim_{N\to+\infty}\int_{0<t_1<\cdots<t_i\leq N} \boldsymbol{\lambda}_N^{+\mathsf{T}} dF_{(\tau_1,\ldots,\tau_n)}(t_1,\ldots,t_n) = \\
&= \lim_{N\to+\infty}\int_{0<t_1<\cdots<t_i\leq N}\left(\sum_{j=1}^{k}\sum_{i=1}^{i:t_i\leq N}\lambda(t_i)\,\eta_j^{N-i}\,\mathbf{t}_i^{\mathsf{T}}\,\boldsymbol{\alpha}_j\boldsymbol{\beta}_j^{\mathsf{T}}\right)dF_{(\tau_1,\ldots,\tau_n)}(t_1,\ldots,t_n),
\end{aligned} \tag{27}$$

and we observe that by Theorem 4 and by the first hypothesis, for every sequence of positive real numbers $(t_i)_{i\geq 1}$ such that $\lim_{i\to+\infty} t_i = +\infty$ and $t_1 < t_2 < \cdots < t_i < \ldots$, we have that

$$\lim_{N\to+\infty}\left(\sum_{j=1}^{k}\sum_{i=1}^{i:t_i\leq N}\lambda(t_i)\,\eta_j^{N-i}\,\mathbf{t}_i^{\mathsf{T}}\,\boldsymbol{\alpha}_j\boldsymbol{\beta}_j^{\mathsf{T}}\right) = \sum_{j=1}^{k}\frac{\lambda_\infty}{1-\eta_j}\mathbf{t}_\infty^{\mathsf{T}}\boldsymbol{\alpha}_j\boldsymbol{\beta}_j^{\mathsf{T}}. \tag{28}$$

The limit in the last term of Formula (27) requires a result of Lebesgue convergence theorem type but with varying measures. For the purpose of applying Theorem 5, we introduce the adequate context and notations and then we will apply the referred theorem.

Consider the space $X = [0, +\infty[^{\aleph_0}$ defined to be the space of infinite sequences of numbers in $[0, +\infty[$, that is,

$$X = \{t = (t_1, \dots, t_i, \dots) : \forall i \geq 1, t_i \in [0, +\infty[\}.$$

Recall that with the metric d given by

$$\forall t = (t_1, \dots, t_i, \dots), t' = (t'_1, \dots, t'_i, \dots) \in X,\ d(t, t') := \sum_{i=1}^{+\infty} \frac{\min(1, |t_i - t'_1|)}{2^i},$$

X is a metric space, locally compact, separable and complete (see, for instance, in [59], pp. 9–10). We will consider $X = [0, +\infty[^{\aleph_0}$ endowed with the Borel σ-algebra $\mathcal{B}(X)$ generated by the family $\mathcal{P}_f$ given by

$$\mathcal{P}_f = \left\{ A_{i_1} \times A_{i_2} \times \cdots \times A_{i_p} \ : \ p \geq 1,\ A_{i_1} \in \mathcal{B}([0, +\infty[) \right\},$$

with $\mathcal{B}([0, +\infty[)$ the Borel σ-algebra of $[0, +\infty[$. We now take $(\tau_i)_{i \geq 0}$ the sequence of the jump times of the process represented in Formula (A17). First, we define the sequence of measures $(\mu_n)_{n \geq 1}$ where for each $n \geq 1$ we have that μ_n is defined on the measurable space $([0, +\infty[^n, \mathcal{B}([0, +\infty[^n))$ by considering, for $A_1 \times A_2 \times \cdots A_n$ with $A_i \in \mathcal{B}([0, +\infty[)$, that

$$\mu_n(A_1 \times A_2 \times \cdots A_n) = \mathbb{P}[\tau_1 \in A_1, \dots, \tau_n \in A_n] = \int_{t_1 \in A_1, \dots, t_n \in A_n} dF_{(\tau_1, \dots, \tau_n)}(t_1, \dots, t_n). \quad (29)$$

Being so, μ_n is the probability joint law of $(\tau_1, \dots, \tau_n)$ and the last integral in the last term of Formula (27) is exactly an integration with respect to the measure μ_n. As a consequence of Formula (29), the sequence $(\mu_n)_{n \geq 1}$ verifies the compatibility conditions of Kolmogorov extension theorem (see [60], p. 46) and so there is a probability measure $\boldsymbol{\mu}$, defined on $(X, \mathcal{B}(X))$, having as finite dimensional distributions the measures of the sequence $(\mu_n)_{n \geq 1}$.

Now, for each $n \geq 1$, we can consider $\widetilde{\mu}_n$ the extension of μ_n to the measurable space $(X, \mathcal{B}(X))$ in the following way:

$$\forall A \in \mathcal{B}(X)\ \ \widetilde{\mu}_n(A) = \int_{\{t = (t_1, \dots, t_i, \dots) \in A \,:\, t_1, \dots, t_n \in [0, +\infty[\}} dF_{(\tau_1, \dots, \tau_n)}(t_1, \dots, t_n). \quad (30)$$

In fact, with this definition the restriction of $\widetilde{\mu}_n$ to $\mathcal{B}([0, +\infty[^n)$ is exactly μ_n. An important observation is the following. Consider $A := A_{i_1} \times A_{i_2} \times \cdots \times A_{i_p} \in \mathcal{P}_f$. Then, for $m \geq i_p$ we have that

$$\begin{aligned} \widetilde{\mu}_m(A) &= \int_{\{t = (t_1, \dots, t_i, \dots) \in A \,:\, t_1, \dots, t_m \in [0, +\infty[\}} dF_{(\tau_1, \dots, \tau_m)}(t_1, \dots, t_m) = \\ &= \int_{\{t = (t_1, \dots, t_i, \dots) \in A \,:\, t_1, \dots, t_{i_p} \in [0, +\infty[\}} dF_{(\tau_1, \dots, \tau_{i_p})}(t_1, \dots, t_{i_p}) = \\ &= \widetilde{\mu}_{i_p}(A) = \mu_{i_p}(A) = \boldsymbol{\mu}(A), \end{aligned} \quad (31)$$

thus showing that for every $A \in \mathcal{P}_f$ the sequence $(\widetilde{\mu}_m(A))_{m \geq 1}$ converges to $\boldsymbol{\mu}(A)$. Now, by Theorem 2.2 in [59] (p. 17), as $\mathcal{P}_f$ is a π-system and every open set in the metric space (X, d) is a countable union of elements of $\mathcal{P}_f$, we have that the sequence $(\widetilde{\mu}_m)_{m \geq 1}$ converges weakly to $\boldsymbol{\mu}$. In order to apply Theorem 5 to compute the limit, we may consider two approaches to deal with the fact that $\Lambda_N^{+\top}$ is a vector of finite dimension k. Either we proceed component wise or we consider norms. Let us follow the second path. Define, for integer N, and some constant M,

$$f_N(\boldsymbol{t}) = f_N(t_1,\ldots,t_i,\ldots) := \sum_{i=1}^{i:t_i\leq N} \lambda(t_i)\, \eta_j^{N-i}\, \mathbf{t}_i^{\mathsf{T}}\, \boldsymbol{\alpha}_j \boldsymbol{\beta}_j^{\mathsf{T}},$$

and also,

$$g_N(\boldsymbol{t}) \equiv g := \left\| \sum_{j=1}^{k} \frac{\lambda_\infty}{1-\eta_j} \mathbf{t}_\infty^{\mathsf{T}} \boldsymbol{\alpha}_j \boldsymbol{\beta}_j^{\mathsf{T}} \right\| + M,$$

in such a way that $\|f_N(\boldsymbol{t})\| \leq g$; such choice of M is possible as a consequence of Formula (28). We can verify that the sequence $(f_N)_{N\geq 1}$ converges continuously to a function f by using Theorem 4.1.1 in [22] (p. 373). In fact, let us consider a sequence $(\boldsymbol{t}_N)_{n\geq 1}$ converging to some $(\boldsymbol{t}^\infty = (t_1^\infty,\ldots,t_i^\infty,\ldots)$ in the metric space $(\mathbf{X},d)$. With $(\boldsymbol{t}_N = (t_1^N,\ldots,t_i^N,\ldots)$ we surely have that $\lim_{N\to+\infty} t_i^N = t_i^\infty$ for all $i \geq 1$. As a consequence of the continuity of λ and of Theorem 4.1.1 in [22] (p. 373), we have that

$$\lim_{N\to+\infty} f_N(\boldsymbol{t}_N) = \lim_{N\to+\infty} \sum_{i=1}^{i:t_i^N\leq N} \lambda(t_i^N)\, \eta_j^{N-i}\, \mathbf{t}_i^{\mathsf{T}}\, \boldsymbol{\alpha}_j \boldsymbol{\beta}_j^{\mathsf{T}} = \sum_{j=1}^{k} \frac{\lambda(\lim_{i\to+\infty} t_i^\infty)}{1-\eta_j} \mathbf{t}_\infty^{\mathsf{T}} \boldsymbol{\alpha}_j \boldsymbol{\beta}_j^{\mathsf{T}} =: f(\boldsymbol{t}^\infty).$$

It is clear now that the sequences $(f_N)_{N\geq 1}$, $(g_N)_{t\geq 1}$ and $(\widetilde{\mu}_n)_{n\geq 1}$ satisfy together with $\boldsymbol{\mu}$ the hypothesis of Theorem 5 and so the announced result in Formula (25) follows. □

Remark 11 (Alternative proof for the weak convergence of the sequence $(\widetilde{\mu}_n)_{n\geq 1}$)**.** *There is another proof the weak convergence of the sequence $(\widetilde{\mu}_m)_{m\geq 1}$ to $\boldsymbol{\mu}$ that we now present. We proceed by showing that the sequence $(\widetilde{\mu}_n)_{n\geq 1}$ is relatively compact—as a consequence of Prohorov theorem (see [59], pp. 59–63)—because, as we will show next, this sequence is tight. Let an arbitrary $0 < \epsilon < 1$ be given and consider a sequence of positive numbers $(\xi_i)_{i\geq 1}$ such that, by Tchebychev inequality and using the fact that the stopping times τ_i have finite integrals,*

$$\mathbb{P}[\tau_i > \xi_i] \leq \frac{\mathbb{E}[\tau_i]}{\xi_i},$$

in such a way that

$$\sum_{i=1}^{+\infty} \frac{\mathbb{E}[\tau_i]}{\xi_i} < \epsilon.$$

Now consider the Borel set $K_\epsilon = \prod_{i=1}^{+\infty}[0,\xi_i] \subset \mathbf{X}$ which is compact by Tychonov theorem. We now have that

$$\begin{aligned}
\widetilde{\mu}_n(K_\epsilon) &= \int_{\{\boldsymbol{t}=(t_1,\ldots,t_i,\ldots)\in K_\epsilon : t_1,\ldots,t_n\in[0,+\infty[\}} dF_{(\tau_1,\ldots,\tau_n)}(t_1,\ldots,t_n) = \\
&= \int_{\prod_{i=1}^{n}[0,\xi_i]} dF_{(\tau_1,\ldots,\tau_n)}(t_1,\ldots,t_n) = \\
&= \mathbb{P}\left[(\tau_1,\ldots,\tau_n) \in \prod_{i=1}^{n}[0,\xi_i]\right] = \mathbb{P}\left[\bigcap_{i=1}^{n}\{\tau_i \leq \xi_i\}\right] = 1 - \mathbb{P}\left[\bigcup_{i=1}^{n}\{\tau_i > \xi_i\}\right] \geq \\
&\geq 1 - \sum_{i=1}^{n} \frac{\mathbb{E}[\tau_i]}{\xi_i} \geq 1 - \sum_{i=1}^{+\infty} \frac{\mathbb{E}[\tau_i]}{\xi_i} \geq 1 - \epsilon,
\end{aligned}$$

thus showing that the sequence of probability measures $(\widetilde{\mu}_n)_{n\geq 1}$ is tight in the measurable space $(\mathbf{X},\mathcal{B}(\mathbf{X}))$. As said, by Prokhorov's theorem, this implies that the sequence $(\widetilde{\mu}_n)_{n\geq 1}$ is relatively compact, that is, for every subsequence of $(\widetilde{\mu}_n)_{n\geq 1}$, there exists a further subsequence and a probability measure such that this subsequence converges weakly to the said probability measure. Now, as, by construction, the probability measure $\boldsymbol{\mu}$ has, as finite dimensional distributions the *probability measures $(\widetilde{\mu}_n)_{n\geq 1}$ we can say that for $n \geq 1$, the finite dimensional distributions of $\widetilde{\mu}_n$ converge weakly to the finite dimensional distributions of $\boldsymbol{\mu}$. As a consequence, following the observation in [59] (p. 58), the sequence $(\widetilde{\mu}_n)_{n\geq 1}$ converges weakly to $\boldsymbol{\mu}$.*

Remark 12 (Applying Theorem 6)**.** *If we manage to estimate a discrete time Markov chain transition matrix and if we manage to fit some function f—such that $\lim_{t\to+\infty} f(t) = \lambda_\infty$—to the number of new incoming members in the population at a set of non accumulating non-evenly spaced dates (as done with a statistical procedure in [22] or, with a simple fitting in [25]) then, Theorem 6 allows us to get the asymptotic expected number of elements in the transient classes of a sMp having as embedded Markov chain the estimated one.*

4.3. Open Continuous Time Processes from Open Markov Schemes

We may follow the approach of open Markov schemes in [26] and define a process in continuous time after getting a process in random discrete times describing, at least on average, the evolution of the elements in each transient class. Let us briefly recall the main idea. A population model is driven by a Markov chain defined by a sequence of initial distributions given, for $n \geq 1$, by $(\mathbf{q}^n)^\intercal = (q_1^n, q_2^n, \ldots, q_{r_\star}^n)$ and a transition matrix $\mathbf{P} = [p_{ij}], 1 \leq i,j \leq r$. After the first transition, the new values of the proportions in all states, after one transition, can be recovered from $\mathbf{P}^\intercal\mathbf{q} = (\mathbf{q}^\intercal\mathbf{P})^\intercal$ and, after n transitions, by $(\mathbf{P}^{(n)})^\intercal\mathbf{q} = (\mathbf{q}^\intercal\mathbf{P}^{(n)})^\intercal$. We want to account for the evolution of the **expected** number of elements in each class supposing that, at each **random** date τ_k, a random number X_{τ_k} of new elements enters the population. Just after the second cohort enters the population, a first transition occurs in the first cohort driven by the Markov chain law and so on and so forth. Table 1 summarizes this accounting process in which, at each step k, we distribute multinomially the new random arrivals X_{τ_k} according to the probability vector $\mathbf{q}^k$ and the elements in each class are redistributed according to the Markov chain transition matrix $\mathbf{P}$.

Table 1. Accounting of n Markov cohorts each with an initial distribution.

Date	τ_1	τ_2	...	τ_{n-1}	τ_n
τ_1	$\mathbb{E}[X_{\tau_1}](\mathbf{q}^1)^\intercal$	$\mathbb{E}[X_{\tau_1}](\mathbf{q}^1)^\intercal\mathbf{P}$	...	$\mathbb{E}[X_{\tau_1}](\mathbf{q}^1)^\intercal\mathbf{P}^{(n-2)}$	$\mathbb{E}[X_{\tau_1}](\mathbf{q}^1)^\intercal\mathbf{P}^{(n-1)}$
τ_2	–	$\mathbb{E}[X_{\tau_2}](\mathbf{q}^2)^\intercal$	...	$\mathbb{E}[X_{\tau_2}](\mathbf{q}^2)^\intercal\mathbf{P}^{(n-3)}$	$\mathbb{E}[X_{\tau_2}](\mathbf{q}^2)^\intercal\mathbf{P}^{(n-2)}$
...	...	...	...	...	...
τ_n	–	–	–	–	$\mathbb{E}[X_{\tau_n}](\mathbf{q}^n)^\intercal$

At date τ_k, if we suppose that each new set of individuals in the population, a cohort, evolves independently from any one of the already existing sets of individuals but, accordingly, to the same Markov chain model, we may recover the total **expected** number of elements in each class at date τ_k by computing the sum:

$$\overline{\mathbf{K}}_n = \sum_{k=1}^{n} \mathbb{E}[X_{\tau_k}](\mathbf{q}^k)^\intercal\ \mathbf{P}^{(n-k)}. \tag{32}$$

Each vector component corresponds precisely to the **expected** number of elements in each class. In order to further study the properties of $(\overline{\mathbf{K}}_n)_{n\geq 1}$, given the properties of a stochastic process $\mathbb{X} = (X_{\tau_k})_{k\geq 1}$, we will randomize formula (32) by considering, instead, for $n \geq 1$:

$$\mathbf{K}_n = \sum_{k=1}^{n} X_{\tau_k}(\mathbf{q}^k)^\intercal\ \mathbf{P}^{(n-k)}, \tag{33}$$

and we observe that in any case $\mathbb{E}[\mathbf{K}] = \overline{\mathbf{K}}_n$. It is known that if the vector of classification probabilities is constant $\mathbf{c}^k = \mathbf{c}$ and if the $\mathbb{X}$ is an ARMA, ARIMA, or SARIMA process, then the populations in each of the transient classes can be described by a sum of a deterministic trend, plus an ARMA process plus an evanescent process, that is a centered process $(Y_k)_{k\geq 1}$ such that $\lim_{k\to+\infty} \mathbb{E}\left[|Y_k|^2\right] = 0$ (see Theorems 3.1 and 3.2 in [26]).

The step process in continuous time naturally associated with the discrete time one would be then defined by for $t \geq 0$ by

$$\mathbf{K}_t := \sum_{n=0}^{+\infty} \mathbf{K}_n \mathbb{1}_{[\tau_n,\tau_{n+1}[}(t) = \sum_{n=0}^{+\infty} \left(\sum_{k=1}^{n} X_{\tau_k} (\mathbf{q}^k)^\top \, \mathbf{P}^{(n-k)} \right) \mathbb{1}_{[\tau_n,\tau_{n+1}[}(t).$$

In order to study this process we will have to take advantage of the properties of $\mathbb{X}$ and of the family of stopping times $(\tau_k)_{k\geq 0}$. It should be noticed that if the process $\mathbb{X} = (X_t)_{t\geq 0}$ is Poisson distributed and the laws of the sequence $(\tau_k)_{k\geq 0}$ are known and it possible to determine the expected value of $\mathbf{K}_t$ for $t \geq 0$ with a result similar to Theorem 6.

5. Conclusions

In this work, we studied several ways to associate, to an open Markov chain process in discrete time—which is often the sole accessible fruit of observation—a continuous time Markov or semi-Markov process that bears some natural relation with the discrete time process. Furthermore, we expect that association to allow the extension of the study of open populations from the discrete to the continuous time model. For that purpose, we consider three approaches: the first, for the continuous time Markov chains; the second, for the semi Markov case; and the third, for the open Markov schemes (see in [26]). For the semi-Markov case, under the hypothesis that we only observe the influx of new individuals in the population at the times of the random jumps, in the main result we determine the expected value of the vector of parameters of the conditional Poisson distributions in the transient classes when the influx of new members is Poisson distributed. The third approach, dealing with open Markov schemes is similar to the second one whenever we consider a similar context hypothesis, that is, distributed incoming new members of the population with known distributions and observation of this influx of new individuals at the times of the random jumps. In the case of the first approach, that is, for the case of Markov chain in continuous time, we propose a calibration procedure for which the embeddable Markov chains provide optimal solutions. In this case also, the study of open populations models relies on the main result proved for the semi-Markov case approach. Future work encompasses applications to real data and the determination of criteria to assess the quality of the association of the continuous model to the observed discrete time model.

Author Contributions: All authors contributed equally to this work. All authors have read and agreed to the published version of the manuscript.

Funding: For the second author, this work was done under partial financial support of RFBR (Grant n. 19-01-00451). For the first and third author this work was partially supported through the project of the Centro de Matemática e Aplicações, UID/MAT/00297/2020 financed by the Fundação para a Ciência e a Tecnologia (Portuguese Foundation for Science and Technology). The APC was funded by the insurance company Fidelidade.

Acknowledgments: This work was published with finantial support from the insurance company Fidelidade. The authors would like to thank Fidelidade for this generous support and also, for their interest in the development of models for insurance problems in Portugal. The authors express gratitude to Professor Panagiotis C.G. Vassiliou for his enlightening comments on a previous version of this work and to the comments, corrections and questions of the referees, in particular, to the one question that motivated the inclusion of Remark 5.

Conflicts of Interest: The authors declare no conflict of interest.

Appendix A. Some Essential Results on Continuous Time Markov Chains

In this exposition of the most relevant results pertinent to our purposes, we follow mainly the references [29–31]. As this exposition is a mere reminder of needed notions and results, the proofs are omitted unless the result is essential for our purposes.

Definition A1 (Continuous time Markov chain). *Let $\mathcal{I}$ be some **finite** set; for instance, $\Theta = \{\theta_1, \theta_2, \ldots, \theta_r\}$ of Section 2. A stochastic process $(X_t)_{t\geq 0}$ is a **continuous time Markov chain** with state space $\mathcal{I}$ if and only if the following **Markov property** is verified, namely, for all $i_0, i_1, \ldots i_n \in \mathcal{I}$ and $0 = t_0 < t_1 < \cdots < t_n < \cdots$ we have that*

$$\mathbb{P}\left[X_{t_n} = i_n \middle| X_{t_{n-1}} = i_{n-1}, \ldots X_{t_1} = i_1, X_{t_0} = i_0\right] = \\ = \mathbb{P}\left[X_{t_n} = i_n \middle| X_{t_{n-1}} = i_{n-1}\right].$$

We observe that by force of the Markov property in Definition A1 the law of a continuous time Markov chain depends only on the following transition probabilities. Let $\boldsymbol{I}$ be the identity matrix with dimension $\#\mathcal{I}$ the Kronecker's delta be given by

$$\delta_i^j = \begin{cases} 0 & i \neq j \\ 1 & i = j. \end{cases}$$

Definition A2 (Transition probabilities). *Let $\mathcal{I}$ be the state space of $(X_t)_{t\geq 0}$ a continuous time Markov chain. The **transition probabilities** are defined by*

$$\forall i, j \in \mathcal{I},\ s < t,\ p(s,i,t,j) = \mathbb{P}[X_t = j | X_s = i] \text{ and } p(t,i,t,j) = \delta_i^j.$$

*Let $\mathcal{L}(\mathbb{R}^{\#\mathcal{I}})$ be the space of square matrices with coefficients in $\mathbb{R}$. The **transition probability matrix function** $\boldsymbol{P} : \mathbb{R}_+ \times \mathbb{R}_+ \mapsto \mathcal{L}(\mathbb{R}^{\#\mathcal{I}})$ is defined by*

$$\forall i, j \in \mathcal{I},\ s < t,\ \boldsymbol{P}(s,t) = [p(s,i,t,j)]_{i,j\in\mathcal{I}} \text{ and } \boldsymbol{P}(t,t) = \boldsymbol{I}. \tag{A1}$$

Transition probabilities of Markov processes in general satisfy a very important functional equation that results from the Markov property.

Theorem A1 (Chapman-Kolmogorov equations). *Consider a NH-CT-MC as given in Definition A1. Let $\boldsymbol{P}$ its transition probability matrix function as given in Definition A2. We then have*

$$\forall s, u, t,\ 0 \leq s < u < t,\ \boldsymbol{P}(s,t) = \boldsymbol{P}(s,u)\boldsymbol{P}(u,t) \tag{A2}$$

As an application of the celebrated existence theorem of Kolmogorov (in the form exposed in [61], pp. 8–10) we have that, under a set of natural hypothesis, there exists a NH-CT-MC such as the one in Definition A1.

Theorem A2 (On the existence of NH-CT-MC). *Let p_0 be an initial probability over $\mathcal{I}$. Consider a matrix valued function $\boldsymbol{P} : \mathbb{R}_+ \times \mathbb{R}_+ \mapsto \mathcal{L}(\mathbb{R}^{\#\mathcal{I}})$ denoted by $\boldsymbol{P}(s,t) = [p(s,i,t,j)]_{i,j\in\mathcal{I}}$ and satisfying Formulas (A3) and (A4) below, that is,*

1. *For all $s < t$ and for all $i \in \mathcal{I}$*

$$\sum_{j\in\mathcal{I}} p(s,i,t,j) = 1. \tag{A3}$$

2. *Formula (A2) in Theorem A1, namely,*

$$\forall s, u, t,\ s < u < t,\ \boldsymbol{P}(s,t) = \boldsymbol{P}(s,u)\boldsymbol{P}(u,t). \tag{A4}$$

Define, for all $i_0, i_1, \ldots i_n \in \mathcal{I}$ and $0 = t_0 < t_1 < \cdots < t_n < \cdots$, the function

$$\nu_{t_0,t_1,\ldots,t_n}(i_0, i_1, \ldots, i_n) = \\ = p_0(i_0)p(t_0,i_0,t_1,i_1)p(t_1,i_1,t_2,i_2)\cdots p(t_{n-1},i_{n-1},t_n,i_n), \tag{A5}$$

and extend this definition to all possible $t_0, t_1, \ldots, t_n, \ldots$ by considering, with the adequate ordering permutation σ of $\{0, 1, 2, \ldots, \#\mathcal{I}\}$ such that we have $t_{\sigma(0)} < t_{\sigma(1)} < \ldots, < t_{\sigma(n)}$,

$$\nu_{t_{\sigma(0)}, t_{\sigma(1)}, \ldots, t_{\sigma(n)}}(i_0, i_1, \ldots, i_n) = \nu_{t_0, t_1, \ldots t_n}(i_{\sigma^{-1}(0)}, i_{\sigma^{-1}(1)}, \ldots, i_{\sigma^{-1}(n)}). \tag{A6}$$

Then, $(\nu_{t_0, t_1, \ldots, t_n})_{t_0, t_1, \ldots, t_n, n \geq 1}$ is a family of probability measures satisfying the compatibility conditions of Kolmogorov existence theorem and so, there exists $\mathbb{P}$ a probability measure over the canonical probability space $(\Omega, \mathcal{A})$—with $\Omega = \mathcal{I}^{\mathbb{R}_+}$ and $\mathcal{A} = \mathcal{P}(\mathcal{I})^{\mathbb{R}_+}$—such that if the stochastic process $(X_t)_{t \geq 0}$ is denoted by

$$\forall \omega = (i_t)_{t \geq 0} \in \Omega, \; X_t(\omega) = i_t,$$

then,

$$\forall i, j \in \mathcal{I}, \; s < t, \; p(s, i, t, j) = \mathbb{P}[X_t = j | \, X_s = i] \text{ and } p(t, i, t, j) = \delta_i^j, \tag{A7}$$

that is, $(X_t)_{t \geq 0}$ has $\mathbf{P}(s, t) = [p(s, i, t, j)]_{i, j \in \mathcal{I}}$—together with $\mathbf{P}(t, t) = \mathbf{I}$—as its transition probabilities.

A natural and useful way of defining transition probabilities is by means of the transition intensities that act like differential coefficients of transition probability functions.

Definition A3 (Transition intensities). *Let $\mathcal{L}(\mathbb{R}^{\#\mathcal{I}})$ be the space of square matrices with coefficients in $\mathbb{R}$. A function $\mathbf{Q} : \mathbb{R} \mapsto \mathcal{L}(\mathbb{R}^{\#\mathcal{I}})$ denoted by*

$$\mathbf{Q}(t) = [q(t, i, j)]_{i, j \in \mathcal{I}},$$

*is a **transition intensity** iff for almost all $t \geq 0$ it verifies*

(i) $\forall i \in \mathcal{I}, \; t \geq 0, \; q(t, i, i) \leq 0$;
(ii) $\forall i \in \mathcal{I}, \; t \geq 0, \; q(t, i, j) - q(t, i, i) \geq 0$;
(iii) $\forall i \in \mathcal{I} \; \sum_{j \in \mathcal{I}} q(t, i, j) = 0$.

There is a way to write differential equations—the Kolmogorov backward and forward equations—useful for recovering the transition probability matrix from the intensities matrix and to study important properties of these transition probabilities.

Theorem A3 (Backward and Forward Kolmogorov equations). *Suppose that $\mathbf{P}(s, t)$ is continuous at s, that is,*

$$\lim_{t \downarrow 0} \mathbf{P}(0, t) = \mathbf{I} \text{ and } \lim_{t \downarrow s} \mathbf{P}(s, t) = \lim_{t \uparrow s} \mathbf{P}(t, s) = \mathbf{I}. \tag{A8}$$

If there exists $\mathbf{Q}$ such that

$$\begin{aligned} \mathbf{Q}(t) = & \lim_{k+h \to 0_+, k \equiv 0 \vee h \equiv 0} \frac{\mathbf{P}(t-k, t+h) - \mathbf{I}}{k+h} = \lim_{h \downarrow 0, h > 0} \frac{\mathbf{P}(t, t+h) - \mathbf{I}}{h} = \\ = & \lim_{k \downarrow 0, k > 0} \frac{\mathbf{P}(t-k, t) - \mathbf{I}}{k}, \end{aligned} \tag{A9}$$

then we have the backward Kolmogorov (matrix) equation:

$$\frac{\partial}{\partial s} \mathbf{P}(s, t) = -\mathbf{Q}(s) \mathbf{P}(s, t), \; \mathbf{P}(s, s) = \mathbf{I}, \tag{A10}$$

and the forward Kolmogorov (matrix) equation:

$$\frac{\partial}{\partial t} \mathbf{P}(s, t) = \mathbf{P}(s, t) \mathbf{Q}(s), \; \mathbf{P}(t, t) = \mathbf{I}. \tag{A11}$$

Remark A1. *The general theory of Markov processes shows that the condition that* $\mathbf{P}(s,t)$ *is continuous in both s and t is sufficient to ensure the existence of the matrix intensities* $\mathbf{Q}$ *given in Formulas* (A9) *(see [31], p. 232). By means of a change of time Goodman (see [41]) proved that the existence of solutions of Kolmogorov equations is amenable to an application of Caratheodory's existence theorem for differential equations.*

Given transition intensities satisfying an integrability condition there are transition probabilities uniquely associated with these transition intensities.

Theorem A4 (Transition probabilities from intensities). *Let* $\mathbf{Q}$ *be a transition intensity as in Definition A3 such that Theorem A3 holds. Then, we have that*

$$\mathbf{P}(s,t) = \mathbf{I} + \int_s^t \mathbf{Q}(u)\mathbf{P}(u,t)du \text{ and } \mathbf{P}(s,t) = \mathbf{I} + \int_s^t \mathbf{P}(s,u)\mathbf{Q}(u)du. \tag{A12}$$

The existence of a NH-CT-MC can also be guaranteed by a constructive procedure that we now present and that is most useful for simulation.

Remark A2 (Constructive definition). *Given a transition intensity* $\mathbf{Q}$ *define*

$$p^\star(t,i,j) = \begin{cases} \frac{1-\delta_i^j}{-q(t,i,i)} q(t,i,j) & q(t,i,i) \neq 0 \\ \delta_i^j & q(t,i,i) = 0. \end{cases}$$

1. *Let* $X_0 = i$*, according to some initial distribution on* $\mathcal{I}$*; the sequence* $(\tau_n)_{n\geq 0}$ *is defined by induction as follows;* $\tau_0 \equiv 0$.
2. τ_1 *time of first jump with Exponential distribution function:*

$$F_{\tau_1}(t) = \mathbb{P}[\tau_1 \leq t] = 1 - \exp\left(\int_0^t q(u,i,i)du\right),$$

and

$$\mathbb{P}[X_{s_1} = j |\, \tau_1 = s_1,\ X_0 = i] = p^\star(s_1,i,j),$$

and so $X_t = i$ *for* $0 \equiv \tau_0 \leq t < \tau_1$*. We note that this distribution of the stopping time is mandatory as a consequence of a general result on the distribution of sojourn times of a continuous time Markov chain (see Theorem 2.3.15 in [31], p. 221).*

3. *Given that* $\tau_1 = s_1$ *and* $X_{s_1} = j$*,* τ_2 *time of the second jump with Exponential distribution function*

$$F_{\tau_2|\tau_1=s_1}(t) = \mathbb{P}[\tau_2 \leq t \mid \tau_1 = s_1] = 1 - \exp\left(\int_0^t q(u+s_1,j,j)du\right)$$

and

$$\mathbb{P}[X_{s_2} = k |\, \tau_1 = s_1,\ X_0 = i,\ \tau_2 = s_2,\ X_{s_1} = j] = p^\star(s_1+s_2,j,k),$$

and so $X_t = j$ *for* $\tau_1 \leq t < \tau_2$.

The following result ensures that the preceding construction yields the desired result.

Theorem A5 (The continuous time Markov chain). *Let the intensities satisfy condition given by Formula* (A12) *in Theorem A4. Then, given the times* $(\tau_0)_{n\geq 1}$*, we have that with the sequence* $(Y_n)_{n\geq 1}$ *defined by* $Y_n = X_{\tau_n}$*, the process defined by:*

$$X_t = \sum_{n=0}^{+\infty} Y_n \mathbb{1}_{[\tau_n,\tau_{n+1}[}(t) = \sum_{n=0}^{+\infty} X_{\tau_n} \mathbb{1}_{[\tau_n,\tau_{n+1}[}(t) \tag{A13}$$

is a continuous time Markov chain with transition probabilities $\mathbf{P}$ *given by Definition A2 and transition intensities* $\mathbf{Q}$ *given by Definition A3 and Theorem A3.*

Proof. This theorem is stated and proved, in the general case of Markov continuous time Markov processes in [31] (p. 229). □

Lemma A1. *Let* $q : \mathbb{R}_+ \mapsto \mathbb{R}$ *a measurable function integrable over every bounded interval of* $\mathbb{R}_+$*. Then, we have that*

$$\int_s^t \int_{s_1}^t \cdots \int_{s_{n-1}}^t q(s_1)q(s_2)\ldots q(s_n)ds_n \ldots ds_2 ds_1 = \frac{\left(\int_s^t q(u)du\right)^n}{n!},$$

for all $0 \leq s \leq t$*,* $n \geq 1$*.*

Proof. Let us observe that, for $n = 2$, we have that

$$\begin{aligned}\left(\int_s^t q(u)du\right)^2 &= \int_s^t \int_s^t q(v)q(u)dudv = \\ &= \int_s^t \int_s^t \mathbb{I}_{\{u \leq v\}} q(v)q(u)dudv + \int_s^t \int_s^t \mathbb{I}_{\{v \leq u\}} q(v)q(u)dudv.\end{aligned}$$

By induction we have for all $n \geq 1$, and for every permutation $\sigma \in \mathfrak{S}_n$

$$\begin{aligned}&\left(\int_s^t q(u)du\right)^n = \\ &= \sum_{\sigma \in \mathfrak{S}_n} \int_s^t \cdots \int_s^t \mathbb{I}_{\{u_{\sigma(1)} \leq u_{\sigma(2)} \leq \cdots \leq u_{\sigma(n)}\}} q(u_1) \ldots q(u_1) du_n \ldots du_1 = \\ &= n! \int_s^t \cdots \int_s^t \mathbb{I}_{\{u_1 \leq u_2 \leq \cdots \leq u_n\}} q(u_1) \ldots q(u_1) du_n \ldots du_1 = \\ &= \int_s^t \int_{u_1}^t \cdots \int_{u_{n-1}}^t q(u_1)q(u_2) \ldots q(u_n) du_n \ldots du_2 du_1,\end{aligned}$$

as all the integrals in the sum are equal by the symmetry of the integrand function, and then, by Fubini theorem. □

Remark A3 (On a fundamental condition). *The condition on* q *stated in Lemma A1 and reformulated in Formula (7) is the key to the proof of important results. In fact we have that this condition is sufficient to ensure that the associated Markov process has no discontinuities of the second type (see [31], p. 227) and, most important for the goals in this work, that the trajectories of the associated Markov process are step functions, that is, any trajectory has only a finite number of jumps in any compact subinterval of* $[0, +\infty[$*; we will detail this last part of the remark in Theorem A6.*

Under the perspective of our main motivation the following result is crucial.

Theorem A6 (The non accumulation property of the jump times of a Markov chain). *Let the intensities satisfy condition given by the statement of Lemma A1. Then, given the times* $(\tau_n)_{n \geq 1}$*, we have that:*

$$\mathbb{P}\left[\sum_{n=1}^{+\infty} \tau_n = +\infty\right] = 1, \tag{A14}$$

and so the trajectories of the process are step functions.

Proof. Property in Formula (A14) has non immediate proof. We present a proof based on a result in [62] (p. 160), stating that the condition given by:

$$\lim_{h\downarrow 0}\sup_{t,i}\sum_{j\neq i} p(t,i,t+h,j) = 0, \tag{A15}$$

guarantees that the process has a stochastic equivalent that is a step process, meaning that for any trajectory ω the set of jumps of this trajectory has no limit points in the interval $[0,\zeta(\omega)[$, with $\zeta(\omega)$ being the end date of the trajectory. This result is based on a thorough analysis (see [62], pp. 149–159) of the conditions for a Markov process not to have discontinuities of the second type, meaning that the right-hand side and left-hand side limits exists for every date point and every trajectory. Now, with,

$$q(t) := \max_{1\leq i\leq \#\mathcal{I}} |q(t,i,i)|,$$

by virtue of the condition on q in Lemma A1—that is reformulated more precisely in Formula (7) of the statement in Theorem 1—we have that:

$$p(t,i,t+h,j) \leq \sum_{k=1}^{+\infty} \frac{\left(\#\mathcal{I}\ \int_t^{t+h} q(u)du\right)^k}{k!}.$$

Therefore, for almost all $t\in[0,T]$,

$$\lim_{h\downarrow 0}\sup_{t,i}\sum_{j\neq i} p(t,i,t+h,j) = (\#\mathcal{I}-1)\lim_{h\downarrow 0}\sup_{t}\sum_{k=0}^{+\infty} \frac{\left(\#\mathcal{I}\cdot \int_t^{t+h} q(u)du\right)^k}{k!} =$$

$$= (\#\mathcal{I}-1)\lim_{h\downarrow 0}\sup_{t}\sum_{k=0}^{+\infty} \frac{\left(h\cdot\#\mathcal{I}\cdot \frac{1}{h}\int_t^{t+h} q(u)du\right)^k}{k!} = 0,$$

as the series is uniformly convergent and for almost all $t\in[0,T]$,

$$\lim_{h\downarrow 0}\frac{1}{h}\int_t^{t+h} q(u)du = q(t),$$

by Lebesgue's differentiation theorem. □

Remark A4 (Negative properties). *The following negative properties suggest the alternative calibration approach that we propose in Section 3.2. Given* $(X_{\tau_n})_{n\geq 0}$, *the successive states occupied by the process, we observe that*

- *the times* $(\tau_n)_{n\geq 1}$ *are* ***not*** *independent;*
- *the sequence* $(Y_n)_{n\geq 1}$ *defined by* $Y_n = X_{\tau_n}$ *is* ***not*** *a Markov chain.*

Appendix B. Semi-Markov Processes: A Short Review

For the reader's convenience we present a short summary of the most important results semi-Markov processes (sMp), needed in this work, following [63] (pp. 189–200). The main foundational references for the theory of sMp are [32,64,65]. Important developments can be read in [33,66,67]. Among the many works with relevance for applications we refer, for instance, [68–73]. Let us consider a complete probability space $(\Omega,\mathcal{F},\mathbb{P})$. The approach of Markov and semi-Markov processes via kernels if fruitful and so we are lead to the following definitions and results for what we will now follow, mainly, the works in [67] (pp. 7–15) and in [33]. Consider a general measurable state space $(\mathbf{\Theta},\mathcal{A}(\mathbf{\Theta}))$. The σ-algebra $\mathcal{A}(\mathbf{\Theta})$ may be seen as the observable sets of the state space of the process $\mathbf{\Theta}$.

Definition A4 (Semi-Markov transition kernel). *A map $Q : \boldsymbol{\Theta} \times \mathcal{A}(\boldsymbol{\Theta}) \times [0, +\infty[\to [0, 1]$ such that $(x, B, t) \mapsto Q(x, B, t)$ is a **semi-Markov transition kernel** if it satisfies the following properties.*

(i) *$Q(x, \cdot, t)$ is measurable with respect to $\mathcal{A}(\boldsymbol{\Theta}) \times \mathcal{B}([0, +\infty[)$ with $\mathcal{B}([0, +\infty[)$ the Borel σ-algebra of $[0, +\infty[$.*

(ii) *For fixed $t > 0$, $Q(\cdot, \cdot, t) : \boldsymbol{\Theta} \times \mathcal{A}(\boldsymbol{\Theta}) \to [0, 1]$ is a **semistochastic kernel**, that is,*

*(ii.1)For fixed $\boldsymbol{\theta} \in \boldsymbol{\Theta}$ and $t > 0$, the map $Q(\boldsymbol{\theta}, \cdot, t) : \mathcal{A}(\boldsymbol{\Theta}) \to [0, 1]$ is a measure and we have $Q(\boldsymbol{\theta}, \boldsymbol{\Theta}, t) \leq 1$; if $Q(\boldsymbol{\theta}, \boldsymbol{\Theta}, t) = 1$ we have that $Q(\cdot, \cdot, t)$ is a **stochastic kernel**.*

(ii.2)For a fixed $T \in \boldsymbol{\Theta}$ we have that $Q(\cdot, T, t) : \boldsymbol{\Theta} \to [0, 1]$ is measurable with respect to $\mathcal{A}(\boldsymbol{\Theta})$.

(iii) *For fixed $(\boldsymbol{\theta}, T) \in \boldsymbol{\Theta} \times \mathcal{A}(\boldsymbol{\Theta})$ we have that the function $Q(\boldsymbol{\theta}, T, t) : [0, +\infty[\to [0, 1]$ is a nondecreasing function, continuous from the right and such that $Q(\boldsymbol{\theta}, T, 0) = 0$.*

(iv) *$P(\cdot, \cdot) : \boldsymbol{\Theta} \times \mathcal{A}(\boldsymbol{\Theta}) \to [0, 1]$ defined to be: $P(\cdot, \cdot) = Q(\cdot, \cdot, +\infty) = \lim_{t \to +\infty} Q(\cdot, \cdot, t)$ is a stochastic kernel.*

(v) *For any $\boldsymbol{\theta} \in \boldsymbol{\Theta}$ we have that the function defined for $t \in [0, +\infty[$ by $F_{\boldsymbol{\theta}}(t) := Q(\boldsymbol{\theta}, \boldsymbol{\Theta}, t)$ is a probability distribution function.*

Now, consider Q a semi-Markov transition kernel, a continuous time stochastic process $(Y_t)_{t \geq 0}$ defined on this probability space and $\mathbb{F} = (\mathcal{F}_t)_{t \geq 0}$ the natural filtration associated to this process, i.e., $\mathcal{F}_t := \sigma(Y_s : s \leq t)$ is the algebra-σ generated by the variables of the process until time t. We now consider a sequence of random variables $(Z_n)_{n \geq 0}$—taking values in a state space $\boldsymbol{\Theta}$, that for our purposes will, in general, be finite state space $\boldsymbol{\Theta} = \{\boldsymbol{\theta}_1, \boldsymbol{\theta}_2, \ldots, \boldsymbol{\theta}_r\}$ and sometimes an infinite one $\boldsymbol{\Theta} = \{\boldsymbol{\theta}_1, \boldsymbol{\theta}_2, \ldots, \boldsymbol{\theta}_r, \ldots\}$—the sequence being adapted to the filtration $\mathbb{F}$. We consider also $0 \equiv \tau_0 < \tau_1 < \tau_2 < \cdots < \tau_n < \cdots$ an increasing sequence of $\mathbb{F}$-stopping times, denoted by $\mathcal{T}$ and $\Delta_n := \tau_n - \tau_{n-1}$ for $n \geq 1$.

Definition A5 (Markov renewal process). *A two dimensional discrete time process $(Z_n, \Delta_n)_{n \geq 0}$ with state space $\boldsymbol{\Theta} \times [0, +\infty[$ verifying,*

$$\mathbb{P}\left[Z_{n+1} = \boldsymbol{\theta}_j, \Delta_n \leq t \middle| Z_0, \ldots, Z_n, \Delta_1, \Delta_2, \ldots, \Delta_n\right] = \mathbb{P}\left[Z_{n+1} = \boldsymbol{\theta}_j, \Delta_n \leq t \middle| Z_n\right],$$

*for all $\boldsymbol{\theta}_j \in \boldsymbol{\Theta}$, $t \geq 0$ and almost surely that is, an homogeneous two dimensional Markov Chain, is a **Markov renewal process** if its transition probabilities are given by:*

$$Q(\boldsymbol{\theta}, T, t) = \mathbb{P}[Z_{n+1} \in T,\ \Delta_n \leq t | Z_n = \boldsymbol{\theta}].$$

Remark A5 (Markov chains and Markov renewal processes). *The transition probabilities of a Markov renewal process do not depend on the second component; as so, a Markov renewal process is a process of different type of a two dimensional Markov chain process. The first component of a Markov renewal process is a Markov chain, denoted the embedded Markov chain, with transition probabilities given by:*

$$P(\boldsymbol{\theta}, T) = Q(\boldsymbol{\theta}, T, +\infty) = \lim_{t \to +\infty} Q(\boldsymbol{\theta}, T, t) = \mathbb{P}[Z_{n+1} \in T | Z_n = \boldsymbol{\theta}].$$

Definition A6 (Markov renewal times). *The **Markov renewal times** of the Markov renewal process $(\tau_n)_{n \geq 0}$ are defined by*

$$\tau_n = \sum_{k=1}^{n} \Delta_k,$$

and the probability distribution functions $F_{\boldsymbol{\theta}}$ of the Markov renewal times depend on the states of the embedded Markov chain, as, by definition we have

$$F_{\boldsymbol{\theta}}(t) := Q(\boldsymbol{\theta}, \boldsymbol{\Theta}, t) = \mathbb{P}[\Delta_n \leq t | Z_n = \boldsymbol{\theta}].$$

Proposition A1. *Consider a general measurable state space $(\boldsymbol{\Theta}, \mathcal{A}(\boldsymbol{\Theta}))$. Let $\boldsymbol{Q}$ be a semi-Markov transition kernel and $\boldsymbol{P}$ the associated stochastic kernel according to Definition A4. Then, there exists a function $F_{\boldsymbol{\theta}}(\gamma, t)$ such that:*

$$Q(\boldsymbol{\theta}, T, t) = \int_T F_{\boldsymbol{\theta}}(\gamma, t) \boldsymbol{P}(\boldsymbol{\theta}, d\gamma). \tag{A16}$$

Proof. As we have for $\boldsymbol{\theta} \in \boldsymbol{\Theta}$ and $T \in \mathcal{A}(\boldsymbol{\Theta}))$ that $\boldsymbol{P}(\boldsymbol{\theta}, T) = \boldsymbol{Q}(\boldsymbol{\theta}, T, +\infty)$, we may conclude that $\boldsymbol{Q}(\boldsymbol{\theta}, T, +\infty) \leq \boldsymbol{P}(\boldsymbol{\theta}, T)$ and so, the measure $\boldsymbol{Q}(\boldsymbol{\theta}, \cdot, +\infty)$ is absolutely continuous with respect to the probability measure $\boldsymbol{P}(\boldsymbol{\theta}, \cdot)$ on $(\boldsymbol{\Theta}, \mathcal{A}(\boldsymbol{\Theta}))$ and so, by the Radon–Nicodym theorem, there exists a density $F_{\boldsymbol{\theta}}(\gamma, t)$ verifying Formula (A16). □

Remark A6 (Semi-Markov kernel for discrete space state). *In the case of a discrete state space, say $\boldsymbol{\Theta} = \{\boldsymbol{\theta}_1, \boldsymbol{\theta}_2, \dots, \boldsymbol{\theta}_r, \dots\}$, we may consider $\mathcal{A}(\boldsymbol{\Theta}) = \mathcal{P}(\boldsymbol{\Theta})$ the maximal σ-algebra of all the subsets of $\boldsymbol{\Theta}$) and, with this condition, a semi-Markov kernel $\boldsymbol{Q}$ is defined by a matrix function $Q = [q(i,j,t)]_{i,j \geq 1, t \geq 0}$ such that*

(i) *For $i, j \geq 1$ fixed the function $q(i,j,\cdot) : [0, +\infty[\to [0,1]$ is nondecreasing.*
(ii) *For $i \geq 1$ fixed the function $F_i(t) := \sum_{j \geq 1} q(i,j,t)$ is a probability distribution function.*
(iii) *The matrix $\boldsymbol{P} = [p(i,j)]_{i,j \geq 1, t \geq 0}$ with $p(i,j) := q(i,j,+\infty) = \lim_{t \to +\infty} q(i,j,t)$ is a stochastic matrix.*

Definition A7 (Semi-Markov process). *The process $(Y_t)_{t \geq 0}$ is a **semi-Markov process** if:*

(i) *The process admits a representation given, for $t \geq 0$, by*

$$Y_t = \sum_{n=0}^{+\infty} Z_n \mathbb{I}_{[\tau_n, \tau_{n+1}[}(t). \tag{A17}$$

(ii) *For $n \geq 0$ we have that $Z_n = Y_{\tau_n}$.*
(iii) *The process $(Z_n, \tau_n)_{n \geq 0}$ is a **Markov renewal process** (Mrp), that is, it verifies*

$$\begin{aligned} &\mathbb{P}\left[Z_{n+1} = \boldsymbol{\theta}_j, \tau_{n+1} - \tau_n \leq t \,|\, Z_0, \dots, Z_n, \tau_1, \tau_2, \dots . \tau_n\right] = \\ &= \mathbb{P}\left[Z_{n+1} = \boldsymbol{\theta}_j, \tau_{n+1} - \tau_n \leq t \,|\, Z_n\right], \end{aligned} \tag{A18}$$

for all $\boldsymbol{\theta}_j \in \boldsymbol{\Theta}$, $t \geq 0$ and almost surely—as it is a conditional expectation.

Proposition A2 (The sMp as a Markov chain). *The process $(Z_n, \tau_n)_{n \geq 0}$ is a Markov chain with state space $\boldsymbol{\Theta} \times [0, +\infty[$ and with semi-Markov transition kernel given by:*

$$q(i,j,t) := \mathbb{P}\left[Z_{n+1} = \boldsymbol{\theta}_j, \tau_{n+1} - \tau_n \leq t \,|\, Z_n = \boldsymbol{\theta}_i\right]. \tag{A19}$$

Proposition A3 (The embedded Markov chain of the Mrp). *The process $(Z_n)_{n \geq 0}$ is a Markov chain with state space $\boldsymbol{\Theta}$ with transition probabilities given by:*

$$p(i,j) := q(i,j,+\infty) = \mathbb{P}\left[Z_{n+1} = \boldsymbol{\theta}_j \,|\, Z_n = \boldsymbol{\theta}_i\right], \tag{A20}$$

*and is denoted as the **embedded** Markov chain of the Mrp.*

Proposition A4 (The conditional distribution function of the time between two successive jumps). *Let $\boldsymbol{Q} = [q(i,j,t)]_{i,j \in \{1,2,\dots r\}, t \geq 0}$ be the semi-Markov kernel as in Proposition A20. Let the times between successive jumps be $\Delta_n := \tau_n - \tau_{n-1}$ have the conditional distribution function of the time between two successive jumps be given by*

$$F_{ij}(t) := \mathbb{P}\left[\Delta_n \leq t \,|\, Z_n = \boldsymbol{\theta}_i,\ Z_{n+1} = \boldsymbol{\theta}_j\right]. \tag{A21}$$

Then, the semi-Markov kernel verifies,

$$q(i,j,t) := \mathbb{P}\left[Z_{n+1} = \boldsymbol{\theta}_j, \Delta_n \leq t \middle| Z_n = \boldsymbol{\theta}_i\right] = p(i,j)F_{ij}(t), \tag{A22}$$

with $p(i,j)$ as defined in Proposition A3.

Proof. It is a consequence of Proposition A1. □

Remark A7 (Homogeneous Markov chains as semi Markov processes). *Let $(X_t)_{t\geq 0}$ be a homogeneous Markov chain in continuous time with state space $\boldsymbol{\Theta} = \{\boldsymbol{\theta}_1, \boldsymbol{\theta}_2, \dots, \boldsymbol{\theta}_r, \dots\}$ and with—time independent—transition intensities given by $\mathbf{Q}(t) = [q(i,j)]_{i,j\geq 1}$ (see Definition A3). Then, by the well known results on homogeneous Markov chains (see [29] pp. 317, 318) and by the representation given by Formula (A22), we have that*

$$q(t,i,j) = \begin{cases} \frac{q(i,j)}{-q(i,i)}\left(1 - e^{q(i,i)t}\right) & i \neq j, \\ 0 & i = j \text{ or } q(i,i) = 0\,, \end{cases} \tag{A23}$$

is the semi Markov kernel of a sMp. Being so, comparing Formula (A23) with Formulas (A21) and (A22), we can see that the main difference between a sMp and a continuous time Markov process is the fact that in the sMp case the conditional distribution function of the time between two successive jumps depend not only on the initial state of the jump but also on the final state, while in the homogeneous Markov chain case the dependence is only on the initial state of the jump.

Definition A8 (The sojourn time distribution in a state). *The* ***sojourn time distribution*** *in the state $\boldsymbol{\theta}_i \in \boldsymbol{\Theta} = \{\boldsymbol{\theta}_1, \boldsymbol{\theta}_2, \dots, \boldsymbol{\theta}_r, \dots\}$, is defined by:*

$$H_i(t) := \sum_{j=1}^{+\infty} q(i,j,t) = \sum_{j=1}^{+\infty} p(i,j)F_{ij}(t). \tag{A24}$$

Its mean value represent the mean sojourn time in state $\boldsymbol{\theta}_i$ of the sMP $(Y_t)_{t\geq 0}$.

Definition A9 (Regular sMp). *A sMP $(Y_t)_{t\geq 0}$ is* ***regular****, with $N(t)$ the number of jumps of the process in the time interval $]0,t]$ given by:*

$$N(t) := \sup\{n \geq 0 : \tau_n \leq t\}, \tag{A25}$$

defined for $t > 0$ verifies for all $\boldsymbol{\theta}_i \in \boldsymbol{\Theta}$,

$$\mathbb{P}_i[N(t) < +\infty] := \mathbb{P}[N(t) < +\infty | Z_0 = \boldsymbol{\theta}_i] = 1. \tag{A26}$$

Proposition A5 (Jumps times of a regular sMp do not have accumulation points). *Let the sMP $(Y_t)_{t\geq 0}$ be regular. Then, almost surely, $\lim_{n\to+\infty} \tau_n = +\infty$ and, for any $T \in \mathbb{R}_+$ and almost all $\omega \in \Omega$:*

$$\#\{k \geq 1 : \tau_k(\omega) \leq T\} < +\infty. \tag{A27}$$

This means that in every compact time interval $[0,T]$, for almost all $\omega \in \Omega$ there is only a finite number of times $\tau_k(\omega)$ in this interval.

The following fundamental theorem ensures that for sMp with finite state space the sequence of stopping times do not accumulate in a compact interval.

Theorem A7 (A sufficient condition for regularity of a sMp). *Let $\alpha > 0$ and $\beta > 0$ be constants such that or every state $\boldsymbol{\theta}_i$ the sojourn time distribution in this state $H_i(t)$ defined in Definition A8 verifies:*

$$H_i(\alpha) < 1 - \beta.$$

Then, the sMp is regular. In particular, any sMp with a finite state space is regular.

Proof. See in [74] (p. 88). □

Remark A8 (On the estimation of sMp). *The estimation of sMp is dealt, for instance, in [75,76].*

References

1. Vajda, S. The stratified semi-stationary population. *Biometrika* **1947**, *34*, 243–254. [CrossRef] [PubMed]
2. Young, A.; Almond, G. Predicting Distributions of Staff. *Comput. J.* **1961**, *3*, 246–250. [CrossRef]
3. Bartholomew, D.J. A multi-stage renewal process. *J. R. Statist. Soc. Ser. B* **1963**, *25*, 150–168. [CrossRef]
4. Bartholomew, D.J. *Stochastic Models for Social Processes*, 2nd ed.; Wiley Series in Probability and Mathematical Statistics; John Wiley & Sons: London, UK; New York, NY, USA; Sydney, Australia, 1973.
5. Bartholomew, D.J. *Stochastic Models for Social Processes*, 3rd ed.; Wiley Series in Probability and Mathematical Statistics; John Wiley & Sons, Ltd.: Chichester, UK, 1982.
6. Gani, J. Formulae for Projecting Enrolments and Degrees Awarded in Universities. *J. R. Stat. Soc. Ser. A* **1963**, *126*, 400–409. [CrossRef]
7. Bowerman, B.; David, H.T.; Isaacson, D. The convergence of Cesaro averages for certain nonstationary Markov chains. *Stoch. Process. Appl.* **1977**, *5*, 221–230. [CrossRef]
8. Vassiliou, P.C.G. Cyclic behaviour and asymptotic stability of nonhomogeneous Markov systems. *J. Appl. Probab.* **1984**, *21*, 315–325. [CrossRef]
9. Vassiliou, P.C.G. Asymptotic variability of nonhomogeneous Markov systems under cyclic behaviour. *Eur. J. Oper. Res.* **1986**, *27*, 215–228. [CrossRef]
10. Dimitriou, V.A.; Georgiou, A.C. Introduction, analysis and asymptotic behavior of a multi-level manpower planning model in a continuous time setting under potential department contraction. *Commun. Statist. Theory Methods* **2021**, *50*, 1173–1199. [CrossRef]
11. Salgado-García, R. Open Markov Chains: Cumulant Dynamics, Fluctuations and Correlations. *Entropy* **2021**, *23*, 256. [CrossRef]
12. Vassiliou, P.C.G.; Papadopoulou, A.A. Nonhomogeneous semi-Markov systems and maintainability of the state sizes. *J. Appl. Probab.* **1992**, *29*, 519–534. [CrossRef]
13. Papadopoulou, A.A.; Vassiliou, P.C.G. Asymptotic behavior of nonhomogeneous semi-Markov systems. *Linear Algebra Appl.* **1994**, *210*, 153–198. [CrossRef]
14. Vassiliou, P.C.G. Asymptotic Behavior of Markov Systems. *J. Appl. Probab.* **1982**, *19*, 851–857. [CrossRef]
15. Vassiliou, P.C.G. Markov Systems in a General State Space. *Commun. Stat. Theory Methods* **2014**, *43*, 1322–1339. [CrossRef]
16. Vassiliou, P.-C.G. Rate of Convergence and Periodicity of the Expected Population Structure of Markov Systems that Live in a General State Space. *Mathematics* **2020**, *8*, 1021. [CrossRef]
17. Vassiliou, P.-C.G. Non-Homogeneous Markov Set Systems. *Mathematics* **2021**, *9*, 471. [CrossRef]
18. McClean, S.I. A continuous-time population model with Poisson recruitment. *J. Appl. Probab.* **1976**, *13*, 348–354. [CrossRef]
19. McClean, S.I. Continuous-time stochastic models of a multigrade population. *J. Appl. Probab.* **1978**, *15*, 26–37. [CrossRef]
20. McClean, S.I. A Semi-Markov Model for a Multigrade Population with Poisson Recruitment. *J. Appl. Probab.* **1980**, *17*, 846–852. [CrossRef]
21. Papadopoulou, A.A.; Vassiliou, P.C.G. Continuous time nonhomogeneous semi-Markov systems. In *Semi-Markov Models and Applications (Compiègne, 1998)*; Kluwer Academic Publishers: Dordrecht, The Netherlands, 1999; pp. 241–251.
22. Esquível, M.L.; Fernandes, J.M.; Guerreiro, G.R. On the evolution and asymptotic analysis of open Markov populations: application to consumption credit. *Stoch. Models* **2014**, *30*, 365–389. [CrossRef]
23. Guerreiro, G.R.; Mexia, J.A.T.; de Fátima Miguens, M. Statistical approach for open bonus malus. *Astin Bull.* **2014**, *44*, 63–83. [CrossRef]
24. Afonso, L.B.; Cardoso, R.M.R.; Egídio dos Reis, A.D.; Guerreiro, G.R. Ruin Probabilities And Capital Requirement for Open Automobile Portfolios With a Bonus-Malus System Based on Claim Counts. *J. Risk Insur.* **2020**, *87*, 501–522. [CrossRef]
25. Esquível, M.L.; Patrício, P.; Guerreiro, G.R. From ODE to Open Markov Chains, via SDE: an application to models for infections in individuals and populations. *Comput. Math. Biophys.* **2020**, *8*, 180–197. [CrossRef]
26. Esquível, M.; Guerreiro, G.; Fernandes, J. Open Markov chain scheme models. *REVSTAT* **2017**, *15*, 277–297.
27. Esquível, M.L.; Guerreiro, G.R.; Oliveira, M.C.; Corte Real, P. Calibration of Transition Intensities for a Multistate Model: Application to Long-Term Care. *Risks* **2021**, *9*. [CrossRef]
28. Resnick, S.I. *Adventures in Stochastic Processes*; Birkhäuser: Boston, MA, USA, 1992.
29. Rolski, T.; Schmidli, H.; Schmidt, V.; Teugels, J. *Stochastic Processes for Insurance and Finance*; Wiley Series in Probability and Statistics; John Wiley & Sons Ltd.: Chichester, UK, 1999. [CrossRef]
30. Iosifescu, M. *Finite Markov Processes and Their Applications*; Wiley Series in Probability and Mathematical Statistics; John Wiley & Sons, Ltd.: Chichester, UK; Editura Tehnică: Bucharest, Romania, 1980; p. 295.
31. Iosifescu, M.; Tăutu, P. *Stochastic Processes and Applications in Biology and Medicine. I: Theory*; Biomathematics; Editura Academiei RSR: Bucharest, Romania; Springer: Berlin, Germany; New York, NY, USA, 1973; Volume 3, p. 331.

32. Pyke, R. Markov renewal processes: Definitions and preliminary properties. *Ann. Math. Statist.* **1961**, *32*, 1231–1242. [CrossRef]
33. Korolyuk, V.S.; Korolyuk, V.V. Stochastic Models of Systems. In *Mathematics and its Applications*; Kluwer Academic Publishers: Dordrecht, The Netherlands, 1999; Volume 469. [CrossRef]
34. Kingman, J.F.C. The imbedding problem for finite Markov chains. *Probab. Theory Relat. Fields* **1962**, *1*, 14–24. [CrossRef]
35. Johansen, S. The Imbedding Problem for Finite Markov Chains. In *Geometric Methods in System Theory*, 1st ed.; Mayne, D.Q.B.R.W., Ed.; D. Reidel Publishing Company: Dordrecht, The Netherlands; Boston, MA, USA, 1973; Volume 1, Chapter 13, pp. 227–237.
36. Johansen, S. A central limit theorem for finite semigroups and its application to the imbedding problem for finite state Markov chains. *Z. Wahrscheinlichkeitstheorie Verw. Gebiete* **1973**, *26*, 171–190. [CrossRef]
37. Johansen, S. Some Results on the Imbedding Problem for Finite Markov Chains. *J. Lond. Math. Soc.* **1974**, *2*, 345–351. [CrossRef]
38. Fuglede, B. On the imbedding problem for stochastic and doubly stochastic matrices. *Probab. Theory Relat. Fields* **1988**, *80*, 241–260. [CrossRef]
39. Guerry, M.A. On the Embedding Problem for Discrete-Time Markov Chains. *J. Appl. Probab.* **2013**, *50*, 918–930. [CrossRef]
40. Jia, C. A solution to the reversible embedding problem for finite Markov chains. *Stat. Probab. Lett.* **2016**, *116*, 122–130. [CrossRef]
41. Goodman, G.S. An intrinsic time for non-stationary finite Markov chains. *Probab. Theory Relat. Fields* **1970**, *16*, 165–180. [CrossRef]
42. Singer, B. Estimation of Nonstationary Markov Chains from Panel Data. *Sociol. Methodol.* **1981**, *12*, 319–337. [CrossRef]
43. Lencastre, P.; Raischel, F.; Rogers, T.; Lind, P.G. From empirical data to time-inhomogeneous continuous Markov processes. *Phys. Rev. E* **2016**, *93*, 032135. [CrossRef]
44. Ekhosuehi, V.U. On the use of Cauchy integral formula for the embedding problem of discrete-time Markov chains. *Commun. Stat. Theory Methods* **2021**, 1–15. [CrossRef]
45. Coddington, E.A.; Levinson, N. *Theory of Ordinary Differential Equations*; McGraw-Hill Book Company, Inc.: New York, NY, USA; Toronto, ON, Canada; London, UK, 1955.
46. Rudin, W. *Real and Complex Analysis*, 3rd ed.; McGraw-Hill Book Co.: New York, NY, USA, 1987.
47. Kurzweil, J. Ordinary differential equations. In *Studies in Applied Mechanics*; Introduction to the theory of ordinary differential equations in the real domain, Translated from the Czech by Michal Basch; Elsevier Scientific Publishing Co.: Amsterdam, The Netherlands, 1986; Volume 13, p. 440.
48. Teschl, G. Ordinary differential equations and dynamical systems. In *Graduate Studies in Mathematics*; American Mathematical Society: Providence, RI, USA, 2012; Volume 140. [CrossRef]
49. Nevanlinna, F.; Nevanlinna, R. *Absolute Analysis*; Translated from the German by Phillip Emig, Die Grundlehren der mathematischen Wissenschaften, Band 102; Springer: New York, NY, USA; Heidelberg, Germany, 1973.
50. Severi, F.; Scorza Dragoni, G. *Lezioni di analisi. Vol. 3. Equazioni Differenziali Ordinarie e Loro Sistemi, Problemi al Contorno Relativi, Serie Trigonometriche, Applicazioni Geometriche*; Cesare Zuffi: Bologna, Italy, 1951.
51. Dobrušin, R.L. Generalization of Kolmogorov's equations for Markov processes with a finite number of possible states. *Matematicheskii Sbornik* **1953**, *33*, 567–596.
52. Pritchard, D.J. Modeling Disability in Long-Term Care Insurance. *N. Am. Actuar. J.* **2006**, *10*, 48–75. [CrossRef]
53. Kingman, J.F.C. Ergodic properties of continuous-time Markov processes and their discrete skeletons. *Proc. Lond. Math. Soc.* **1963**, *13*, 593–604. [CrossRef]
54. Conner, H. A note on limit theorems for Markov branching processes. *Proc. Am. Math. Soc.* **1967**, *18*, 76–86. [CrossRef]
55. Israel, R.B.; Rosenthal, J.S.; Wei, J.Z. Finding generators for Markov chains via empirical transition matrices, with applications to credit ratings. *Math. Financ.* **2001**, *11*, 245–265. [CrossRef]
56. Guerreiro, G.R.; Mexia, J.A.T. Stochastic vortices in periodically reclassified populations. *Discuss. Math. Probab. Stat.* **2008**, *28*, 209–227. [CrossRef]
57. Feller, W. *An Introduction to Probability Theory and Its Applications. Vol. I*, 3rd ed.; John Wiley & Sons, Inc.: New York, NY, USA; London, UK; Sydney, Australia, 1968.
58. Serfozo, R. Convergence of Lebesgue integrals with varying measures. *Sankhyā Ser. A* **1982**, *44*, 380–402.
59. Billingsley, P. *Convergence of Probability Measures*, 2nd ed.; Wiley Series in Probability and Statistics: Probability and Statistics; John Wiley & Sons, Inc.: New York, NY, USA, 1999. [CrossRef]
60. Durrett, R. *Probability—Theory and Examples*. In *Cambridge Series in Statistical and Probabilistic Mathematics*; Cambridge University Press: Cambridge, UK, 2019; Volume 49. [CrossRef]
61. Skorokhod, A.V. *Lectures on the Theory of Stochastic Processes*; VSP: Utrecht, The Netherlands; TBiMC Scientific Publishers: Kiev, Ukraine, 1996.
62. Dynkin, E.B. *Theory of Markov Processes*; Translated from the Russian by D. E. Brown and edited by T. Köváry, Reprint of the 1961 English translation; Dover Publications, Inc.: Mineola, NY, USA, 2006.
63. Iosifescu, M.; Limnios, N.; Oprişan, G. *Introduction to Stochastic Models*; Applied Stochastic Methods Series; Translated from the 2007 French original by Vlad Barbu; ISTE: London, UK; John Wiley & Sons, Inc.: Hoboken, NJ, USA, 2010. [CrossRef]
64. Pyke, R. Markov renewal processes with finitely many states. *Ann. Math. Statist.* **1961**, *32*, 1243–1259. [CrossRef]
65. Feller, W. On semi-Markov processes. *Proc. Nat. Acad. Sci. USA* **1964**, *51*, 653–659. [CrossRef]
66. Kurtz, T.G. Comparison of semi-Markov and Markov processes. *Ann. Math. Statist.* **1971**, *42*, 991–1002. [CrossRef]

67. Korolyuk, V.; Swishchuk, A. Semi-Markov random evolutions. In *Mathematics and its Applications*; Translated from the 1992 Russian original by V. Zayats and revised by the authors; Kluwer Academic Publishers: Dordrecht, The Netherlands, 1995; Volume 308. [CrossRef]
68. Janssen, J.; de Dominicis, R. Finite non-homogeneous semi-Markov processes: Theoretical and computational aspects. *Insur. Math. Econ.* **1984**, *3*, 157–165. [CrossRef]
69. Janssen, J.; Limnios, N. (Eds.) *Semi-Markov Models and Applications*; Selected papers from the 2nd International Symposium on Semi-Markov Models: Theory and Applications held in Compiègne, December 1998; Kluwer Academic Publishers: Dordrecht, The Netherlands, 1999. [CrossRef]
70. Janssen, J.; Manca, R. *Applied Semi-Markov Processes*; Springer: New York, NY, USA, 2006.
71. Janssen, J.; Manca, R. *Semi-Markov Risk Models for Finance, Insurance and Reliability*; Springer: New York, NY, USA, 2007.
72. Barbu, V.S.; Limnios, N. Semi-Markov chains and hidden semi-Markov models toward applications. In *Lecture Notes in Statistics*; Springer: New York, NY, USA, 2008; Volume 191.
73. Grabski, F. *Semi-Markov Processes: Applications in System Reliability and Maintenance*; Elsevier: Amsterdam, The Netherlands, 2015.
74. Ross, S.M. *Applied Probability Models with Optimization Applications*; Reprint of the 1970 original; Dover Publications, Inc.: New York, NY, USA, 1992.
75. Moore, E.H.; Pyke, R. Estimation of the transition distributions of a Markov renewal process. *Ann. Inst. Stat. Math.* **1968**, *20*, 411. [CrossRef]
76. Ouhbi, B.; Limnios, N. Nonparametric Estimation for Semi-Markov Processes Based on its Hazard Rate Functions. *Stat. Inference Stoch. Process.* **1999**, *2*, 151–173. [CrossRef]

mathematics

MDPI

Article

Partial Diffusion Markov Model of Heterogeneous TCP Link: Optimization with Incomplete Information

Andrey Borisov [1,2,3,4], Alexey Bosov [1,2], Gregory Miller [1,*] and Igor Sokolov [3]

1 Federal Research Center "Computer Science and Control" of the Russian Academy of Sciences, 44/2 Vavilova Str., 119333 Moscow, Russia; ABorisov@frccsc.ru (A.B.); abosov@frccsc.ru (A.B.)
2 Moscow Aviation Institute, 4, Volokolamskoe Shosse, 125993 Moscow, Russia
3 Faculty of Computational Mathematics and Cybernetics, Lomonosov Moscow State University, GSP-1, 1-52 Leninskiye Gory, 119991 Moscow, Russia; ISokolov@cs.msu.ru
4 Moscow Center for Fundamental and Applied Mathematics, Lomonosov Moscow State University, GSP-1, Leninskie Gory, 119991 Moscow, Russia
* Correspondence: GMiller@frccsc.ru

Abstract: The paper presents a new mathematical model of TCP (Transmission Control Protocol) link functioning in a heterogeneous (wired/wireless) channel. It represents a controllable, partially observable stochastic dynamic system. The system state describes the status of the modeled TCP link and expresses it via an unobservable controllable MJP (Markov jump process) with finite-state space. Observations are formed by low-frequency counting processes of packet losses and timeouts and a high-frequency compound Poisson process of packet acknowledgments. The information transmission through the TCP-equipped channel is considered a stochastic control problem with incomplete information. The main idea to solve it is to impose the separation principle on the problem. The paper proposes a mathematical framework and algorithmic support to implement the solution. It includes a solution to the stochastic control problem with complete information, a diffusion approximation of the high-frequency observations, a solution to the MJP state filtering problem given the observations with multiplicative noises, and a numerical scheme of the filtering algorithm. The paper also contains the results of a comparative study of the proposed state-based congestion control algorithm with the contemporary TCP versions: Illinois, CUBIC, Compound, and BBR (Bottleneck Bandwidth and RTT).

Keywords: controllable Markov jump processes; compound Poisson processes; diffusion limits; stochastic control problem with incomplete information; novel queuing models in applications

Citation: Borisov, A.; Bosov, A.; Miller, G.; Sokolov, I. Partial Diffusion Markov Model of Heterogeneous TCP Link: Optimization with Incomplete Information. *Mathematics* **2021**, *9*, 1632. https://doi.org/10.3390/math9141632

Academic Editors: Panagiotis-Christos Vassiliou and Andreas C. Georgiou

Received: 13 June 2021
Accepted: 8 July 2021
Published: 10 July 2021

Publisher's Note: MDPI stays neutral with regard to jurisdictional claims in published maps and institutional affiliations.

Copyright: © 2021 by the authors. Licensee MDPI, Basel, Switzerland. This article is an open access article distributed under the terms and conditions of the Creative Commons Attribution (CC BY) license (https://creativecommons.org/licenses/by/4.0/).

1. Introduction

Despite its age of almost 50 years, the Transmission Control Protocol (TCP) [1] is still an object of permanent modernization and improvement, and this evolution represents a natural perpetual process. The root of this development lies in incessant challenges caused by a wide variety of computer networks, impetuous progress in the communication devices design, and strengthening of requirements to the information transmission [2–4]. Meanwhile, guaranteeing data transfer independent of the hardware platform is the key task of the TCP algorithm; both the stable functioning and effective use of the available channel bandwidth are also the performance characteristics of each specific version of TCP. The congestion control algorithms are responsible for the implementation of all these functions. They use two characteristics as the control actions. The basic one is the congestion window size (cwnd), i.e., the number of packets sent without acknowledgment. A less influential one is the retransmission timeout, i.e., some waiting time for the acknowledgment of the successful packet reception, which excess is treated by the congestion control algorithm as a packet loss.

When most channels were wire channels and had a relatively small capacity and queue waiting time *"Additive Increase–Multiple Decrease"* (AIMD) congestion control rule demonstrated good performance. This presumed a linear growth of the cwnd between two successive packet losses when the cwnd abruptly decreased in a jump-like manner. The effectiveness of this strategy for such channels was transparent. First, the small channel capacity gave a chance to reach a bandwidth limit linearly without losses for a rather short time. Second, wired hops were so reliable that the fact of a sudden packet loss presumed congestion at some "bottleneck" almost surely. Therefore, the loss indicated the necessity to reduce the sending rate. This simple reason was a base to develop such *loss-based* versions of TCP as Tahoe, New Reno, etc. [5].

In the case of the "long fat" channels (ones with huge capacity and long queue waiting times), AIMD-based versions of TCP turned out to be ineffective: they underused the channel bandwidth significantly. In the case of the channels with high capacity, the linear growth does not allow for the congestion window to quickly achieve values close to the available bandwidth. Plus, a loss of at least one packet decreases the data transferring speed even more. In addition, if a channel includes a wireless hop, facts of single packet losses are not an explicit congestion indicator. The *round-trip time* (RTT) parameter starts to play a remarkable role in the congestion control algorithm, and this brings to the variety of the TCP versions: delay-sensitive, hybrid loss-delay, bandwidth estimation-based, etc. [2]. All the modifications make the congestion control algorithm more tolerant to packet losses: after each loss, it decreases cwnd not multiplicative but more sparingly. At the same time, the cwnd growth speed is more aggressive to reach the channel bandwidth faster. The bandwidth value is unknown but estimated given all past statistics of the channel functioning. The algorithm probes more or less gentle cwnd enlargement to give a chance to use all channel resources. Hence, the typical cwnd curve between two packet losses demonstrates a concave [6] or mixed concave-convex character [7].

The ubiquitous application of wireless technologies in computer networks is a challenge to TCP protocol performance and claims its subsequent enhancement. Jitter and periodical signal fading in the wireless channel hops are extra sources of uncertainty of the channel real throughput. These physical phenomena affect both the new mathematical models of the channel functioning and the congestion control algorithms.

Mathematical models of computer network traffic are also developed intensively. With no goal to present a comprehensive overview of these models, we only mention their major classes

- Markov and hidden Markov models [8–11],
- queuing systems [12,13],
- models, based on the fluid or diffusion approximation of jump processes [14–16],
- network calculus models [17–19],
- models involving selfsimilar processes [20–22],
- concurrent models and games [23–25], etc.

Generally speaking, a prospective mathematical model of a channel should satisfy the conditions below.

1. A model should describe the data transferring process adequately.
2. A model should represent a trade-off between a complicated object with many parameters, their uncertainty along with the uncertainty introduced by the external disturbances, and simplicity.
3. A model should operate with the same collection of statistical information as the one available in the real channel.
4. A model should provide a possibility to simulate the collection of recent "concurrent" versions of TCP.
5. The chosen model presumes the presence of the developed mathematical framework for the solution to the complex of all the analysis, estimation/identification and optimization/control problems. Availability of both the theoretical solution to the problems above and their efficient numerical realization is strongly encouraged.

The aim of the paper is two-fold. First, this is a presentation of a new mathematical model of the TCP link functioning based on the heterogeneous (wired/wireless) channel. It represents a controllable, partially observable stochastic dynamic system. The system state describes the status of the modeled TCP link and expresses it via the controllable *Markov jump process* (MJP) with a finite-state space. This space can be chosen arbitrarily depending on the desired detailing of the link description. Below in this paper, we consider four possible channel states:

- e_1: the channel is idle,
- e_2: the channel is loaded moderately,
- e_3: congestion in the wired segment,
- e_4: signal fading in the wireless hop.

Looking rather simple, this model admits successful description of such a problematic link phenomenon as congestion in a channel "bottleneck" and the carrier radio signal fading.

The observations included into the model correspond to those available to a TCP control algorithm on the sending side. Two observable processes describe the flow of packet losses and the flow of timeouts. They are represented by controllable Cox processes with intensity that depends both on the control and unobserved link state. The third observation is a flow of the acknowledgments concerning the successful packet reception on the receiving node. The flow is expressed in terms of a *compound Poisson processes* (CPP). Its first component represents a counting process of acknowledgment reception moments, and the second one registers corresponding individual values of *the Round-Trip Time* (RTT).

In the paper, we control the TCP varying the cwnd value only; however, the proposed model allows other control parameters, e.g., RTO (retransmission timeout). We also demonstrate how the proposed mathematical model can describe various contemporary versions of the TCP: Illinois, CUBIC, BBR, and Compound.

The second aim of the paper is presentation of a new TCP prototype version. Its mathematical background is both the solution to the optimal MJP state control under complete information, and the solution to the optimal MJP state filtering given the diffusion and counting observations. The performance of the proposed prototype is demonstrated on the complex of the numerical experiments.

The paper is organized as follows. Section 2 contains a detailed description of the TCP link mathematical model in terms of the controllable stochastic observation system, along with the optimization problem of data transmission through this link.

One can enhance the use of the channel resources in terms of the optimal stochastic control with incomplete information. However, this approach promises complications during its realization: starting from the proof of the optimal solution existence and concluding by bulky numerical algorithms of its realization. Hence, we propose a rather simple suboptimal solution to the problem along with its effective numerical implementation.

To develop the TCP prototype, we need a substantial mathematical framework, which is introduced in Section 3:

- Section 3.1 contains the solution to the optimal MJP control problem with instant geometric control constraints and complete information [26],
- Section 3.2 introduces a diffusion approximation for the high-frequency CPP describing the packet acknowledgment flow [27],
- Section 3.3 presents a solution to the optimal MJP state filtering problem given both counting and diffusion observations with state-dependent noise [28],
- Section 3.4 contains a numerical algorithm for the optimal filtering realization [28].

In general, the articles [26–28] represent a formal, detailed mathematical background of all applied inferences presented in this paper. We use it in Section 4 to develop a new congestion control algorithm as follows. At the first stage, we calculate a high-precision channel state estimate based on the available observations discretized by time. At the second stage, we apply a separation principle: the obtained filtering estimate replaces the actual MJP state during the process of the optimal control synthesis with the complete information.

The aim of Section 5 is two-fold. First, it demonstrates the potential of the proposed mathematical model to describe various versions of the TCP: classic AIMD congestion control scheme and TCP Illinois (Section 5.1), TCP CUBIC (Section 5.2), TCP Compound (Section 5.3), TCP BBR (Section 5.4).

Second, the section contains the comparison of the proposed state-based TCP with versions mentioned above: Section 5.5 highlights some details of the numerical realization of the proposed TCP version, and Section 5.6 represents the summary of the performed numerical experiments. Section 6 contains concluding remarks.

2. Problem of Optimal Data Transmission through TCP Channel

On the canonical Wiener-Poisson space with filtration $(\Omega, \mathcal{F}, \mathcal{P}, \{\mathcal{F}_t\})$ [29,30] we consider the following controllable stochastic system, describing the TCP link functioning

$$X_t = X_0 + \int_0^t A(u_s)X_s ds + \alpha_t, \tag{1}$$

$$Y_t = \int_0^t B(u_s)X_s ds + \beta_t, \tag{2}$$

$$Z_t = \int_0^t C(u_s)X_s ds + \gamma_t, \tag{3}$$

$$\{(\tau_n, V_n)\}_{n\in\mathbb{N}}. \tag{4}$$

Here the TCP link state X_t is a controllable finite-state MJP with values in the set $\mathbb{S}^N \triangleq \{e_1, ..., e_n\}$ formed by unit coordinate vectors of the Euclidean space $\mathbb{R}^N$. The initial value X_0 has a known distribution π, $A(u) = \|A^{ij}(u)\|_{i,j=\overline{1,N}}$ is a controllable transition intensity matrix and α_t is a $\mathcal{F}_t$-adapted martingale with the quadratic characteristic [31]

$$\langle \alpha, \alpha \rangle_t = \int_0^t \Big(\operatorname{diag}(A(u_s)X_s) - A(u_s)\operatorname{diag}(X_s) - \operatorname{diag}(X_s)A^\top(u_s) \Big) ds.$$

The link state is unobservable, and the complex of observations $(Y_t, Z_t, \{(\tau_n, V_t)\})$ includes three components.

- Y_t is a counting process (flow) of packet losses described by its martingale representation (2): β_t is an $\mathcal{F}_t$-adapted martingale with the quadratic characteristic

$$\langle \beta, \beta \rangle_t = \int_0^t B(u_s)X_s ds,$$

  $B(u) \triangleq \mathrm{row}(B^1(u), \dots, B^N(u))$ represents the collection of the loss intensities of the flow given the conditions $X_t = e_n$, $n = \overline{1,N}$.
- Z_t is a counting process (flow) of packet timeouts described by its martingale representation (3): γ_t is an $\mathcal{F}_t$-adapted martingale with the quadratic characteristic

$$\langle \gamma, \gamma \rangle_t = \int_0^t C(u_s)X_s ds,$$

  $C(u) \triangleq \mathrm{row}(C^1(u), \dots, C^N(u))$ represents the collection of the timeout intensities of the flow given the conditions $X_t = e_n$, $n = \overline{1,N}$.
- $\{(\tau_n, V_t)\}$ is a flow of successful packet acknowledgments: here τ_n stands for the time instant of the n-th acknowledgment arrival and V_t does for the specific RTT of the n-th acknowledgment. It represents controllable *compound Poisson process* (CPP) with the intensity driven by the Markov state X_t: the predictable measure generated by $\{(\tau_n, V_t)\}$ conditioned by the MJP state X takes the form

$$\mu_p(\omega, dt, dv) = \lambda(u_t)\, \mathrm{diag}(X_{t-})\Lambda(u_t, v)dtdv.$$

Here $\lambda(u_t) \triangleq \mathrm{row}(\lambda^1(u_t), \dots, \lambda^N(u_t))$ is a vector-valued function with continuous positive components, its nth component represent conditional intensity of acknowledgment arrivals given $X_t = e_n$; $\Lambda(u_t, v) \triangleq \mathrm{col}(\Lambda^1(u_t, v), \dots, \Lambda^N(u_t, v))$ is a vector-valued function with continuous components, its nth component represent conditional *probability density function* (pdf) with respect to v given $X_t = e_n$ for each fixed u_t.

All martingale terms in the processes X, Y, Z and (τ, V) are strongly orthogonal.

The control u_t represents a current size of the congestion window, i.e., portion of packets which can be instantly transmitted. The set of admissible control contains all $\mathcal{O}_t$-predictable processes ($\mathcal{O}_t \triangleq \sigma\{Y_s, Z_s, (\tau_n, V_n) : s, \tau_n \in [0, t]\}$ stands for a natural filtration induced by all observations available up to the moment t) with the geometric constraint:

$$u_s \in \mathsf{U} \triangleq [\underline{u}, \overline{u}] \subset \mathbb{R}_+ \quad \mathcal{P}\text{-a.s. for all } s \geqslant 0. \tag{5}$$

The intensity of acknowledgment arrivals is much more than all the state transition, packet loss and timeout ones:

$$\min_{n,u} \lambda^n(u) \gg \max_{n,u}(|A^n(u)|, B^n(u), C^n(u)).$$

The performance criterion

$$J(U) \triangleq \mathsf{E}\left\{\psi X_T + \int_0^T (\phi(u_s) - u_s\xi)X_s ds\right\} \to \max_U \tag{6}$$

represents an average profit for the transmitted information, which should be maximized. Here

- $\psi \triangleq \mathrm{row}(\psi^1, \dots, \psi^N)$ is a vector of conditional gains given the terminal state X_T,
- $\phi(u_s) \triangleq \mathrm{row}(\phi^1(u_s), \dots, \phi^N(u_s))$ includes strictly concave components, which represent conditional instant gains for the transmitted information given the current link state X_s,
- $\xi \triangleq \mathrm{row}(\xi^1, \dots, \xi^N)$ is a vector of specific transmission expenses per information unit in each link state.

The problem under consideration is challenging. First, in general, optimal control problems of stochastic jump processes with incomplete information are rather complicated [31–34]. Their proper statement and solution depends on the answer to several auxiliary questions/problems: the martingale one [35], the one of strong solution existence and uniqueness and the one of measurable control selection (see [36] and references within). Without positive answers to the questions, we cannot use the martingale theory [35,37] to express optimal control in terms of either variation inequalities (dynamic programming equation as the preferable outcome) or stochastic maximum principle. Please note that negative answers presumes only impossibility to use the mathematical tools mentioned above. Apparently, the control problem can be modified slightly to provide its solution existence which can be found involving other still undiscovered frameworks.

Second, both the dynamic programming equation and stochastic maximum principle have forward-backward form which complicates synthesis of the optimal control in the explicit form. The authors of [36] have solved the analogous problem of the MJP state (1) control observing the flow of packet losses (2) only. The theoretical optimal solution has been characterized both via the dynamic programming equation and the maximum principle. At the same time, the authors have presented a numerical realization of the obtained result only for the case when the transition intensity matrix of the MJP is independent of the control (i.e., the state is uncontrollable), and control affects the intensity of the losses only. Despite the restrictive conditions the obtained practical results have looked rather prospective: the optimal policy has demonstrated piecewise concave nature similar to the modern versions of TCP: Illinois [6], CUBIC without probe phase [38,39], Compound [40,41] etc.

Third, the essential weak points of the optimal control implementation are its poor robustness relating to the imprecise knowledge of the control system characteristics and small perturbations of the synthesized control to its performance. This means that either control system parameters slightly misspecified towards its unknown nominal, or "instrumental errors" in control caused by imperfection of its numerical realization could nullify gain of the sophisticated optimal control in comparison with a stable suboptimal algorithm.

Fourth, the flow of packet acknowledgments has high intensity and hence leads to a high-frequency control, which is resource intensive.

Keeping in mind all arguments above we avoid the direct solution to the optimal stochastic control problem (6) of the MJP (1) state given the observations (2), (3) and (4) including the martingale problem and the ones of the solution existence and uniqueness. Instead of this we use solutions to a complex of adjacent problems and propose a suboptimal control algorithm of high performance.

3. Mathematical Background

As a basis of the proposed suboptimal control algorithm, we use the following arguments and mathematical results. We derive the algorithm basing on the following mathematical results and reasons.

1. The solution to the optimal stochastic control of the MJP (1) state with the complete information does exist and can be defined as a solution to the equation of dynamic programming [26].
2. The high frequency allows us to approximate the observable controlled CPP (4) by a drifting Brownian motion [42] with the parameters modulated by the MJP state [27]. We can describe the distribution of the diffusion approximation via some moment characteristics only, and this fact leads to robustness of the subsequent state filtering algorithm towards the imprecise knowledge of the specific distribution of compound Poisson process jumps.
3. The conversion of high-frequency acknowledgment flow to a diffusion process gives a possibility to use the solution to the optimal MJP (1) state filtering problem given the "diffusion" and counting observations [43]. This is extension of the Wonham filter [44] to the case of the diffusion observations with state-dependent noises. Under rather mild identifiability conditions the optimal filtering estimate coincides with the exact MJP state.
4. The dynamic programming equation corresponding to the control problem with complete information mentioned at item 1, represents the system of ordinary differential equations with well-developed methods of numerical solution. By contrast, the equations of the generalized Wonham filter [43] require design of special numerical procedures similar to [28].
5. To complete the control synthesis, we postulate *a separation principle*. This means we put the state filtering estimate mentioned at items 3, 4 into the control strategy defined at item 1.

3.1. Optimal Control Strategy with Complete Information

Let us consider the controllable MJP (1) which should be optimized with respect to the optimality criterion (6) where the set $\mathcal{U}$ of all admissible controls U includes all $\mathcal{O}_t$-predictable processes with the geometric constraint (5).

Let us define the Bellman function $V(t,x) : [0,T] \times \mathbb{S}^N \to \mathbb{R}$:

$$\mathbf{B}(t,x) \triangleq \sup_{U \in \mathcal{U}} \mathsf{E}\left\{ \psi X_T + \int_t^T (\phi(u_s) - u_s\xi(s))X_s ds \middle| X_t = x \right\}. \tag{7}$$

Obviously, the function $\mathbf{B}(t,x)$ can be presented in the form $\mathbf{B}(t,x) = \eta^\top(t)x$, where $\eta(t) \triangleq \mathrm{col}(\eta^1(t),\dots,\eta^N(t)) = \mathrm{col}(\mathbf{B}(t,e_1),\dots,\mathbf{B}(t,e_N))$ is a vector-valued function.

Theorem 1. *The assertions below are true [26].*

1. *The function $\eta(t)$ is the unique solution to the Cauchy problem*

$$\begin{cases} \dot{\eta}^n(t) = \max_{u\in \mathsf{U}} \left[\sum_{j=1}^N A^{jn}(u)\eta^j(t) + \xi^n(u) \right], & n = \overline{1,N}, \quad 0 \leqslant t < T, \\ \eta^n(T) = \psi^n, & n = \overline{1,N}. \end{cases} \tag{8}$$

2. *There exists a Borel function $\widehat{u}_t(x) : [0,T] \times \mathbb{S}^N \to \mathsf{U}$, such that*

$$\widehat{u}_t(x) \in \mathop{\mathrm{Argmax}}_{u\in\mathsf{U}} \left[\sum_{n=1}^N \left(\sum_{j=1}^N A^{jn}(u)\eta^j(t) + \xi^n(u) \right) x^n \right] \tag{9}$$

for any $(t,s) \in [0,T] \times \mathbb{S}^N$.

3. *The random process $\widehat{U}_t \triangleq \widehat{u}_t(X_{t-})$ is an optimal control strategy for the problem (1), (6).*
4. *The optimal value of criterion (6) has the form $\max_{U\in\mathsf{U}} J(U) = J(\widehat{U}) = \eta^\top(0)\pi$; moreover, supremum in (7) is attained for any $(t,x) \in [0,T] \times \mathbb{S}^N$ at the strategy $\{\widehat{U}_s,\ s \in [t,T]\}$.*

The theorem establishes the base of the practical control realization. Indeed, all variants of possible optimal controls (9) can be calculated and stored in advance via solution to (8), before the control synthesis. The synthesis itself represents the selection of suitable control from the set of possible ones using the "current" MJP state X_{t-}.

3.2. Diffusion Approximation of High-Frequency Counting Observations

Use of the "genuine" acknowledgments flow (4) to synthesize the control leads to discontinuous one with high frequency. Its calculation may be resource intensive: each new-coming acknowledgment triggers the control recalculation algorithm. The contemporary TCP versions are exactly like this, but they are relatively simple, so not too "costly".

Once we consider (4) discretized by time with some appropriate time increment, we can see the probability distribution of the observation increments look like mixtures of some Gaussians due to *the central limit theorem for renewal-reward processes* (CLTRRP). In this subsection we answer two questions. First, we determine characteristics of these mixtures. Second, we form recommendations how to choose time increment value to provide appropriate closeness of the real discretized observation distribution to the theoretical mixture above.

First, to perceive the nature of diffusion approximation, we investigate the CPPs with a fixed control $u \in \mathsf{U}$. We consider a collection of the CPPs $\{(\tau_n^j, V_n^j)\}_{\substack{n\in\mathbb{N}, j=\overline{1,N},\\ u\in\mathsf{U}}}$ with the predictable measures $\{\mu_p^j(dt,dv)\}_{\substack{s>0,\\ u\in\mathsf{U}}}$:

$$\mu_p^j(dt,dv) \triangleq \lambda^j(u)\Lambda^j(u,v)ds dv.$$

Probabilistically they correspond to initial CPP $\{(\tau_n, V_n)\}$ staying in the "single mode": $X_t \equiv e_j$ and a fixed control value $u_t \equiv u$. Each CPP generates a stochastic measure

$$\mu^j(\omega,dt,dv) \triangleq \sum_{n\in\mathbb{N}} \delta_{(\tau_n^j(\omega),V_n^j(\omega))}(dt,dv).$$

Keeping in mind the specific form of the predictable measures μ_p^j, we can compute the moment characteristics for one jump of the CPPs:

$$m_\tau^j \triangleq \mathsf{E}\left\{\tau_1^j\right\} = \frac{1}{\lambda^j(u)}, \qquad m_V^j \triangleq \mathsf{E}\left\{V_1^j\right\} = \int_{\mathbb{R}} v\Lambda^j(u,v)dv, \tag{10}$$
$$\sigma_\tau^j \triangleq \sqrt{\mathrm{var}(\tau_1^j)} = \frac{1}{\lambda^j(u)}, \qquad \sigma_V^j \triangleq \sqrt{\mathrm{var}(V_1^j)} = \sqrt{\int_{\mathbb{R}} v^2\Lambda^j(u,v)dv - \left(m_V^j\right)^2},$$
$$\kappa^j \triangleq \mathrm{cov}(\tau_1^j, V_1^j) = 0.$$

We investigate the asymptotic behavior of the distribution of the two-dimensional random process

$$\Theta_t^j \triangleq \begin{bmatrix} \int_{[0,t]\times\mathbb{R}} \mu^j(ds,dv) \\ \int_{[0,t]\times\mathbb{R}} v\mu^j(ds,dv) \end{bmatrix} = \begin{bmatrix} \sum_{n\in\mathbb{N}} \mathbf{I}(t-\tau_n^j) \\ \sum_{n\in\mathbb{N}} V_n\mathbf{I}(t-\tau_n^j) \end{bmatrix} \tag{11}$$

when $t \to \infty$. The first component represents the total number of acknowledgments received at the sender over the time interval $[0,t]$, the second component, in turn, stands for the corresponding cumulative RTT value. The author of [42] proved a version of CLTRRP:

$$\frac{1}{\sqrt{\lambda^j(u)t}}\left(\Theta_t^j - \begin{bmatrix} \lambda^j(u)t \\ m_V^j\lambda^j(u)t \end{bmatrix}\right) \xrightarrow{Law} \mathcal{N}\left(\begin{bmatrix} 0 \\ 0 \end{bmatrix}, \begin{bmatrix} 1 & m_V^j \\ m_V^j & (m_V^j)^2 + (\sigma_V^j)^2 \end{bmatrix}\right) \tag{12}$$

as $t \to \infty$. In other words, for rather huge t

$$\frac{1}{\sqrt{t}}\Theta_t^j \simeq \mathcal{N}\left(\begin{bmatrix} (\lambda^j(u))^{\frac{3}{2}}\sqrt{t} \\ m_V^j(\lambda^j(u))^{\frac{3}{2}}\sqrt{t} \end{bmatrix}, \begin{bmatrix} \lambda^j(u) & \lambda^j(u)m_V^j \\ \lambda^j(u)m_V^j & \lambda^j(u)\left[(m_V^j)^2 + (\sigma_V^j)^2\right] \end{bmatrix}\right).$$

Let us complicate the model, mixing the CPPs $\{(\tau_n^j, V_n^j)\}_{\substack{n\in\mathbb{N}, j=\overline{1,N},\\ u\in\mathsf{U}}}$ above with probabilities $\pi = \mathrm{col}(\pi^1,\dots,\pi^N)$

$$\begin{bmatrix} \overline{\tau}_n \\ \overline{V}_n \end{bmatrix} = \sum_{j=1}^N X_0^j \begin{bmatrix} \tau_n^j \\ V_n^j \end{bmatrix}. \tag{13}$$

Here $X_0 \triangleq \mathrm{col}(X_0^1,\dots,X_0^N) \in \mathbb{S}^N$ is an $\mathcal{F}_0$-measurable random vector, independent of $\{(\tau_n^j, V_n^j)\}_{\substack{n\in\mathbb{N}, j=\overline{1,N},\\ u\in\mathsf{U}}}$; $X_0 \sim \pi_0$. It is easy to verify that the predictable measure generated by $\{(\overline{\tau}_n, \overline{V}_n)\}$, conditioned by X_0, takes the form

$$\overline{\mu}_p(\omega,dt,dv) = \lambda(u_t)\,\mathrm{diag}(X_0)\Lambda(u_t,v)dtdv.$$

Please note that the mixed CPP (13) represents a specific case of the observations (4) with "single mode" MJP X: $A(u) \equiv 0$, $X_0 \sim \pi$.

Making inferences as above we can conclude that for rather huge t

$$\begin{aligned} &\frac{1}{\sqrt{t}}\begin{bmatrix} \sum_{k\in\mathbb{N}} \mathbf{I}(t-\overline{\tau}_k) \\ \sum_{k\in\mathbb{N}} \overline{V}_k\mathbf{I}(t-\overline{V}_k) \end{bmatrix} \\ &\simeq \sum_{j=1}^N \pi^j\mathcal{N}\left(\begin{bmatrix} (\lambda^j(u))^{\frac{3}{2}}\sqrt{t} \\ m_V^j(\lambda^j(u))^{\frac{3}{2}}\sqrt{t} \end{bmatrix}, \begin{bmatrix} \lambda^j(u) & \lambda^j(u)m_V^j \\ \lambda^j(u)m_V^j & \lambda^j(u)\left[(m_V^j)^2 + (\sigma_V^j)^2\right] \end{bmatrix}\right). \end{aligned} \tag{14}$$

Therefore, given some MJP state X_s distribution (conditional or unconditional) at the time instant s and a constant control $u_q \equiv u \in \mathsf{U}$, $q \in [s, s+h)$ we assume that the cumulative observation increment over the interval $[s, s+h)$ is distributed approximately in the following way

$$
\frac{1}{\sqrt{h}}\begin{bmatrix} \sum\limits_{k\in\mathbb{N}} \mathbf{I}(t-\tau_k)\mathbf{I}(\tau_k-s) \\ \sum\limits_{k\in\mathbb{N}} V_k\mathbf{I}(t-\tau_k)\mathbf{I}(\tau_k-s) \end{bmatrix} \\ \simeq \sum_{j=1}^{N}\widehat{X}_s^j\mathcal{N}\left(\begin{bmatrix} (\lambda^j(u))^{\frac{3}{2}}\sqrt{t} \\ m_V^j(\lambda^j(u))^{\frac{3}{2}}\sqrt{t} \end{bmatrix}, \begin{bmatrix} \lambda^j(u) & \lambda^j(u)m_V^j \\ \lambda^j(u)m_V^j & \lambda^j(u)\left[(m_V^j)^2+(\sigma_V^j)^2\right] \end{bmatrix}\right). \tag{15}
$$

By analogy with (15) for the cumulative process, corresponding to the acknowledgment flow (4)

$$
Q_t \triangleq \begin{bmatrix} \sum\limits_{k\in\mathbb{N}} \mathbf{I}(t-\tau_k) \\ \sum\limits_{k\in\mathbb{N}} V_k\mathbf{I}(t-\tau_k) \end{bmatrix} \tag{16}
$$

we propose the following approximate diffusion model

$$
Q_t = \int_0^t D(u_s)X_s ds + \int_0^t \sum_{n=1}^{N} e_n^\top X_s E_n^{\frac{1}{2}}(u_s)dW_s, \tag{17}
$$

where

$$
D(u) \triangleq \begin{bmatrix} (\lambda^1(u))^{\frac{3}{2}} & (\lambda^2(u))^{\frac{3}{2}} & \dots & (\lambda^N(u))^{\frac{3}{2}} \\ m_V^1(\lambda^1(u))^{\frac{3}{2}} & m_V^2(\lambda^2(u))^{\frac{3}{2}} & \dots & m_V^N(\lambda^N(u))^{\frac{3}{2}} \end{bmatrix},
$$

$$
E_n(u) \triangleq \begin{bmatrix} \lambda^n(u) & \lambda^n(u)m_V^{u,n} \\ \lambda^n(u)m_V^{u,n} & \lambda^n(u)((m_V^{u,n})^2+(\sigma_V^{u,n})^2) \end{bmatrix}
$$

Model (17) gives a chance both to solve the MJP state filtering problem given the diffusion and counting observations and develop corresponding algorithms of the numerical solution to the filtering problem.

By contrast with weak convergence in (12), any convergence in (15) is absent. First, *the right-hand side* (RHS) of (15) contains the mathematical expectation which is increasing function of t. Second, we determine (15) under hypothesis that the MJP state X remains unchanged over the discretization interval: $X_q \equiv X_s$, $q \in [s, s+t)$. In the general case, the probability of MJP state transition increases to 1 when the interval length t increases infinitely.

Use of the time-discretized observations (4) at the first stage of the control synthesis–MJP state filtering–presumes calculation of likelihood ratios for the single Gaussian modes and their mixtures. Therefore, the filtering performance depends on both the "theoretical" pdf (15) and the closeness of real distribution of the observation increments to (15).

We form recommendations for appropriate choice of the time interval for discretization of (4). On the one hand, the length should provide the appropriate performance of the diffusion approximation (15), when there is no MJP state transitions over the time interval. On the other hand, the interval length should be small enough to guarantee small probability of those state transitions.

In the CLT the closeness of the limit distribution and the pre-limit one is described by the Berry–Esseen inequality in terms of either the uniform metric or the total variation one [45–47]. By contrast, we are interested in closeness of the corresponding PDFs, and the appropriated results are valid for the case of the "classic" CLT, not for CLTRRP.

We propose some heuristic technique choose the discretization interval length, basing on a performance criterion of the distribution approximation.

We refer to the "single mode" processes Θ_t^j and construct the processes

$$
\overline{\Theta}_h^j \triangleq \left(\sqrt{\Theta_h^{j,1}}\right)^+ \frac{1}{\sigma_V^j}\left(\Theta_h^{j,2} - m_V^j\Theta_h^{j,1}\right). \tag{18}
$$

From the definition one can conclude that $\overline{\Theta}_h^j$ represents the normalized sum of the random number of independent equally distributed normalized random summands. We investigate closedness of its distribution to the standard Gaussian one depending on time h.

Below in the filtering algorithm we operate with various likelihood ratios calculated via the pdfs, hence we need to characterize a distance between the pre-limit pdf and its limit one. The precise distance is difficult to calculate, and we must turn to some upper bound of this quantity.

Let $\mu(dx)$ be some positive measure on $(\mathbb{R}, \mathcal{B}(\mathbb{R}))$, and there exist both the pdf $\frac{dP_a}{d\mu}$ of the pre-limit distribution and the limit one $\frac{dP_\ell}{d\mu}$. Then the relative approximation error takes the form

$$\Delta(x) \triangleq \frac{\left|\frac{dP_a}{d\mu} - \frac{dP_\ell}{d\mu}\right|}{\frac{dP_\ell}{d\mu}}(x),$$

and its average

$$\int \Delta(x) \frac{dP_\ell}{d\mu}(x)\mu(dx) = Var(P_a, P_\ell)$$

coincides with *the total variation distance* (TVD) between P_a and P_ℓ.

We use the notation $P^j(x,h) \triangleq \mathcal{P}\left\{\overline{\Theta}_h^j \leqslant x\right\}$ for the pre-limit distribution function, $P_n^j(x)$ stands for the distribution function of the normalized sum of n independent equally distributed normalized random summands with the pdf $\Lambda^j(u)$, and $\Phi(x) \triangleq \int_{-\infty}^{x} \frac{1}{\sqrt{2\pi}} e^{-\frac{z^2}{2}} dz$ does for the distribution function of the standard Gaussian random value. From the total probability formula, it follows that

$$P^j(x,h) = e^{-\lambda^j(u)h}\left(\mathbf{I}(x) + \sum_{n \in \mathbb{N}} \frac{(\lambda^j(u)h)^n}{n!} P_n^j(x)\right), \tag{19}$$

where $\mathbf{I}(x)$ is the Heaviside function.

Proposition 1. *For $\lambda^j(u)h \geqslant \frac{3+\sqrt{13}}{2}$ an approximate upper bound of $Var(P^j, \Phi)$ can be written as*

$$J^j(h) = e^{-\lambda^j(u)h}\left(2 + C_1\left(2\Phi(-3) + \left(\frac{1}{\sqrt{1 - \frac{3}{\sqrt{\lambda^j(u)h}}}} + \frac{1}{\sqrt{1 + \frac{3}{\sqrt{\lambda^j(u)h}}}}\right)\right)\right), \tag{20}$$

where $C_1 = C_1(\Lambda^j(u,\cdot))$ is some parameter.

Proof. From (19) and the results of [48] (Theorem 1.1) and [49] (Theorem 2.6) the following inequalities are true

$$Var(P^j, \Phi) \leqslant e^{-\lambda^j(u)h}\left(2 + \sum_{n \in \mathbb{N}} \frac{(\lambda^j(u)h)^2}{n!} Var(P_n^j, \Phi)\right) \leqslant e^{-\lambda^j(u)h}\left(2 + C_1 \sum_{n \in \mathbb{N}} \frac{(\lambda^j(u)h)^2}{\sqrt{n}n!}\right), \tag{21}$$

where $C_1 = C_1(\Lambda^j(u,\cdot))$ is some parameter (see [48,49] for details).

Under the Proposition conditions the approximation of the Poisson distribution by the Gaussian one is valid

$$\sum_{n\in\mathbb{N}} \frac{(\lambda^j(u)h)^2}{\sqrt{n}n!} \approx \int_1^{\infty} \frac{1}{\sqrt{x}} \frac{1}{\sqrt{2\pi\lambda^j(u)h}} e^{\frac{(x-\lambda^j(u)h)^2}{2\lambda^j(u)h}} dx$$
$$\leqslant 2\Phi(-3) + \int\limits_{\lambda^j(u)h-3\sqrt{\lambda^j(u)h}}^{\lambda^j(u)h+3\sqrt{\lambda^j(u)h}} (ax+b)\frac{1}{\sqrt{x}} \frac{1}{\sqrt{2\pi\lambda^j(u)h}} e^{\frac{(x-\lambda^j(u)h)^2}{2\lambda^j(u)h}} dx, \tag{22}$$

where

$$\begin{cases} a \triangleq \frac{\sqrt{\lambda^j(u)h-3\sqrt{\lambda^j(u)h}}-\sqrt{\lambda^j(u)h+3\sqrt{\lambda^j(u)h}}}{6\sqrt{\lambda^j(u)h}\sqrt{(\lambda^j(u)h)^2-9\lambda^j(u)h}}, \\ b \triangleq \frac{1}{\sqrt{\lambda^j(u)h-3\sqrt{\lambda^j(u)h}}} - \frac{\lambda^j(u)h-3\sqrt{\lambda^j(u)h}-\sqrt{\lambda^j(u)h-9}}{6\sqrt{\lambda^j(u)h+3\sqrt{\lambda^j(u)h}}}. \end{cases} \tag{23}$$

Coefficients a and b above correspond to a piecewise linear majorant for $y(x) = \frac{1}{\sqrt{x}}$ over the interval $[1, +\infty)$ (see Figure 1).

We can calculate the last integral analytically

$$\sum_{n\in\mathbb{N}} \frac{(\lambda^j(u)h)^2}{\sqrt{n}n!} \lesssim 2\Phi(-3) + (1-2\Phi(-3))(a\lambda^j(u)h+b) \leqslant 2\Phi(-3) + a\lambda^j(u)h + b$$
$$= 2\Phi(-3) + \frac{1}{2\sqrt{\lambda^j(u)h}}\left(\frac{1}{\sqrt{1-\frac{3}{\sqrt{\lambda^j(u)h}}}} + \frac{1}{\sqrt{1+\frac{3}{\sqrt{\lambda^j(u)h}}}}\right). \tag{24}$$

Using the RHS of (24) in (21) we obtain the approximate upper bound (20). This ends the sketch of the proof of the Proposition. □

To characterize the distance between the Q_t (16) increment distribution and its diffusion approximation (17) we should take into account the chance of the MJP transition during the discretization interval. Let us suppose $X_t^u = e_j$, then, taking into account (20), the upper bound of $Var(P^u, \Phi | X_t = e_j)$ can be obtained by the total probability formula:

$$Var(P, \Phi | X_t = e_j) \leqslant \mathcal{J}^j(u,h) \triangleq$$
$$\triangleq e^{(A^{jj}(u)-\lambda^j(u))h}\left(2 + C_1\left(2\Phi(-3) + \left(\frac{1}{\sqrt{1-\frac{3}{\sqrt{\lambda^j(u)h}}}} + \frac{1}{\sqrt{1+\frac{3}{\sqrt{\lambda^j(u)h}}}}\right)\right)\right) + 2\left(1 - e^{A^{jj}(u)h}\right). \tag{25}$$

The second summand in (25) answers the chance the MJP can leave the state e_j during the time interval with probability $1 - e^{A^{jj}(u)h}$, and the multiplier 2 is the upper bound of the TVD for any distributions.

To take into account the statistical uncertainty of the current state X_t^u, we must consider the following averaged criterion:

$$\mathbf{J}(u, p_1, \ldots, p_N, h) \triangleq \sum_{j=1}^{N} p_j \mathcal{J}^j(u,h), \tag{26}$$

which describes the guaranteeing estimate of distribution distance for the case of the fixed control $u \in \mathrm{U}$ and $X_t^u \sim \mathrm{col}(p_1, \ldots, p_N)$.

From the practical point of view, the "rational" value of the time increment h can be chosen following to the one of policies:

1. Numerical analysis of the values $\mathcal{J}^j(u,h)$ for various (j,u,h) for the choice of an appropriate value for h.
2. Solution to the individual minimax problems

$$\mathcal{J}^j(u,h) \to \min_{h:\lambda^j(u)h \geqslant \frac{3+\sqrt{13}}{2},\, u\in \mathsf{U}} \max_{u\in\mathsf{U}}, \qquad j=\overline{1,N}$$

with subsequent choice of the maximal h from the set of the individual solutions.

3. Solution to the general minimax problem

$$\mathbf{J}(u,p_1,\dots,p_N,h)) \to \min_{h:\lambda^j(u)h \geqslant \frac{3+\sqrt{13}}{2},\, u\in\mathsf{U},\, j=\overline{1,N}} \max_{\substack{u\in\mathsf{U},\\ (p_1,\dots,p_N)\in\Pi}} .$$

In this paper, we use the first policy as the most economical one.

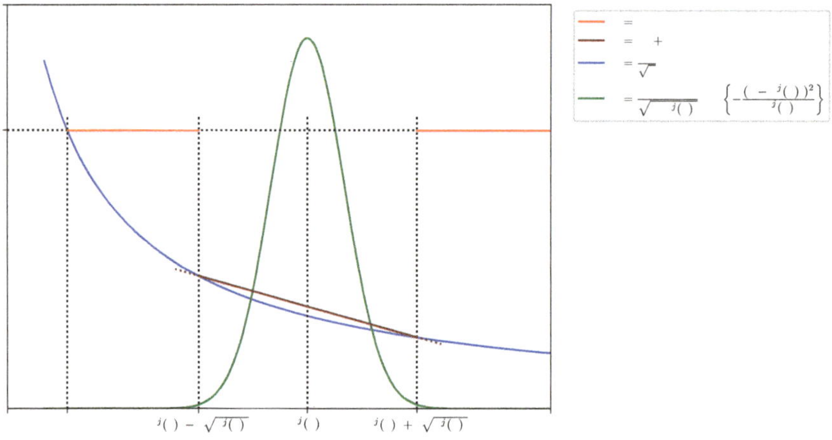

Figure 1. The function $y=\frac{1}{\sqrt{x}}$ and its piecewise linear majorant against the Gaussian.

3.3. Optimal Filtering of MJP State Given Counting and Diffusion Observations

In this section, we investigate MJP state (1) filtering problem given counting (2), (3) and diffusion observations (17). Without loss of generality to simplify the presentation and subsequent analysis of the solution to the MJP filtering problem we must introduce below the additional assumptions.

1. The control u_t represents an observable nonrandom *cádlág*-process.
2. The noises in Q_t are uniformly nondegenerate [50], i.e., $\min_{\substack{1\leqslant n\leqslant N,\\ u\in\mathsf{U}}} E_n(u) > \alpha I$ for some $\alpha > 0$.
3. The processes $K_{ij}(u_t) \triangleq \mathbf{I}_{\{\mathbf{0}\}}(E_i(u_t)-E_j(u_t))$, $i,j=\overline{1,N}$ has a finite local variation (here and below $\mathbf{0}$ stands for a zero matrix of appropriate dimensionality); $K(u_t) \triangleq \|K_{ij}(u_t)\|_{i,j=\overline{1,N}}$ is the corresponding $N\times N$-dimensional matrix-valued function.
The optimal filtering problem is to find *a Conditional Mathematical Expectation* (CME) $\widehat{X}^u_t \triangleq \mathsf{E}\{X^u_t|\mathcal{O}_{t+}\}$, where $\mathcal{O}_t \triangleq \sigma\{Y_s, Z_s, Q_s, s\in[0,t]\}$ is a natural flow of σ-algebras generated by the observations (2), (3) and (17).

The noise intensity in the observations (17) depends on the estimated state X, and this fact prevents to apply the known results of the optimal nonlinear filtering [37]. To overcome this obstacle, we use a special transformation of available diffusion observations [28]. Here we present a sketch of this transformation.

The Ito rule gives a possibility to obtain the observable quadratic characteristics of Q:

$$\langle Q, Q\rangle_t = \int_0^t \sum_{n=1}^N e_n^\top X_s E_n(u_s)ds. \tag{27}$$

We use the normalized diffusion observations

$$\overline{Q}_t \triangleq \int_0^t \left(\frac{d\langle Q, Q\rangle_s}{ds}\right)^{-\frac{1}{2}} dQ_s. \tag{28}$$

as the first block component of the transformed observations. The model of this process is the following

$$\overline{Q}_t = \int_0^t \overline{D}(u_s)X_s ds + \overline{W}_s, \tag{29}$$

where $\overline{D}(u) \triangleq \sum_{n=1}^N E_n^{-\frac{1}{2}}(u)D(u)\operatorname{diag} e_n$, and $\overline{W}_t$ is a standard Wiener process of appropriate dimensionality.

The quadratic characteristics $\langle Q, Q\rangle$ contains essential statistical information which should be included in the estimation algorithm. This process is a linear transformation of the estimated MJP state.

It is easy to verify that

$$F(u_t, X_t) \triangleq \frac{d\langle Q, Q\rangle_s}{ds}\bigg|_{s=t+} = \sum_{n=1}^N e_n^\top X_t E_n(u_t),$$

however, result of the direct derivation is a matrix-valued function with the excess dimensionality. All its statistical information is included in the complete preimage of F:

$$F = F(u, x) \xrightarrow{F^{-1}} \{e_n \in \mathbb{S}^N : E_n(u) = F\}.$$

In [28] we explain in detail how to reduce the "rough" process F to the N-dimensional "compressed" process H_t, which has the model

$$H_t = L(u_t)X_t, \tag{30}$$

where $L(u_t)$ is an $N \times N$-dimensional matrix-valued function with cádlág components; its rows are orthogonal and contains 0 or 1 only.

One can rewrite the process H_t as a cumulative sum of the jumps occurred at some nonrandom (or O_t-predictable) moments τ (the term H_t^D) and one, which accumulates jumps at the random (totally inaccessible) moments (the term H_t^R):

$$H_t = \underbrace{L(u_0)X_0 + \sum_{\tau \leqslant t} \Delta L(u_\tau)X_\tau}_{\triangleq H_t^D} + \underbrace{\int_0^t L(u_s)dX_s}_{\triangleq H_t^R}.$$

The process H_t^D represents the second block component of the transformed diffusion observations. To obtain the third component we must express H_t^R through the equivalent complex of the counting processes $G_t = \operatorname{col}(G_t^1, \dots, G_t^N)$:

$$G_t \triangleq \int_0^t (I - \operatorname{diag} H_{s-})dH_s - H_t^D.$$

The components of the process have the following properties.

1. Each component G_t^n has the martingale representation

$$G_t^n = \int_0^t \mathbf{1}\Gamma_n(u_s)X_s ds + \int_0^t (1 - L_n(u_s)X_{s-})L_n(u_s)d\alpha_s^u, \tag{31}$$

where α_t^u is the martingale from the state representation (1), $L_n(u) \triangleq e_n^\top L(u)$ and

$$\Gamma^n(u) \triangleq \mathrm{diag}(L_n(u))\Lambda^\top(u)(I - \mathrm{diag}(L_n(u))).$$

2. $[G^n, G^m]_t \equiv 0$ for any $n \neq m$, and $\langle G^n, G^n\rangle_t = \int_0^t \mathbf{1}\Gamma_n(u_s)X_s ds$.

Below we present a stochastic system for the CME $\widehat{X}_t$ along with its properties.

Proposition 2. *The following assertions are true.*

1. *The CME $\widehat{X}$ is the unique strong solution to the stochastic system*

$$\begin{aligned}
\widehat{X}_t = {} & \left((H_0^D)^\top L(u_0)\pi_0\right)^+ \mathrm{diag}(H_0^D)L(u_0)\pi_0 + \int_0^t \Lambda^\top(u_s)\widehat{X}_s ds + \int_0^t \widehat{k}_s \overline{D}^\top(u_s)d\omega_s \\
& + \sum_{n=1}^N \int_0^t \left(\Gamma_n(u_s) - \mathbf{1}\Gamma_n(u_s)\widehat{X}_{s-} I\right)\widehat{X}_{s-}\left(\mathbf{1}\Gamma_n(u_s)\widehat{X}_{s-}\right)^+ d\nu_s^n \\
& + \int_0^t \widehat{k}_s B^\top(u_s)(B(u_s)\widehat{X}_{s-})^+ d\widehat{\beta}_s + \int_0^t \widehat{k}_s C^\top(u_s)(C(u_s)\widehat{X}_{s-})^+ d\widehat{\gamma}_s \\
& + \sum_{\tau \leqslant t}\left(\left((\Delta H_\tau^D)^\top \Delta L(u_\tau)\widehat{X}_{\tau-}\right)^+ \mathrm{diag}(\Delta H_\tau^D)L(u_\tau) - I\right)\widehat{X}_{\tau-},
\end{aligned} \tag{32}$$

where

$$\begin{aligned}
\widehat{k}_t &\triangleq \mathrm{diag}\,\widehat{X}_t - \widehat{X}_t(\widehat{X}_t)^\top = \mathsf{E}\left\{(\widehat{X}_t - X_t)(\widehat{X}_t - X_t)^\top | O_{t+}\right\}, \\
\omega_t &\triangleq \int_0^t (d\overline{Q}_s - \overline{D}(u_s)\widehat{X}_s ds), \\
\nu_t^n &\triangleq \int_0^t (dG_s^n - \mathbf{1}\Gamma_n(u_s)\widehat{X}_{s-}ds), \quad n = \overline{1,N}, \\
\widehat{\beta}_t &\triangleq \int_0^t (dY_s - B(u_s)\widehat{X}_{s-}ds), \\
\widehat{\gamma}_t &\triangleq \int_0^t (dZ_s - C(u_s)\widehat{X}_{s-}ds).
\end{aligned}$$

2. *The estimate of the maximum a posteriori probability (MAP) $\widetilde{X}_t = e_n$: $n \in \operatorname{Argmax}_{1\leqslant m\leqslant N} e_m^\top \widehat{X}_t$ minimizes the $\mathcal{L}_1$-criterion, i.e., $\widetilde{X}_t \in \operatorname{Argmin}_{\overline{X}_t} \mathsf{E}\{\|\overline{X}_t - X_t\|_1\}$.*
3. *If $E_n(u) \neq E_m(u)$ for any $n \neq m$ almost everywhere on $[0,t]$, then $\widehat{X}_t = X_t$ $\mathcal{P}$-a.s.*

The validity of items 1 and 3 in Proposition 2 can be proved by complete analogy with [28] (Theorem 1, Corollary 1), meanwhile the one of item 2 is proved in [51].

The theoretical assertions above are also meaningful from the practical point of view for subsequent design of the suboptimal control of MJP state under incomplete information. First, the CME $\widehat{X}_t$ represents a solution to some closed finite-dimensional stochastic system, by contrast with the general case of the optimal filtering problem [37]. Second, the paths of the CME $\widehat{X}_t$ usually are piecewise continuous functions with values in Π, meanwhile the MJP X state trajectories are $\mathcal{P}$-a.s. piecewise constant functions with values in $\mathbb{S}^N$. Therefore, we cannot directly substitute the state X by its estimate $\widehat{X}$, imposing the separation principle to this control problem. The CME $\widehat{X}$ can be easily transformed into the MAP estimate $\widetilde{X}$ with the paths with the same properties as the ones of X. Assertion 2 of Proposition indicates that the proposed MAP estimate is also $\mathcal{L}_1$-optimal. Third, if the observation system satisfies the identifiability conditions (see Assertion 3 of Proposition) then the MJP state can be restored *exactly* given the indirect noisy observations. This crucial property gives a chance to reduce the initial control problem with incomplete information to the one with complete information. Obviously, any numerical realization of the filtering estimate leads to some approximation errors, nevertheless Assertion 3 allows one to hope that the small filtering errors cause acceptable control performance.

At the same time, results of Proposition 2 are difficult for the direct application. First, due to the approximation of the acknowledgment flow (4) by the diffusion model (17), the former one is valid and can be effectively applied only for the observation increments over the time interval of significant length (see Section 3.2). Second, the process H_t, playing the key role in the estimation, is not observable directly, and represents a result of some stochastic limit passage since it is based on the quadratic characteristic $\langle Q, Q\rangle$. Due to the boundedness from below of the diffusion observation time increment, direct calculation of H_t looks impossible. In the next subsection, basing on the time-discretized diffusion observations we present a special numerical algorithm of the nonlinear filtering together with its performance characteristics.

3.4. Numerical Realization of Filtering Algorithm

To construct the numerical algorithm of the MJP state filtering given the combination of both the diffusion and counting observations we consider a time-invariant version of the observation system (1), (3), (2), (17) given the observations discretized by time with the time increment $h > 0$ ($t_r \triangleq rh,\ r \in \mathbb{N}$):

$$X_t = X_0 + \int_0^t AX_s ds + \alpha_t, \tag{33}$$

$$\mathcal{Y}_r = \int_{t_{r-1}}^{t_r} BX_s ds + (\beta_{t_{r-1}} - \beta_{t_r}), \tag{34}$$

$$\mathcal{Z}_r = \int_{t_{r-1}}^{t_r} CX_s ds + (\gamma_{t_{r-1}} - \gamma_{t_{r-1}}), \tag{35}$$

$$\mathcal{Q}_r = \int_{t_{r-1}}^{t_r} DX_s ds + \int_{t_{r-1}}^{t_r} \sum_{n=1}^{N} e_n^\top X_s E_n^{\frac{1}{2}} dW_s, \tag{36}$$

and $\mathcal{O}_r \triangleq \sigma\{\mathcal{Y}_n, \mathcal{Z}_n, \mathcal{Q}_n,\ n \leqslant r\}$ is a natural filtration generated by the discretized observations.

An assumption that coefficients A, B, C, D and E are constant, is not too restrictive in practice because below we will construct the MJP control which will be constant during the time discretization intervals. Please note that the discretized observations $\mathcal{Y}_r$, $\mathcal{Z}_r$ and $\mathcal{Q}_r$ are conditionally independent given $\mathcal{F}_{t_r}^X \vee \mathcal{O}_{r-1}$ due to the properties of the Wiener-Poisson canonical space and the result of [50] (Lemma 7.5). Specifically, the distribution of $\mathcal{Y}_r$, $\mathcal{Z}_r$ and $\mathcal{Q}_r$ depends on the random vector $\eta_r = \mathrm{col}(\eta_r^1, \dots, \eta_r^N) = \int_{t_{r-1}}^{t_r} X_s ds$ is a random vector composed of the occupation times of the state X in each state e_n during the interval $[t_{r-1}, t_r]$. Then

- conditional distribution of $\mathcal{Y}_r$ given $\mathcal{F}_{t_r}^X \vee \mathcal{O}_{r-1}$ is the Poisson one with the parameter $B\eta_r$,
- conditional distribution of $\mathcal{Z}_r$ given $\mathcal{F}_{t_r}^X \vee \mathcal{O}_{r-1}$ is the Poisson one with the parameter $C\eta_r$,
- conditional distribution of $\mathcal{Q}_r$ given $\mathcal{F}_{t_r}^X \vee \mathcal{O}_{r-1}$ is the Gaussian one with the mean $D\eta_r$ and covariance matrix $\sum_{n=1}^N \eta_r^n E_n$.

Below in the presentation we use the following notations:

- $\overline{A} \triangleq \max_{n=\overline{1,N}} |A_{nn}|$;
- $\mathcal{D} \triangleq \{u = \mathrm{col}(u^1, \dots, u^N) :\ u^n \geqslant 0,\ \sum_{n=1}^N u^n = h\}$ is an $(N-1)$-dimensional simplex in the space $\mathbb{R}^M$; $\mathcal{D}$ is a distribution support of the vector v_r;
- $\Pi \triangleq \{\pi = \mathrm{col}(\pi^1, \dots, \pi^N) :\ \pi^n \geqslant 0,\ \sum_{n=1}^N \pi^n = 1\}$ is a "probabilistic simplex" formed by the possible values of π;
- N_r^X is a random number of the state X_t transitions, occurred on the interval $[t_{r-1}, t_r]$,
- $\rho^{k,\ell,q}(du)$ is a conditional distribution of the vector $X_{t_r}^\ell \mathbf{I}_{\{q\}}(N_r^X) v_r$ given $X_{t_{r-1}} = e_k$, i.e., for any $\mathcal{G} \in \mathcal{B}(\mathbb{R}^M)$ the following equality is true:

$$\mathsf{E}\left\{\mathbf{I}_{\mathcal{G}}(v_r)\mathbf{I}_{\{q\}}(N_r^X) X_{t_r}^\ell | X_{t_{r-1}} = e_k\right\} = \int_{\mathcal{G}} \rho^{k,\ell,q}(du);$$

- $\|\alpha\|_K^2 \triangleq \alpha^\top K\alpha$, $\langle\alpha,\beta\rangle_K \triangleq \alpha^\top K\beta$;
- $\mathcal{N}(q,m,K) \triangleq (2\pi)^{-M/2}\det^{-1/2}K\exp\left\{-\frac{1}{2}\|y-m)\|^2_{K^{-1}}\right\}$ is an M-dimensional Gaussian *probability density function* (pdf) with the expectation m and nondegenerate covariance matrix K;
- $\mathcal{P}(n,a) \triangleq e^{-a}\frac{a^n}{n!}$ is a Poisson distribution with the parameter a;
- $\Upsilon^{k,j,s}(y,z,q) \triangleq \int_{\mathcal{D}} \mathcal{P}(y,Bv)\mathcal{P}(z,Cv)\mathcal{N}(q,Dv,\sum_{i=1}^N v^i E_i)\rho^{k,j,s}(dv)$.

Below is an assertion introducing the calculation algorithm of the MJP state given the discretized observations $\widehat{\mathcal{X}}_r \triangleq \mathsf{E}\{X_{t_r}|\mathcal{O}_r\}$.

Proposition 3. *The filtering estimate $\widehat{\mathcal{X}}_r$ can be calculated be the following recursive algorithm*

$$\widehat{\mathcal{X}}_r^j = \frac{\sum_{n_1=1}^N \widehat{\mathcal{X}}_r^{n_1} \sum_{s_1=0}^\infty \Upsilon^{n_1,j,s_1}(\mathcal{Y}_r,\mathcal{Z}_r,\mathcal{Q}_r)}{\sum_{n_2,j_2=1}^N \widehat{\mathcal{X}}_r^{n_2} \sum_{s_2=0}^\infty \Upsilon^{n_2,j_2,s_2}(\mathcal{Y}_r,\mathcal{Z}_r,\mathcal{Q}_r)}, \qquad j=\overline{1,N}, \tag{37}$$

and initial condition

$$\widehat{\mathcal{X}}_0 = \pi_0. \tag{38}$$

Proof of Proposition 3 can be performed similarly to [28] (Lemma 3).

To construct a numerically realizable algorithm we must restrict the sums both in the numerator and denominator of (37)

$$\overline{\mathcal{X}}_r^j(S) = \frac{\sum_{n_1=1}^N \overline{\mathcal{X}}_r^{n_1} \sum_{s_1=0}^S \Upsilon^{n_1,j,s_1}(\mathcal{Y}_r,\mathcal{Z}_r,\mathcal{Q}_r)}{\sum_{n_2,j_2=1}^N \overline{\mathcal{X}}_r^{n_2} \sum_{s_2=0}^S \Upsilon^{n_2,j_2,s_2}(\mathcal{Y}_r,\mathcal{Z}_r,\mathcal{Q}_r)}, \qquad j=\overline{1,N}, \tag{39}$$

and obtain the analytical approximation of the Sth order.

We present some summands Υ of the low order s:

$$\Upsilon^{k,j,0}(y,z,q) = \delta_{kj}e^{A_{kk}h}\mathcal{P}(y,B^k h)\mathcal{P}(y,C^k h)\mathcal{N}(q,hD^k,hE_k),$$

$$\begin{aligned}\Upsilon^{k,j,1}(y,z,q) = {} & (1-\delta_{kj})A_{jk}e^{A_{jj}h} \\ & \times \int_0^h e^{(A_{kk}-A_{jj})v}\mathcal{P}(y,B^k v + B^j(h-v))\mathcal{P}(z,C^k v + C^j(h-v)) \\ & \times \mathcal{N}(q,vD^k+(h-v)D^j,vE_k+(h-v)E_j)dv,\end{aligned}$$

$$\begin{aligned}& \Upsilon^{k,j,2}(y,z,q) \\ & = \sum_{i:i\neq k,i\neq j} A_{ik}A_{ji}e^{A_{jj}h}\int_0^h\int_0^{h-v^k} e^{(A_{kk}-A_{ii})v^k+(A_{ii}-A_{jj})v^j}\mathcal{P}(y,B^k v^i + B^i v^i + B^j(h-v^k-v^j)) \\ & \times \mathcal{P}(z,C^k v^i + C^i v^i + C^j(h-v^k-v^j)) \\ & \times \mathcal{N}(q,v^k D^k + v^i D^i + (h-v^k-v^i)D^j, v^k E_k + v^i E_i + (h-v^k-v^i)E_j)dv^i dv^k,\end{aligned}$$

where D^k is the kth column of the matrix D. Other summands are also determined by the total probability formula and have complicated form. Obviously, the integrals above cannot be calculated analytically, and we approximate them by some integral sums

$$\widetilde{\Upsilon}^{k,j,s}(y,z,q) \triangleq \sum_{\ell}^{L} \mathcal{P}(y, Bv_\ell)\mathcal{P}(z, Cv_\ell)\mathcal{N}(q, Dv_\ell, \sum_{i=1}^{N} v_\ell^i E_i)\varrho_\ell^{kj}, \quad (40)$$

where $\{v_\ell\}_{\ell=\overline{1,L}} \subset \mathcal{D}$ is a collection of points, and $\{\varrho_\ell^{kj}\}_{\ell=\overline{1,L}}$ are corresponding weights, such that $\sum_{j=1}^{N}\sum_{\ell=1}^{L} \varrho_\ell^{kj} \leqslant 1$. Therefore, we calculate the filtering estimate by the recursion

$$\widetilde{\mathcal{X}}_r^j(S) = \frac{\sum\limits_{n_1=1}^{N} \widetilde{\mathcal{X}}_r^{n_1} \sum\limits_{s_1=0}^{S} \Upsilon^{n_1,j,s_1}(\mathcal{Y}_r, \mathcal{Z}_r, \mathcal{Q}_r)}{\sum\limits_{n_2,j_2=1}^{N} \widetilde{\mathcal{X}}_r^{n_2} \sum\limits_{s_2=0}^{S} \Upsilon^{n_2,j_2,s_2}(\mathcal{Y}_r, \mathcal{Z}_r, \mathcal{Q}_r)}, \qquad j = \overline{1,N}, \quad (41)$$

and refer it as *the numerical approximation of the Sth order, corresponding to a chosen numerical integration scheme.*

Let us fix a time instant t, and consider the asymptotic performance of approximation (41) as $h \to 0$. The performance index is $\sup_{\pi\in\Pi} \mathsf{E}_\pi\left\{\|\widetilde{X}_r - \widehat{X}_r\|_1\right\}$, i.e., an average of the $\mathcal{L}_1$-norm of the filtering error calculated at the step r for the worst initial distribution of the MJP.

Proposition 4. *If the condition*

$$\max_{k=\overline{1,N},(y,z)\in\mathbb{Z}_+^2} \sum_{j=1}^{N} \int_{\mathbb{R}^2} \left| \sum_{s=0}^{S} \widetilde{\Upsilon}^{n_1,j,s_1}(y,z,q) - \Upsilon^{n_1,j,s_1}(y,z,q) \right| dq < \delta,$$

holds, then for small enough h

$$\sup_{\pi\in\Pi} \mathsf{E}_\pi\left\{\|\widetilde{X}_{t/h} - \widehat{X}_{t/h}\|_1\right\} \leqslant 2t\left(2\overline{A}\frac{(\overline{A}h)^S}{(S+1)!} + \frac{\delta}{h}\right). \quad (42)$$

Proof of Proposition 4 can be performed similarly to [28] (Lemma 4, Theorem 2). The first term in (42) characterizes the error of the analytical approximation: formula (39) takes into account at most S possible state transitions occurred during the time discretization interval $[t_{r-1}, t_r]$. The second term in (42) describes an impact of numerical integration error to the overall performance of the filtering approximation. We can deduce that the effective choice of the integration scheme should provide the equal contribution of both summands in (42).

For the numerical study we choose the analytical approximation of the 1st order realized by the middle-point scheme:

$$\widetilde{\Upsilon}^{k,j,0}(y,z,q) = \Upsilon^{k,j,0}(y,z,q),$$

$$\widetilde{\Upsilon}^{k,j,1}(y,z,q) = (1-\delta_{kj})A_{jk}e^{\frac{h}{2}(A_{jj}+A_{jj})}h\mathcal{P}(y, \tfrac{h}{2}(B^k+B^j))\mathcal{P}(z, \tfrac{h}{2}(C^k+C^j))\mathcal{N}(q, \tfrac{h}{2}(D^k+D^j), \tfrac{h}{2}(E_k+E_j)).$$

4. State-Based Modification of TCP

In this section, we describe a TCP channel mathematical model we later use for simulation of some modern TCP versions and their comparison with the state-based optimal control policy. The model we use here is in general following the one of [52]. The main distinctive characteristic of this model is the channel state allocation: we use three states to describe the wire channel condition and add one extra state to cover the issues of the wireless connection. This allocation presents a reasonable trade-off between a comprehensive connection state model taking into account all possible features (including the data flows from every channel user, the current packet distribution in all the channel hops and buffers' queues, and signal quality in the wireless channel segment) and the feasibility of the mathematical modeling.

Thus, we suppose that the link state from a sender to a receiver is described by a controllable MJP X_t (1) with four possible states:

- e_1 is assigned for low channel load,
- e_2 is for moderate load,
- e_3 is for wired segment congestion,
- e_4 is for signal fading in the wireless segment.

The intensity matrix $A(u) = \|A^{ij}(u)\|_{i,j=\overline{1,4}}$ is defined based on the following assumptions: the link has a single bottleneck device, which remains the same during the whole transmission, this bottleneck device uses Random Early Detection (RED) queuing discipline [53], its buffer capacity is Q, and the RED threshold of guaranteed packet rejection is W'' ($W'' \leqslant Q$).

We also assume that the wireless connection quality does not dependent on the data flow, hence the intensities $A^{\cdot 4}$ and $A^{4\cdot}$ corresponding to the transitions from/into the state e_4 are independent of the control u_s. Furthermore, the direct transitions between the e_1 and e_3 without passing through the e_2 are assumed impossible, i.e., $A^{13} = A^{31} \equiv 0$.

The controllable components of $A(u_t)$ have the form

$$\begin{aligned}
A^{21}(u_t) &= \begin{cases} A_0^{21} + \dfrac{C^{21}}{U_{bdp} - u_t}, & \text{if} \quad u_t < U_{bdp}, \\ \overline{A}, & \text{otherwise}; \end{cases} \\
A^{12}(u_t) &= A_0^{12} + C^{12} \max(U_{bdp} - u_t, 0); \\
A^{32}(u_t) &= \begin{cases} A_0^{32} + \dfrac{C^{32}}{W'' - u_t}, & \text{if} \quad u_t < W'', \\ \overline{A}, & \text{otherwise}; \end{cases} \\
A^{23}(u_t) &= A_0^{23} + C^{23} \max(W'' - u_t, 0),
\end{aligned}$$

where U_{bdp} is the control, which corresponds to the bandwidth-delay product (BDP), in other words—the maximum window size yielding throughput equal to channel bandwidth. The constant $\overline{A}$ is a level of intensity which guarantees the state transition during the forthcoming RTT.

The dependence of $A^{ji}(u_t)$ on control u_t is straightforward. In the state e_1, the number of packets in the link is less than U_{bdp}; and in the state e_2 the "bottleneck" buffer begins to fill. The inverse proportionality of $A^{21}(u_t)$ on u_t and guaranteeing intensity $\overline{A}$ provides the increasing probability of $e_1 \to e_2$ transition as u_t approaches to U_{bdp} and guarantees the transition when the threshold U_{bdp} is reached. The constant additive term A_0^{21} stands for a chance of the $e_1 \to e_2$ transition under low control values $u < U_{bdp}$, which are probable due to the external flows. When u_t decreases to levels less than U_{bdp}, the probability of backward transition $e_2 \to e_1$ increases linearly due to the constant flow processing rate. The transition intensities $e_2 \leftrightarrows e_3$ act the same way, but with a different threshold, namely W''.

The conditional intensities of the acknowledgment arrivals $\lambda^j(u)$ depend on the control u and, according to (10), are inversely proportional to the average time between the acknowledgment arrivals:

$$\lambda^j(u) = \frac{1}{m_\tau^j(u)}.$$

We assume that if no packets are lost, then during each RTT cycle, the sender receives back the acknowledgments for all the packets currently being sent into the network; hence we assume that the following relation is valid:

$$m_\tau^j(u) = \frac{m_V^j(u)}{u}.$$

The average RTT for each state $m_V^j(u)$ is assumed to be a sum of the following components:

- constant propagation delay, δ_0,
- average queuing delay caused by external data, flows $m_{V,ext}^j$,
- average queuing delay caused by the data flow under control, $u \cdot m_{V,self}^j$.

Summing up the assumptions, we have the following relation for the conditional intensity of the acknowledgment arrivals:

$$\lambda^j(u) = \frac{u}{\delta_0 + m_{V,ext}^j + u m_{V,self}^j}. \quad (43)$$

The counting processes for loss (2) and timeouts (3) can now be defined as thinned versions of the acknowledgment flow with following conditional intensities:

$$\begin{aligned} B^j(u) &= B_0^j + \lambda^j(u) P_l(u), \\ C^j(u) &= \lambda^j(u) P_{to}^j. \end{aligned} \quad (44)$$

Here P_{to}^j denotes the conditional probabilities of a timeout in the corresponding states. For the states $e_{1,2,3}$, which are related to the wired part of the link, we assume that the only cause for a timeout is a temporary communication hardware fault; and hence the probabilities for these states are constant and equal to each other: $P_{to}^1 = P_{to}^2 = P_{to}^3$. In the state e_4, the timeouts follow the wireless carrier signal fading; hence the probability of a timeout P_{to}^4 is different but still independent of the control u.

The packet loss conditional probabilities, on the contrary, are the functions of the control u. If the control value is less than the RED threshold $u < W''$, then

$$P_l^1(u) = P_0, \quad P_l^2(u) = P_0 + \max\left(\frac{U_t - W'}{W'' - W'}(P_1 - P_0), 0\right), \quad P_l^3(u) = 1, \quad P_l^4(U_t) = P_l^4,$$

where P_0 is the probability of a packet loss in the wired segment during its propagation through the media, W' is the lower RED threshold ($W' < W''$). If the threshold of guaranteed packet loss is exceeded, then the loss is inevitable, thus $P^j(u) = 1$ for any j, if $u \geq W''$.

To conclude the definition of the loss and timeout intensities, it remains to mention that the additive terms B_0^j in the loss intensity $B(u)$ stand for the losses caused by the external flows.

5. Comparative Study with Modern Versions of TCP

We have completely described the observation system (1)–(4) and its parameters' dependence on the control u. Let $\mathcal{O}_t \triangleq \{Y_s, Z_s, Q_s, 0 \leqslant s \leqslant t\}$ be the natural filtration generated by the observations available up to the moment t. Generally speaking, any $\mathcal{O}_t$-predictable nonnegative control U_t is admissible to (1)–(4).

In this section, we present the control processes, which describe the modern versions of TCP in terms of the presented model of channel state and observations. We also present here a state-based TCP control modification, which is based on the optimal state filtering and optimal control strategy. The section will be concluded by a comparative analysis of the TCP versions' performance.

In what follows we will assume that the constant values U_{bdp}, W'', δ_0 and $m_{V,self}^j$ are selected so as to comply with the link of $C = 100$ Mbps capacity, propagation delay of $\delta_0 = 0.1$ s, bottleneck queue limit of $Q = 100$ packets, and $MSS = 1000$ bytes:

$$U_{bdp} = \frac{10^6\, C\, \delta_0}{8\, MSS} = 1250, \quad W'' = U_{bdp} + Q = 1350, \quad m_{V,self}^j = \frac{8\, MSS}{10^6\, C} = 8 \cdot 10^{-5}.$$

5.1. AIMD Scheme and TCP Illinois

In [52] we presented an AIMD type control u_t policy, which remains the same for the present channel model:

$$\begin{cases} u_t = u_0 + \int_0^t \mathbf{I}_{[\underline{W}, W_t^{th})}(u_{s-}) \frac{u_{s-}}{r_{s-}} ds + \int_0^t \mathbf{I}_{[W_t^{th}, +\infty)}(u_{s-}) \frac{\alpha_s}{r_{s-}} ds - \int_0^t \beta_s u_{s-} dY_s + \int_0^t (\underline{W} - u_{s-}) dZ_s, \\ W_t^{th} = W_0^{th} + \int_0^t \left(\frac{1}{2} u_{s-} - W_{s-}^{th} \right) dY_s, \end{cases} \quad (45)$$

where

- $\mathbf{I}_S(u)$ is an indicator function equal to one, if $u \in S$, and zero otherwise,
- $\underline{W}$ is the minimal window size,
- W_t^{th} is a threshold actuating congestion avoidance phase,
- r_t is the exponential smoothing estimate of RTT,
- α_t and β_t are $\mathcal{O}_t$-predictable coefficients of additive increase and multiplicative decrease.

The first term in (45) describes the slow start mode, the second and the third stand for the linear increase and the multiplicative decrease in the congestion avoidance phase, and the fourth provides the window rollback to the minimal value $\underline{W}$ and return to the slow start mode when a timeout event occurs.

In the case $\alpha_t \equiv 1$ and $\beta_t \equiv 0.5$ Equation (45) represents the New Reno algorithm. The Illinois concave control policy is defined by convex α_t and increasing linear β_t functions of the average queuing delay $d_a = \sum_{j=1}^{4} (m_{V,ext}^j + u m_{V,self}^j) e_j^T X_t$:

$$\alpha_t(d_a) = \begin{cases} \alpha_{max} & \text{if } d_a \leqslant d_1 \\ \frac{\kappa_1}{\kappa_2 + d_a} & \text{otherwise,} \end{cases}$$
$$\beta_t(d_a) = \begin{cases} \beta_{min} & \text{if } d_a \leqslant d_2 \\ \kappa_3 + \kappa_4 d_a & \text{if } d_2 < d_a < d_3 \\ \beta_{max} & \text{otherwise,} \end{cases} \quad (46)$$

The parameters κ_i and d_i and other details of the Illinois control scheme can be found in [6]. It should be noted that the most important parameters are the maximum and minimum additive increase and multiplicative decrease coefficients, which for the standard implementation are set to $[\alpha_{min}, \alpha_{max}] = [0.3, 10]$, $[\beta_{min}, \beta_{max}] = [0.125, 0.5]$. In Figure 2, we present the simulation results for the Illinois TCP control policy for these standard parameters. The upper plot presents the channel parameters' dynamics, including RTT (in red), losses (black triangles), and timeouts (red crosses). The filling color indicates the channel states: white for idle, green for moderate load, red for congestion in the wired segment, and grey for the wireless segment signal fading. The lower plot shows the control dynamics and the critical thresholds: U_{bdp}, which corresponds to the channel bandwidth-delay product and buffer overflow low bound $U_{bdp} + W''$.

One can notice that by processing only the RTT information, the algorithm succeeds in the determination of the U_{bdp} and becomes much more prudent once the bottleneck buffer starts to fill. This results in long periods of relatively high transmission rates without buffer overflows and rare losses. Nevertheless, during the intervals, when the channel is idle, the control values growth speed is insufficient, which results in underuse of the channel resources and, in the end, in lower average transmission rate.

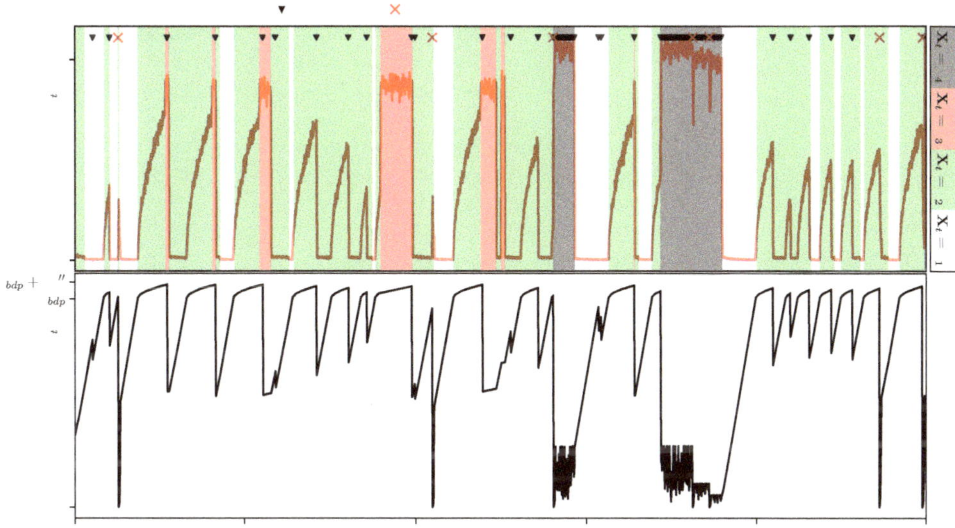

Figure 2. TCP channel simulation example for Illinois control algorithm.

5.2. TCP CUBIC

In contrast with TCP Illinois, this version of TCP does not rely on RTT observations most of the time. Instead, it considers the control value, at which a loss occurred last time,

$$W_t^{max} = \int_0^t (u_{s-} - W_{s-}^{max})\, dY_s,$$

as the highest network use control and tends to form a plateau in the close region to this point. To that end, it keeps counting the time since the last loss or timeout,

$$T_t^{loss} = t - \int_0^t t\, dY_s - \int_0^t t\, dZ_s,$$

and sets the control according to a cubic function of T_t^{loss} forming two regions: a concave region to reach the last maximum control value of W_t^{max}, and then a convex region of network probing, where the control growth speed becomes higher as the time without loss increases. Upon the loss event, the control is reduced according to a constant multiplicative decrease coefficient β, and when a timeout occurs, the control is reset to a minimal window size $\underline{W}$. Summing up, the TCP CUBIC control can be represented as follows:

$$u(t) = W_t^{max} + C\left(T_t^{loss} - \left(\frac{W_t^{max}(1-\beta)}{C}\right)^{\frac{1}{3}}\right)^3 - \int_0^t \beta u_{s-} dY_s + \int_0^t (\underline{W} - u_{s-})\, dZ_s, \quad (47)$$

where C is a constant fixed to determine the aggressiveness of control growth: with higher C values (for example, $C = 4.0$), CUBIC tends to be more aggressive, which can be quite useful in high BDP networks.

In Figure 3, we present the simulation results for the TCP CUBIC control with multiplicative decrease coefficient $\beta = 0.9$ and scale constant $C = 4.0$. It should be noted that this simulation is based on a more precise model of the protocol described in [38] and takes into account such details as TCP-friendly region and fast convergence heuristics. These details were not reflected in Equation (47) to avoid unnecessary complications. As in the previous Figure, the upper plot presents the channel dynamics (RTT, losses, timeouts, and state), and the lower plot shows the dynamics of the control.

One can see that TCP CUBIC manages to keep the control close to the desired U_{bdp} value, allowing fast recovery after losses. At the same time, the probing phase, which is symmetrical to the recovery phase, is too aggressive, and the average throughput would benefit from longer "plateau" periods. Another advantage, which must be mentioned, is the ability to adjust to dramatic changes in the media: in contrast with TCP Illinois, the CUBIC protocol keeps the control at low values throughout the whole period of wireless signal degradation, which results in fewer losses.

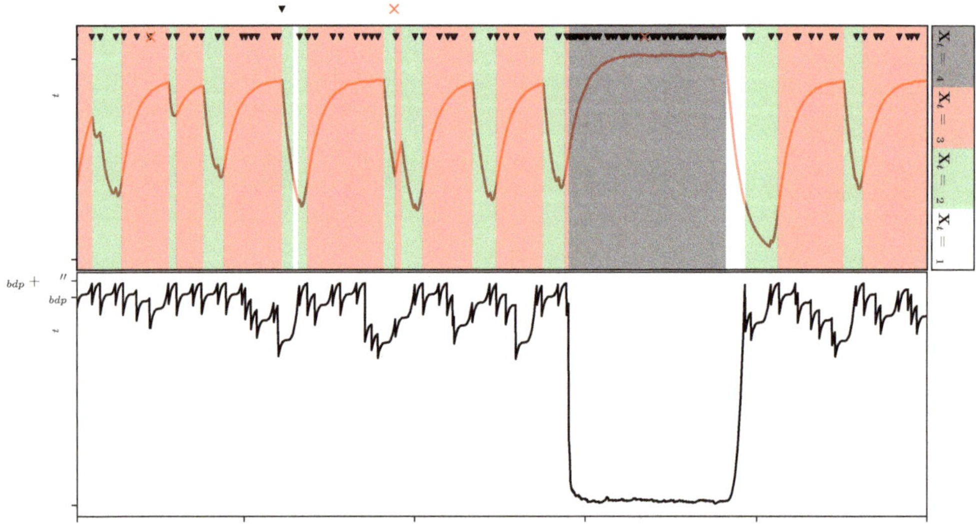

Figure 3. TCP channel simulation example for CUBIC control algorithm.

5.3. TCP Compound

The TCP Compound algorithm tries to benefit both from the loss-based and congestion-based approach. To that end, the authors enhance the standard AIMD congestion avoidance scheme with an additional component, which allows faster growth on an idle channel when standard AIMD control underuses the resources [40]. When the congestion is detected, the window is adjusted to avoid packet losses. To estimate the congestion, the TCP Compound scheme compares the estimated number of backlogged packets (bottleneck queue size) d_t with a known threshold value γ. The estimate of the queue size is computed as follows:

$$d_t = u_t \left(1 - \frac{V_t^{min}}{V_t}\right),$$

where V_t is current, and V_t^{min} is a minimum registered RTT value.

The entire TCP Compound control scheme can be represented by the following expression:

$$\begin{aligned} u_t = u_0 &+ \int_0^t \mathbf{I}_{[\underline{W}, W_t^{th})}(u_{s-}) \frac{u_{s-}}{r_{s-}} ds \\ &+ \int_0^t \mathbf{I}_{[W_t^{th}, +\infty)}(u_{s-}) \left(\mathbf{I}_{[0,\gamma)}(d_{s-}) u_t^{\kappa} \frac{\alpha}{r_{s-}} - \mathbf{I}_{[\gamma,+\infty)}(d_{s-}) \zeta d_{s-} \right) ds \\ &- \int_0^t \beta u_{s-} dY_s + \int_0^t (\underline{W} - u_{s-}) dZ_s, \end{aligned} \quad (48)$$

where

- $\mathbf{I}_{[\underline{W}, W_t^{th})}(u_{s-})$ is the slow start indicator,
- $\mathbf{I}_{[0,\gamma)}(d_{s-})$ is the congestion indicators,
- α, β, κ, ζ are tunable protocol parameters.

In (48), the first term describes the slow start mode, the second term reflects the growth phase and correction upon congestion detection, the third stands for the multiplicative decrease, and the fourth provides the window rollback and return to the slow start mode when a timeout event occurs.

In Figure 4, we present the simulation results for the TCP Compound protocol with standard parameter values: $\alpha = \beta = 0.125$, $\kappa = 0.75$, $\zeta = 1.0$. The backlog estimate threshold value for congestion indication is set to $\gamma = 80$. The upper plot presents the channel dynamics (RTT, losses, timeouts, and state), and the lower plot shows the dynamics of the control (in black) and the estimated backlog size d_t (in blue). The figure illustrates the correction of the control when the backlog size estimate reaches the threshold and high control values when the bottleneck buffer queue is assumed empty. It should be noted that TCP Compound, such as the Illinois version, fails to quickly adapt to the wireless signal degradation, demonstrating high instability and a big number of losses during this channel state.

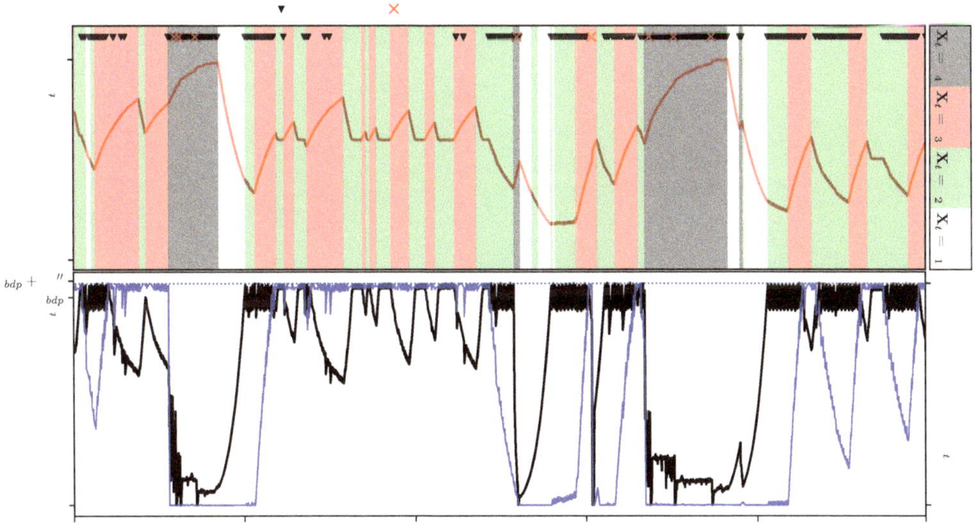

Figure 4. TCP channel simulation example for Compound control algorithm.

5.4. TCP BBR

The TCP BBR algorithm is purely delay-based [54]. It is designed with the idea of maintaining the total data in the channel equal to the BDP. At this load, a connection runs with the highest throughput and lowest delay. The BDP value is estimated as a product of *RTprop*—round-trip propagation time and *BtlBw*—bottleneck bandwidth or delivery rate. An estimate for the propagation time is the minimum registered RTT over a long time:

$$RTprop_t = min\{RTT_s\}, \quad s \in [t - W_R, t],$$

where W_R typically varies from tens of seconds to minutes. To estimate the delivery rate, BBR calculates the ratio of the portion of data delivered to the time elapsed from the delivery start. Since this ratio is calculated for every acknowledgment received, it is natural to take the data "inflight" at the moment the packet was sent as a portion and the RTT of this acknowledgment as the time elapsed from the delivery start. The estimated delivery rate then is a maximum of such ratios taken over a period W_B equal to 6–10 RTTs:

$$BtlBw_t = max\left\{\frac{u(t - RTT_t)}{RTT_t}\right\}, \quad s \in [t - W_B, t].$$

The main problem of this approach is that the propagation time and the delivery rate cannot be observed at the same time. Indeed, the bottleneck buffer must be empty to observe RTT values close to the propagation time and, to observe the capacity of the channel, it must be overfilled. This problem is solved by two modes of the steady-state regime: ProbeBW and ProbeRTT. In ProbeBW, the algorithm cycles through eight phases with the following pacing gain values: $p_t = (5/4, 3/4, 1, 1, 1, 1, 1, 1)$. The length of each phase is equal to the current estimate of the propagation time $RTprop_t$. Thus, the capacity of the channel is achieved by a periodical increase of the sending rate followed by a rollback for the queue drain. ProbRTT is turned on when the value of $RTprop_t$ is not updated for a long time. In this mode, the transmission barely stops for a short time to fully drain the queue. Simulation experiments show that in the present model, the last mode is redundant since BBR manages to maintain a very precise estimate of the propagation delay spending the whole time in ProbBW mode. Plus, we excluded from consideration the Startup and Drain modes since they are usually very short.

Thus, finally, the BBR control is defined as follows:

$$u_t = RTprop_t \cdot BtlBw_t \cdot p_t^T e[(t/RTprop_t)\,\%\,8 + 1], \tag{49}$$

where $e[k] \in \mathbb{R}^8$ is a vector with unity on k-th place and zeros on all others, and % is the modulo operator.

In Figure 5, we present the simulation results for the TCP BBR protocol. The upper plot presents the channel dynamics (RTT, losses, timeouts, and state), and the lower plot shows the dynamics of the control (in black) and the estimate of the BDP control equal to $RTprop_t \cdot BtlBw_t$ (in blue). One can notice that this estimate is quite precise, nevertheless, the channel is congested almost the whole time. This means that the BBR algorithm is too aggressive for the channel at hand parameters: the bottleneck buffer size is not enough to accommodate the periodical 25% sending rate increase.

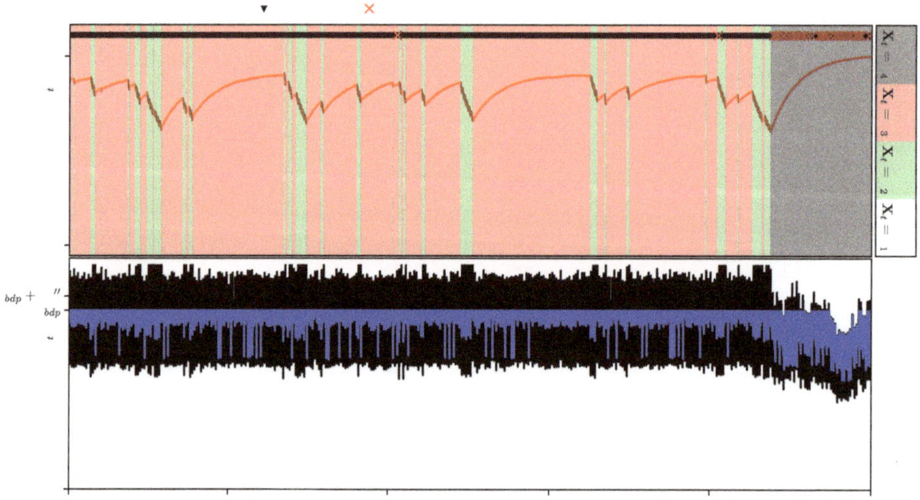

Figure 5. TCP channel simulation example for BBR control algorithm.

5.5. State-Based TCP

To obtain the state-based TCP control strategy, the optimization problem (6) needs to be solved for some predefined gains (instantaneous and terminal) and transmission expenses.

It is natural to bind the transmission expense function $\xi = (\xi^1, \dots, \xi^4)^T$ with the intensity of losses, which we aim to minimize, hence set

$$\xi^j(u) = k^j B^j(u), \tag{50}$$

where $k^1, \dots, k^4$ are coefficients, which reflect the gravity of losses in particular channel states.

We take the same instantaneous gain, as in [36]:

$$\phi^j(u) = -\frac{a^j}{m^j_V(u)\,u} = -\frac{a^j}{\delta_0 + m^j_{V,ext} + u m^j_{V,self}}, \tag{51}$$

where $a^1, \dots, a^4$ are coefficients, which define the utility of the traffic, depending on the channel state.

Analyzing the behavior of the TCP versions described earlier in the present paper, we may conclude that the most beneficial in terms of the throughput and losses is the state e_2 (moderate load). Hence, it is natural to design the state-based version with the goal of spending most of the time in this state. Terminal gains ψ^j, satisfying the condition $max\{\psi^j\} = \psi^2$, would reflect this idea.

In Figure 6 (left), we present a solution to the problem (6) with transmission expenses and instantaneous gains given by (50)–(51) with $k = (10^{-4}, 10, 10^2, 1)^T$ and $a = (100, 100, 1, 100)^T$. The terminal gains are $\psi = -10^6 \cdot (2, 1, 2, 4)^T$, and the right bound of the observation interval is set to a rather small value of the propagation delay $T = \delta_0 = 0.1$ so that the impact of the terminal gains on the criterion would be more valuable. The controls for the states e_1 (idle), e_2 (moderate load), e_3 (congestion), e_4 (wireless signal fading) are given in grey, green, red, and black colors, respectively.

One can observe that the optimal control we obtained is almost constant. This is a very useful property in terms of the scalability of the results. Indeed, the control strategy equal to the mean of the optimal controls

$$\overline{u}_t = \frac{1}{T}\int_0^T u_s\,ds, \tag{52}$$

does not depend on the interval, where the original optimization problem (6) was defined.

In Figure 6 (right), we present three plots, which illustrate the behavior of state occupation probabilities of the channel X_t with constant controls (52) given three different initial states: $X_0 = e_1$, $X_0 = e_2$, $X_0 = e_3$. The color scheme is the same: grey, green, red, and black lines show the occupation probabilities for respectively e_1, e_2, e_3, e_4 states. With solid lines, we show the probabilities obtained as a result of the Kolmogorov equation solution, and with dotted lines, we show the same probabilities obtained through the Monte-Carlo sampling (with 1000 trajectories). One can see that even on a bigger time interval ($T = 5$ s), the goal of the state-based control is achieved: from any given initial condition, the channel manages to revert to (or maintain) the most favorable state e_2.

In Figure 7, we present the simulation results for the state-based control policy. The upper plot presents the channel dynamics (RTT, losses, timeouts, and state), and the lower plot shows the optimal channel estimate $\widehat{X}_t$ in the form of a stack plot: the height of the white/green/red/grey area at a certain point of time corresponds to the conditional probability of state idle/moderate load/congestion/wireless signal fading. This plot demonstrates that the quality of the estimates is good and that the hidden channel state may be adequately revealed based on the available information. In the lower plot of Figure 7, we also show the dynamic of the control

$$\widehat{u}_t = \overline{u}_t^T \widehat{X}_t,$$

where $\overline{u}_t$ is given by (52). One can see that even on a larger interval, the main property of the proposed control strategy remains: the channel spends most of the time in the state e_2, which results in better throughput and fewer losses.

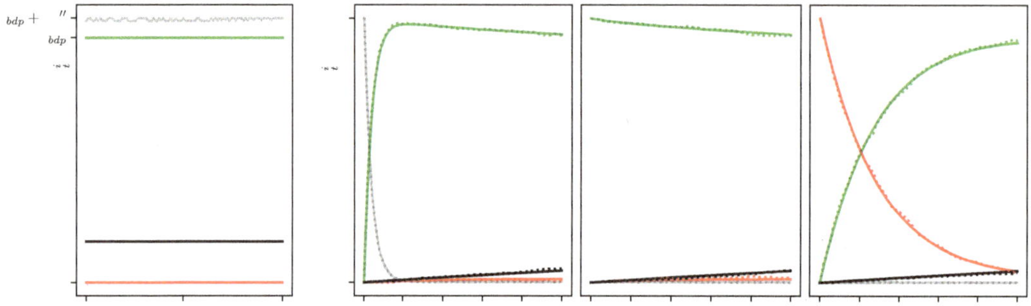

Figure 6. State-based control (**left**) and state occupation probabilities for three initial states: $X_0 = e_1$, $X_0 = e_2$, $X_0 = e_3$ (**right**).

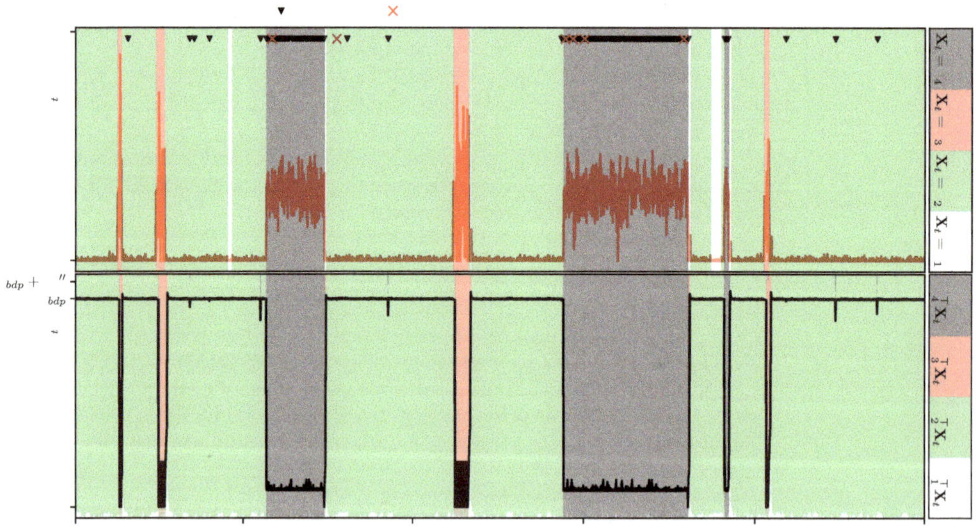

Figure 7. TCP channel simulation example for state-based control algorithm.

5.6. Comparison

To compare the performance of the TCP control schemes discussed above, we use statistical modeling. The performance metrics, namely the average throughput (a measure of bandwidth usage effectiveness) and the loss percentage (a measure of predisposition to congestions, which affect other users), are calculated on samples long enough to make the variance negligible. This way is preferable in comparison with taking the average on a bunch of short-term samples since it diminishes the effect of transient phases: initial probing for available channel characteristics, which is implemented differently but is an essential part of all TCP protocol versions.

On samples of 10^6 seconds, we compare the state-based control with TCP Illinois, CUBIC, Compound, and BBR versions. To make the comparison fairer, we variate, where available, the parameters of TCP control algorithms to achieve better performance. For TCP Cubic, we take three values of multiplicative decrease coefficient $\beta \in \{0.7, 0.8, 0.9\}$; for TCP Compound, we consider nine values of the backlog estimate threshold $\gamma \in \{10, 20, 30, 40, 50, 60, 70, 80, 90\}$. Other parameters of the protocol are the same as they were defined in Sections 5.2 and 5.3 since they have little or negative effect on the performance.

For the state-based version described in Section 5.5, one can tune the protocol behavior by choosing different optimization criteria (6). Nevertheless, since, in our case, the optimal control is constant, instead of the variation of the coefficients of the transmission expenses (50) and instantaneous gain (51), we can directly manipulate these constant values assigned for the channel states. The experiments show that changing controls for states e_1, e_2, e_3, which correspond to the wired part of the transmission channel, makes the performance worse. At the same time, the variation of the control for the state e_4 (wireless signal fading) can bring value; hence we consider four cases: $u_t^4 \in \{20, 50, 100, 200\}$.

The simulation results are summarized in Figure 8, where we present the average throughput and loss percentage and are detailed in Table 1, where one can also find the control algorithm parameters and state occupation times.

One can immediately observe the same occupation time value for the state e_4, which is an indirect indicator of the sufficiency of the chosen simulation sample length: since the transition to and from the state of wireless signal fading does not depend on the control values, the limit probability for the corresponding state should be the same.

The highest occupation time for the state e_2 of moderate channel load is demonstrated by the state-based control. In addition, it can be confirmed that this allows this control algorithm to demonstrate better performance: for the case of $u_t^4 = 20$, the losses are minimal, and the average throughput is second best. It should be noted that the best throughput value demonstrated by the BBR protocol is only possible at the cost of huge losses. This is a characteristic feature of this control algorithm on shallow buffers [55]: it is too aggressive for a channel with chosen characteristics, and a small buffer cannot accommodate frequent 25% speed jumps.

The last thing, which is worth mentioning, is the ability of the state-based protocol to be tuned specifically for the cases of wireless channel issues. Depending on the application, it may try to maintain the maximal possible transmission rate at a cost of huge losses, or, vice versa, drop the speed and wait for the connection to restore to the full speed.

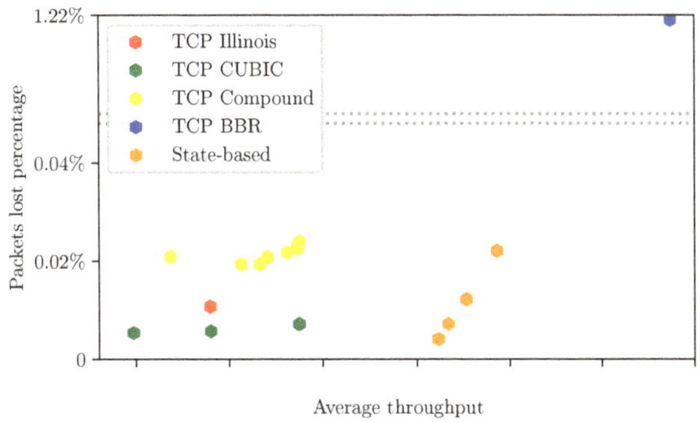

Figure 8. Performance of TCP control versions.

Table 1. Performance metrics.

Protocol	Parameter	Throughput	% loss	e_1	e_2	e_3	e_4
Illinois		63.97	0.011	15.3%	37.7%	24.8%	22.2%
CUBIC	$\beta = 0.7$	59.85	0.005	25.7%	31.2%	20.9%	22.2%
CUBIC	$\beta = 0.8$	63.99	0.006	17.7%	30.8%	29.3%	22.2%
CUBIC	$\beta = 0.9$	68.74	0.007	8.9%	24.2%	44.7%	22.2%
Compound	$\gamma = 10$	61.81	0.021	14.3%	42.6%	20.9%	22.2%
Compound	$\gamma = 20$	65.63	0.019	10.7%	43.4%	23.7%	22.2%
Compound	$\gamma = 30$	66.61	0.019	9.6%	39.9%	28.3%	22.2%
Compound	$\gamma = 40$	67.02	0.021	9.0%	31.1%	37.7%	22.2%
Compound	$\gamma = 50$	68.08	0.022	8.5%	26.8%	42.5%	22.2%
Compound	$\gamma = 60$	68.63	0.022	8.3%	24.1%	45.4%	22.2%
Compound	$\gamma = 70$	68.68	0.023	8.3%	22.8%	46.7%	22.2%
Compound	$\gamma = 80$	68.76	0.024	8.3%	22.9%	46.6%	22.2%
Compound	$\gamma = 90$	68.77	0.024	8.3%	22.8%	46.7%	22.2%
BBR		88.65	1.219	0.7%	8.2%	68.9%	22.2%
State-based	$u_t^4 = 20$	76.15	0.004	1.9%	74.2%	1.7%	22.2%
State-based	$u_t^4 = 50$	76.68	0.007	1.8%	74.3%	1.7%	22.2%
State-based	$u_t^4 = 100$	77.64	0.012	1.7%	74.4%	1.7%	22.2%
State-based	$u_t^4 = 200$	79.29	0.022	1.6%	74.5%	1.7%	22.2%

6. Conclusions

The class of controllable Markov jump processes equipped by the stochastic analysis framework represents an effective tool for the description of a TCP governed communication connection. The hidden channel state is described by a Markov jump process with a finite-state space, characterizing both the current channel load and physical "health status". The state equation admits both to include various types of existing congestion control algorithms (Illinois, CUBIC, Compound, BBR, etc.) and to incorporate some novelties.

The available observations represent the Markov jump processes, namely the Cox processes of the packet losses and timeouts and compound Poisson processes of the packet reception acknowledgments.

The available mathematical framework admits designing the complete technological chain of the TCP congestion control optimization, namely:

- to describe properly the congestion control problem as the stochastic control one,
- to solve the problem above in the case of complete information under the admissible controls with geometric constraints,
- to simplify the mathematical model of available observations, replacing the high-frequency packet acknowledgments flow by its diffusion limit,
- to solve the connection state filtering by the available observations and obtain high-precision state estimates,
- to design effective numerical algorithms for the filtering and control problems solution,
- to apply the separation principle and the loop of congestion control synthesis, using the connection state estimates instead of their exact values.

The result of this optimization represents the proposed state-based version of TCP. The paper contains a comparative analysis of the proposed algorithm against the other contemporary TCP versions and demonstrates its advantages.

The potential of the controllable Markov jump processes for the description of the transport and applied layer communication protocols is far from being exhausted. In perspective, one can use it both for the enhancement of the existing protocols (see, e.g., multi-path TCP [56]) and for the development of new ones (see, e.g., "TCP-free" protocols such as QUIC [57]).

In conclusion, we should also note that the mathematical potential of Markov chains/ Markov jump processes allows designing complete technological chains "mathematical model-properly formulated mathematical problem-theoretical solution-efficient numerical algorithm" to solve many applied problems of the analysis, estimation, and control in such areas as biology [58–60], epidemiology [61–63], inventory control [64], mathematical finance [65], insurance [66,67], etc.

Author Contributions: Conceptualization, A.B. (Andrey Borisov), I.S.; methodology, A.B. (Andrey Borisov), G.M.; software, G.M.; validation, A.B. (Alexey Bosov); formal analysis and investigation, A.B. (Andrey Borisov), G.M.; writing—original draft preparation, A.B. (Andrey Borisov), G.M.; writing—review and editing, A.B. (Alexey Bosov), I.S.; visualization, G.M.; supervision, A.B. (Alexey Bosov), I.S. All authors have read and agreed to the published version of the manuscript.

Funding: The work of Andrey Borisov, Alexey Bosov, and Gregory Miller was partially supported by the Russian Foundation of Basic Research (RFBR Grant No. 19-07-00187-A).

Institutional Review Board Statement: Not applicable.

Informed Consent Statement: Not applicable.

Data Availability Statement: Data sharing is not applicable to this article.

Conflicts of Interest: The authors declare no conflict of interest.

Abbreviations

The following abbreviations are used in this manuscript:

BBR	Bottleneck Bandwidth and RTT
BDP	bandwidth-delay product
CLTRRP	central limit theorem
CLTRRP	central limit theorem for renewal-reward processes
CME	conditional mathematical expectation
CPP	compound Poisson process
cwnd	congestion window size
MAP	maximum a posteriori probability
MJP	Markov jump process
pdf	probability density function
RHS	right-hand side
RTO	retransmission timeout
RTT	round-trip time
TCP	Transmission Control Protocol
TVD	the total variation distance

References

1. Cerf, V.; Kahn, R. A Protocol for Packet Network Intercommunication. *IEEE Trans. Commun.* **1974**, *22*, 637–648. [CrossRef]
2. Al-Saadi, R.; Armitage, G.; But, J.; Branch, P. A Survey of Delay-Based and Hybrid TCP Congestion Control Algorithms. *IEEE Commun. Surv. Tutor.* **2019**, *21*, 3609–3638. [CrossRef]
3. Polese, M.; Chiariotti, F.; Bonetto, E.; Rigotto, F.; Zanella, A.; Zorzi, M. A Survey on Recent Advances in Transport Layer Protocols. *IEEE Commun. Surv. Tutor.* **2019**, *21*, 3584–3608. [CrossRef]
4. Mishra, A.; Sun, X.; Jain, A.; Pande, S.; Joshi, R.; Leong, B. The Great Internet TCP Congestion Control Census. In *Abstracts of the 2020 SIGMETRICS/Performance Joint International Conference on Measurement and Modeling of Computer Systems*; Association for Computing Machinery: New York, NY, USA, 2020; pp. 59–60. [CrossRef]
5. Sikdar, B.; Kalyanaraman, S.; Vastola, K. Analytic models for the latency and steady-state throughput of TCP Tahoe, Reno, and SACK. *IEEE/ACM Trans. Netw.* **2003**, *11*, 959–971. [CrossRef]
6. Liu, S.; Başar, T.; Srikant, R. TCP-Illinois: A Loss- and Delay-Based Congestion Control Algorithm for High-Speed Networks. *Perform. Eval.* **2008**, *65*, 417–440. [CrossRef]
7. Wang, J.; Wen, J.; Han, Y.; Zhang, J.; Li, C.; Xiong, Z. CUBIC-FIT: A High Performance and TCP CUBIC Friendly Congestion Control Algorithm. *IEEE Commun. Lett.* **2013**, *17*, 1664–1667. [CrossRef]
8. Altman, E.; Avrachenkov, K.; Barakat, C. TCP in Presence of Bursty Losses. In Proceedings of the 2000 ACM SIGMETRICS International Conference on Measurement and Modeling of Computer Systems, Santa Clara, CA, USA, 18–21 June 2000; Association for Computing Machinery: New York, NY, USA, 2000; pp. 124–133, [CrossRef]

9. Ephraim, Y.; Merhav, N. Hidden Markov processes. *IEEE Trans. Inf. Theory* **2002**, *48*, 1518–1569. [CrossRef]
10. Borisov, A.V.; Miller, G.B. Analysis and filtration of special discrete-time Markov processes. II: Optimal filtration. *Autom. Remote Control* **2005**, *66*, 1125–1136. [CrossRef]
11. Hasslinger, G.; Hohlfeld, O. The Gilbert-Elliott Model for Packet Loss in Real Time Services on the Internet. In Proceedings of the 14th GI/ITG Conference-Measurement, Modelling and Evalutation of Computer and Communication Systems, Dortmund, Germany, 31 March–2 April 2008; pp. 269–283.
12. Kleinrock, L. *Queueing Systems: Volume 2: Computer Applications*; John Wiley & Sons: New York, NY, USA, 1976.
13. Bertsekas, D.; Gallager, R. *Data Networks*; Prentice-Hall: Hoboken, NJ, USA, 1992.
14. Misra, V.; Gong, W.B.; Towsley, D. Fluid-Based Analysis of a Network of AQM Routers Supporting TCP Flows with an Application to RED. *SIGCOMM Comput. Commun. Rev.* **2000**, *30*, 151–160. [CrossRef]
15. Kushner, H. *Heavy Traffic Analysis of Controlled Queueing and Communication Networks*; Springer: New York, NY, USA, 2001.
16. Whitt, W. *Stochastic-Process Limits. An Introduction to Stochastic-Process Limits and their Application to Queues*; Springer: New York, NY, USA, 2002.
17. Le Boudec, J.Y.; Thiran, P. *Network Calculus: A Theory of Deterministic Queuing Systems for the Internet*; Springer: Berlin/Heidelberg, Germany, 2001.
18. Jiang, Y.; Liu, Y. *Stochastic Network Calculus*; Springer: London, UK, 2008.
19. Fidler, M.; Rizk, A. A Guide to the Stochastic Network Calculus. *IEEE Commun. Surv. Tutor.* **2015**, *17*, 92–105. [CrossRef]
20. Leland, W.; Taqqu, M.; Willinger, W.; Wilson, D. On the self-similar nature of Ethernet traffic (extended version). *IEEE/ACM Trans. Netw.* **1994**, *2*, 1–15. [CrossRef]
21. Crovella, M.; Bestavros, A. Self-similarity in World Wide Web traffic: Evidence and possible causes. *IEEE/ACM Trans. Netw.* **1997**, *5*, 835–846. [CrossRef]
22. Park, K.; Willinger, W. *Self-Similar Network Traffic and Performance Evaluation*; John Wiley & Sons: New York, NY, USA, 2000.
23. Altman, E.; Boulogne, T.; El-Azouzi, R.; Jiménez, T.; Wynter, L. A Survey on Networking Games in Telecommunications. *Comput. Oper. Res.* **2006**, *33*, 286–311. [CrossRef]
24. Habachi, O.; El-azouzi, R.; Hayel, Y. A Stackelberg Model for Opportunistic Sensing in Cognitive Radio Networks. *IEEE Trans. Wirel. Commun.* **2013**, *12*, 2148–2159. [CrossRef]
25. Liu, K.J.R.; Wang, B. *Cognitive Radio Networking and Security: A Game-Theoretic View*, 1st ed.; Cambridge University Press: Cambridge, UK, 2010.
26. Miller, B.M.; Miller, G.B.; Semenikhin, K.V. Methods to design optimal control of Markov process with finite state set in the presence of constraints. *Autom. Remote Control* **2011**, *72*, 323–341. [CrossRef]
27. Borisov, A.V. Robust Filtering Algorithm for Markov Jump Processes with High-Frequency Counting Observations. *Autom. Remote Control* **2020**, *81*, 575–588. [CrossRef]
28. Borisov, A.; Sokolov, I. Optimal Filtering of Markov Jump Processes Given Observations with State-Dependent Noises: Exact Solution and Stable Numerical Schemes. *Mathematics* **2020**, *8*, 506. [CrossRef]
29. Ishikawa, Y.; Kunita, H. Malliavin calculus on the Wiener-Poisson space and its application to canonical SDE with jumps. *Stoch. Process. Their Appl.* **2006**, *116*, 1743–1769. [CrossRef]
30. Borisov, A.; Miller, G.; Stefanovich, A. Controllable Markov Jump Processes. I. Optimum Filtering Based on Complex Observations. *J. Comput. Syst. Sci. Int.* **2018**, *57*, 890–906. [CrossRef]
31. Elliott, R.J.; Moore, J.B.; Aggoun, L. *Hidden Markov Models: Estimation and Control*; Springer: New York, NY, USA, 1995.
32. Fleming, W.; Rishel, R.; Rishel, R.; Collection, K.M.R. *Deterministic and Stochastic Optimal Control*; Applications of Mathematics; Springer: Berlin/Heidelberg, Germany, 1975.
33. Davis, M. *Markov Models & Optimization*; Chapman & Hall/CRC Monographs on Statistics & Applied Probability; Chapman & Hall/CRC: London, UK, 1993.
34. Cohen, S.; Elliott, R. *Stochastic Calculus and Applications*; Probability and Its Applications; Springer: New York, NY, USA, 2015.
35. Jacod, J.; Shiryaev, A. *Limit Theorems for Stochastic Processes*; Grundlehren der Mathematischen Wissenschaften; Springer: Berlin/Heidelberg, Germany, 2013.
36. Miller, B.M.; Avrachenkov, K.; Stepanyan, K.V.; Miller, G.B. Flow control as stochastic optimal control problem with incomplete information. In Proceedings of the INFOCOM 2005 24th Annual Joint Conference of the IEEE Computer and Communications Societies, Miami, FL, USA, 13–17 March 2005; pp. 1328–1337. [CrossRef]
37. Liptser, R.; Shiryaev, A. *Theory of Martingales*; Mathematics and its Applications; Springer: Dortrecht, The Netherlands, 1989.
38. Ha, S.; Rhee, I.; Xu, L. CUBIC: A New TCP-Friendly High-Speed TCP Variant. *SIGOPS Oper. Syst. Rev.* **2008**, *42*, 64–74. [CrossRef]
39. Kato, T.; Haruyama, S.; Yamamoto, R.; Ohzahata, S. mpCUBIC: A CUBIC-like Congestion Control Algorithm for Multipath TCP. In *Trends and Innovations in Information Systems and Technologies*; Rocha, Á., Adeli, H., Reis, L.P., Costanzo, S., Orovic, I., Moreira, F., Eds.; Springer International Publishing: Cham, Switzerlands, 2020; pp. 306–317. [CrossRef]
40. Tan, K.; Song, J.; Zhang, Q.; Sridharan, M. A Compound TCP Approach for High-Speed and Long Distance Networks. In Proceedings of the IEEE INFOCOM 2006 25th IEEE International Conference on Computer Communications, Barcelona, Catalunya, Spain, 23–29 April 2006; pp. 1–12. [CrossRef]
41. Oda, H.; Hisamatsu, H.; Noborio, H. Compound TCP+: A Solution for Compound TCP Unfairness in Wireless LAN. *J. Inf. Process.* **2013**, *21*, 122–130. [CrossRef]

42. Smith, W. Regenerative Stochastic Processes. *Proc. R. Soc. Lond. Ser. A Math. Phys. Sci.* **1955**, *232*, 6–31. [CrossRef]
43. Borisov, A.V. Wonham Filtering by Observations with Multiplicative Noises. *Autom. Remote Control* **2018**, *79*, 39–50. [CrossRef]
44. Wonham, W.M. Some Applications of Stochastic Differential Equations to Optimal Nonlinear Filtering. *J. Soc. Ind. Appl. Math. Ser. A Control* **1964**, *2*, 347–369. [CrossRef]
45. Borovkov, A. *Asymptotic Methods in Queuing Theory*; John Wiley & Sons: Hoboken, NJ, USA, 1984.
46. Gut, A. *Stopped Random Walks: Limit Theorems and Applications*; Springer Series in Operations Research and Financial Engineering; Springer: New York, NY, USA, 2009.
47. Fischer, H. *A History of the Central Limit Theorem: From Classical to Modern Probability Theory*; Sources and Studies in the History of Mathematics and Physical Sciences; Springer: New York, NY, USA, 2010.
48. Bobkov, S.G.; Chistyakov, G.P.; Götze, F. Berry-Esseen bounds in the entropic central limit theorem. *Probab. Theory Relat. Fields* **2014**, *159*, 435–478. [CrossRef]
49. Bally, V.; Caramellino, L. Asymptotic development for the CLT in total variation distance. *Bernoulli* **2016**, *22*, 2442–2485. [CrossRef]
50. Liptser, R.; Shiryaev, A. *Statistics of Random Processes II: Applications*; Springer: Berlin/Heidelberg, Germany, 2001.
51. Borisov, A.V. Application of optimal filtering methods for on-line monitoring of queueing network states. *Autom. Remote Control* **2016**, *77*, 277–296. [CrossRef]
52. Borisov, A.V.; Bosov, A.V.; Miller, G.B.; Stefanovich, A.I. Optimization of TCP Algorithm for Wired–Wireless Channels Based on Connection State Estimation. In Proceedings of the 2019 IEEE 58th Conference on Decision and Control (CDC), Nice, France, 11–13 December 2019; pp. 728–733. [CrossRef]
53. Floyd, S.; Jacobson, V. Random early detection gateways for congestion avoidance. *IEEE/ACM Trans. Netw.* **1993**, *1*, 397–413. [CrossRef]
54. Cardwell, N.; Cheng, Y.; Gunn, C.S.; Yeganeh, S.H.; Jacobson, V. BBR: Congestion-Based Congestion Control. *Commun. ACM* **2017**, *60*, 58–66. [CrossRef]
55. Claypool, S.; Claypool, M.; Chung, J.; Li, F. Sharing but not Caring-Performance of TCP BBR and TCP CUBIC at the Network Bottleneck. In Proceedings of the 4th IARIA International Conference on Advances in Computation, Communications and Services (ACCSE), Nice, France, 28 July–2 August 2019; pp. 74–81.
56. Pokhrel, S.R.; Panda, M.; Vu, H.L. Analytical Modeling of Multipath TCP Over Last-Mile Wireless. *IEEE/ACM Trans. Netw.* **2017**, *25*, 1876–1891. [CrossRef]
57. Kharat, P.; Kulkarni, M. Modified QUIC protocol with congestion control for improved network performance. *IET Commun.* **2021**, *15*, 1210–1222. [CrossRef]
58. Krogh, A.; Brown, M.; Mian, I.; Sjölander, K.; Haussler, D. Hidden Markov Models in Computational Biology: Applications to Protein Modeling. *J. Mol. Biol.* **1994**, *235*, 1501–1531. [CrossRef] [PubMed]
59. Huelsenbeck, J.; Larget, B.; Swofford, D. A compound Poisson process for relaxing the molecular clock. *Genetics* **2000**, *154*, 1879–1892. [CrossRef] [PubMed]
60. Karchin, R.; Cline, M.; Mandel-Gutfreund, Y.; Karplus, K. Hidden Markov models that use predicted local structure for fold recognition: Alphabets of backbone geometry. *Proteins Struct. Funct. Bioinform.* **2003**, *51*, 504–514. [CrossRef]
61. Cauchemez, S.; Carrat, F.; Viboud, C.; Valleron, A.; Böelle, P. A Bayesian MCMC approach to study transmission of influenza: Application to household longitudinal data. *Stat. Med.* **2004**, *22*, 3469–3487. [CrossRef]
62. Allen, L.J.S. An Introduction to Stochastic Epidemic Models. In *Mathematical Epidemiology*; Brauer, F., van den Driessche, P., Wu, J., Eds.; Springer: Berlin/Heidelberg, Germany, 2008; pp. 81–130. [CrossRef]
63. Gómez, S.; Arenas, A.; Borge-Holthoefer, J.; Meloni, S.; Moreno, Y. Discrete-time Markov chain approach to contact-based disease spreading in complex networks. *EPL (Europhys. Lett.)* **2010**, *89*, 38009. [CrossRef]
64. Papadopoulos, C.T.; Li, J.; O'Kelly, M.E. A classification and review of timed Markov models of manufacturing systems. *Comput. Ind. Eng.* **2019**, *128*, 219–244. [CrossRef]
65. Ang, A.; Timmermann, A. Regime Changes and Financial Markets. *Annu. Rev. Financ. Econ.* **2012**, *4*, 313–337. [CrossRef]
66. Paulsen, J. Risk theory in a stochastic economic environment. *Stoch. Process. Appl.* **1993**, *46*, 327–361. [CrossRef]
67. Christiansen, M. Multistate models in health insurance. *AStA Adv. Stat. Anal.* **2012**, *96*, 155–186. [CrossRef]

mathematics

MDPI

Article

Evaluating the Efficiency of Off-Ball Screens in Elite Basketball Teams via Second-Order Markov Modelling

Nikolaos Stavropoulos [1], Alexandra Papadopoulou [2] and Pavlos Kolias [2,*]

[1] Laboratory of Evaluation of Human Biological Performance, School of Physical Education and Sport Science, Aristotle University of Thessaloniki, 57001 Thessaloniki, Greece; nstavrop@phed.auth.gr

[2] Section of Statistics and Operational Research, Department of Mathematics, Aristotle University of Thessaloniki, 54124 Thessaloniki, Greece; apapado@math.auth.gr

* Correspondence: pakolias@math.auth.gr; Tel.: +30-2315-31-7627

Abstract: In basketball, the offensive movements on both strong and weak sides and tactical behavior play major roles in the effectiveness of a team's offense. In the literature, studies are mostly focused on offensive actions, such as ball screens on the strong side. In the present paper, for the first time a second-order Markov model is defined to evaluate players' interactions on the weak side, particularly for exploring the effectiveness of tactical structures and off-ball screens regarding the final outcome. The sample consisted of 1170 possessions of the FIBA Basketball Champions League 2018–2019. The variables of interest were the type of screen on the weak side, the finishing move, and the outcome of the shot. The model incorporates partial non-homogeneity according to the time of the execution (0–24″) and the quarter of playtime, and it is conditioned on the off-ball screen type. Regarding the overall performance, the results indicated that the outcome of each possession was influenced not only by the type of the executed shot, but also by the specific type of screen that took place earlier on the weak side of the offense. Thus, the proposed model could operate as an advisory tool for the coach's strategic plans.

Keywords: basketball; Markov chain; second order; off-ball screens; performance

Citation: Stavropoulos, N.; Papadopoulou, A.; Kolias, P. Evaluating the Efficiency of Off-Ball Screens in Elite Basketball Teams via Second-Order Markov Modelling. *Mathematics* **2021**, *9*, 1991. https://doi.org/10.3390/math9161991

Academic Editors: Panagiotis-Christos Vassiliou and Andreas C. Georgiou

Received: 8 July 2021
Accepted: 17 August 2021
Published: 20 August 2021

Publisher's Note: MDPI stays neutral with regard to jurisdictional claims in published maps and institutional affiliations.

Copyright: © 2021 by the authors. Licensee MDPI, Basel, Switzerland. This article is an open access article distributed under the terms and conditions of the Creative Commons Attribution (CC BY) license (https://creativecommons.org/licenses/by/4.0/).

1. Introduction

Basketball is a team sport that is constantly evolving due to the changes in regulations, the faster pace, the increasing physical abilities of the players, and the upgrading of training methods. Offensive movements and players' tactical cooperation play major roles in both individual and team performance concerning offense [1,2]. The most frequent offensive movement between two players on the strong side is the ball screen. Ball screens are important coordinated movements used in offense, providing enhanced strategy on the court [3]. During the action of ball screen, one player is the screener, who blocks the defensive movements of the opponents from an appropriate area, and the other is the ball handler, who creates opportunities by either passing to the screener-cutter (roll or pop out to the basket) or becoming the cutter by executing a shot himself [4,5]. Previous studies have indicated that the effectiveness of the screen is affected by time-related characteristics, such as the offense's remaining time, the type of screen and the area of execution [1]. In addition, coordinated movements on the weak side are also extremely important for the overall offensive performance of each team. According to previous findings, the most common offensive tactics used on the weak side are the off-ball screens [6]. The continuous movements and screen types on the weak side are crucial factors in allowing advantageous positions while executing the shots. Previous results in NCAA basketball league have shown that the winning teams had approximately 11 off-ball screens less than the losing teams [7].

Statistical and stochastic modelling has already been applied to model performance in basketball. The most common approach is to apply linear or generalized linear regression models to box-score data, while considering each individual player's statistics and overall team statistics [8]. Furthermore, researchers have applied quantile regression methods, which can provide more specific descriptions of the relationships between key performance indicators and the outcome of a basketball game compared to multiple regression [9]. These approaches lack the detail of the evolution of the match, as they mainly focus on overall performance. Other studies have used discriminant analysis to obtain the most dominant factors that could potentially lead a team to victory in both the Basketball World Cup and domestic leagues [10–12]. Play-by-play data have more recently been included in basketball research and expanded the traditional use of summary statistics of tournaments. Such data can be used for more detailed illustrations of the evolution of basketball matches [13].

Markov models are useful for modelling play-by-play data, as they effectively describe the evolution of future successive possession by each of the two competing teams. One of the earliest attempts was the application of a Markov chain with state space consisting of the team that had the possession, how the possession was taken, and the points scored during the previous possession. This model could derive the progression of the basketball match over time [14]. In NCAA, a combination of logistic regression and Markov modelling has been used to evaluate the rankings of the teams and predict the final standings of the tournament [15]. Furthermore, researchers have applied a Markov model to simulate a basketball match in the NBA and forecast the outcome of the match and the points scored, based on the transition matrix [13]. This model captured the non-homogeneity of a basketball match, which was mainly observed in the first and last minutes of playtime, and provided a more detailed state space, including time, the difference in points and characteristics of the teams. Basketball formations also play a crucial role in the overall performance of a team, as different positions exhibit different characteristics and should optimally cooperate with the rest of the team. Markov chain modelling has been used to compare the offensive and defensive performances of different formations, and the performance of these formations over time [16]. Finally, Markov chains have been used as modelling tools in various other domains, such as manpower planning, finance, healthcare, biology, and others [17–28].

The class of high-order Markov chains is an essential stochastic tool, which fits more adequately when the phenomena under investigation incorporate longer dependencies. One of the earliest studies with high-order Markov chains applied them in manpower systems, and they presented a considerably better fit compared to first-order Markov models [29]. A major problem of high-order Markov chains is the great number of the parameters that must be estimated, which increases geometrically according to the order of the model. Raftery, in 1985, was the first to propose a high-order Markov model, called the mixture transition distribution (MTD) model, where each transition probability is a weighted linear combination of the previous transition probabilities [30]. In this formulation, one can estimate a smaller number of parameters by solving a linear system, as in the well-known Yule–Walker system of equations found in time-series analysis. The limiting distribution of high-order Markov chains was studied in [31]. Ching and his colleagues extended Raftery's model by introducing variability into the transition probability matrices, and proved that, given some mild conditions, the proposed model has a stationary distribution [32]. More recently, in the field of the mixture transition probability models for high-order Markov chains, the G-inhomogeneous Markov system was introduced, and its asymptotic behavior, under assumptions easily met in practice, was studied [33]. Applications of high-order Markov chains can be found in various domains, such as DNA analysis [34,35], analysis of wind speed [36], and manpower planning [29].

To our knowledge, there exist limited studies concerning basketball screens on the weak side of the court and their influences on a game's outcome. The purpose of the present study was to develop a second-order Markov modelling framework that would evaluate the characteristics of off-ball screens that positively affect the finishing move and the outcome of the offensive movement, thus improving the performance of the team. Apart from the overall performance, the aim of the current paper is to examine how time, expressed either as the quarter of play or as the time clock (0–24 s), could influence the transition probabilities from screens and finishing moves to outcomes. Section 2 presents the main methodological tools adopted in this paper. More specifically, Section 2.1 presents the theoretical background and definitions of high-order Markov chains. Section 2.2 presents the description of the data and the measured variables, embeds the second-order Markov theory in the basketball context, where the state space and the basic parameters of the Markov modelling are provided. Section 3 provides the results of the analysis. Section 4 discusses the obtained results from a basketball viewpoint and finally, the conclusions are provided in Section 5.

2. Modelling Framework

2.1. Second-Order Markov Modelling

A first-order Markov chain $\{X_n\}$, $n = 0, 1, \ldots,$ with state space $V = \{1, 2, \ldots, m\}$ is a discrete stochastic process, in which the transition to the next state is governed only by the current state of the process and it is independent of the past states. This property, called *Markovian*, could be written as

$$P(X_{n+1} = j | X_n = i,\ X_{n-1} = i_{n-1},\ \ldots, X_0 = i_0) = P(X_{n+1} = j | X_n = i) = p_{ij}(n),$$

where $i, j, i_0, \ldots, i_{n-1} \in V$ and $\sum_{i=1}^{m} p_{ij}(n) = 1,\ p_{ij}(n) \geq 0$. The matrix $\boldsymbol{P}(n)$, which contains the probabilities $p_{ij}(n)$ is called the transition probability matrix. If the probabilities $p_{ij}(n)$ are independent of time, i.e., $p_{ij}(n) = p_{ij},\ \forall\ n \in \mathbb{N}$, then the Markov chain is called *time homogeneous*. If we consider a first-order Markov chain, then the k-order Markov chain (X_n) with state space $S = \{1, 2, \ldots, M\}$, where the states of S are k-tuples of the elements of V, is a discrete stochastic process, for which the *k-order Markovian* property holds:

$$P(X_{n+1} = j | X_n = i,\ \ldots,\ X_0 = i_0) = P(X_{n+1} = j | X_n = i,\ \ldots,\ X_{n-k+1} = i_{n-k+1}),$$

and the number of states is equal to $M = (m-1)m^k$. In general, the transition probability matrix of the high-order Markov chain will contain many zero cells, as it is impossible to transition to states where the past observations do not overlap. To present the transition probabilities in a more elegant way, we can use the *reduced* transition probability matrix, which contains only the non-zero probabilities [37]. For example, the reduced transition probability matrix for a second-order Markov chain with state space $S = \{1, 2\}$ is presented in Table 1. Note that in a second-order Markov chain, the subscript of the probabilities contain three states, where the first two refer to past states and the last one to the next state.

Table 1. Transition probability matrix of a second-order two-state Markov chain in reduced form.

		X_t	
X_{t-2}	X_{t-1}	1	2
1	1	p_{111}	p_{112}
2	1	p_{211}	p_{212}
1	2	p_{121}	p_{122}
2	2	p_{221}	p_{222}

By using this technique, we can transform any Markov chain of order n to a first-order model, by appropriately changing the state space and keeping all the n-tuples. The high-order Markov chains are, in general, more efficient as they acquire memory and can capture longer dependencies compared to the first order; however, the number of parameters increases with geometric growth with respect to the order. This leads to computational problems while estimating all the parameters. Some alternative specifications of the n-order model have been proposed, which reduce the set of parameters by applying linear dependencies between the n-step probabilities [30]. These MTD models are, in general, more practical to estimate, however the assumption of dependent transition probabilities may not be necessary, especially when we are dealing with short-term correlations. In the basketball context, the outcome of each possession could be influenced by two preceding events, namely, the type of screen on the weak side of the court and the finishing action. Thus, a second-order Markov chain could be more feasible for estimation, as the number of varying parameters is reasonable for direct estimations of the transition probabilities. Hence, the model could examine the relationship between those past movements and the final outcome of the offense. In this scenario, the transition probabilities $p_{kij}(n)$ denote the probability that the Markov chain will transition to state j, while currently it is at state i at time n and the previous state was k. With inclusion of the second-order transition probabilities, we can arrange the non-homogeneous second-order transition matrix $\boldsymbol{P}(n)$, which is the basic parameter of the process. The maximum likelihood estimates (MLE) for the transition probabilities of a second-order Markov chain are given by

$$\hat{p}_{kij}(n) = \frac{N(k, i(n) \to j)}{\sum_{x \in M} N(k, i(n) \to x)},$$

where $N(k, i(n) \to j)$ denotes the number of transitions from the pair (k, i) to state j, starting from the position n. Please note that if we assume that the transition probabilities are time-invariant, that is $\boldsymbol{P}(n) = \boldsymbol{P}$, then the MLE estimates for the transition probabilities are given by

$$\hat{p}_{kij} = \frac{N(k, i \to j)}{\sum_{x \in M} N(k, i \to x)},$$

where $(k, i \to j)$ denotes the number of transitions from the pair (k, i) to state j.

2.2. Basketball Modelling

In the context of basketball, assume that $\{X_n\}$ is a discrete first-order Markov chain that denotes the current event taking place during the offense. The events that happen are the screen type (TS), the finishing move type (TF) and the outcome (O). Hence, the process takes values in the three-dimensional state space, which is $V = \{TS,\ TF, O\}$. For example, consider the scenario where a team obtains possession and screens outside the paint with a staggered screen and the player that gets the ball shoots from inside the paint with a lay-up and scores a 2-pt shot; then, the associated transitions of this scenario will be, *"Staggered screen outside the paint, 0, 0 → 0, Lay-up, 0 → 0,0, Successful 2-pt shot"*.

To model the successive events during each offense, we have used a sample of 1170 possessions by 16 competing teams of the FIBA Basketball Champions League 2018–2019. The recordings of the possessions were made using the "SportScout" video-analysis software. The possessions were observed by three assistant coaches, with at least 5 years of experience in professional basketball. Cohen's kappa (κ) correlation coefficient was used to assess the inter-rater reliability. The values obtained displayed a high degree of agreement (κ_{min} = 0.91). For each possession, the events were recorded, as well as the time of the shot clock (T) and the quarter of playtime (Q1–Q4). The levels of each of the recorded variables are presented in Table 2. The possible outcomes consisted of successful and unsuccessful 2- and 3-pt shots and possession change, which includes turnovers, steals, blocks, offensive fouls, and the violation of the 24 s duration of offense.

The screen types were defined using standard basketball terminology. More specifically, two consecutive screens for a player, in the same direction away from the ball were defined as a staggered screen. A flare screen was defined as a screen set at the elbow of the free throw line where the player fades out on the weak side. Screen the screener occurs when an offensive player sets a screen and, at the same time, receives a screen from a teammate. To pass on the side and set a screen for a player in the opposite direction was described as a screen away. Down screen is a screen where an offensive player sets himself in a position away from the ball. Back screen occurs when an offensive player stands behind the defensive player with his back toward the basket. Single- and double-staggered screens were combined into one category, as well as the single- and double- high-cross screens. Examples of screen types under consideration are presented in Figure 1.

Table 2. Recorded variables, levels, and coding indices.

Variables	Levels	Level Coding
Type of screen (TS)	Staggered screen	TS1
	Flare screen	TS2
	Screen the screener	TS3
	Back screen	TS4
	Down screen	TS5
	High cross screen	TS6
	Screen away	TS7
Type of finishing move (TF)	Dunk	TF1
	Lay-up	TF2
	2-pt shot	TF3
	3-pt shot	TF4
	None	TF5
Outcome (O)	Successful 2-pt shot	O1
	Missed 2-pt shot	O2
	Successful 3-pt shot	O3
	Missed 3-pt shot	O4
	Possession change	O5
Time (T)	0–8 s	T1
	8–24 s	T2
Quarter (Q)	First Quarter	Q1
	Second Quarter	Q2
	Third Quarter	Q3
	Fourth Quarter	Q4

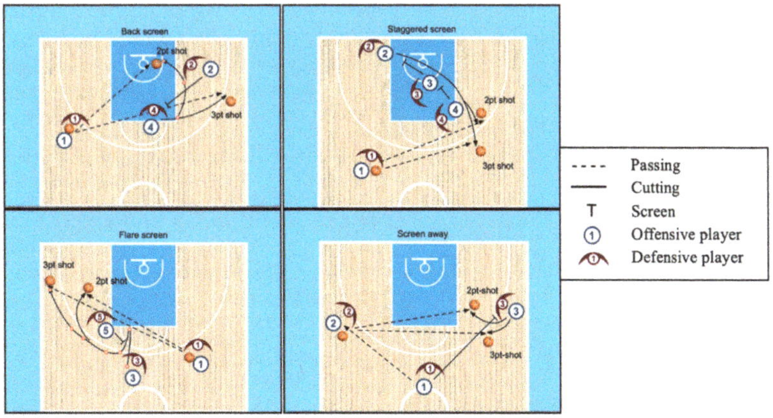

Figure 1. Execution of indicative offensive screens, back screens (**top left**), staggered screens (**top right**), flare screens (**bottom left**) and screen away (**bottom right**).

We shall note here that not all transitions were observed, for example if the finishing move was a middle-range shot (2-pt), the only possible outcomes would be either a successful or unsuccessful 2-pt shot. For the first-order Markov chain, the possible transitions between states are presented in Table 3. Apparently, the Markov chain exhibits periodic behavior with period $d = 3$, as each screen is always followed by a finishing move and each finishing move is only followed by the outcome of the possession.

Table 3. Possible transitions between the states in the first-order Markov chain.

	TS1	TS2	TS3	TS4	TS5	TS6	TS7	TF1	TF2	TF3	TF4	TF5	O1	O2	O3	O4	O5
TS1	0	0	0	0	0	0	0	X	X	X	X	X	0	0	0	0	0
TS2	0	0	0	0	0	0	0	X	X	X	X	X	0	0	0	0	0
TS3	0	0	0	0	0	0	0	X	X	X	X	X	0	0	0	0	0
TS4	0	0	0	0	0	0	0	X	X	X	X	X	0	0	0	0	0
TS5	0	0	0	0	0	0	0	X	X	X	X	X	0	0	0	0	0
TS6	0	0	0	0	0	0	0	X	X	X	X	X	0	0	0	0	0
TS7	0	0	0	0	0	0	0	X	X	X	X	X	0	0	0	0	0
TF1	0	0	0	0	0	0	0	0	0	0	0	0	X	X	0	0	X
TF2	0	0	0	0	0	0	0	0	0	0	0	0	X	X	0	0	X
TF3	0	0	0	0	0	0	0	0	0	0	0	0	X	X	0	0	X
TF4	0	0	0	0	0	0	0	0	0	0	0	0	0	0	X	X	X
TF5	0	0	0	0	0	0	0	0	0	0	0	0	0	0	0	0	X
O1	X	X	X	X	X	X	X	0	0	0	0	0	0	0	0	0	0
O2	X	X	X	X	X	X	X	0	0	0	0	0	0	0	0	0	0
O3	X	X	X	X	X	X	X	0	0	0	0	0	0	0	0	0	0
O4	X	X	X	X	X	X	X	0	0	0	0	0	0	0	0	0	0
O5	X	X	X	X	X	X	X	0	0	0	0	0	0	0	0	0	0

TS: Type of screen, TF: Type of finishing move, O: Outcome, X: Non-zero probability.

It is of interest to examine whether the process $\{X_n\}$ incorporates memory, i.e., a higher-order Markov model would provide a more adequate fit. In relation to basketball, a coach may assume that the outcome of a possession does not only depend on the type of the executed shot, but on the previous characteristics of the phase, such as the type of screen, as it could probably alter the evolution of the possession and provide more space and freedom for a well-executed shot. Hence, we would like to test the null hypothesis that the process is of order r = 1 versus the alternative hypothesis, r = 2. For testing this hypothesis, we used the likelihood ratio test (LRT). The likelihood ratio (LR) is given by

$$LR = -2(LL_1 - LL_2),$$

where LL_1 and LL_2 denote the log-likelihood of models of order 1 and order 2, respectively. The log-likelihood ratio is an essential tool for the comparison of two competing Markov models [38] and can be used to evaluate well-known goodness-of-fit metrics, such as the AIC and BIC [39]. The likelihood ratio asymptotically follows a chi-squared distribution with degrees of freedom (df) equal to the difference of degrees of freedom of the two models, thus it can provide a *p*-value that can lead to the rejection of the null hypothesis, if it is smaller than a predefined cut-off value α (commonly α is set to 0.05). Adopting the notations of a previous work, where the authors assessed the order of a Markov chain applied in DNA sequences [40], one can formulate the likelihood ratio for two competing Markov models by

$$LR = -2\left(\sum_{a_2,a_3} n_{a_2,a_3} \log\left(\frac{n_{a_2,a_3}}{n_{a_2}}\right) - \sum_{a_1,a_2,a_3} n_{a_1,a_2,a_3} \log\left(\frac{n_{a_1,a_2,a_3}}{n_{a_1 a_2}}\right)\right),$$

where

$$n_{a_1,a_2,a_3} = \sum_{k=1}^{n-2} I(X_k = a_1, X_{k+1} = a_2, X_{k+2} = a_3),$$

$$n_{a_2,a_3} = \sum_{k=1}^{n-1} I(X_k = a_2, X_{k+1} = a_3),$$

denote the number of observed triplets and pairs of $a_1, a_2, a_3 \epsilon S$, respectively. Also, we note that the ratios $\frac{n_{a_2,a_3}}{n_{a_2}}$ and $\frac{n_{a_1,a_2,a_3}}{n_{a_1 a_2}}$ are the empirical estimators of the transition probabilities, e.g., $\hat{p}_{a_2,a_3}$ and $\hat{p}_{a_1,a_2,a_3}$, respectively. The LR could be simplified as

$$LR = -2\left(\sum_{a_2,a_3} n_{a_2,a_3}\log(\hat{p}_{a_2,a_3}) - \sum_{a_1,a_2,a_3} n_{a_1,a_2,a_3}\log(\hat{p}_{a_1,a_2,a_3})\right).$$

In general, a Markov chain of order r with state space S has $(|S|-1)|S|^r$ varying parameters. However, in our case, the number of varying parameters would be less, since in the basketball context the transition probability matrix prohibited some transitions. For example, when the offensive player shoots a 3-pt shot, the possible transitions would not include any other outcome, apart from a successful or missed 3-pt shot. More specifically, the numbers of estimated transition probabilities were 82 and 354 for the first- and second-order Markov chain, respectively. The likelihood ratio value was calculated equal to 395.242, which resulted in $p < 0.001$, therefore the likelihood ratio test indicated to reject the null hypothesis, in favor of r = 2.

The results of the significant relationships between the three components (screen type, finishing move and outcome) lead to establishing a model that includes second order dependencies, therefore a second-order Markov chain is proposed to study the effect of screen type and finishing move on the outcome of the possession. The state space $S = \{(TS, TF), (TF, O), (O, TS)\}$ of the second-order Markov chain consists of the ordered pairs of events that belong in the state-space V of the first-order Markov chain. The transition probabilities are presented in Table 4, in reduced form. Several considerations were made regarding the time, as a parameter that influences the frequency of specific off-ball screens and outcomes. First, the off-ball screen possessions were designated into two categories, 0–8 s and 8–24 s, according to the shot clock time at the time of the finishing move. For each subsample, the transition probabilities were estimated and the asymptotic probability vectors were also estimated. Second, we differentiated the offensive movements between the first three quarters and the last quarter of the game, where in the last quarter, as the pace of the game increases, the losing team can make a comeback.

Table 4. Transition probability matrix of the second-order Markov chain in reduced form.

		O (X_t)			
TS (X_{t-2})	**TF** (X_{t-1})	**O1**	**O2**	...	**O5**
TS1	TF1	$p_{TS1\ TF1\ O1}$	$p_{TS1\ TF1\ O2}$	...	$p_{TS1\ TF1\ O5}$
TS2	TF1	$p_{TS2\ TF1\ O1}$	$p_{TS2\ TF1\ O2}$	...	$p_{TS2\ TF1\ O5}$
TS3	TF1	$p_{TS3\ TF1\ O1}$	$p_{TS3\ TF1\ O2}$	...	$p_{TS3\ TF1\ O5}$
TS4	TF1	$p_{TS4\ TF1\ O1}$	$p_{TS4\ TF1\ O2}$	...	$p_{TS4\ TF1\ O5}$
TS5	TF1	$p_{TS5\ TF1\ O1}$	$p_{TS5\ TF1\ O2}$	...	$p_{TS5\ TF1\ O5}$
TS6	TF1	$p_{TS6\ TF1\ O1}$	$p_{TS6\ TF1\ O2}$	...	$p_{TS6\ TF1\ O5}$
TS7	TF1	$p_{TS7\ TF1\ O1}$	$p_{TS7\ TF1\ O2}$	...	$p_{TS7\ TF1\ O5}$
TS1	TF2	$p_{TS1\ TF2\ O1}$	$p_{TS1\ TF2\ O2}$	...	$p_{TS1\ TF2\ O5}$
TS2	TF2	$p_{TS2\ TF2\ O1}$	$p_{TS2\ TF2\ O2}$	...	$p_{TS2\ TF2\ O5}$
TS3	TF2	$p_{TS3\ TF2\ O1}$	$p_{TS3\ TF2\ O2}$	...	$p_{TS3\ TF2\ O5}$
TS4	TF2	$p_{TS4\ TF2\ O1}$	$p_{TS4\ TF2\ O2}$	...	$p_{TS4\ TF2\ O5}$
TS5	TF2	$p_{TS5\ TF2\ O1}$	$p_{TS5\ TF2\ O2}$	...	$p_{TS5\ TF2\ O5}$
TS6	TF2	$p_{TS6\ TF2\ O1}$	$p_{TS6\ TF2\ O2}$	...	$p_{TS6\ TF2\ O5}$
TS7	TF2	$p_{TS7\ TF2\ O1}$	$p_{TS7\ TF2\ O2}$	...	$p_{TS7\ TF2\ O5}$
⋮	⋮	⋮	⋮	...	⋮
TS1	TF5	$p_{TS1\ TF5\ O1}$	$p_{TS1\ TF5\ O2}$	...	$p_{TS1\ TF5\ O5}$
TS2	TF5	$p_{TS2\ TF5\ O1}$	$p_{TS2\ TF5\ O2}$	...	$p_{TS2\ TF5\ O5}$

Table 4. *Cont.*

TS (X_{t-2})	TF (X_{t-1})	O (X_t)			
		O1	O2	...	O5
TS3	TF5	$p_{TS3\ TF5\ O1}$	$p_{TS3\ TF5\ O2}$	...	$p_{TS3\ TF5\ O5}$
TS4	TF5	$p_{TS4\ TF5\ O1}$	$p_{TS4\ TF5\ O2}$	...	$p_{TS4\ TF5\ O5}$
TS5	TF5	$p_{TS5\ TF5\ O1}$	$p_{TS5\ TF5\ O2}$	...	$p_{TS5\ TF5\ O5}$
TS6	TF5	$p_{TS6\ TF5\ O1}$	$p_{TS6\ TF5\ O2}$	...	$p_{TS6\ TF5\ O5}$
TS7	TF5	$p_{TS7\ TF5\ O1}$	$p_{TS7\ TF5\ O2}$	...	$p_{TS7\ TF5\ O5}$

3. Results

3.1. Overall Performance

The screens with the highest frequency were staggered screens (41%), followed by flare screens (15%). Furthermore, offensive players decided to execute their offense by using 3-pt shots (57%), followed by making use of 2-pt shots (22%); conversely, the lay-up frequency appeared lower (10%). The estimated transition probabilities of the first-order Markov chain showed that, on average, the probability of a successful outcome was less when compared to a missed attempt. Except for dunks, where the success was assured, 2-pt shots showed the highest probability of success ($p = 0.48$), followed by 3-pt shots ($p = 0.41$) and lay-ups ($p = 0.35$). Lay-ups also showed the highest probability of a possession change, caused by a block, turnover, or foul ($p = 0.27$). Table 5 presents the second-order transition probabilities between finishing moves and successful 2- or 3-point shots, conditional on screen type. Schematically, Figure 2 visualizes the relationship between the pairs: screen type/finishing move and screen type/outcome. Lay-ups were mainly enhanced by back screens, as it was found that the succession of back screens and lay-ups results in 0.78 probability of scoring a 2-pt shot. Most 2-pt shots were successfully executed, when the preceding off-ball screen was flare, staggered or down screen. Concerning 3-pt shots, the two types of screens where the outcome was optimal, were the high-cross and screen the screener.

Table 5. Overall transition probability estimates between finishing moves and screens to successful shots.

Screen Type (X_{t-2})	Finishing Move (X_{t-1})	Successful Shot (X_t)
Staggered	Lay-up	0.27
Flare [1]	Lay-up	0.17
Screen the screener	Lay-up	0.44
Back screen	Lay-up	0.78
Down screen	Lay-up	0.45
High cross	Lay-up	0.47
Screen away	Lay-up	0.15
Staggered	2-pt shot	0.51
Flare [1]	2-pt shot	0.56
Screen the screener	2-pt shot	0.30
Back screen	2-pt shot	0.47
Down screen	2-pt shot	0.50
High cross	2-pt shot	0.39
Screen away	2-pt shot	0.48
Staggered	3-pt shot	0.41
Flare [1]	3-pt shot	0.34
Flare [2]	3-pt shot	0.33
Screen the screener	3-pt shot	0.52
Back screen	3-pt shot	0.29
Down screen	3-pt shot	0.42
High cross	3-pt shot	0.67
Screen away	3-pt shot	0.40

[1]: inside the paint, [2]: perimeter area.

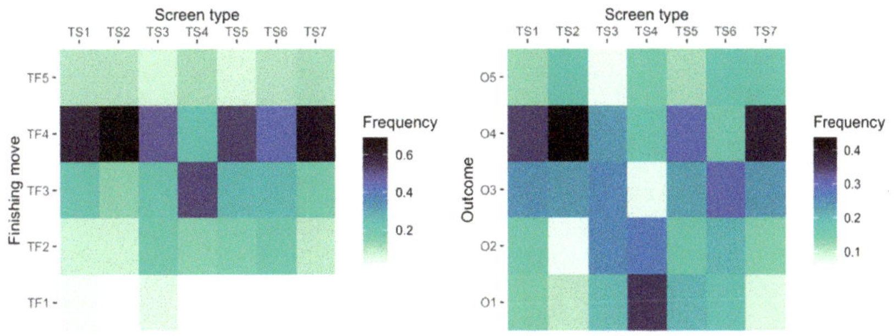

Figure 2. Heatmap of the relationship between screen type/finishing move and screen type/outcome.

3.2. Time Comparison

Table 6 presents the transition probabilities of the second-order Markov chain, that were estimated separately for the outcomes that took place in the first eight seconds of the possession and the last sixteen seconds of the possession. Screen the screener enhanced both 2- and 3-pt shots during the 0–8 s compared to the last seconds of the possession. High-cross screens were effectively beneficial for 3-pt shots during the interval 0–8 s, on the other hand screen-away was associated with well-executed 3-pt shots in the last seconds of the possession. Table 7 presents the comparison of the successful outcomes between the first to third and fourth quarters of playtime. In general, in the last quarter, the efficiency of the most frequently used screens was elevated, as the probabilities to successfully execute a shot were greater compared to the first three quarters. Flare screens inside the perimeter area and high-cross screen in the last quarter, according to our data, guaranteed the outcome of a 2- and 3-pt shots, respectively.

Table 6. Transition probabilities between finishing moves and screens to successful shots.

		Successful Outcome (X_t)	
Screen Type (X_{t-2})	**Finishing Move (X_{t-1})**	**T1**	**T2**
Down screen	Lay-up	0.50	0.43
Staggered	2-pt shot	0.58	0.47
Flare	2-pt shot	0.33	0.50
Screen the screener	2-pt shot	1.00	0.67
Back screen	2-pt shot	0.36	0.56
Down screen	2-pt shot	0.66	0.47
Screen away	2-pt shot	0.55	0.42
Staggered	3-pt shot	0.56	0.38
Flare [1]	3-pt shot	0.35	0.33
Screen the screener	3-pt shot	0.77	0.36
Down screen	3-pt shot	0.52	0.37
High cross	3-pt shot	1.00	0.66
Screen away	3-pt shot	0.17	0.44

[1]: inside the paint.

The asymptotic probabilities of the most-used pairs of screens/finishing moves and finishing moves/outcomes are presented in Table 8. Regarding screens and finishing moves, the most frequent pairs in the court were staggered screens followed by 3-pt shots and flare screens with 3-pt shots. Concerning finishing moves and outcomes, most 3-pt shots were unsuccessful, while the successful and missed 2-pt shots had the same frequency.

Table 7. Comparison of the probability estimates between finishing moves and screens to successful outcomes in the first three vs. last quarter.

		Successful Outcome (X_t)	
Screen Type (X_{t-2})	**Finishing Move (X_{t-1})**	**Q1–Q3**	**Q4**
Staggered	Lay-up	0.30	0.20
Down screen	Lay-up	0.20	0.50
Staggered	2-pt shot	0.52	0.40
Flare [1]	2-pt shot	0.33	1.00
Down screen	2-pt shot	0.30	0.50
Staggered	3-pt shot	0.38	0.59
Flare [1]	3-pt shot	0.37	0.14
Screen the screener	3-pt shot	0.52	0.50
Back screen	3-pt shot	0.20	0.50
Down screen	3-pt shot	0.37	0.58
High cross	3-pt shot	0.63	1.00
Screen away	3-pt shot	0.34	0.69

[1]: inside the paint.

Table 8. Asymptotic probabilities of frequent screens, finishing moves and outcomes.

TF-O	*p*	TS-TF	*p*
TF4-O4	0.33	TS1-TF4	0.24
TF4-O3	0.23	TS2-TF4	0.09
TF3-O1	0.11	TS7-TS4	0.08
TF3-O2	0.11	TS1-TF3	0.08
TF5-O5	0.05	TS5-TF4	0.07
TF2-O1	0.04	TS6-TF4	0.04
TF2-O2	0.04		
TF2-O5	0.03		
TF4-O5	0.01		
TF1-O1	0.01		

Probabilities lower than 0.01 were excluded.

4. Discussion

The aim of the present study is to develop a second-order Markov modelling framework that would evaluate the efficiency of off-ball screens that positively affect the finishing move and the outcome. Relevant literature regarding the strong side of the offense have indicated that screens on the strong side were beneficial for the offense [41]; however, limited studies were conducted concerning the weak side. The outcome of every action in the basketball context depends on several factors, such as the type of defense, the characteristics of the players involved, the scoreboard, and the finishing moves and the screen types on the strong and weak side. The present paper, focusing on offensive actions, attempts to investigate the decision taken by the players on the weak side of the offense. While executing weak side offensive movements, it was found that the two screens that had the highest frequency were staggered screens, followed by flare screens. This occurs because in the first type of screen, there are two consecutive screens in the same direction for a teammate away from the ball. A stagger screen creates more space and allows the cutter to rub the defensive player on the first or second screen for a middle range or 3-pt shot. Conversely, flare screens create clear out situations on the perimeter for a 2-pt or 3-pt shot. According to [42], which undertook an analysis of basketball at the Olympic Games, the findings showed that the successful or unsuccessful 2-pt or 3-pt shots are the most important indicators for winning teams.

Our findings also revealed that the players, during off-ball screens, decide to execute their offense more by using 3-pt shots, followed by making use of a middle-range 2-pt shot; conversely, the lay-up option was not frequent. This is in line with [10], in which research at the World Cup 2019 pointed out that winning teams were more successful on their 3-pt shot attempts, on equally competitive teams. Regarding the effectiveness in the variations of executing the off-ball screens and finishing the offense, greater success is observed in using back screens and lay-ups, followed by flare screens and 2-pt shot; whereas the combination of high-cross screens and 3-pt shots was advantageous. Back screens are movements that take place on the "back" of the defensive player while playing defense on the weak side area. Offensive players, executing this type of screen use back door movements to receive the ball for a lay-up. However, although the combination of back screens and lay-ups lead to higher probability of successfully executing the offense, only a few instances of the above actions were observed in the court. Furthermore, flare screen is a collaboration in which the screener sets up a screen at the corner of the free-throw line and the cutter, instead of moving towards the basket, takes the screen and fades out to open space, away from the ball, for a middle-range shot. Moreover, a cross screen appears when a player cuts to the opposite side of the floor to set a screen for a teammate. Predominantly, this happens at the top of the key and gets a player who was on the strong side of the floor open for a quick 3-pt shot. For the execution of this screen, the coach can use two power forwards players, and additionally, a guard one with a center. The results agree with [16], which presented those different formations wherein the players achieved different effectiveness while leading to a basket.

The present study, by applying a second-order Markov model, demonstrated interesting findings, which confirmed that the offense is influenced by specific screens on the weak side of the court. Using staggered screens, it was shown that when the players executed 2-pt shots, they were led to more successful outcomes. This type of screen can be used inside the paint, where the cutter can go into the corner for a 3-pt shot, whenever the attacking player can go out on different sides of the perimeter for a 3-pt shot. The flare screen, executed either inside or outside the perimeter area, provided equal results with regards to successful 3-pt shots. The above combinations could be interpreted by the arrival of American players in European basketball, indicating that the European basketball has become more unrestrained, such as the NBA. Finally, the Markovian model also predicted that a successful offensive combination is a down screen followed by the execution of a 2-pt shot. The latter combination is probably explained by the fact that the attacks take place inside the paint, as the down screen is made to release mainly the taller players and make a flash movement towards the ball, leading to a better position while leading to a basket.

Concerning the shot clock, the results indicated that specific screen types, such as screen the screener and high-cross, that occur rapidly before the set-play of the offense at the top of the key area during the transition game led to more successful offensive movements in the first 8 s of the possession. On the other hand, during the interval 8–24 s of the offense, the players achieved greater mobility, thus they used screen away to provide the perimeter shooter with an optimal area to execute the 3-pt shot. This result confirms previous findings, which showed that defenders have more fatigue during the last seconds of the offense, thus the resulting offensive screen could be successful [43]. The results in the last quarter, showed that 3-pt shots were positively influenced by high-cross, staggered, screen away, back, and down screens. This can be explained by the fact that in the last minutes of playtime, the players using the aforementioned screens aim to optimize their final score. Previous studies suggested that possession effectiveness was found to be elevated by using different tactical strategies during the last minutes of playtime [41]. On the contrary, in the first three quarters, staggered screens, which consist of two consecutive screens from different offensive players, provided the opportunity to a teammate to receive the ball for an easy lay-up or 2-pt shot. In general, in the last quarter, the efficiency of the most frequently-used screens was elevated, as the probabilities to successfully execute a shot were greater, compared to the first three quarters.

5. Conclusions

In the recent years, the study of performance indicators and their use in the strategy of basketball teams to maximize performance has been the subject of extended research. Via second-order Markov modelling, this paper provided insights into the behaviors and interactions of the players using the screens, and the final attempt of the shots on the weak side. In conclusion, attacks away from the ball are movements without prior verbal signals in which players must perform a specific screen with great speed and accuracy. It is worth noting that this study provides useful information for coaches who may have the opportunity to use it in training programs aimed at the individual improvement of players, and also to improve and maximize the team's offense. We suggest further research that could bring about advances in play, including the area of execution or screen, as a covariate that would influence the outcome of the offense, the cutting movements, or the characteristics of the line-ups on the weak side of the offensive team. In addition, a semi-Markov model could provide a more detailed picture of the offense, incorporating sojourn times between offensive movements, if appropriate data were available. By knowing the strengths and weaknesses of the attack, the coach can have a complete picture of the offense on both sides and adjust the preparation for the next movement to succeed in a basketball game.

Author Contributions: Conceptualization, N.S., A.P. and P.K.; Methodology, N.S., A.P. and P.K.; Software, A.P. and P.K.; Validation, N.S., A.P. and P.K.; Formal analysis, N.S., A.P. and P.K.; Investigation, N.S., A.P. and P.K.; Data curation, N.S., A.P. and P.K.; Writing—original draft preparation, N.S., A.P. and P.K.; Writing—review and editing, N.S., A.P. and P.K.; Visualization, N.S., A.P. and P.K.; Supervision, N.S. and A.P. All authors have read and agreed to the published version of the manuscript.

Funding: This research received no external funding.

Data Availability Statement: The data presented in this study are available on request from the corresponding author. The data are not publicly available due to privacy restrictions.

Acknowledgments: The authors greatly acknowledge the comments and suggestions of the three anonymous referees, which improved the quality and the presentation of the current paper.

Conflicts of Interest: The authors declare no conflict of interest.

References

1. Gómez, M.; Battaglia, O.; Lorenzo, A.; Lorenzo, J.; Jimenez, S.L.; Sampaio, J. Effectiveness during ball screens in elite basketball games. *J. Sports Sci.* **2015**, *33*, 1844–1852. [CrossRef] [PubMed]
2. Lamas, L.; Junior, D.D.R.; Santana, F.; Rostaiser, E.; Negretti, L.; Ugrinowitsch, C. Space creation dynamics in basketball offence: Validation and evaluation of elite teams. *Int. J. Perform. Anal. Sport* **2011**, *11*, 71–84. [CrossRef]
3. Vaquera, A.; García-Tormo, J.V.; Ruano, M.G.; Morante, J. An exploration of ball screen effectiveness on elite basketball teams. *Int. J. Perform. Anal. Sport* **2016**, *16*, 475–485. [CrossRef]
4. Lamas, L.; Santana, F.; Heiner, M.; Ugrinowitsch, C.; Fellingham, G. Modeling the Offensive-Defensive Interaction and Resulting Outcomes in Basketball. *PLoS ONE* **2015**, *10*, e0144435. [CrossRef] [PubMed]
5. Calvo, J.L.; García, A.M.; Navandar, A. Analysis of mismatch after ball screens in Spanish professional basketball. *Int. J. Perform. Anal. Sport* **2017**, *17*, 555–562. [CrossRef]
6. Bazanov, B.; Võhandu, P.; Haljand, R. Factors influencing the teamwork intensity in basketball. *Int. J. Perform. Anal. Sport* **2006**, *6*, 88–96. [CrossRef]
7. Conte, D.; Tessitore, A.; Gjullin, A.; MacKinnon, D.; Lupo, C.; Favero, T. Investigating the game-related statistics and tactical profile in NCAA division I men's basketball games. *Biol. Sport* **2018**, *35*, 137–143. [CrossRef]
8. Kubatko, J.; Oliver, D.; Pelton, K.; Rosenbaum, D.T. A Starting Point for Analyzing Basketball Statistics. *J. Quant. Anal. Sports* **2007**, *3*. [CrossRef]
9. Zhang, S.; Gomez, M.; Yi, Q.; Dong, R.; Leicht, A.; Lorenzo, A. Modelling the Relationship between Match Outcome and Match Performances during the 2019 FIBA Basketball World Cup: A Quantile Regression Analysis. *Int. J. Environ. Res. Public Health* **2020**, *17*, 5722. [CrossRef]
10. Stavropoulos, N.; Kolias, P.; Papadopoulou, A.; Stavropoulou, G. Game related predictors discriminating between winning and losing teams in preliminary, second and final round of basketball world cup 2019. *Int. J. Perform. Anal. Sport* **2021**, *21*, 383–395. [CrossRef]
11. Ibáñez, S.J.; Sampaio, J.; Sáenz-López, P.; Giménez, J.; Janeira, M.A. Game statistics discriminating the final outcome of junior world basketball championship matches (Portugal 1999). *J. Hum. Mov. Stud.* **2003**, *45*, 1–20.
12. Sampaio, J.; Janeira, M. Statistical analyses of basketball team performance: Understanding teams' wins and losses according to a different index of ball possessions. *Int. J. Perform. Anal. Sport* **2003**, *3*, 40–49. [CrossRef]
13. Vračar, P.; Štrumbelj, E.; Kononenko, I. Modeling basketball play-by-play data. *Expert Syst. Appl.* **2016**, *44*, 58–66. [CrossRef]

14. Shirley, K. A Markov model for basketball. In *New England Symposium for Statistics in Sports*; Boston, MA, USA, 2007; p. 82. Available online: http://www.nessis.org/nessis07/Kenny_Shirley.pdf (accessed on 8 June 2021).
15. Kvam, P.; Sokol, J.S. A logistic regression/Markov chain model for NCAA basketball. *Nav. Res. Logist.* **2006**, *53*, 788–803. [CrossRef]
16. Kolias, P.; Stavropoulos, N.; Papadopoulou, A.; Kostakidis, T. Evaluating basketball player's rotation line-ups performance via statistical markov chain modelling. *Int. J. Sports Sci. Coach.* **2021**. [CrossRef]
17. Young, A.; Vassiliou, P.-C.G. A non-linear model on the promotion of staff. *J. R. Stat. Soc. Ser. A* **1974**, *137*, 584–595. [CrossRef]
18. Vassiliou, P.-C.G. A Markov chain model for wastage in manpower systems. *J. Oper. Res. Soc.* **1976**, *27*, 57–70. [CrossRef]
19. Vassiliou, P.-C.G. On the Limiting Behaviour of a Nonhomogeneous Markov Chain Model in Manpower Systems. *Biometrika* **1981**, *68*, 557. [CrossRef]
20. Tsantas, N.; Vassiliou, P.-C.G. The non-homogeneous Markov system in a stochastic environment. *J. Appl. Probab.* **1993**, *30*, 285–301. [CrossRef]
21. Georgiou, A.; Vassiliou, P.-C. Cost models in nonhomogeneous Markov systems. *Eur. J. Oper. Res.* **1997**, *100*, 81–96. [CrossRef]
22. Dimitriou, V.; Georgiou, A.; Tsantas, N. The multivariate non-homogeneous Markov manpower system in a departmental mobility framework. *Eur. J. Oper. Res.* **2013**, *228*, 112–121. [CrossRef]
23. D'Amico, G.; Di Biase, G.; Manca, R. Income inequality dynamic measurement of Markov models: Application to some European countries. *Econ. Model.* **2012**, *29*, 1598–1602. [CrossRef]
24. Faddy, M.J.; McClean, S. Analysing data on lengths of stay of hospital patients using phase-type distributions. *Appl. Stoch. Models Bus. Ind.* **1999**, *15*, 311–317. [CrossRef]
25. McClean, S.; Millard, P. Where to treat the older patient? Can Markov models help us better understand the relationship between hospital and community care? *J. Oper. Res. Soc.* **2007**, *58*, 255–261. [CrossRef]
26. Garg, L.; McClean, S.; Meenan, B.; Millard, P. Non-homogeneous Markov models for sequential pattern mining of healthcare data. *IMA J. Manag. Math.* **2008**, *20*, 327–344. [CrossRef]
27. Almagor, H. A Markov analysis of DNA sequences. *J. Theor. Biol.* **1983**, *104*, 633–645. [CrossRef]
28. Waterman, M.S. Introduction to Computational Biology: Maps, Sequences and Genomes. *Biometrika* **1998**, *54*, 398. [CrossRef]
29. Vassiliou, P.-C.G. A High Order Non-linear Markovian Model for Promotion in Manpower Systems. *J. R. Stat. Soc. Ser. A* **1978**, *141*, 86. [CrossRef]
30. Raftery, A.E. A model for high-order Markov chains. *J. R. Stat. Soc. Ser. B* **1985**, *47*, 528–539. [CrossRef]
31. Adke, S.R.; Deshmukh, S.R. Limit distribution of a high order Markov chain. *J. R. Stat. Soc. Ser. B* **1988**, *50*, 105–108. [CrossRef]
32. Ching, W.K.; Fung, E.S.; Ng, M.K. A higher-order Markov model for the Newsboy's problem. *J. Oper. Res. Soc.* **2003**, *54*, 291–298. [CrossRef]
33. Vassiliou, P.-C.G.; Moysiadis, T.P. G-Inhomogeneous Markov Systems of High Order. *Methodol. Comput. Appl. Probab.* **2010**, *12*, 271–292. [CrossRef]
34. Raftery, A.; Tavaré, S. Estimation and modelling repeated patterns in high order Markov chains with the mixture transition distribution model. *J. R. Stat. Soc. Ser. C* **1994**, *43*, 179–199. [CrossRef]
35. Garden, P.W. Markov analysis of viral DNA/RNA sequences. *J. Theor. Biol.* **1980**, *82*, 679–684. [CrossRef]
36. Shamshad, A.; Bawadi, M.; Wanhussin, W.; Majid, T.A.; Sanusi, S. First and second order Markov chain models for synthetic generation of wind speed time series. *Energy* **2005**, *30*, 693–708. [CrossRef]
37. Berchtold, A.; Raftery, A. The Mixture Transition Distribution Model for High-Order Markov Chains and Non-Gaussian Time Series. *Stat. Sci.* **2002**, *17*, 328–356. [CrossRef]
38. Katz, R. On Some Criteria for Estimating the Order of a Markov Chain. *Technometrics* **1981**, *23*, 243. [CrossRef]
39. Tong, H. Determination of the order of a Markov chain by Akaike's information criterion. *J. Appl. Probab.* **1975**, *12*, 488–497. [CrossRef]
40. Menéndez, M.L.; Pardo, L.; Pardo, M.C.; Zografos, K. Testing the Order of Markov Dependence in DNA Sequences. *Methodol. Comput. Appl. Probab.* **2008**, *13*, 59–74. [CrossRef]
41. Gómez, M.A.; Lorenzo, A.; Ibañez, S.J.; Sampaio, J. Ball possession effectiveness in men's and women's elite basketball according to situational variables in different game periods. *J. Sports Sci.* **2013**, *31*, 1578–1587. [CrossRef]
42. Leicht, A.S.; Gómez, M.A.; Woods, C.T. Explaining Match Outcome During the Men's Basketball Tournament at The Olympic Games. *J. Sports Sci. Med.* **2017**, *16*, 468–473. [PubMed]
43. Bourbousson, J.; Sève, C.; McGarry, T. Space–time coordination dynamics in basketball: Part 2. The interaction between the two teams. *J. Sports Sci.* **2010**, *28*, 349–358. [CrossRef] [PubMed]

mathematics

MDPI

Article

Discrete Time Hybrid Semi-Markov Models in Manpower Planning

Brecht Verbeken * and Marie-Anne Guerry

Department of Business Technology and Operations, Vrije Universiteit Brussel, Pleinlaan, 2, 1050 Brussels, Belgium; Marie-Anne.Guerry@vub.be
* Correspondence: brecht.verbeken@vub.be

Abstract: Discrete time Markov models are used in a wide variety of social sciences. However, these models possess the memoryless property, which makes them less suitable for certain applications. Semi-Markov models allow for more flexible sojourn time distributions, which can accommodate for duration of stay effects. An overview of differences and possible obstacles regarding the use of Markov and semi-Markov models in manpower planning was first given by Valliant and Milkovich (1977). We further elaborate on their insights and introduce hybrid semi-Markov models for open systems with transition-dependent sojourn time distributions. Hybrid semi-Markov models aim to reduce model complexity in terms of the number of parameters to be estimated by only taking into account duration of stay effects for those transitions for which it is useful. Prediction equations for the stock vector are derived and discussed. Furthermore, the insights are illustrated and discussed based on a real world personnel dataset. The hybrid semi-Markov model is compared with the Markov and the semi-Markov models by diverse model selection criteria.

Keywords: semi-Markov model; Markov model; hybrid semi-Markov model; manpower planning

Citation: Verbeken, B.; Guerry, M.-A. Discrete Time Hybrid Semi-Markov Models in Manpower Planning. *Mathematics* **2021**, *9*, 1681. https://doi.org/10.3390/math9141681

Academic Editor: Panagiotis-Christos Vassiliou

Received: 14 June 2021
Accepted: 13 July 2021
Published: 16 July 2021

Publisher's Note: MDPI stays neutral with regard to jurisdictional claims in published maps and institutional affiliations.

Copyright: © 2021 by the authors. Licensee MDPI, Basel, Switzerland. This article is an open access article distributed under the terms and conditions of the Creative Commons Attribution (CC BY) license (https://creativecommons.org/licenses/by/4.0/).

1. Introduction

Manpower planning is a key aspect of modern human resources management. The principal aim of manpower planning is the development of plans dealing with future human resource requirements. In this way, an effective manpower planning policy can avoid future shortages and excesses of staff members. Such an imbalance between the actual and the required staff is highly undesirable because it would lead to higher costs and/or less profits. Since manpower planning itself is concerned with the description and prediction of large groups of employees, whose behaviour can be unpredictable at the individual level, it is only natural to study aggregated data, where statistical patterns may appear. So, it is no surprise that the use of mathematical models for manpower planning can be traced back to at least 1779 when Rowe used a career-modeling plan in the Royal Marines [1].

Since the 1960s and the dawn of the computer age, such models have become an essential tool for the modern manager. Pioneering work concerning mathematical approaches for manpower planning was carried out by Vajda [2,3] and Bartholomew [4,5], whereas Almond and Young [6] were the first to study a real world application of an open homogeneous Markov chain model. Since then, various other manpower planning model approaches have been considered. In the work of Vassiliou [7], the non-homogeneous Markov system was introduced. This idea was expanded upon by Vasilliou et al. [8]. Other work regarding non-homogeneous discrete time (semi-)Markov models includes the works of Papadopoulou [9] and Dimitriou et al. [10] as well as continuous time (semi-)Markov models by McClean et al. [11,12], Papadopoulou et al. [13] and Mehlmann [14]. It is important to remark that the scope of those models is not limited to humans [7], as is the case in manpower planning, but that it can be any biological being or object. Some examples of other populations modeled by this class of stochastic processes [15] include ecological modeling [16] and biological Markov population models [17] and financial applications [18]. It is remarkable that, until recently, discrete time homogeneous semi-Markov models were somewhat neglected in manpower planning.

One of the assumptions of a Markov model is that the length of time a person stays in a state S_i before going to another state S_j only depends on the state S_i itself. Moreover, the waiting time distribution, often called the sojourn time distribution, exhibits the memoryless property. Which means that it does not account for possible duration of stay effects. In this case, the sojourn time distribution is in fact a geometric distribution. However, in practice those assumptions may pose an unrealistic limitation. An alternative model that may solve those problems is a semi-Markov model, which can be viewed as a natural extension of a Markov model. In recent years, the use of discrete time semi-Markov models became more and more popular in various fields such as reliability and survival analysis [19], DNA analysis [20,21], disability insurance [22], credit risk [23–26], and wind speed and tornado modeling [27,28]. Moreover, insights regarding discrete time semi-Markov models contribute to the use of continuous time semi-Markov models [29].

Markov models and semi-Markov models both have advantages: Markov models are less complex and more transparent. In the manpower planning context, for example, this makes a classical Markov model easier to interpret and understand for a manager. Semi-Markov models, on the other hand, allow capturing duration of stay effects due to their more general sojourn distributions. This provides motivation to build hybrid models that incorporate the best of both approaches. In the previous work of Guédon, so-called hidden hybrid semi-Markov chains are presented that combine Markovian states with semi-Markovian states [30]. Since it is possible that, for a particular state, some of the transitions are Markovian while other transitions are semi-Markovian [22], the present paper introduces the concepts of Markovian transitions and semi-Markovian transitions. In this way, Markovian and semi-Markovian transitions are a further refinement of Markovian and semi-Markovian states.

Furthermore, both Markov and semi-Markov models require longitudinal data for their parameter estimation. In practice, however, longitudinal data are often left truncated or right censored, which may lead to estimation problems [31], especially in a semi-Markovian context, where more general sojourn time distributions are allowed. Previous works [11,32] suggest alternative approaches [23,33] to deal with this drawback, such as restricting the analyses to the items for which there is complete information, artificially truncating the data or using adapted formulas for the estimation of the parameters.

In this paper, we discuss the advantages and disadvantages of Markov and semi-Markov manpower planning models in Section 2. In Section 3, we present the so-called hybrid semi-Markov model, which uses a mix of Markov (geometric) and more general (Weibull) sojourn time distributions, offering some advantages: the hybrid semi-Markov model allows for capturing duration of stay effects where useful and reduces the number of parameters to estimate, where possible. In this way, the hybrid semi-Markov model enables one to improve on the semi-Markov model in case the amount of available data is limited. Finally, in Section 4, we use a real world personnel dataset to illustrate our insights. The hybrid semi-Markov model is compared with the Markov model as well as with the semi-Markov model based on several criteria.

2. Markov and Semi-Markov Manpower Planning Models

To model a manpower system, one has to account for three different types of flows: the incoming flows (recruitments), the internal flows between the different personnel categories and the outgoing flows (wastage). We consider $G + 1$ states, given by G personnel categories and one absorbing state W, corresponding to the wastage. First of all, the classical Markov model [4] will be discussed; afterwards, a semi-Markov model for manpower planning based on [19] will be proposed. An interesting reference regarding (semi-)Markov processes is [34]. The discussion on the classical Markov model (in Section 2.1) and the semi-Markov model (in Section 2.2) contributes in defining the new hybrid semi-Markov model in Section 3.

2.1. Markov Model

All models in this section are Markov processes and generalizations thereof, such as semi-Markov processes. However, all models have their limitations and are subjected to restrictions. In this setting, one of the assumptions we make is the so-called Markov property, which states that the probability of reaching a future state is independent of the past states and only depends on the present state. For a second-order Markov chain, this probability of entering a state at time $t+1$ also depends on the state at time $t-1$. To assess the Markov property, we will use Equation (1) below, which tests a first-order against a second-order Markov chain. The use of a classical Markov model without meeting the first-order assumption may lead to false conclusions and incorrect analysis results. An extensive discussion about the often overlooked need to check for the Markov property can be found in [35]. For a given stochastic process $\{X_t\}_t$ with $G+1$ states $\{S_1, \cdots, S_{G+1}\}$ and data over a time horizon $[0, T]$, we will use the following χ^2 goodness of fit test to verify the first-order assumption, as described in [35,36],

$$\chi_e^2 = \sum_{i \in \mathcal{G}} \sum_{j \in \mathcal{G}} \sum_{l \in \mathcal{G}} n_{ij} \frac{(\widehat{p_{ijl}} - \widehat{p_{jl}})^2}{\widehat{p_{jl}}} \tag{1}$$

with index set $\mathcal{G} = \{1, 2, \cdots, G, G+1\}$ and $n_{ij} = \sum_{t=0}^{T-1} \sum_{k=0}^{m} n_{ij}(t,k)$, where $n_{ij}(t,k)$ is the number of persons that are at time t in the state S_i with grade seniority k and at time $t+1$ in state S_j and m is the maximal grade seniority observed in the database. $\widehat{p_{jl}}$ is the maximum likelihood estimator of the transition probability p_{jl} with $N_j(t) = \sum_{i \in \mathcal{G}} \sum_{k=0}^{m} n_{ij}(t-1,k)$ being the number of persons in state S_j at time t , where $\widehat{p_{ijl}}$ is the maximum likelihood estimator of the transition probability p_{ijl} and where $n_{ijl}(t,k)$ is the number of persons that are at time t in the state S_i with grade seniority k at time $t+1$ in the state S_j and at time $t+2$ in the state S_l:

$$\widehat{p_{jl}} = \frac{\sum_{t=0}^{T-1} \sum_{k=0}^{m} n_{jl}(t,k)}{\sum_{t=0}^{T-1} N_j(t)}. \tag{2}$$

$$p_{jl} = \Pr(X_t = S_l | X_{t-1} = S_j) \tag{3}$$

$$\widehat{p_{ijl}} = \frac{\sum_{t=0}^{T-2} \sum_{k=0}^{m} n_{ijl}(t,k)}{\sum_{t=0}^{T-2} \sum_{k=0}^{m} n_{ij}(t,k)}. \tag{4}$$

$$p_{ijl} = \Pr(X_{t+2} = S_l | X_{t+1} = S_j, X_t = S_i) \tag{5}$$

Only non-zero $\widehat{p_{jl}}$ are taken into account for computing χ_e^2. Under the assumption that the Markov property is satisfied, i.e., that we are looking for a Markov chain of order 1, the test statistic χ_e^2 has a χ^2-distribution with $(G+1)^3$ degrees of freedom. If this assumption holds, we can proceed with the classical Markov approach, in which transition probabilities are assumed to be equal for individuals within a category.

The use of time homogeneous Markov chains in manpower planning is well-known (see, for example, [4]) . Given the G states corresponding to different personnel categories $S_1, \cdots, S_G$ and a wastage state $W = S_{G+1}$, one can define a Markov process $\{X_t\}_t$ on those states with transition probabilities p_{jl} that can be estimated by Equation (2). If we denote the stock vector at time t by $\mathbf{N}(t) = (N_1(t), N_2(t), \cdots, N_G(t), W(t))$, and write $\mathbf{R}(t) = (R_1(t), R_2(t), \cdots, R_G(t), 0)$ for the recruitment vector at time t, then we obtain the prediction equation [4] for the stocks at time $t+1$:

$$\mathbf{N}(t+1) = \mathbf{N}(t) \cdot \mathbf{P} + \mathbf{R}(t+1), \tag{6}$$

where **P** is the matrix with elements $\widehat{p_{jl}}$.

Due to their simplicity, time homogeneous Markov chain models are used in a wide variety of domains and applications. As there are relatively few parameters to estimate in a time homogeneous Markov chain model, they are not too data demanding. However, on the other hand, they cannot be used to account for duration of stay effects and they are less flexible due to the so-called memoryless property, which implies that their sojourn time distributions are geometrical distributed by construction. This shortcoming is accounted for in semi-Markov models.

2.2. Semi-Markov Model

Again, consider a system with a finite number of states $\{S_1, \cdots, S_G, S_{G+1}\}$ and let us denote the set of indices by $\mathcal{G} = \{1, 2, \cdots, G, G+1\}$. Furthermore, let T_n and J_n denote, respectively, the time of the n-th transition and the state occupied after the n-th transition. A semi-Markov process is equivalent to a Markov renewal process [37] and is completely determined by an initial distribution $\delta = (\delta_1, \cdots \delta_G, \delta_{G+1})$ and a discrete semi-Markov kernel $\mathbf{q} = (q_{ij}(k) : i, j \in \mathcal{G}, k \in \mathbb{N})$ where

$$q_{ij}(k) = \Pr(J_{n+1} = S_j, T_{n+1} - T_n = k \mid J_n = S_i). \tag{7}$$

It can be shown that $\{J_n\}_n$ itself is a Markov chain via

$$p_{ij}^{\infty} = \Pr(J_{n+1} = S_j \mid J_n = S_i), \tag{8}$$

i.e., p_{ij}^{∞} is the probability, starting from S_i, that the next state will eventually be S_j, regardless of the duration time. We write $\mathbf{P}^{\infty} = (p_{ij}^{\infty} : i, j \in \mathcal{G})$ for the associated transition matrix. This allows for the following decomposition:

$$q_{ij}(k) = p_{ij}^{\infty} f_{ij}(k) \tag{9}$$

where $\mathbf{f} = (f_{ij}(k) : i, j \in \mathcal{G}, k \in \mathbb{N})$ consists of the sojourn time distributions, conditioned by the next state to be visited:

$$f_{ij}(k) = \Pr(T_{n+1} - T_n = k \mid J_n = S_i, J_{n+1} = S_j) \tag{10}$$

A few remarks are in order at this point. First of all, only actual transitions are accounted for, in the sense that transitions to the same state are prohibited, so that $p_{ii}^{\infty} = 0$ for every $i \in \mathcal{G}$. Furthermore, instantaneous transitions are not allowed either: the chain has to spend at least one unit of time in a state, which corresponds to $f_{ij}(0) = q_{ij}(0) = 0$ for every $i, j \in \mathcal{G}$.

The main difference in regard to the Markov chain model is the fact that the sojourn time distributions **f** can be any discrete distribution, incorporating the possible duration of stay effects. Note that a Markov chain with transition matrix $\mathbf{P} = (p_{ij} : i, j \in \mathcal{G})$ itself can be viewed as a semi-Markov chain with geometrically distributed sojourn times for which

$$q_{ij}(k) = \begin{cases} p_{ij}\, p_{ii}^{k-1} & \text{if } i \neq j \text{ and } k \in \mathbb{N}_0 \\ 0 & \text{elsewhere.} \end{cases} \tag{11}$$

In order to use this framework for a manpower planning model, one starts in the same way as in the case of a Markov chain model with dividing the population in $G+1$ states and determining the corresponding stock vector $\mathbf{N}(t)$. In contrast with the Markov chain model, we incorporate the grade seniority of the employees in our model. Instead of a vector $\mathbf{N}(t)$ consisting of the total number of people in each personnel category at time t, every entry of $\mathbf{N}(t)$ corresponds to a vector of a certain length m containing the number of employees with seniority l, with $1 \leq l \leq m$. This disaggregation of the entries of $\mathbf{N}(t)$ results in a matrix, whose columns will be denoted by $\mathbf{N}(t,k)$ as in Figure 1. So, the first column, $\mathbf{N}(t,0)$, corresponds to the employees with grade seniority 0 at time t, the second column, $\mathbf{N}(t,1)$, corresponds to the employees with grade seniority 1 at time t, ... up to the $m+1$-th column that corresponds to the employees with grade seniority m at time t , where m is the maximal grade seniority observed in the database. We will call this matrix the seniority based stock matrix. Note that $N_i(t,k)$ corresponds to the number of employees in state S_i with grade seniority k at time t and that $\sum_{k=0}^{m} N_i(t,k) = N_i(t)$ for each $i \in \mathcal{G}$ and every $t \in \mathbb{N}$. The vectors $\mathbf{N}(t,k)$ enable the expression of the prediction equation for the stock vector as in Theorem 2. An equivalent approach is presented in [8], where the semi-Markov system is transformed into a Markov system. While the present paper considers a separate vector $\mathbf{N}(t,k)$ for each grade seniority k, in [8], this information is gathered into one vector.

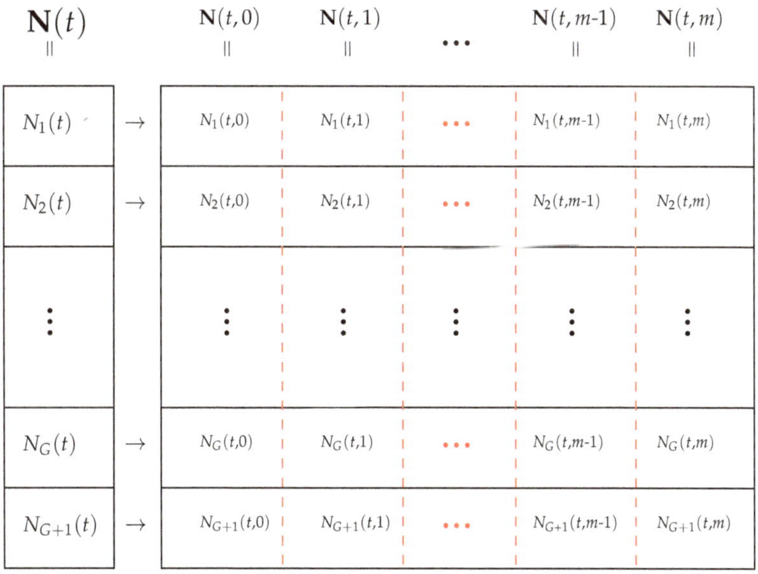

Figure 1. The seniority based stock matrix, consisting of columns $\mathbf{N}(t,k)$.

Now, we can estimate a discrete semi-Markov kernel $\mathbf{q}$ using the maximum likelihood estimator [19]:

$$\widehat{q_{ij}(k)} = \frac{\sum_{t=0}^{T-1} n_{ij}(t,k)}{n_i} \tag{12}$$

where $n_i = \sum_{j \neq i} \sum_{t=0}^{T-1} \sum_{k=0}^{m} n_{ji}(t,k)$, i.e., the total number of visits to state i. Furthermore, we can use this $\mathbf{q}$ to calculate the grade seniority transition matrices $\mathbf{P}(k) = (P_{ij}(k) : i,j \in \mathcal{G})$, the one-step ahead transition matrix for group members with grade seniority k, is defined by:

$$P_{ij}(k) = \Pr(J_{n+1} = j, T_{n+1} - T_n = k \mid J_n = i, T_{n+1} - T_n > k-1) \tag{13}$$

In practice, $\mathbf{P}(k)$ can be calculated in the following way.

Theorem 1. *For all k such that* $\sum_{h \in \mathcal{G}} \sum_{m=0}^{k-1} q_{ih}(m) \neq 1$ *we have*

$$P_{ij}(k) = \frac{q_{ij}(k)}{1 - \sum_{h \in \mathcal{G}} \sum_{m=0}^{k-1} q_{ih}(m)} \tag{14}$$

Proof.

$$\begin{aligned}P_{ij}(k) &= \Pr(J_{n+1} = j, T_{n+1} - T_n = k \mid J_n = i, T_{n+1} - T_n > k-1) \\ &= \frac{\Pr(J_{n+1} = j, T_{n+1} - T_n = k \mid J_n = i)}{\Pr(T_{n+1} - T_n > k-1 \mid J_n = i)} \\ &= \frac{\Pr(J_{n+1} = j, T_{n+1} - T_n = k \mid J_n = i)}{1 - \Pr(T_{n+1} - T_n \leq k-1 \mid J_n = i)} \\ &= \frac{q_{ij}(k)}{1 - \sum_{h \in \mathcal{G}} \sum_{m=0}^{k-1} q_{ih}(m)}\end{aligned}$$

□

Combining all of the above, we arrive at:

Theorem 2. *For a semi-Markov system, the prediction equation for the stock vector at time* $t+1$ *is as follows:*

$$\mathbf{N}(t+1) = \sum_{k=0}^{m} \Big(\mathbf{N}(t,k) \cdot \mathbf{P}(k)\Big) + \mathbf{R}(t+1), \tag{15}$$

where $\mathbf{N}(t,k)$ *is the stock vector of people with grade seniority* k *at time* t, $\mathbf{R}(t+1)$ *is the recruitment vector at time* $t+1$, $\mathbf{P}(k)$ *is the one-step ahead transition matrix for people with grade seniority* k *and* m *is the maximum of all grade seniorities.*

At first glance, it would seem that the semi-Markov model is a preferable model due to its more flexible sojourn time distributions and its greater generality. However, to build a semi-Markov model, one has to estimate more parameters, such that a sufficiently long time series of data may be necessary to avoid problems with overfitting [38]. This may limit the utility of semi-Markov modeling in manpower planning as a data horizon of, for example, less than ten years may be insufficient for the realization of some transitions and so for the required data for estimating the semi-Markov kernel $\mathbf{q}$.

3. Hybrid Semi-Markov Model

In Section 2.2, we note that a Markov chain with transition matrix $\mathbf{P} = (p_{ij} : i,j \in \mathcal{G})$ can be viewed as a semi-Markov chain, i.e., a semi-Markov chain can be considered as an extension of a Markov chain, where more general and flexible sojourn time distributions are allowed. However, in practice, it can be difficult to decide which approach is more adequate to model the manpower system in question. Due to its greater generality, the semi-Markov chain may look as the most preferable model at first sight. However, in practice, the data requirements to result in accurate parameter estimates may limit the utility of semi-Markov models in manpower planning [38]. For these reasons, the presented hybrid semi-Markov model examines, for each pair of states (S_i, S_j), whether the transition from S_i to S_j can be considered as a Markov transition or should be modeled as a semi-Markov transition. In order to make an adequate choice for a particular transition from S_i to S_j between a Markov and a semi-Markov approach, one can use a technique which was introduced in [22] and which is briefly discussed below.

The semi-Markov hypothesis is tested at the level of the sojourn time distributions f_{ij}. A transition from S_i to S_j can be considered Markovian if its corresponding sojourn time f_{ij} is geometrically distributed. Under the geometrical hypothesis, the equality $f_{ij}(2) = f_{ij}(1)(1 - f_{ij}(1))$ holds and a significant deviation of $f_{ij}(1)(1 - f_{ij}(1)) - f_{ij}(2)$ from zero has to be seen as evidence to the contrary, i.e., evidence in favor of a (more general) sojourn time distribution. The test statistic, as introduced in [22], is given by:

$$\widehat{S}_{ij} = \frac{\sqrt{n_{ij}}(\widehat{f}_{ij}(1) * (1 - \widehat{f}_{ij}(1)) - \widehat{f}_{ij}(2))}{\sqrt{\widehat{f}_{ij}(1)(1 - \widehat{f}_{ij}(1))^2(2 - \widehat{f}_{ij}(1))}} \tag{16}$$

where $n_{ij} = \sum_{t=0}^{T-1} \sum_{k=0}^{m} n_{ij}(t,k)$ denotes the observed total number of transitions from S_i to S_j and where $\widehat{f_{ij}}(k)$ is the maximum likelihood estimator of the probability $f_{ij}(k)$ (see [19]):

$$\widehat{f_{ij}}(k) = \frac{\sum_{t=0}^{T-1} n_{ij}(t,k)}{n_{ij}}. \tag{17}$$

Under the geometrical hypothesis H_0, the test statistic $\widehat{S}_{ij}$ is asymptotically normally distributed.

Note that for a system with $G+1$ states, this test has to be run $(G+1)*G$ times as this test permits us to make a decision about the sojourn time distribution for each f_{ij} individually, which allows for a so-called hybrid semi-Markov model—a semi-Markov model that incorporates the sojourn time distributions of the classical Markov model for those pairs (S_i, S_j) where geometric sojourn time distributions may be assumed and that enables the use of more general sojourn time distributions for those pairs (S_i, S_j) where necessary. This approach can be seen as a further generalization of techniques used in [22,39], where the same criterion was used to make a decision about the sojourn time distributions at the level of the states instead of the transitions. Since the sojourn time distribution is determined per pair (S_i, S_j), and hence for each possible transition, the hybrid semi-Markov model is based on transition-dependent sojourn time distributions. In this way, we can construct a model that unites the best of the Markovian and (pure) semi-Markovian worlds, as we will only have to estimate extra parameters of the sojourn time distributions if those parameters might improve the goodness of fit.

Previous studies concerning semi-Markov models often used the discrete Weibull distribution [19] whenever the geometrical hypothesis is rejected. The choice for the discrete Weibull distribution is motivated by the fact that the discrete Weibull distribution can be viewed as a more flexible generalization of the geometric distribution [40]:

$$\textbf{CMF}\ dweibull(k,\alpha,\beta) = 1 - \alpha^{(k+1)^{\beta}} \tag{18}$$

$$\textbf{CMF}\ geometric(k,p) = 1 - (1-p)^{(k+1)} \tag{19}$$

so $geometric(k,p) = dweibull(k, 1-p, 1)$.

Note that, in the semi-Markov setting, the prediction equation of the stock vector (Equation (15)) is nothing more than a generalization of the prediction equation of the stock vector in the Markov setting (Equation (1)), as in the latter case the $\mathbf{P}(k)$ are equal for all k. So, to arrive at the prediction equation for the stock vector of the hybrid semi-Markov model, one can recycle Equation (15), where $P_{ij}(k)$ will be dependent on k due to the sojourn time distributions associated with the (S_i, S_j) for which the Markov hypothesis does not hold.

A procedure to decide on whether to use a Markov model, a semi-Markov model or a hybrid semi-Markov model is graphically represented in Figure 2.

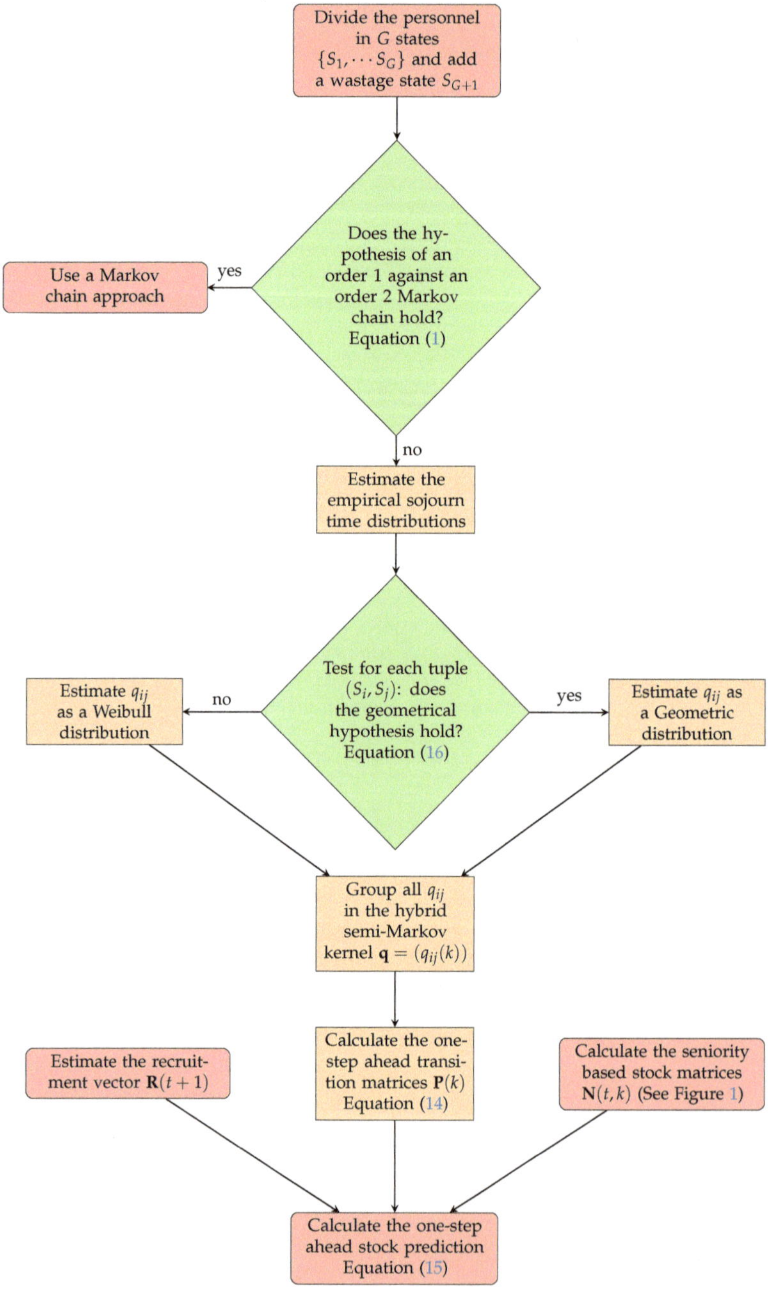

Figure 2. Decision flowchart for the hybrid semi-Markov model.

4. Application

4.1. Data Handling

The subject of this research is modeling (a subsystem of) the academic staff of the Vrije Universiteit Brussel (VUB). An anonymized personnel database including the career paths of all academic staff at the VUB between 1999 and 2013 was at our disposal for this study. The aim is to estimate the number of teaching staff in the various grades for the near future. In our study, we have chosen to avoid left censoring issues: since the analyzed data contain only a limited number of data lines where left censoring is involved, we did not take into account the first observed state of an employee in case it was subjected to left censoring. We corrected for right censoring in computing the estimations of the parameters [41].

After extensive data cleansing, we obtained the career paths of 1585 relevant employees. Only data from 1999 to 2012 were included to avoid look-ahead bias as we aim to estimate the number of teaching staff in 2013. Concerning the division of the personnel in G states, we opted for the common hierarchical academic ranking structure in Belgium as in Table 1.

Table 1. Personnel categories in our manpower system.

State		
S_1	Doctor-assistent	(lecturer with a PhD)
S_2	Docent	(assistent professor)
S_3	Hoofddocent	(associate professor)
S_4	Hoogleraar	(full professor)

Furthermore, we included an additional state, state S_5, which corresponds to wastage in our system. Contrary to most applications in the literature, we did not consider the wastage state to be an absorbing state as it regularly happens during academic careers that people who leave their universities are employed again later on. This happens in our dataset for 371 cases. The observed transitions between the states in our system are visualized in Figure 3.

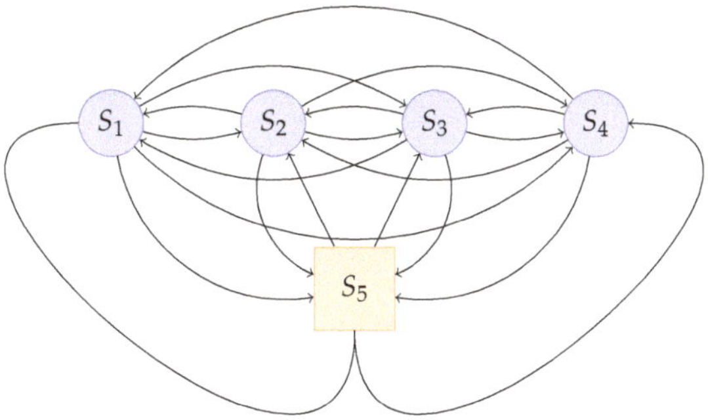

Figure 3. Graph of the states and state transitions.

First of all, the Markov property (Equation (1)) was assessed. Defining the level of significance at $\alpha = 0.05$, the null hypothesis states that the Markov property is met. As we consider five states in our subsystem, it follows that the test statistic χ_e^2 has a χ^2-distribution with 5^3 degrees of freedom under the null hypothesis. We obtained $\chi_e^2 = 4984.911$, which means that we reject the zero hypothesis at the significance level $\alpha = 0.05$. These findings let us conclude that the whole system, consisting of five states, does not satisfy the Markov property.

4.2. Parameter Estimation and Modeling

We now use the same data as in the previous section to estimate the empirical sojourn time distributions $\widehat{f_{ij}}$ according to Equation (17) with the aid of the R package SMM [41] and apply the test statistic $\widehat{S}_{ij}$ (Equation (16)) to each tuple of states (S_i, S_j). The results are summarized in Table 2.

Table 2. Values of test statistic $\widehat{S}_{ij}$.

	S_1	S_2	S_3	S_4	S_5
S_1	/	0.12	−0.00	0	−3.76
S_2	0.49	/	0.83	0	−1.22
S_3	0.17	0.09	/	−1.87	−1.68
S_4	0	0	0	/	−1.10
S_5	2.05	1.12	0.08	0	/

Under the geometrical hypothesis H_0, these test statistics $\widehat{S}_{ij}$ are asymptotically normally distributed. At a significance level of $\alpha = 0.05$, we reject the null hypothesis if and only if $|\widehat{S}_{ij}| > 1.96$. This means we have to reject the geometrical hypothesis for the sojourn time distributions f_{15} and f_{51}. Using the R package SMM [41], we estimated all f_{ij}s as parametric distributions: f_{15} and f_{51} as Weibull distributions and the other f_{ij}s as geometric distributions. We now consider three different models:

- **M**, a classical Markov model as in Section 2.1;
- **SMW**, a semi-Markov model as in Section 2.2, where all f_{ij}s are Weibull distributions;
- **HSM**, the hybrid semi-Markov model as in Section 3 with the f_{ij}s as described above.

4.3. Comparison of the Different Models

We used Equation (15) to predict the stock vector in 2013 starting from the stock vector in 2012 for the three models mentioned above, as a first indication of the performance of those models. We took the factual recruitment vector for $\mathbf{R}(t+1)$. The forecasts, including the standard deviations [42], are summarized in Table 3.

Table 3. Model predictions of the stock vector in 2013. The standard deviations are within brackets.

	Model Predictions						
	M		**SMW**		**HSM**		**Actual Stocks in 2013**
S_1	235.32	(8.01)	191.54	(8.85)	206.22	(8.45)	229
S_2	292.77	(7.81)	246.68	(9.81)	298.07	(8.05)	304
S_3	97.06	(4.65)	96.94	(6.56)	97.28	(4.68)	96
S_4	58.84	(2.86)	64.34	(4.79)	58.89	(2.86)	64

It is immediately obvious, looking at Table 3, that the **SMW** model is the worst predictor of the stock vector for the first two personnel categories in the setting above. Other prediction results are more similar. In what follows, **M**, **SMW** and **HSM** are compared based on several model selection criteria such as AIC and BIC. Afterwards, we used the likelihood ratio test statistic to state a final model preference [43].

First, we analyzed the goodness of fit of our different models using the AIC and BIC according to the formulas below [44],

$$\begin{aligned} AIC &= 2n - 2l(M_i) \\ BIC &= n\ \ln(\kappa) - 2l(M_i) \end{aligned} \tag{20}$$

where n corresponds to the number of estimated parameters in the model M_i, $l(M_i)$ is the log-likelihood function for M_i and κ corresponds to the total number of observations, which is the number of observed transitions in our case.

We obtained the following values for the log-likelihood function:

$$\begin{aligned} l(\mathbf{M}) &= -3723.73, \\ l(\mathbf{SMW}) &= -3912.79, \\ l(\mathbf{HSM}) &= -3682.55 \end{aligned}$$

It is immediately apparent from the equations above that **SMW** is an unfeasible model, as it has the most parameters but the worst fit of our three models. We now proceed to calculate the AIC and BIC of the three models in question. The results are summarized in Table 4.

Table 4. AIC and BIC values.

	Selection Criteria	
	AIC	BIC
M	7487	7624
SMW	7906	8179
HSM	7409	7559

The hybrid semi-Markov model **HSM** has the lowest BIC and AIC values, which means that it outperforms both the semi-Markov model **SMW** and the Markov model **M** with regard to the goodness of fit. Furthermore it is remarkable that the semi-Markov model **SMW** turns out to be the model with the worst fit of the three models concerning the AIC, BIC or even the values of the log-likelihood function itself. This may sound counter-intuitive at first as this model is the most flexible model of the three. We theorize that this is probably due to the more demanding data requirements needed to estimate a higher amount of parameters, which can lead to problems with overfitting.

At last, in order to make a final choice between the models above, one can assess the goodness of fit between the Markov model **M** and the hybrid semi-Markov model **HSM** by means of the likelihood ratio test for nested models as $\mathbf{M} \subset \mathbf{HSM}$ [44]. For two nested statistical models $M_1 \subset M_2$, the likelihood ratio test statistic is given by:

$$\lambda_{LR} = -2[l(M_1) - l(M_2)] \tag{21}$$

where $l(M_1)$ and $l(M_2)$ are the values of the log-likelihood function for M_1 and M_2, respectively. This test statistic is, under the zero hypothesis, i.e., that the more simple model is in fact the true model, asymptotically χ^2 distributed with d degrees of freedom, where d is the number of additional parameters in the more complex model.

We now proceed to use the likelihood ratio test to assess the goodness of fit between the two remaining models of interest: **M** and **HSM**. We arrive at the following value for the test statistic λ_{LR}.

$$\lambda_{LR} = 82.36 \tag{22}$$

As **HSM** adds two additional parameters to **M**, it follows that the test statistic λ_{LR} has a χ^2-distribution with two degrees of freedom under the null hypothesis. We obtained $\lambda_{LR} = 82.36$, which means that we reject the zero hypothesis at the significance level $\alpha = 0.05$ in favor of the alternative hypothesis, i.e., that **HSM** is the better model, which is consistent with the AIC and BIC values in Table 4. Hence, for illustrative purposes, our three models can be ranked according to their goodness of fit as **HSM**, **M** and finally **SMW**.

5. Conclusions

In this paper, we use a discrete time semi-Markov framework to model an open manpower system. At first sight, such a model might appear to be the preferable model as it is not only a more flexible model in nature but also enables us to account for duration of stay effects. However, such a model does not always show to be superior in an empirical context due to the fact that more parameters have to be estimated, which necessitates the availability of a vast amount of data and which may lead to overfitting in the absence of enough data. Therefore, we introduce a hybrid semi-Markov model, that is a semi-Markov model in which Markov sojourn time distributions are used for those transitions (S_i, S_j) where it is not useful to account for duration of stay effects and in which Weibull distributed sojourn times are used for those transitions (S_i, S_j) where the geometrical hypothesis does not hold. Hence, the hybrid semi-Markov model takes the duration of stay effect into account only for those transitions where it can contribute to the improvement of the goodness of fit. In this way, the hybrid semi-Markov combines the best of both worlds by capturing duration of stay effects where useful and reduces the number of parameters to estimate, where possible. Finally we used a real world personnel dataset to illustrate our insights and made a comparison between the Markov model, the semi-Markov model and the hybrid semi-Markov model.

The authors view the use of this specific dataset as one of the most important limitations of this research, as alternative or richer databases may exhibit other characteristics which could lead to other model choices. In addition, future research may focus on the use of other non-Weibull distributions or might explore the possibilities of a hybrid semi-Markov model in a non-homogeneous context.

Author Contributions: Conceptualization, B.V.; methodology, B.V. and M.-A.G.; validation, B.V. and M.-A.G.; data curation, B.V.; writing—original draft preparation, B.V. and M.-A.G.; writing—review and editing, B.V. and M.-A.G.; visualization, B.V.; supervision, M.-A.G. All authors have read and agreed to the published version of the manuscript.

Funding: This research received no external funding.

Data Availability Statement: The data are not publicly available due to privacy restrictions.

Acknowledgments: The authors would like to thank the reviewers for their remarks and valuable suggestions.

Conflicts of Interest: The authors declare no conflict of interest.

References

1. Jones, E. An actuarial problem concerning the Royal Marines. *J. Staple Inn. Actuar. Soc.* **1946**, *6*, 38–42. [CrossRef]
2. Vajda, S. The Stratified Semi-Stationary Population. *Biometrika* **1947**, *34*, 243–254. [CrossRef]
3. Vajda, S. *Mathematics of Manpower Planning*; Wiley: Chichester, UK, 1978.
4. Bartholomew, D.J. The Statistical Approach to Manpower Planning. *Statistician* **1971**, *20*, 3–26. [CrossRef]
5. Bartholomew, D.J. A Multi-Stage Renewal Process. *J. R. Stat. Soc. Ser. B (Methodol.)* **1963**, *25*, 150–168. [CrossRef]
6. Young, A.; Almond, G. Predicting Distributions of Staff. *Comput. J.* **1961**, *3*, 246–250. [CrossRef]
7. Vassiliou, P.C.G. Asymptotic behavior of Markov systems. *J. Appl. Probab.* **1982**, *19*, 851–857. [CrossRef]
8. Vassiliou, P.C.G.; Papadopoulou, A.A. Non-homogeneous semi-Markov systems and maintainability of the state sizes. *J. Appl. Probab.* **1992**, *29*, 519–534. [CrossRef]
9. Papadopoulou, A.A. Economic Rewards in Non-homogeneous Semi-Markov Systems. *Commun. Stat. Theory Methods* **2004**, *33*, 681–696. [CrossRef]
10. Dimitriou, V.; Georgiou, A.; Tsantas, N. The multivariate non-homogeneous Markov manpower system in a departmental mobility framework. *Eur. J. Oper. Res.* **2013**, *228*, 112–121. [CrossRef]
11. McClean, S.; Montgomery, E.; Ugwuowo, F. Non-homogeneous continuous-time Markov and semi-Markov manpower models. *Appl. Stoch. Model. Data Anal.* **1997**, *13*, 191–198. [CrossRef]
12. McClean, S. Continuous-Time Stochastic Models of a Multigrade Population. *J. Appl. Probab.* **1978**, *15*, 26–37. [CrossRef]
13. Papadopoulou, A.; Vassiliou, P.C. Continuous time non homogeneous semi-Markov systems. In *Semi-Markov Models and Applications*; Janssen, J., Limnios, N., Eds.; Springer: Boston, MA, USA, 1999; pp. 241–251. [CrossRef]
14. Mehlmann, A. Semi-Markovian Manpower Models in Continuous Time. *Appl. Probab. Probab.* **1979**, *16*, 416–422. [CrossRef]

15. Esquível, M.L.; Krasii, N.P.; Guerreiro, G.R. Open Markov Type Population Models: From Discrete to Continuous Time. *Mathematics* **2021**, *9*, 1496. [CrossRef]
16. Moore, A.D. The semi-Markov process: A useful tool in the analysis of vegetation dynamics for management. *J. Environ. Manag.* **1990**, *30*, 111–130. [CrossRef]
17. Cohen, J.E. Markov population processes as models of primate social and population dynamics. *Theor. Popul. Biol.* **1972**, *3*, 119–134. [CrossRef]
18. Guerreiro, G.R.; Mexia, J.T.; Miguens, M.F. A Model for Open Populations Subject to Periodical Re-Classifications. *J. Stat. Theory Pract.* **2010**, *4*, 303–321. [CrossRef]
19. Barbu, V.S.; Limnios, N. Semi-Markov chains and hidden semi-Markov models toward applications: Their use in reliability and DNA analysis. In *Lecture Notes in Statistics*; Springer Science & Business Media: Berlin/Heidelberg, Germany, 2009; Volume 191.
20. Papadopoulou, A.A. Some Results on Modeling Biological Sequences and Web Navigation with a Semi Markov Chain. *Commun. Stat. Theory Methods* **2013**, *42*, 2853–2871. [CrossRef]
21. Kolias, P.; Papadopoulou, A. Investigating some attributes of periodicity in DNA sequences via semi-Markov modelling. *arXiv* **2019**, arXiv:stat.AP/1907.03119.
22. Stenberg, F.; Silvestrov, D.; Manca, R. Semi-Markov reward models for disability insurance. *Theory Stoch. Process.* **2006**, *12*, 239–254.
23. Vasileiou, A.; Vassiliou, P.C. An inhomogeneous semi-Markov model for the term structure of credit risk spreads. *Adv. Appl. Probab.* **2006**, *38*, 171–198. [CrossRef]
24. Vassiliou, P.C.; Vasileiou, A. Asymptotic behaviour of the survival probabilities in an inhomogeneous semi-Markov model for the migration process in credit risk. *Linear Algebra Its Appl.* **2013**, *438*, 2880–2903. [CrossRef]
25. D'Amico, G.; Janssen, J.; Manca, R. Valuing credit default swap in a non-homogeneous semi-Markovian rating based model. *Comput. Econ.* **2007**, *29*, 119–138. [CrossRef]
26. D'Amico, G.; Janssen, J.; Manca, R. Initial and final backward and forward discrete time non-homogeneous semi-Markov credit risk models. *Methodol. Comput. Appl. Probab.* **2010**, *12*, 215–225. [CrossRef]
27. D'Amico, G.; Manca, R.; Corini, C.; Petroni, F.; Prattico, F. Tornadoes and related damage costs: Statistical modelling with a semi-Markov approach. *Geomat. Nat. Hazards Risk* **2016**, *7*, 1600–1609. [CrossRef]
28. D'Amico, G.; Petroni, F.; Prattico, F. Wind speed modeled as an indexed semi-{M}arkov process. *Environmetrics* **2013**, *24*, 367–376. [CrossRef]
29. Wu, B.; Maya, B.I.G.; Limnios, N. Using Semi-Markov Chains to Solve Semi-Markov Processes. *Methodol. Comput. Appl. Probab.* **2020**, 1–13. [CrossRef]
30. Guédon, Y. Hidden hybrid Markov/semi-Markov chains. *Comput. Stat. Data Anal.* **2005**, *49*, 663–688. [CrossRef]
31. McClean, S.I.; Gribbin, J.O. Estimation for incomplete manpower data. *Appl. Stoch. Model. Data Anal.* **1987**, *3*, 13–25. [CrossRef]
32. Kalbfleisch, J.D.; Prentice, R.L. *The Statistical Analysis of Failure Time Data*; John Wiley & Sons: Hoboken, NJ, USA, 2011; Volume 360. [CrossRef]
33. McClean, S.; Gribbin, O. A non-parametric competing risks model for manpower planning. *Appl. Stoch. Model. Data Anal.* **1991**, *7*, 327–341. [CrossRef]
34. Howard, R.A. *Dynamic Probabilistic Systems: Markov Models*; Courier Corporation: North Chelmsford, UK, 2012; Volume 1.
35. Bickenbach, F.; Bode, E. *Markov or Not Markov-This Should Be a Question*; Technical Report; Kiel Working Paper; Kiel Institute for the World Economy (IfW): Kiel, Germany, 2001.
36. Anderson, T.W.; Goodman, L.A. Statistical Inference about Markov Chains. *Ann. Math. Stat.* **1957**, *28*, 89–110. [CrossRef]
37. Vassiliou, P.C. Non-Homogeneous Semi-Markov and Markov Renewal Processes and Change of Measure in Credit Risk. *Mathematics* **2021**, *9*, 55. [CrossRef]
38. Valliant, R.; Milkovich, G. Comparison of Semi-Markov and Markov Models in a Personnel Forecasting Application. *Decis. Sci.* **1977**, *8*, 465–477. [CrossRef]
39. D'Amico, G.; Petroni, F.; Prattico, F. Semi-Markov Models in High Frequency Finance: A Review. *arXiv* **2013**, arXiv:q-fin.ST/1312.3894.
40. Nakagawa, T.; Yoda, H. Relationships Among Distributions. *IEEE Trans. Reliab.* **1977**, *26*, 352–353. [CrossRef]
41. Barbu, V.; Bérard, C.; Cellier, D.; Sautreuil, M.; Vergne, N. SMM: An R Package for Estimation and Simulation of Discrete-time semi-Markov Models. *R J.* **2018**, *10*, 226. [CrossRef]
42. Papadopoulou, A.; Vassiliou, P.C.G. On the Variances and Convariances of the Duration State Sizes of Semi-Markov Systems. *Commun. Stat. Theory Methods* **2014**, *43*, 1470–1483. [CrossRef]
43. Udom, A.U.; Ebedoro, U.G. On multinomial hidden Markov model for hierarchical manpower systems. *Commun. Stat. Theory Methods* **2021**, *50*, 1370–1386. [CrossRef]
44. Koch, K. *Parameter Estimation and Hypothesis Testing in Linear Models*, 2nd ed.; Springer: Berlin/Heidelberg, Germany, 1999. [CrossRef]

mathematics

Article

On State Occupancies, First Passage Times and Duration in Non-Homogeneous Semi-Markov Chains

Andreas C. Georgiou [1,*], Alexandra Papadopoulou [2], Pavlos Kolias [2], Haris Palikrousis [2] and Evanthia Farmakioti [2]

[1] Quantitative Methods and Decision Analytics Lab, Department of Business Administration, University of Macedonia, 54636 Thessaloniki, Greece
[2] Department of Mathematics, Aristotle University of Thessaloniki, 54124 Thessaloniki, Greece; apapado@math.auth.gr (A.P.); pakolias@math.auth.gr (P.K.); palihar7@gmail.com (H.P.); evanthiafarmakioti93@gmail.com (E.F.)
* Correspondence: acg@uom.edu.gr

Abstract: Semi-Markov processes generalize the Markov chains framework by utilizing abstract sojourn time distributions. They are widely known for offering enhanced accuracy in modeling stochastic phenomena. The aim of this paper is to provide closed analytic forms for three types of probabilities which describe attributes of considerable research interest in semi-Markov modeling: (a) the number of transitions to a state through time (Occupancy), (b) the number of transitions or the amount of time required to observe the first passage to a state (First passage time) and (c) the number of transitions or the amount of time required after a state is entered before the first real transition is made to another state (Duration). The non-homogeneous in time recursive relations of the above probabilities are developed and a description of the corresponding geometric transforms is produced. By applying appropriate properties, the closed analytic forms of the above probabilities are provided. Finally, data from human DNA sequences are used to illustrate the theoretical results of the paper.

Keywords: semi-Markov modeling; occupancy; first passage time; duration; non-homogeneity; DNA sequences

Citation: Georgiou, A.C.; Papadopoulou, A.; Kolias, P.; Palikrousis, H.; Farmakioti, E. On State Occupancies, First Passage Times and Duration in Non-Homogeneous Semi-Markov Chains. *Mathematics* **2021**, *9*, 1745. https://doi.org/10.3390/math9151745

Academic Editor: Alexander Zeifman

Received: 20 May 2021
Accepted: 21 July 2021
Published: 24 July 2021

Publisher's Note: MDPI stays neutral with regard to jurisdictional claims in published maps and institutional affiliations.

Copyright: © 2021 by the authors. Licensee MDPI, Basel, Switzerland. This article is an open access article distributed under the terms and conditions of the Creative Commons Attribution (CC BY) license (https://creativecommons.org/licenses/by/4.0/).

1. Introduction

Human populations can be divided into categories (states and classes) taking into account some of their basic characteristics, such as place of residence, social class or rank in a hierarchy system. People usually move from a category to another category in a probabilistic manner and a person's history contains a sequence of sojourn times in the various categories and a set of transitions that have taken place. These are the basic parameters that construct a semi-Markov chain (SMC), according to which a mathematical model can be developed for the study of those systems [1,2]. These systems do not necessarily have to include humans, instead, they can describe any potential system characterized by and composed of historical observations, such as stay times in situations as well as transitions from one category to another. If, for the study of a population system, we reside on a Markov chain, we assume that the probability of transition from one category in another does not depend on the length of stay. Nonetheless, this time dependence is, in some cases, desirable to include in the process since it provides additional useful information. In this case, the transitions of such a system are not merely described by a typical Markov chain procedure and Semi-Markov models are introduced as the stochastic tools that provide a more rigorous framework accommodating a greater variety of applied probability models [3–5]. Various applications of semi-Markov processes include manpower planning, credit risk, word sequencing and DNA analysis [6–14].

In addition to semi-Markov processes, the non-homogeneous semi-Markov system (NHSMS) was defined, introducing a class of broader stochastic models [15,16] that provide a more general framework to describe the complex semantics of the system involved. Semi-Markov systems, which deploy a number of Markov chains evolving in parallel, are mostly applied in manpower planning, where the most important issues pertain to the evolution, control and asymptotic behavior [17–19]. In the last two decades, there has been an extended body of literature regarding the theory and results about NHMS [20–29]. The dynamic characteristics of the semi-Markov systems influence the number of times the chain occupies a state, of how long it takes to leave a state as well as the probability of first passage to a state. Therefore, in order to accompany the basic parameters of the semi-Markov chain and to enhance the modeling framework, additional attributes of critical interest are the occupancy, first passage time and duration probabilities, which are described as follows

1. *Occupancy probabilities.* These probabilities describe the distribution of the random variables that define the number of times the SMC has visited a specific state during an arbitrary time interval.
2. *First passage time probabilities.* These are the probabilities that describe the transition from a state to a different state for the first time. The properties of the first passage time probabilities have been investigated for Markov processes and some specific types of semi-Markov processes [30–35]. Details for the first passage time probabilities have been also presented for various stochastic processes [36].
3. *Duration probabilities.* These probabilities describe the distribution of random variables that define the time needed for the SMC to transfer to a different state.

DNA sequences are usually studied using probabilistic models, as nucleotide appearances are inter-correlated and attempts to use Markov models to model them have been reported [10,37]. One of the earliest studies applied a Markov model on the nucleotide alphabet $\{A, C, G, T\}$ to estimate the transition probability matrix and the number of doublets and triplets [38]. Several statistics have been proposed to test the dependency order of the sequence, e.g., the Markov order, such as the phi-divergent statistics and conditional mutual information [39–41]. More advances in the subject include hidden-Markov models that are able to model different regions of DNA sequences [42]. Word occurrences are also of interest in DNA analysis [43]. Previous studies have examined the distribution, moments and properties of successive word occurrences [44,45]. Papadopoulou has provided some examples of semi-Markov models on modeling biological sequences [46]. Furthermore, algorithmic applications for estimating the first passage time probabilities in genomic sequences have been reported [47].

The aim of this study is to provide insight on the actual mechanism of the recursive relations of the probabilities mentioned above. Section 2 presents the basic parameters of a SMC, the interval transition probabilities and the entrance probabilities. Section 3 presents the main results of the paper, that is, the closed analytic solutions for the occupancy, duration and first passage time probabilities. The final section applies these theoretical results to human genome DNA strands. For the first illustration, the aim is to find the corresponding probabilities between nucleotide words and their symmetric complements by using the analytic form of the first passage time probabilities. Finally, for the second illustration, the frequency of the dinucleotide GC is examined for two distinct DNA sequences, using the occupancy probabilities.

2. Basic Framework

We can consider the semi-Markov chain $\{X_t\}_{t\geq 1}$ with state space $S = \{1, 2, \ldots, N\}$ as a discrete stochastic process in which the successive states are defined by the transition probability matrix and the sojourn time in each state is described by a random variable conditioned on the current and the next state to be transitioned into. Thus, during the transition times, the process is equivalent to a Markov process. We call this Markovian process the *embedded* process. Let transition probabilities $p_{ij}(t)$ be the probability of a SMC provided that it entered state i during its last transition at time t to transition to state j in the next transition. The transition probabilities should satisfy the same equations of a

Markovian process, that is, $p_{ij} \geq 0, \forall i,j \in S$ and $\sum_{j=1}^{N} p_{ij} = 1, \forall i \in S$. When the process enters state i at time t, we assume that this state determines the next transition to state j, which occurs according to the transition probabilities. However, before making the transition from state i to state j and after the next state j is selected, the chain holds in state i for time τ_{ij}. The sojourn time τ_{ij} is a positive random variable with density function $h_{ij}(\cdot)$, which is called the function of sojourn time to transition from state i to state j. Thus, $Prob[\tau_{ij} = m] = h_{ij}(m)$, for $m = 1, 2, .., $ and $i,j \in S$. We assume that the mean values of the distributions of sojourn times are finite and $h_{ij}(0) = 0$. In matrix notation, the basic parameters of the semi-Markov chain are the sequence of transition matrices $\{\boldsymbol{P}(t)\}_{t=0}^{\infty}$ and the sequence of sojourn time matrices $\{\boldsymbol{H}(m)\}_{m=1}^{\infty}$. The probabilities of the *waiting times* $w_i(t,m)$ are defined as follows:

$$w_i(t,m) = \sum_{j=1}^{N} p_{ij}(t) h_{ij}(m) = Prob[\tau_i = m|t],$$

where τ_i is the holding time of the SMC in state i. The *core matrix* of the SMC connects the transition probabilities and the sojourn times and it is defined as follows:

$$\boldsymbol{C}(t,m) = \{c_{ij}(t,m)\}_{ij \in S} = \boldsymbol{P}(t) \circ \boldsymbol{H}(m).$$

The operator $\{\circ\}$ denotes the element-wise product of matrices (Hadamard product). Using the *core* matrix, we define $q_{ij}(k|t,n)$, which is the joint probability that the SMC will be in state j at time $t+n$ and that it has made k transitions during the time interval $(t, t+n]$, given that at time t the process has entered state i. In order to calculate the probability $q_{i,j}(k|t,n)$, we distinguish two cases. First, we consider that during the time interval $(t, t+n]$ the number of transitions is zero. Then, in order for the process at time $t+n$ to be in state j, given that no transitions were made, it must be that the states i,j are the same. Secondly, assume that the SMC makes the first transition to state r at time $t+m,\ 0 < m < n$. Then, in the time interval $(t, t+m]$, we have one transition to state r and, in the remaining time interval $(t+m, t+n]$, we have the remaining $k-1$ transitions, with a final transition to state j. Thus, the resulting formula is as follows:

$$q_{ij}(k|t,n) = \delta_{ij}\delta(k)^{>}w_i(t,n) + \sum_{r=1}^{N}\sum_{m=0}^{n} c_{ir}(t,m) q_{rj}(k-1|t+m, n-m).$$

where ${}^{>}w_i(t,n) = \sum_{k=n+1}^{\infty} w_i(t,k)$ indicates the survival function of $w_i(t,n)$ and $\delta(k) = 1$ if k is zero, otherwise it is zero. If we are not interested in counting the number of transitions up to the final state j, we can deduce the following recursive relationship.

$$q_{ij}(t,n) = \delta_{ij}^{>} w_i(t,n) + \sum_{r=1}^{N}\sum_{m=0}^{n} c_{ir}(t,m) q_{rj}(t+m, n-m).$$

We also define the quantity $e_{i,j}(k|t,n)$, which is the probability that the SMC enters state j at time $t+n$ and the total number of transitions in the time interval $(t, t+n]$ is k, given that the SMC has entered state i at the initial position. Here, we can distinguish two cases. First, we assume that the number of transitions in the time interval $(t, t+n]$ is zero. Then, to enter in state j at time $t+n$, the states i and j must be the same since state i was entered at the initial time. For the second case, suppose that the SMC at time $t+m,\ 0 < m < n$ makes its first transition to state r. Then, at the time interval $(t, t+m]$ we have a transition to state r and, at the time interval $(t+m, t+n]$, we have the remaining $k-1$ transitions, with the final transition to state j. These facts result in the following recursive relationship.

$$e_{ij}(k|t,n) = \delta_{ij}\delta(n)\delta(k) + \sum_{r=1}^{N}\sum_{m=0}^{n} c_{ir}(t,m) e_{rj}(k-1|t+m, n-m).$$

If we are not interested in the number of transitions up to the final state j, we can reduce the recursive relationship to the quantity $e_{ij}(t,n)$, which are the probabilities that the SMC will enter state j at time n, provided that, at the initial position at time t, the SMC has entered state i. The equation for calculating the probabilities $e_{ij}(t,n)$ is given by the following.

$$e_{ij}(t,n) = \delta_{ij}\delta(n) + \sum_{r=1}^{N}\sum_{m=0}^{n} c_{ir}(t,m)e_{rj}(t+m,n-m).$$

The interval transition probabilities and entrance probabilities are connected by the following relationship.

$$q_{ij}(k|t,n) = \sum_{m=0}^{n} e_{ij}(k|t,m)^{>}w_j(t+m,n-m).$$

3. Theoretical Results: Analytic Solutions of the Recursive Equations

3.1. First Passage Time

The first passage times provide a measure of how long it takes to reach a given state from another. We can think of first passage times either in terms of transitions or of time or both. Thus, let $f_{ij}(k|t,n)$ be the probability that k transitions and time n will be required for the first passage from state i to state j given that the SMC entered state i at time t. Applying a probabilistic argument, we can provide the following recursive formula.

$$f_{ij}(k|t,n) = \sum_{r\neq j}^{N}\sum_{m=0}^{n} c_{ir}(t,m)f_{rj}(k-1|t+m,n-m) + \delta(k-1)c_{ij}(t,n). \tag{1}$$

The first term of equation (1) corresponds to the case where $k>1$ and the SMC makes a transition to some state r different from j at time $t+m$ and then makes a first passage from r to j in $k-1$ transitions during the interval $(t+m,n-m]$. The term is summed over all states and holding times that could describe the first transition. The second term corresponds to the case where $k=1$ and the process moves directly to state j at time $t+n$. If we are not interested in counting the transitions, then the recursive formula of the probabilities $f_{ij}(t,n)$ is provided by the following.

$$f_{ij}(t,n) = \sum_{r\neq j}^{N}\sum_{m=0}^{n} c_{ir}(t,m)f_{rj}(t+m,n-m) + c_{ij}(t,n). \tag{2}$$

Theorem 1. *For each non-homogeneous SMC with discrete state space* $S = 1,2,\ldots,N$, *a sequence of transition probability matrices* $\{\mathbf{P}(t)\}_{t=0}^{\infty}$ *and a sequence of sojourn time matrices* $\{\mathbf{H}(m)\}_{m=1}^{\infty}$, *the probability matrices of first passage times* $\mathbf{F}(k|t,n) = \{f_{ij}(k|t,n)\}_{i,j\in S}$ *are given by the following relationships:*

1. $\mathbf{F}(1|t,n) = \mathbf{C}(t,n)$, *for every* n.
2. $\mathbf{F}(k|t,n) = \mathbf{0}$, *if* $k>n$ *or* $k=0$.
3. $\mathbf{F}(k|t,n) = \sum_{m_1=1}^{n-k+1} * \sum_{m_2=1+m_1}^{n-k+2} * \ldots * \sum_{m_{k-1}=1+m_{k-2}}^{n-1} \prod_{r=0}^{k-1}{}^{\{\mathbf{B}\}}\mathbf{C}(t+m_{k-r-1}, m_{k-r}-m_{k-r-1})$, *for each* $1<k\leq n$,

where $\mathbf{B} = \mathbf{U}-\mathbf{I}$, $\mathbf{U} = \{u_{ij}=1\}_{i,j\in S}$, $\mathbf{I}$ *is the* $N\times N$ *identity matrix and*

$$\prod_{r=0}^{k-1}{}^{\{\mathbf{B}\}}\mathbf{C}(s+m_{k-r-1}, m_{k-r}-m_{k-r-1}) =$$
$$= \mathbf{C}(s,m_1)\{\mathbf{C}(s+m_1,m_2-m_1)\{\ldots\{\mathbf{C}(s+m_{k-1},n-m_{k-1})\circ\mathbf{B}\}\circ\mathbf{B}\}\ldots\}\circ\mathbf{B}\}.$$

Proof. Appendix A.1. □

3.2. Duration

Transitions of a SMC can be divided into two categories: virtual and real. The first category refers to transitions made from one state to the same state, while the second category refers to transitions from one state to a different state. Based on those two categories, one can define the duration as the number of transitions or the time required for the SMC to leave the initial state and to move to a different state, i.e., a real transition to take place for the first time and not a virtual one. Therefore, it is of interest to study the duration probability $d_i(k|t,n)$ defined as the probability that the SMC moves for the first time to a different state that the initial one after n time units and k transitions during the interval $(t, t+n]$, given that the process entered state i at time t. We note here that out of the total k transitions in the above case, $k-1$ transitions are virtual and one transition is real. The duration probabilities for $k \leq n$ are provided by the following.

$$d_i(k|t,n) = \sum_{m=0}^{n} c_{ii}(t,m) d_i(k-1|t+m, n-m) + \delta(k-1)(\mathrm{w}_i(t,n) - c_{ii}(t,n)). \quad (3)$$

In the case that $k > n$ or $k = 0$, then $d_i(k|t,n) = 0$. The rationale of this relationship can be deconstructed into two parts. In the first part, we can assume that the SMC has at least one virtual intermediate transition, while it starts from state i at time t, holds at the state i for m time units and finally transfers to state i again. At this point, the associated probability is $d_i(k-1|t+m, n-m)$. In the second scenario, we assume that the SMC makes no transition up to time $t+n$. Therefore, the chain holds at state i for exactly n time units and then moves to a state j different than i. Thus, the duration defined in the present measures how long it takes to leave a given state.

Theorem 2. *For each non-homogeneous SMC with discrete state space $S = 1, 2, \ldots, N$, a sequence of transition probability matrices $\{\mathbf{P}(t)\}_{t=0}^{\infty}$ and a sequence of sojourn time matrices $\{\boldsymbol{H}(m)\}_{m=1}^{\infty}$, the duration probability matrices $\mathbf{D}(k|t,n) = diag\{d_i(k|t,n)\}_{i \in S}$ are provided by the following relationships:*

1. $\mathbf{D}(1|t,n) = [\mathbf{W}(t,n) - \mathbf{C}(t,n) \circ \mathbf{I}]$, *for every* n.
2. $\mathbf{D}(k|t,n) = \mathbf{0}$, *if* $k > n$ *or* $k = 0$.
3. $\mathbf{D}(k|t,n) = \sum_{m_1=1}^{n-k+1} * \sum_{m_2=1+m_1}^{n-k+2} * \ldots * \sum_{m_{k-1}=1+m_{k-2}}^{n-1} (\mathbf{C}(t,m_1) \circ \mathbf{I})(\mathbf{C}(t+m_1, m_2-m_1) \circ \mathbf{I})$ $\ldots (\mathbf{C}(t+m_{k-2}, m_{k-1}-m_{k-2}) \circ \mathbf{I})(\mathbf{W}(t+m_{k-1}, n-m_{k-1}) - \mathbf{C}(t+m_{k-1}, n-m_{k-1}) \circ \mathbf{I})$, *for each* $1 < k \leq n$,

where $\mathbf{W}(t,n) = diag\{w_i(t,n)\}_{i \in S}$.

Proof. Appendix A.2. □

3.3. Occupancy

We define $v_{ij}(t,n)$ to be the number of times the SMC makes transitions to a state j in time interval of length equal to n, provided that in the initial time t the SMC had entered state i. If the initial state is the same as j, that is when $i = j$, then the initial state is not counted in $v_{ij}(t,n)$. We call the quantity $v_{ij}(t,n)$ as the *occupancy measure* of state j at time $t+n$, provided that the SMC entered state i at time t. Clearly, the quantity $v_{ij}(t,n)$ is a discrete random variable. We define as $\omega_{ij}(\cdot|t,n)$ the probability mass distribution of $v_{ij}(t,n)$, which is $\omega_{ij}(x|t,n) = Prob[v_{ij}(t,n) = x]$. The recursive relationship of the occupancy probabilities is given by the following:

$$\begin{aligned} \omega_{ij}(x|t,n) = & \sum_{\substack{r=1 \\ r \neq j}}^{N} \sum_{m=0}^{n} c_{ir}(t,m) \omega_{rj}(x|t+m, n-m) + \\ & + \sum_{m=0}^{n} c_{ij}(t,m) \omega_{jj}(x-1|t+m, n-m) + \delta(x) > w_i(t,n), \end{aligned} \quad (4)$$

where $i,j \in S, n = 0,1,\ldots,$ and $x = 0,1,\ldots.$

Assumption 1. *In what follows, we assume that the embedded Markov chain is homogeneous, i.e.,* $\{\mathbf{P}(t)\}_{t=0}^{\infty} = \mathbf{P}$*, for each* t*.*

Considering the above assumption, one can use the double geometric transform of the occupancy probabilities as follows.

$$\omega_{ij}^{gg}(y|z) = \sum_{x=0}^{\infty}\sum_{n=0}^{\infty}\omega_{ij}(x|n)z^n y^x.$$

Moreover, from the Equation (4), we can write the double geometric transform of the occupancy probabilities as follows.

$$\omega_{ij}^{gg}(y|z) = \sum_{r=1}^{N} c_{ir}^{g}(z)\omega_{rj}^{gg}(y|z) - (1-y)c_{ij}^{g}(z)\omega_{jj}^{gg}(y|z) + {}^{>}w_i^{g}(z).$$

In matrix notation, we can use the previous results to obtain the following [3]:

$$\mathbf{\Omega}^{gg}(y|z) = \frac{1}{1-z}\mathbf{U} - \frac{1-y}{1-z}[\mathbf{I} - \mathbf{C}^{g}(z)]^{-1}\mathbf{C}^{g}(z)\left(y\mathbf{I} + (1-y)[\mathbf{I} - \mathbf{C}^{g}(z)]^{-1} \circ \mathbf{I}\right)^{-1},$$

where $\mathbf{U}$ is the unit matrix, $\mathbf{\Omega}^{gg}(y|z) = \left\{\omega_{ij}^{gg}(y|z)\right\}_{i,j\in S}$ is the double geometric transform of $\mathbf{\Omega}(x|n) = \{\omega_{ij}(x|n)\}_{i,j\in S}$ and $\mathbf{C}^{g}(z) = \left\{c_{ij}^{g}(z)\right\}_{i,j\in S}$.

The occupancy probabilities are connected with the corresponding homogeneous first passage time probabilities through the following relationship.

$$\omega_{ij}(x|n) = \delta(x)^{>}f_{ij}(n) + \sum_{m=0}^{n} f_{ij}(m)\omega_{jj}(x-1|n-m).$$

Using the double geometric transform, we can present the occupancy probabilities in matrix form according to the geometric transforms of the first passage time probabilities:

$$\mathbf{\Omega}^{gg}(y|z) = {}^{>}\mathbf{F}^{g}(z) + y\mathbf{F}^{g}(z)\left[{}^{>}\mathbf{F}^{g}(z) \circ \mathbf{I}\right][\mathbf{I} - y(\mathbf{F}^{g}(z) \circ \mathbf{I})]^{-1},$$

which could be further simplified by using ${}^{>}f_{ij}^{g}(z) = \frac{1-f_{ij}^{g}(z)}{1-z}$ (Appendix B.1) resulting in matrix notation in (Appendix B.2).

$$\mathbf{\Omega}^{gg}(y|z) = \frac{1}{1-z}\mathbf{U} - \frac{1-y}{1-z}\mathbf{F}^{g}(z)[\mathbf{I} - y\mathbf{F}^{g}(z) \circ \mathbf{I}]^{-1}.$$

We now provide Theorem 3 and Lemma 1 that will be used to prove the main Theorem 4 of the occupancy probabilities with respect to the core matrix.

Theorem 3. *For a SMC with core matrix* $\mathbf{C}(\cdot)$*, we have the following:*

$$\begin{aligned}\mathbf{\Omega}^{g}(z|n) &= (z-1)\textstyle\sum_{j=1}^{n-1}\left[\mathbf{C}(j) + \left[\sum_{i=2}^{j}\left(\mathbf{C}(i-1) + \sum_{k=1}^{i-2}\mathbf{S}_i(k,m_k)\right)\mathbf{C}(j+1-i)\right][\mathbf{\Omega}^{g}(z|n-j)] \circ \mathbf{I}\right] \\ &+z\left[\mathbf{C}(n) + \textstyle\sum_{j=2}^{n}\left(\mathbf{C}(j-1) + \sum_{k=1}^{j-2}\mathbf{S}_j(k,m_k)\right)\mathbf{C}(n+1-j)\right] + \\ &+\left[\textstyle\sum_{j=2}^{n}\left(\mathbf{C}(j-1) + \sum_{k=1}^{j-2}\mathbf{S}_j(k,m_k)\right)^{>}\mathbf{W}(n+1-j) + {}^{>}\mathbf{W}(n)\right],\end{aligned}$$

where $\mathbf{S}_i(k,m_k) = \sum_{m_k=2}^{i-k}\sum_{m_{k-1}=1+m_k}^{i-k+1}\cdots\cdots\sum_{m_1=1+m_2}^{i-1}\prod_{r=-1}^{k-1}\mathbf{C}(m_{k-r-1} - m_{k-r}), \forall i,j \in S$ *and* $n = 0,1,2,\ldots$ *Please note that the* (j,r) *element of* $\mathbf{S}_i(k,m_k)$ *is the probability of moving from state* j *to state* r *after* $i-1$ *time units and* k *intermediate transitions during the interval* $(t, t+i-1]$ *for every* t *due to the time-homogeneity assumption.*

Proof. Appendix A.3. □

Lemma 1. *The product* $\boldsymbol{\Omega}^g(z|n) \circ \mathbf{I}$ *is equal to the following:*

$$\boldsymbol{\Omega}^g(z|n) \circ \mathbf{I} = -(z-1)\sum_{j=1}^{n-1}\left[\left[\sum_{i=1}^{j}\mathbf{a}_{1i}^{-1}\mathbf{C}(j+1-i)\right]\circ\mathbf{I}\right][\boldsymbol{\Omega}^g(z|n-j)\circ\mathbf{I}]$$
$$-z\sum_{j=1}^{n}\left[\mathbf{a}_{1j}^{-1}\mathbf{C}(n+1-j)\right]\circ\mathbf{I}+\sum_{j=1}^{n}\left[-\mathbf{a}_{1j}^{-1>}\mathbf{W}(n+1-j)\right]\circ\mathbf{I},$$

$\forall i,j \in S$ *and* $n = 0,1,2,\ldots$, *where*
$-\mathbf{a}_{1i}^{-1} = \mathbf{C}(i-1) + \sum_{k=1}^{i-2}\mathbf{S}_i(k,m_k)$.

Proof. Appendix A.4. □

We now provide Theorem 4, which describes the analytic solutions of the occupancy probabilities. In order to facilitate the presentation and proof of Theorem 4, we begin with some aggregate notation. Let the following be the case:

$$\mathbf{A}_j = \mathbf{C}(j) + \sum_{i=2}^{j}\left(\mathbf{C}(i-1)+\sum_{k=1}^{i-2}\mathbf{S}_i(k,m_k)\right)\mathbf{C}(j+1-i),$$

$$\mathbf{B}_{n,j} = \left[\sum_{w=2}^{n-j}\left[\left(\mathbf{C}(w-1)+\sum_{k=1}^{w-2}\mathbf{S}_w(k,m_k)\right)^{>}\mathbf{W}(n-j+1-w)\right]\circ\mathbf{I}+{}^{>}\mathbf{W}(n-j)\right]\circ\mathbf{I},$$

$$\mathbf{M}_u = -\left[\mathbf{C}(u-1)+\sum_{i=2}^{u-1}\left(\mathbf{C}(i-1)+\sum_{k=1}^{i-2}\mathbf{S}_i(k,m_k)\right)\mathbf{C}(u-i)\right]\circ\mathbf{I}+\sum_{k=1}^{u-2}(-1)^{k+1}\mathbf{R}_u(k,m_k),$$

$$\mathbf{M}'_u = \left[\mathbf{C}(u-1)+\sum_{i=2}^{u-1}\left(\mathbf{C}(i-1)+\sum_{k=1}^{i-2}\mathbf{S}_i(k,m_k)\right)\mathbf{C}(u-i)\right]\circ\mathbf{I},$$

$$\mathbf{M}''_u = \left[\mathbf{C}(u-1)+\sum_{i=2}^{u-1}\left(\mathbf{C}(i-1)+\sum_{k=1}^{i-2}\mathbf{S}_i(k,m_k)\right)\mathbf{C}(u-i)\right]\circ\mathbf{I}+\left[\sum_{k=1}^{u-2}(k+1)(-1)^k\mathbf{R}_u(k,m_k)\right],$$

$$\mathbf{M}'''_u = \left[\mathbf{C}(u-1)+\sum_{i=2}^{u-1}\left(\mathbf{C}(i-1)+\sum_{k=1}^{i-2}\mathbf{S}_i(k,m_k)\right)\mathbf{C}(u-i)\right]\circ\mathbf{I}-\left[\sum_{k=1}^{u-2}(k+2)(-1)^k\mathbf{R}_u(k,m_k)\right],$$

$$\mathbf{E}_n = \sum_{j=2}^{n}\left(\mathbf{C}(j-1)+\sum_{k=1}^{j-2}\mathbf{S}_j(k,m_k)\right)^{>}\mathbf{W}(n+1-j)+^{>}\mathbf{W}(n),$$

$$\mathbf{F}_{x,u} = x(x-1)\sum_{k=x-3}^{u-2}\left[\prod_{r=-1}^{x-4}(k-r)\right](-1)^{(k-x+3)}\mathbf{R}_u(k,m_k) - x\left[\sum_{k=x-2}^{u-2}\left[\prod_{r=-1}^{x-3}(k-r)\right](-1)^{(k-x+2)}\mathbf{R}_u(k,m_k)\right],$$

$$\mathbf{G}_{u,n,j} = \mathbf{C}(n-j+1-u)\circ\mathbf{I}+\sum_{w=2}^{n-j+1-u}\left[\left(\mathbf{C}(w-1)+\sum_{k=1}^{w-2}\mathbf{S}_w(k,m_k)\right)\mathbf{C}(n-j+2-u-w)\right]\circ\mathbf{I},$$

$$\mathbf{H}_{x,u} = x\sum_{k=x-2}^{u-2}\left[\prod_{r=-1}^{x-3}(k-r)\right](-1)^{k-(x-2)}\mathbf{R}_u(k,m_k) - \sum_{k=x-1}^{u-2}\left[\prod_{r=-1}^{x-2}(k-r)\right](-1)^{k-(x-1)}\mathbf{R}_u(k,m_k),$$

$$\mathbf{Q}_{u,n,j} = \sum_{w=2}^{n-j+1-u}\left[\left(\mathbf{C}(w-1)+\sum_{k=1}^{w-2}\mathbf{S}_w(k,m_k)\right)^{>}\mathbf{W}(n-j+2-u-w)\right]\circ\mathbf{I}+\left[^{>}\mathbf{W}(n-j+1-u)\right]\circ\mathbf{I},$$

where

$$\mathbf{R}_u(k,m_k) = \sum_{m_k=2}^{u-k} \sum_{m_{k-1}=1+m_k}^{u-k+1} \cdots \sum_{m_1=1+m_2}^{u-1} \prod_{r=-1}^{k-1} \left[\sum_{i=1}^{m_{k-r-1}-m_{k-r}} \left(-\mathbf{a}_{1i}^{-1}\right) \mathbf{C}(m_{k-r-1} - m_{k-r} + 1 - i) \right] \circ \mathbf{I},$$

$$\mathbf{S}_i(k,m_k) = \sum_{m_k=2}^{i-k} \sum_{m_{k-1}=1+m_k}^{i-k+1} \cdots \sum_{m_1=1+m_2}^{i-1} \prod_{r=-1}^{k-1} \mathbf{C}(m_{k-r-1} - m_{k-r}),$$

and

$$-\mathbf{a}_{1i}^{-1} = \mathbf{C}(i-1) + \sum_{k=1}^{i-2} \mathbf{S}_i(k,m_k).$$

Theorem 4. *For a SMC with core matrix* $\mathbf{C}(\cdot)$*, by adopting the above notations, we have that the following:*

$$\boldsymbol{\Omega}(0|n) = -\sum_{j=1}^{n-1} \mathbf{A}_j \left[\mathbf{B}_{n,j} + \sum_{u=2}^{n-j} \mathbf{M}_u \mathbf{Q}_{u,n,j} \right] + \mathbf{E}_n,$$

$$\boldsymbol{\Omega}(1|n) = \sum_{j=1}^{n-1} \mathbf{A}_j \left[\mathbf{B}_{n,j} - \mathbf{G}_{1,n,j} - \sum_{u=2}^{n-j} [\mathbf{M}_u + \mathbf{G}_{u,n,j}] - 2\sum_{u=2}^{n-j} \mathbf{M}_u''' \mathbf{Q}_{u,n,j} \right],$$

$$\boldsymbol{\Omega}(2|n) = \sum_{j=1}^{n-1} \mathbf{A}_j \left[2\mathbf{G}_{1,n,j} + \sum_{u=2}^{n-j} \sum_{k=1}^{u-2} \left[(-2k-4)(-1)^k \mathbf{R}_u(k,m_k) \mathbf{G}_{u,n,j} \right] - 4\sum_{u=2}^{n-j} \mathbf{M}_u' \mathbf{G}_{u,n,j} \right.$$
$$\left. +2\sum_{u=2}^{n-j} \mathbf{M}_u' \mathbf{Q}_{u,n,j} - \sum_{u=2}^{n-j} \sum_{k=1}^{u-2} (k+1)(k+2)(-1)^{k-1} \mathbf{R}_u(k,m_k) \mathbf{Q}_{u,n,j} \right],$$

$$\boldsymbol{\Omega}(3|n) = \sum_{j=1}^{n-1} \mathbf{A}_j \left[6\sum_{u=2}^{n-j} \mathbf{M}_u'' \mathbf{G}_{u,n,j} - 3\sum_{u=2}^{n-j} \sum_{k=1}^{u-2} k(k+1)(-1)^{k+1} \mathbf{R}_u(k,m_k) \mathbf{G}_{u,n,j} \right.$$
$$\left. -\sum_{u=2}^{n-j} (k-1)k(k+1)(-1)^{k-2} \mathbf{R}_u(k,m_k) \mathbf{Q}_{u,n,j} + 3\sum_{u=2}^{n-j} \sum_{k=1}^{u-2} k(k+1)(-1)^{k-1} \mathbf{R}_u(k,m_k) \mathbf{Q}_{u,n,j} \right],$$

and

$$\boldsymbol{\Omega}(x|n) = \sum_{j=1}^{n-1} \left[\mathbf{A}_j \sum_{u=2}^{n-j} [\mathbf{F}_{x,u} + \mathbf{G}_{u,n,j} + \mathbf{H}_{x,u} \mathbf{Q}_{u,n,j}] \right], \quad \forall\, x \geq 4.$$

Proof. Appendix A.5. □

4. Illustration

In this section we will accompany the theoretical results of the paper with two applications related to DNA sequences. It is known that a DNA strand consists of a sequence of adenine (A), guanine (G), cytosine (C) and thymine (T), which are the four nucleotides. We assume that a DNA sequence could be described by a homogeneous discrete SMC $\{X_t\}_{t=0}^{\infty}$ with state space $S = \{w_1, w_2, \ldots, w_N\}$, where w_i, $i = 1, 2, \ldots, N$ is a specific word that is a combination of the letters of the DNA alphabet $S = \{A, C, G, T\}$ with length l and t denoting the position of the word inside the sequence.

4.1. Inverted Repeats

The main focus of the following approach is the appearance of specific words formed from the alphabet A, C, G, T and their symmetric complements (inverted repeats). Inverted repeats are commonly found in eukaryotic genomes [48]. The presence of inverted repeats could form DNA cruciforms that have been shown to play an important role in the regulation of natural processes involving DNA. The cruciform structures are important for various biological processes, including replication, regulation of gene expression and nucleosome structure. They have also been implicated in the development of diseases including cancer, Werner's syndrome and others [49].

For each DNA word w, there exists a reversed complement of the word w'. For example, the word $w = ACG$ has the word $w' = CGT$ as an inverted repeat. The main question that we will attempt to address by applying the analytic relationships derived earlier is the following: Given that the SMC entered at the initial position in the word w, we want to estimate the probability of the reversed complement word w' appearing for the first time after a certain range of letters n. We define the distance, d, between two words as the number of letters between the first letter of the initial word that has appeared and the first letter of the following word that subsequently appears. For the sake of simplicity, we consider only the scenario where $d > l$. The DNA sequence that was used for this illustration is the first chromosome of the human genome consisting of 248,956,422 base-pairs that are publicly available from the website of the National Center for Biotechnology Information (NCBI) [50].

For the first illustration, three words of length $l = 7$ were chosen that have been previously shown to exhibit different distances between them and their inverted complements [51]. The words were $w_1 = GGCTCAC$, $w_2 = ATATATG$ and $w_3 = CCACAAT$. For each word, the state space of the SMC consisted of the word and its reversed complement, e.g., $S = \{w_i, w_i'\}$. First, the basic parameters of the SMC were estimated, namely the transition probability matrix and the sequence of sojourn times. The sojourn time was defined as the distance, i.e., the number of nucleotides that occur between each word and its inverted repeat. The transition matrix and the empirical distribution of the sojourn times were estimated using the empirical estimators. The sequence of the core matrices was calculated as the Hadamard product of the transition matrix with the sequence of the sojourn time matrices. For each word $w \in S$, the first passage time probability was calculated between the word w and its reversed complement w' according to the proposed analytic relationship (Theorem 1). For a maximum distance, ($n = 1000$), the highest first passage time probabilities of the three words and their inverted repeats, along with the corresponding distances are illustrated in Figure 1. Concretely, the first passage time probabilities were calculated for the human Chromosome 1, aiming to estimate the most probable distances between words and their symmetrical complements. More specifically, as presented in Figure 1, we have noted that, for the first passage time probabilities, we have $argmax(f_{w_1 w_1'}) = 210$, $argmax(f_{w_2 w_2'}) = 10$ and $argmax(f_{w_3 w_3'}) = 132$ approximating the numerical results of previous studies with corresponding values for the arguments $210, 15$ and 133 for the three words, respectively [51]. This highlights the fact that specific DNA words exhibit different behaviors and the distance between them and their inverted repeats demonstrates variability.

4.2. CpG Islands

Usually, in vertebrate DNA sequences, the dinucleotide CG occurs less frequently than expected [52]. For the second illustration, we considered CpG islands, which are genomic regions that contain an elevated number of the dinucleotide CG. The human genome contains approximately 30 thousand CpG islands. The APRT gene is an example of a CpG region and it was used for this analysis [53]. This gene provides instructions for making an enzyme called adenine phosphoribosyltransferase (APRT). APRT contains approximately 2500 nucleotides and it had been shown to include an elevated amount of the dinucleotide GC [54]. We modeled the sequence of this DNA region as a homogeneous SMC with state space containing all the two-letter words from the DNA alphabet. The transition probability matrix and the sojourn times were estimated using the empirical estimators. The occupancy distribution $\omega_{GCGC}(x|n)$ for a fixed length of $n = 100$ was calculated using the analytic relationship from Theorem 4 in order to estimate the occupancy distribution of specific words up to a specified sequence length. For comparison, we also applied the model to an intron sequence of human's phosphodiesterase gene (PDEA) [55]. The two sequences are publicly available from the NCBI. The occupancy probabilities are presented in Figure 2 up to length $n = 50$. It is confirmed that the number of occupancies of the dinucleotide GC will be greater in the CpG island compared to the intron sequence. As expected, the occupancy probabilities applied on the two sequences indicated that the occurrences of GCs were more frequent in the CpG sequence.

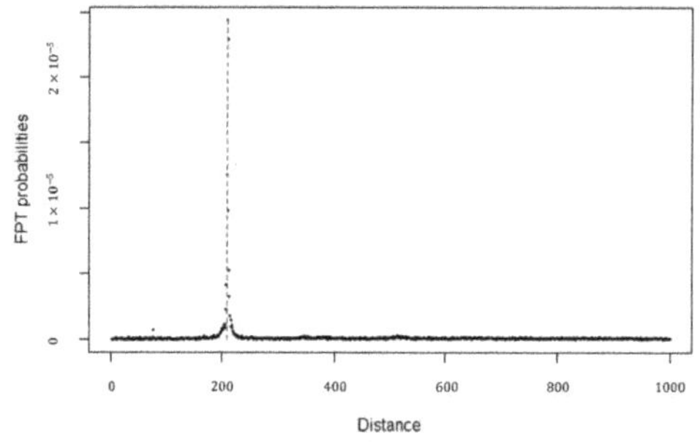

(**a**) $w_1 = GGCTCAC$

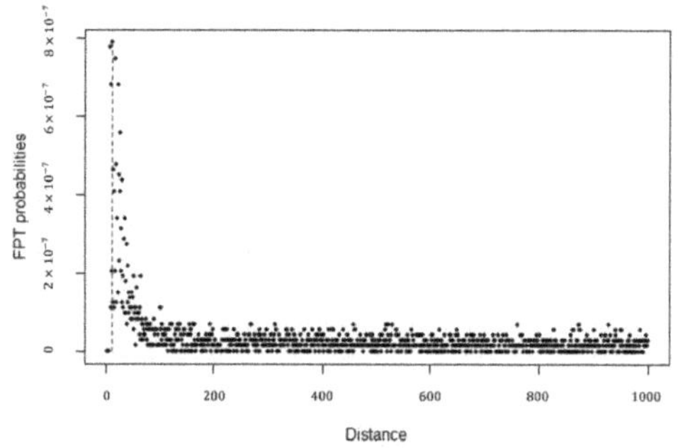

(**b**) $w_2 = ATATATG$

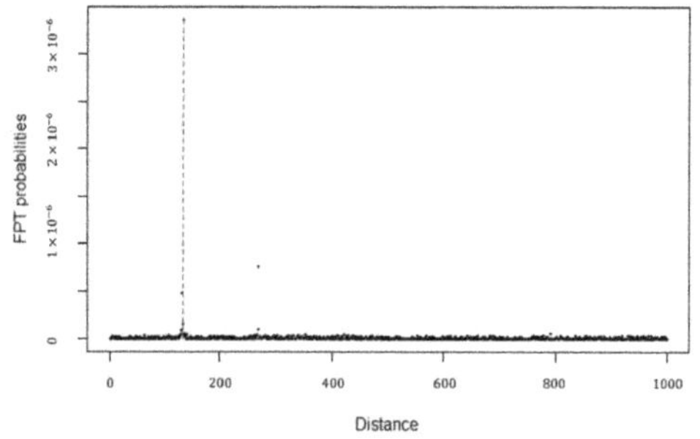

(**c**) $w_3 = CCACAAT$

Figure 1. First passage time (FPT) probabilities for distance $n \leq 1000$.

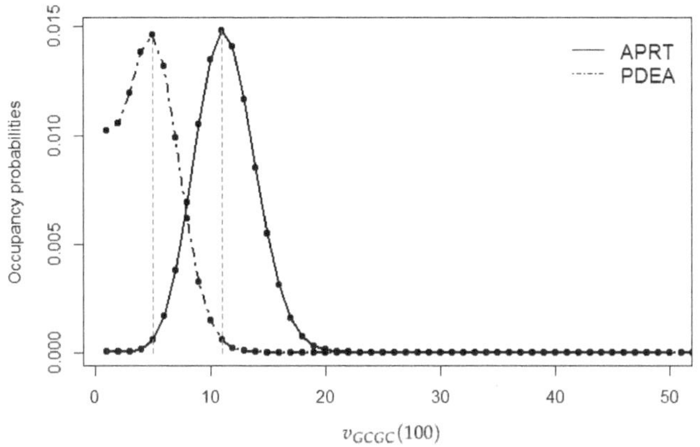

Figure 2. Occupancy probabilities of APRT and PDEA genes.

5. Concluding Remarks

In this article, three classes of important probabilities of a semi-Markov process, namely the first passage time, the occupancy and the duration probabilities were defined and their closed analytic forms were proved by using the basic parameters of the process. The study of the first passage time probability provides information regarding the distribution of the time elapsed to reach a state from another for the first time, either in terms of transitions or time. The second category of duration probabilities provides information about the distribution of the number of virtual transitions taking place before an actual transition to a different state occurs. Finally, the third class of probabilities provides insight information regarding the distribution of the number of times the SMC makes transitions to some state in a time interval of a given length. We provided analytic forms on the actual behavior of the recursive relations of the aforementioned probabilities and included these results into specific propositions and theorems.

The analytical results were accompanied with two illustrations on human genome DNA strands which are often studied using probabilistic modeling and, specifically, Markovian models. Although, in the relevant literature, there exist several algorithmic approaches analyzing the occupancy and appearance of words in DNA sequences, the results of the illustration section strongly suggest that the proposed modeling framework could also be used for the investigation of the structure of genome sequences.

Of course nothing comes without limitations and motivation for further research. For example, additional research effort could aim towards high-order dependencies since DNA sequences often show long-range correlations. This could result in a more coherent modeling approach. Furthermore, additional parameters could be included in the model, for example the length of sequence or specific mutations, resulting in more realistic representations regarding the different structures of complex genome of humans and other organisms. Finally, the proposed model could be applied in completely different contexts, such as natural language processing, linguistics, text similarity and anomaly detection, i.e., areas of machine learning that appear to be amongst the most popular areas in the last decade in data science and stochastic modeling.

Author Contributions: Conceptualization, A.C.G., A.P. and P.K.; Data curation, P.K.; Formal analysis, P.K.; Investigation, A.P. and P.K.; Methodology, A.C.G., A.P., H.P. and E.F.; Software, P.K.; Supervision, A.C.G. and A.P.; Validation, A.C.G.; Visualization, P.K.; Writing—original draft, A.P., P.K., H.P. and E.F.; Writing—review & editing, A.C.G., A.P. and P.K. All authors have read and agreed to the published version of the manuscript.

Funding: This research received no external funding.

Institutional Review Board Statement: Not applicable.

Informed Consent Statement: Not applicable.

Data Availability Statement: Publicly available datasets were analyzed in this study. This data can be found here: https://www.ncbi.nlm.nih.gov/nuccore/CM000663, https://www.ncbi.nlm.nih.gov/gtr/genes/353/, https://www.ncbi.nlm.nih.gov/nuccore/1059792111.

Acknowledgments: The authors greatly acknowledge the comments and suggestions of the three anonymous referees, which improved the content and the presentation of the current paper.

Conflicts of Interest: The authors declare no conflicts of interest.

Appendix A. Proofs

Appendix A.1. Proof of Theorem 1

The results for (1) and (2) are obvious. For the third part, we used the matrix notation of the first passage time probabilities:

$$\mathbf{F}(k|t,n) = \sum_{m=1}^{n-k+1} \mathbf{C}(t,m)\{\mathbf{F}(k-1|t+m,n-m) \circ \mathbf{B}\} + \delta(k-1)\mathbf{C}(t,m),$$

with $\mathbf{F}(k|t,n) = 0$ if $k > n$ or $k = 0$. For $k = 1$ and $m = m_i$ we have shown the results for the case where $k > 1$ can be proved by induction. Thus, we assume that this result holds for $k-1$ and we will show that it also holds for each $k \leq n$. Here we note that the recursive relationship of the first passage time probabilities could be reformulated as follows.

$$f_{ij}(k|t,n) = \sum_{m_1=1}^{n-k+1} \sum_{x_1 \neq j} c_{ix_1}(t,m_1) \left\{ \sum_{m_2=1+m_1}^{n-k+2} \sum_{x_2 \neq j} c_{x_1x_2}(t+m_1,m_2-m_1) \right\}$$
$$\left\{ \cdots \left\{ \sum_{m_{k-1}=1+m_{k-2}}^{n-1} \sum_{x_{k-1} \neq j} c_{x_{k-2}x_{k-1}}(t+m_{k-2},m_{k-1}-m_{k-2}) c_{x_{k-1}j}(t+m_{k-1},n-m_{k-1}) \right\} \right\} \cdots$$
$$+ \delta(k-1)c_{ij}(t,n).$$

Using matrix notation, we can express the previous relationship as the following.

$$\mathbf{F}(k|t,n) = \sum_{m_1=1}^{n-k+1} \mathbf{C}(t,m_1) \left\{ \sum_{m_2=1+m_1}^{n-k+2} \mathbf{C}(t+m_1,m_2-m_1) \right\}$$
$$\left\{ \cdots \left\{ \sum_{m_{k-1}=1+m_{k-2}}^{n-1} \mathbf{C}(t+m_{k-2},m_{k-1}-m_{k-2})\{\mathbf{C}(t+m_{k-1},n-m_{k-1}) \circ \mathbf{B}\} \right\} \circ \mathbf{B} \right\} \cdots \circ \mathbf{B}$$

for $0 < k \leq n$.

The initial conditions are $\mathbf{F}(k|t,n) = 0$ for $k > n$ or $k = 0$ and $\mathbf{F}(1|t,n) = \mathbf{C}(t,n)$. By using the following notation:

$$\sum_{m_1=1}^{n-k+1} \left\{ \sum_{m_2=1+m_1}^{n-k+2} \left\{ \cdots \left\{ \sum_{m_{k-1}=1+m_{k-2}}^{n-1} \right.\right.\right. = \sum_{m_1=1}^{n-k+1} * \sum_{m_2=1+m_1}^{n-k+2} * \cdots * \sum_{m_{k-1}=1+m_{k-2}}^{n-1},$$

we obtain the following.

$$\mathbf{F}(k|t,n) = \sum_{m=1}^{n-k+1} \mathbf{C}(t,m) \left\{ \left\{ \sum_{m_1=1}^{n-m-k+2} * \sum_{m_2=1+m_1}^{n-m-k+3} * \cdots * \sum_{m_{k-2}=1+m_{k-3}}^{n-m-1} \mathbf{C}(t,m_1) \right\} \right\}$$
$$\{\mathbf{C}(t+m_1,m_2-m_1)\{\ldots\{\mathbf{C}(t+m+m_{k-2},n-m-m_{k-2}) \circ \mathbf{B}\} \circ \mathbf{B}\} \ldots\} \circ \mathbf{B}.$$

By the appropriate substitution of the time indices and by the definition of the following operation $\prod_{r=1}^{2}{}^{\{\mathbf{B}\}}\mathbf{A}_r = \mathbf{A}_1 *_B \mathbf{A}_2 = \mathbf{A}_2(\mathbf{A}_1 \circ \mathbf{B})$ for the matrices $\mathbf{A}_1, \mathbf{A}_2, \mathbf{B}$, we obtain the desired result. □

Appendix A.2. Proof of Theorem 2

The results for (1) and (2) are obvious. For the third part, we used induction. By using matrix notation on the recursive relationship, it holds that, for $k = 2$, we have the following.

$$\mathbf{D}(2|t,n) = \sum_{m_1=1}^{n-1} (\mathbf{C}(t,m_1) \circ \mathbf{I})(\mathbf{W}(t+m_1, n-m_1) - \mathbf{C}(t+m_1, n-m_1) \circ \mathbf{I})$$

Now assume that the relationship hold for $k-1$, which is the following.

$$\begin{aligned}\mathbf{D}(k-1|t+m,n-m) = & \sum_{m_1=1}^{n-m-k+2} * \sum_{m_2=1+m_1}^{n-m-k+3} * \ldots * \sum_{m_{k-2}=1+m_{k-3}}^{n-m-1} (\mathbf{C}(t+m,m_1) \circ \mathbf{I}) \\ & (\mathbf{C}(t+m+m_1, m_2-m_1) \circ \mathbf{I}) \ldots (\mathbf{C}(t+m+m_{k-3}, m_{k-2}-m_{k-3}) \circ \mathbf{I}) \\ & (\mathbf{W}(t+m+m_{k-2}, n-m-m_{k-2}) - \mathbf{C}(t+m+m_{k-2}, n-m-m_{k-2}) \circ \mathbf{I}).\end{aligned}$$

Therefore, the following obtains.

$$\begin{aligned}\mathbf{D}(k|t,n) = & \sum_{m=1}^{n-k+1} * \sum_{m_1=1}^{n-m-k+2} * \ldots * \sum_{m_{k-2}=1+m_{k-3}}^{n-m-1} (\mathbf{C}(t+m,m_1) \circ \mathbf{I}) \\ & (\mathbf{C}(t+m+m_1, m_2-m_1) \circ \mathbf{I}) \ldots (\mathbf{C}(t+m+m_{k-3}, m_{k-2}-m_{k-3}) \circ \mathbf{I}) \\ & (\mathbf{W}(t+m+m_{k-2}, n-m-m_{k-2}) - \mathbf{C}(t+m+m_{k-2}, n-m-m_{k-2}) \circ \mathbf{I}).\end{aligned}$$

By appropriately substituting the time indices with $m'_0 = 0$, $m'_1 = m$, $m'_2 = m + m_1$, $\ldots\ m'_i = m + m_{i-1}, \ldots, m'_{k-1} = m + m_{k-2}, i = 1, 2, \ldots, k-1$, where $1 + m_{i-1} \leq m_i \leq n - m - k + i + 1$, we obtain the following:

$$\begin{aligned}\mathbf{D}(k|t,n) = & \sum_{m'_1=1}^{n-k+1} * \sum_{m'_2=1+m'_1}^{n-k+2} * \ldots * \sum_{m'_{k-1}=1+m'_{k-2}}^{n-1} (\mathbf{C}(m'_1) \circ \mathbf{I})(\mathbf{C}(m'_2 - m'_1) \circ \mathbf{I})(\mathbf{C}(m'_3 - m'_2) \circ \mathbf{I}) \\ & \ldots \left(\mathbf{C}\left(m'_{k-1} - m'_{k-2}\right) \circ \mathbf{I}\right)\left(\mathbf{W}\left(n - m'_{k-1}\right) - \mathbf{C}\left(n - m'_{k-1}\right) \circ \mathbf{I}\right),\end{aligned}$$

which results in the stated relationship. □

Appendix A.3. Proof of Theorem 3

Assuming homogeneity in time, Equation (4) is provided by the following:

$$\begin{aligned}\omega_{ij}(x|n) = & \sum_{\substack{r=1 \\ r \neq j}}^{N} \sum_{m=0}^{n} c_{ir}(m)\omega_{rj}(x|n-m) + \\ & + \sum_{m=0}^{n} c_{ij}(m)\omega_{jj}(x-1|n-m) + \delta(x)^{>} w_i(n),\end{aligned} \tag{A1}$$

where $i, j \in S, n = 0, 1, \ldots$ and $x = 0, 1, \ldots$. Equation (A1) can be written as follows.

$$\omega_{ij}(x|n) = \sum_{r=1}^{N} \sum_{m=0}^{n} c_{ir}(m)[\omega_{rj}(x|n-m)(1-\delta_{rj}) + \omega_{rj}(x-1|n-m)\delta_{rj}] + \delta(x)^{>} w_i(n). \tag{A2}$$

Equation (A2) in matrix notation is the following.

$$\mathbf{\Omega}(x|n) = \sum_{m=1}^{n} \mathbf{C}(m)[\mathbf{\Omega}(x|n-m) \circ (\mathbf{U} - \mathbf{I}) + \mathbf{\Omega}(x-1|n-m) \circ \mathbf{I}] + \delta(x)^{>} \mathbf{W}(n).$$

By applying the geometric transform to the above, we obtain the following:

$$\mathbf{\Omega}^g(z|n) = \sum_{m=1}^{n} \mathbf{C}(m)\mathbf{\Omega}^g(z|n-m) + (z-1)\sum_{m=1}^{n} \mathbf{C}(m)[\mathbf{\Omega}^g(z|n-m) \circ \mathbf{I}] +^{>} \mathbf{W}(n),$$

with initial condition $\mathbf{\Omega}^g(z|0) = \mathbf{I}$. Following the methodology of Vassiliou and Papadopoulou (1992), we derive the result of the Theorem 3. [15]

Appendix A.4. Proof of Lemma 1

By using the Hadamard product on Theorem 3, we have the following.

$$\begin{aligned}\mathbf{\Omega}^g(z|n) \circ \mathbf{I} = &-(z-1)\sum_{j=1}^{n-1}\left[\left[\sum_{i=1}^{j} \mathbf{a}_{1i}^{-1}\mathbf{C}(j+1-i)\right][\mathbf{\Omega}^g(z|n-j) \circ \mathbf{I}]\right] \circ \mathbf{I} \\ &- z\sum_{j=1}^{n}\left[\mathbf{a}_{1j}^{-1}\mathbf{C}(n+1-j)\right] \circ \mathbf{I} - \sum_{j=1}^{n}\left[\mathbf{a}_{1j}^{-1>}\mathbf{W}(n+1-j)\right] \circ \mathbf{I}.\end{aligned}$$

By using the following property:

$$(\mathbf{A}(\mathbf{B} \circ \mathbf{I})) \circ \mathbf{I} = (\mathbf{A} \circ \mathbf{I})(\mathbf{B} \circ \mathbf{I}).$$

we obtain the following:

$$\left[\left[\sum_{i=1}^{j} \mathbf{a}_{1i}^{-1}\mathbf{C}(j+1-i)\right][\mathbf{\Omega}^g(z|n-j) \circ \mathbf{I}]\right] \circ \mathbf{I} = \left[\left[\sum_{i=1}^{j} \mathbf{a}_{1i}^{-1}\mathbf{C}(j+1-i)\right] \circ \mathbf{I}\right][\mathbf{\Omega}^g(z|n-j) \circ \mathbf{I}].$$

which completes the proof. □

Appendix A.5. Proof of Theorem 4

An early version of the proof of Theorem 4 can be found in [56]. We analytically present here all necessary steps of the proof. Using the equations provided by the results of Theorem 3 and by substituting $\mathbf{\Omega}^g(z|n) \circ \mathbf{I}$ with the result found in Lemma 1, we can obtain the analytic relation for the geometric transforms of $\mathbf{\Omega}^g(z|n)$, which is as follows:

$$\begin{aligned}\mathbf{\Omega}^g(z|n) = (z-1)\sum_{j=1}^{n-1}\mathbf{A}_j&\left[\begin{array}{c} z\mathbf{G}_{1,n,j} + z\sum_{u=2}^{n-j}\left[(z-1)\mathbf{M}'_u + \sum_{k=1}^{u-2}(z-1)^{k+1}\mathbf{R}_u(k,m_k)\right]\mathbf{G}_{u,n,j} \\ +\mathbf{Q}_{1,n,j} + \sum_{u=2}^{n-j}\left((z-1)\mathbf{M}'_u + \sum_{k=1}^{u-2}(z-1)^{k+1}\mathbf{R}_u(k,m_k)\right)\mathbf{Q}_{u,n,j}\end{array}\right] + \\ &+ z\mathbf{A}_n + \mathbf{E}_n,\end{aligned} \tag{A3}$$

where

$$\mathbf{A}_j = \mathbf{C}(j) + \sum_{i=2}^{j}\left(\mathbf{C}(i-1) + \sum_{k=1}^{i-2}\mathbf{S}_i(k,m_k)\right)\mathbf{C}(j+1-i),$$

$$\mathbf{M}'_u = \left[\mathbf{C}(u-1) + \sum_{i=2}^{u-1}\left(\mathbf{C}(i-1) + \sum_{k=1}^{i-2}\mathbf{S}_i(k,m_k)\right)\mathbf{C}(u-i)\right] \circ \mathbf{I},$$

$$\mathbf{E}_n = \sum_{j=2}^{n}\left(\mathbf{C}(j-1) + \sum_{k=1}^{j-2}\mathbf{S}_j(k,m_k)\right)^{>}\mathbf{W}(n+1-j) +^{>} \mathbf{W}(n),$$

$$\mathbf{G}_{u,n,j} = \mathbf{C}(n-j+1-u) \circ \mathbf{I} + \sum_{w=2}^{n-j+1-u}\left[\left(\mathbf{C}(w-1) + \sum_{k=1}^{w-2}\mathbf{S}_w(k,m_k)\right)\mathbf{C}(n-j+2-u-w)\right] \circ \mathbf{I},$$

$$\mathbf{Q}_{u,n,j} = \sum_{w=2}^{n-j+1-u}\left[\left(\mathbf{C}(w-1) + \sum_{k=1}^{w-2}\mathbf{S}_w(k,m_k)\right)^{>}\mathbf{W}(n-j+2-u-w)\right] \circ \mathbf{I} + [^{>}\mathbf{W}(n-j+1-u)] \circ \mathbf{I}.$$

Then, by applying properties of the inverse geometric transforms by using the equation $\mathbf{\Omega}(x|n) = \frac{1}{x!}\frac{d^{(x)}}{dz^x}\mathbf{\Omega}^g(z|n)\Big|_{z=0}$ and by repeatedly taking the derivatives of $\mathbf{\Omega}^g(z|n)$ with respect to z, we obtain the result of the Theorem 5 for $x \geq 1$.

Finally, for the special case where $x = 0$, by substituting $z = 0$ in expression (A3), we obtain the following:

$$\boldsymbol{\Omega}(0|n) = -\sum_{j=1}^{n-1} \mathbf{A}_j \left[\mathbf{B}_{n,j} + \sum_{u=2}^{n-j} \mathbf{M}_u \mathbf{Q}_{u,n,j} \right] + \mathbf{E}_n,$$

where the following results.

$$\mathbf{B}_{n,j} = \left[\sum_{w=2}^{n-j} \left[\left(\mathbf{C}(w-1) + \sum_{k=1}^{w-2} \mathbf{S}_w(k, m_k) \right) {}^{>}\mathbf{W}(n-j+1-w) \right] \circ \mathbf{I} + {}^{>}\mathbf{W}(n-j) \right] \circ \mathbf{I},$$

$$\mathbf{M}_u = -\left[\mathbf{C}(u-1) + \sum_{i=2}^{u-1} \left(\mathbf{C}(i-1) + \sum_{k=1}^{i-2} \mathbf{S}_i(k, m_k) \right) \mathbf{C}(u-i) \right] \circ \mathbf{I} + \sum_{k=1}^{u-2} (-1)^{k+1} \mathbf{R}_u(k, m_k).$$

Appendix B

Appendix B.1

$$\begin{aligned} {}^{>}f_{ij}(n) = 1 - f_{ij}(n) \Rightarrow {}^{>}f_{ij}^{g}(z) = \textstyle\sum_{n=0}^{\infty} {}^{>}f_{ij}(n) z^n = \sum_{n=0}^{\infty} (1 - f_{ij}(n)) z^n = \\ = \textstyle\sum_{n=0}^{\infty} z^n - \sum_{n=0}^{\infty} f_{ij}(n) z^n = \frac{1}{1-z} - \sum_{n=0}^{\infty} \left(\sum_{m=0}^{n} f_{ij}(m) \right) z^n = \\ = \textstyle\frac{1}{1-z} - \sum_{m=0}^{\infty} \sum_{n-m=0}^{\infty} f_{ij}(m) z^m z^{n-m} = \frac{1}{1-z} - \frac{f_{ij}^{g}(z)}{1-z} = \frac{1-f_{ij}^{g}(z)}{1-z}. \end{aligned}$$

Appendix B.2

$$\begin{aligned} \omega_{ij}^{gg}(y|z) &= {}^{>}f_{ij}^{g}(z) + y f_{ij}^{g}(z) \frac{{}^{>}f_{jj}^{g}(z)}{\left(1 - y f_{jj}^{g}(z)\right)} \\ &= \frac{1 - f_{ij}^{g}(z)}{1-z} + \frac{y f_{ij}^{g}(z)}{\left(1 - y f_{jj}^{g}(z)\right)} \frac{1 - f_{jj}^{g}(z)}{1-z} \\ &= \frac{\left(1 - f_{ij}^{g}(z)\right)\left(1 - y f_{jj}^{g}(z)\right) + y f_{ij}^{g}(z)\left(1 - f_{jj}^{g}(z)\right)}{(1-z)\left(1 - y f_{jj}^{g}(z)\right)} \\ &= \frac{1 - y f_{jj}^{g}(z) - f_{ij}^{g}(z) + y f_{jj}^{g}(z) f_{ij}^{g}(z) + y f_{ij}^{g}(z) - y f_{ij}^{g}(z) f_{jj}^{g}(z)}{(1-z)\left(1 - y f_{jj}^{g}(z)\right)} \\ &= \frac{\left(1 - y f_{jj}^{g}(z)\right) - f_{ij}^{g}(z) + y f_{ij}^{g}(z)}{(1-z)\left(1 - y f_{jj}^{g}(z)\right)} \\ &= \frac{\left(1 - y f_{jj}^{g}(z)\right) - (1-y) f_{ij}^{g}(z)}{(1-z)\left(1 - y f_{jj}^{g}(z)\right)}. \end{aligned}$$

References

1. Pyke, R. Markov renewal processes with finitely many states. *Ann. Math. Stat.* **1961**, *32*, 1243–1259. [CrossRef]
2. Cinlar, E. *Introduction to Stochastic Processes*; Courier Corporation: Chelmsford, MA, USA, 2013.
3. Howard, R.A. *Dynamic Probabilistic Systems: Semi-Markov and Decision Processes*; Dover Publications: Mineola, NY, USA, 2007; Volume 2.
4. McClean, S.I. A semi-Markov model for a multigrade population with Poisson recruitment. *J. Appl. Probab.* **1980**, *17*, 846–852. [CrossRef]
5. McClean, S.I. Semi-Markov models for manpower planning. In *Semi-Markov Models*; Springer: Berlin/Heidelberg, Germany, 1986; pp. 283–300.

6. D'Amico, G.; Di Biase, G.; Janssen, J.; Manca, R. *Semi-Markov Migration Models for Credit Risk*; Wiley Online Library: Hoboken, NJ, USA, 2017.
7. Vassiliou, P.-C.G. Non-Homogeneous Semi-Markov and Markov Renewal Processes and Change of Measure in Credit Risk. *Mathematics* **2021**, *9*, 55. [CrossRef]
8. Janssen, J.; Manca, R. *Applied Semi-Markov Processes*; Springer Science & Business Media: Berlin/Heidelberg, Germany, 2006.
9. Janssen, J. *Semi-Markov Models: Theory and Applications*; Springer Science & Business Media: Berlin/Heidelberg, Germany, 2013.
10. Schbath, S.; Prum, B.; de Turckheim, E. Exceptional motifs in different Markov chain models for a statistical analysis of DNA sequences. *J. Comput. Biol.* **1995**, *2*, 417–437. [CrossRef]
11. De Dominicis, R.; Manca, R. Some new results on the transient behaviour of semi-Markov reward processes. *Methods Oper. Res.* **1986**, *53*, 387–397.
12. Vasileiou, A.; Vassiliou, P.-C.G. An inhomogeneous semi-Markov model for the term structure of credit risk spreads. *Adv. Appl. Probab.* **2006**, *38*, 171–198. [CrossRef]
13. Vassiliou, P.-C.G.; Vasileiou, A. Asymptotic behaviour of the survival probabilities in an inhomogeneous semi-Markov model for the migration process in credit risk. *Linear Algebra Appl.* **2013**, *438*, 2880–2903. [CrossRef]
14. Vassiliou, P.-C.G. Semi-Markov migration process in a stochastic market in credit risk. *Linear Algebra Appl.* **2014**, *450*, 13–43. [CrossRef]
15. Vassiliou, P.-C.G.; Papadopoulou, A. Non-homogeneous semi-Markov systems and maintainability of the state sizes. *J. Appl. Probab.* **1992**, *29*, 519–534. [CrossRef]
16. Vassiliou, P.-C.G. Asymptotic behavior of Markov systems. *J. Appl. Probab.* **1982**, *19*, 851–857. [CrossRef]
17. Dimitriou, V.; Georgiou, A.C. Introduction, analysis and asymptotic behavior of a multi-level manpower planning model in a continuous time setting under potential department contraction. *Commun. Stat. Theory Methods* **2021**, *50*, 1173–1199. [CrossRef]
18. Papadopoulou, A.; Vassiliou, P.-C.G. Asymptotic behavior of nonhomogeneous semi-Markov systems. *Linear Algebra Appl.* **1994**, *210*, 153–198. [CrossRef]
19. Papadopoulou, A.; Vassiliou, P.-C.G. On the variances and convariances of the duration state sizes of semi-Markov systems. *Commun. Stat. Theory Methods* **2014**, *43*, 1470–1483. [CrossRef]
20. Vassiliou, P.-C.G. Markov systems in a general state space. *Commun. Stat. Theory Methods* **2014**, *43*, 1322–1339. [CrossRef]
21. Dimitriou, V.A.; Georgiou, A.C.; Tsantas, N. The multivariate non-homogeneous Markov manpower system in a departmental mobility framework. *Eur. J. Oper. Res.* **2013**, *228*, 112–121. [CrossRef]
22. Symeonaki, M. Theory of fuzzy non homogeneous Markov systems with fuzzy states. *Qual. Quant.* **2015**, *49*, 2369–2385. [CrossRef]
23. Tsaklidis, G.; Vassiliou, P.-C.G. Asymptotic periodicity of the variances and covariances of the state sizes in non-homogeneous Markov systems. *J. Appl. Probab.* **1988**, *25*, 21–33. [CrossRef]
24. Vassiliou, P.-C.G. The evolution of the theory of non-homogeneous Markov systems. *Appl. Stoch. Model. Data Anal.* **1997**, *13*, 159–176. [CrossRef]
25. Vassiliou, P.-C.G.; Georgiou, A.C. Asymptotically attainable structures in nonhomogeneous Markov systems. *Oper. Res.* **1990**, *38*, 537–545. [CrossRef]
26. Ugwuowo, F.I.; McClean, S.I. Modelling heterogeneity in a manpower system: A review. *Appl. Stoch. Model. Bus. Ind.* **2000**, *16*, 99–110. [CrossRef]
27. Symeonaki, M.; Stamatopoulou, G. Describing labour market dynamics through Non Homogeneous Markov System theory. In *Demography of Population Health, Aging and Health Expenditures*; Springer: Berlin/Heidelberg, Germany, 2020; pp. 359–373.
28. Ossai, E.; Uche, P. Maintainability of departmentalized manpower structures in Markov chain model. *Pac. J. Sci. Technol.* **2009**, *2*, 295–302.
29. Guerry, M.A.; De Feyter, T. Optimal recruitment strategies in a multi-level manpower planning model. *J. Oper. Res. Soc.* **2012**, *63*, 931–940. [CrossRef]
30. Hunter, J.J. Stationary distributions and mean first passage times of perturbed Markov chains. *Linear Algebra Appl.* **2005**, *410*, 217–243. [CrossRef]
31. Hunter, J.J. Simple procedures for finding mean first passage times in Markov chains. *Asia-Pac. J. Oper. Res.* **2007**, *24*, 813–829. [CrossRef]
32. Hunter, J.J. The computation of the mean first passage times for Markov chains. *Linear Algebra Appl.* **2018**, *549*, 100–122. [CrossRef]
33. Yao, D.D. First-passage-time moments of Markov processes. *J. Appl. Probab.* **1985**, *22*, 939–945. [CrossRef]
34. Zhang, X.; Hou, Z. The first-passage times of phase semi-Markov processes. *Stat. Probab. Lett.* **2012**, *82*, 40–48. [CrossRef]
35. Pitman, J.; Tang, W. Tree formulas, mean first passage times and Kemeny's constant of a Markov chain. *Bernoulli* **2018**, *24*, 1942–1972. [CrossRef]
36. Redner, S. *A Guide to First-Passage Processes*; Cambridge University Press: Cambridge, UK, 2001.
37. Waterman, M.S. *Introduction to Computational Biology: Maps, Sequences and Genomes*; CRC Press: Boca Raton, FL, USA, 1995.
38. Almagor, H. A Markov analysis of DNA sequences. *J. Theor. Biol.* **1983**, *104*, 633–645. [CrossRef]
39. Menéndez, M.; Pardo, L.; Pardo, M.; Zografos, K. Testing the order of Markov dependence in DNA sequences. *Methodol. Comput. Appl. Probab.* **2011**, *13*, 59–74. [CrossRef]
40. Skewes, A.D.; Welch, R.D. A Markovian analysis of bacterial genome sequence constraints. *PeerJ* **2013**, *1*, e127. [CrossRef]

41. Papapetrou, M.; Kugiumtzis, D. Markov chain order estimation with conditional mutual information. *Phys. A: Stat. Mech. Appl.* **2013**, *392*, 1593–1601. [CrossRef]
42. Boys, R.J.; Henderson, D.A.; Wilkinson, D.J. Detecting homogeneous segments in DNA sequences by using hidden Markov models. *J. R. Stat. Soc. Ser. C* **2000**, *49*, 269–285. [CrossRef]
43. Reinert, G.; Schbath, S.; Waterman, M.S. Probabilistic and statistical properties of words: An overview. *J. Comput. Biol.* **2000**, *7*, 1–46. [CrossRef]
44. Robin, S.; Daudin, J.J. Exact distribution of word occurrences in a random sequence of letters. *J. Appl. Probab.* **1999**, *36*, 179–193. [CrossRef]
45. Schbath, S. An overview on the distribution of word counts in Markov chains. *J. Comput. Biol.* **2000**, *7*, 193–201. [CrossRef] [PubMed]
46. Papadopoulou, A. Some Results on Modeling Biological Sequences and Web Navigation with a Semi Markov Chain. *Commun. Stat. Theory Methods* **2013**, *42*, 2853–2871. [CrossRef]
47. Ricciardi, L.; Crescenzo, A.; Giorno, V.; Nobile, A. An outline of theoretical and algorithmic approaches to first passage time problems with applications to biological modeling. *Math. Jpn.* **1999**, *50*, 247–322.
48. Lavi, B.; Levy Karin, E.; Pupko, T.; Hazkani-Covo, E. The prevalence and evolutionary conservation of inverted repeats in proteobacteria. *Genome Biol. Evol.* **2018**, *10*, 918–927. [CrossRef] [PubMed]
49. Brázda, V.; Laister, R.C.; Jagelská, E.B.; Arrowsmith, C. Cruciform structures are a common DNA feature important for regulating biological processes. *BMC Mol. Biol.* **2011**, *12*, 1–16. [CrossRef]
50. Homo Sapiens Chromosome 1, GRCh38.p13 Primary Assembly. Available online: https://www.ncbi.nlm.nih.gov/nuccore/CM000663 (accessed on 17 December 2020).
51. Tavares, A.H.; Pinho, A.J.; Silva, R.M.; Rodrigues, J.M.; Bastos, C.A.; Ferreira, P.J.; Afreixo, V. DNA word analysis based on the distribution of the distances between symmetric words. *Sci. Rep.* **2017**, *7*, 1–11. [CrossRef] [PubMed]
52. Gardiner-Garden, M.; Frommer, M. CpG islands in vertebrate genomes. *J. Mol. Biol.* **1987**, *196*, 261–282. [CrossRef]
53. APRT adenine phosphoribosyltransferase. Available online: https://www.ncbi.nlm.nih.gov/gtr/genes/353/ (accessed on 17 December 2020).
54. Broderick, T.P.; Schaff, D.A.; Bertino, A.M.; Dush, M.K.; Tischfield, J.A.; Stambrook, P.J. Comparative anatomy of the human APRT gene and enzyme: Nucleotide sequence divergence and conservation of a nonrandom CpG dinucleotide arrangement. *Proc. Natl. Acad. Sci. USA* **1987**, *84*, 3349–3353. [CrossRef] [PubMed]
55. Homo sapiens Human Phosphodiesterase (PDEA) Gene. Available online: https://www.ncbi.nlm.nih.gov/nuccore/1059792111 (accessed on 17 December 2020).
56. Farmakioti, E. Probabilities of State Occupancies in Semi-Markov Chains. Master's Thesis, Aristotle University of Thessaloniki, Thessaloniki, Greece, 2018.

mathematics

MDPI

Article

Sequential Interval Reliability for Discrete-Time Homogeneous Semi-Markov Repairable Systems

Vlad Stefan Barbu [1], Guglielmo D'Amico [2,*] and Thomas Gkelsinis [1]

[1] Laboratory of Mathematics Raphaël Salem, University of Rouen-Normandy, UMR 6085, Avenue de l'Université, BP. 12, F76801 Saint-Étienne-du-Rouvray, France; barbu@univ-rouen.fr (V.S.B.); thomas.gkelsinis@univ-rouen.fr (T.G.)

[2] Department of Economics, University "G. d'Annunzio" of Chieti-Pescara, 66013 Pescara, Italy

* Correspondence: g.damico@unich.it

Abstract: In this paper, a new reliability measure, named sequential interval reliability, is introduced for homogeneous semi-Markov repairable systems in discrete time. This measure is the probability that the system is working in a given sequence of non-overlapping time intervals. Many reliability measures are particular cases of this new reliability measure that we propose; this is the case for the interval reliability, the reliability function and the availability function. A recurrent-type formula is established for the calculation in the transient case and an asymptotic result determines its limiting behaviour. The results are illustrated by means of a numerical example which illustrates the possible application of the measure to real systems.

Keywords: semi-Markov; reliability; transient analysis; asymptotic analysis

Citation: Barbu, V.S.; D'Amico, G.; Gkelsinis, T. Sequential Interval Reliability for Semi-Markov Repairable Systems. *Mathematics* **2021**, *9*, 1997. https://doi.org/10.3390/math9161997

Academic Editors: Panagiotis-Christos Vassiliou and Andreas C. Georgiou

Received: 10 July 2021
Accepted: 19 August 2021
Published: 20 August 2021

Publisher's Note: MDPI stays neutral with regard to jurisdictional claims in published maps and institutional affiliations.

Copyright: © 2021 by the authors. Licensee MDPI, Basel, Switzerland. This article is an open access article distributed under the terms and conditions of the Creative Commons Attribution (CC BY) license (https://creativecommons.org/licenses/by/4.0/).

1. Introduction

This paper is concerned with reliability indicators for semi-Markov systems. As it is well known (see, e.g., [1–6]), semi-Markov processes represent an important modelling tool for practical problems in reliability, survival analysis, financial mathematics, and manpower planning, among other applied domains. The attractiveness of these processes comes from the fact that the sojourn time in a state can be arbitrarily distributed, as compared to Markov processes, where the sojourn time in a state is constrained to be geometrically or exponentially distributed.

Several researchers have investigated the reliability measures of semi-Markov processes. Examples of discrete-time semi-Markov processes with the associated reliability measures and statistical topics can be found in, e.g., [7–10], who proposed a semi-Markov chain usage model in discrete time and provided analytical formulas for the mean and variance of the single-use reliability of the system. The evaluation of reliability indicators for continuous-time semi-Markov processes and statistical inference can be found in [11–15]. The readers interested in solving numerically continuous-time semi-Markov processes by using discrete-time semi-Markov processes for solving continuous ones are referred to [16–19].

In the present work, we propose a new measure for analysing the performance of a system, called the sequential interval reliability (SIR). This generalises the notion of interval reliability, as it is introduced in [20] for discrete-time semi-Markov processes and further studied in [21,22]. In line with the work of [23], we are also interested in a general definition that takes into account the dependence on what is called the final backward. It is worth mentioning that interval reliability was first introduced and studied for continuous-time semi-Markov systems in [24,25]. In those contributions, the interval reliability was expressed in terms of a system of integral equation.

This measure computes the probability that a system is in a working state during a sequence of non-overlapping intervals. This type of measure is of importance in several applications: in reliability, when a system has to perform during consequent time periods; in extreme value theory, where we can be interested in the occurrence of an extreme event during several time periods; in energy studies, where we are interested, for instance, in the electricity consumption that is greater or below a certain threshold; and in financial modelling, in order to create advanced credit scoring models, etc.

This article is structured as follows: in the next section, we introduce the basic semi-Markov notions and notations and we also give the corresponding measures of reliability. The main object of our study, namely sequential interval reliability, is introduced in Section 3. Then, we first perform transient analysis, providing a recurrence formula for computing the SIR. Second, we furnish an asymptotic result, as a time of interest that extends to infinity. A numerical example is provided in Section 4, illustrating some aspects of our theoretical work.

2. Discrete-Time Semi-Markov Processes and Reliability Measures

Let us consider a random system with finite state space $E = \{1, \ldots, s\}$, $s < \infty$ and let $(\Omega, \mathcal{A}, \mathbb{P})$ be a probability space. We assume that the evolution in time of the system is governed by a stochastic process $Z = (Z_k)_{k\in\mathbb{N}}$, defined on $(\Omega, \mathcal{A}, \mathbb{P})$ with values in E; in other words, Z_k gives the state of the system at time k. Let $T = (T_n)_{n\in\mathbb{N}}$, defined on $(\Omega, \mathcal{A}, \mathbb{P})$ with values in $\mathbb{Z}$, be the successive time points when state changes in $(Z_k)_{k\in\mathbb{N}}$ occur (the jump times) and let $J = (J_n)_{n\in\mathbb{N}}$, defined on $(\Omega, \mathcal{A}, \mathbb{P})$ with values in E, be the successively visited states at these time points. We denote by $X = (X_n)_{n\in\mathbb{N}^*}$ the successive sojourn times in the visited states, i.e., $X_{n+1} = S_{n+1} - S_n$, $n \in \mathbb{N}$. The relation between the process Z and the process J of the successively visited states is given by $Z_k = J_{N(k)}$, or, equivalently, $J_n = Z_{T_n}, n, k \in \mathbb{N}$, where:

$$N(k) := \max\{n \in \mathbb{N} \mid T_n \leq k\} \tag{1}$$

is the discrete-time counting process of the number of jumps in $[0, k] \subset \mathbb{N}$.

Definition 1 (Semi-Markov chain SMC and Markov renewal chain MRC)**.** *If we have:*

$$\mathbb{P}(J_{n+1} = j, T_{n+1} - T_n = k | J_n = i, J_{n-1}, \ldots, J_0, T_n, \ldots, T_0) = \mathbb{P}(J_{n+1} = j, T_{n+1} - T_n = k | J_n = i), \tag{2}$$

then $Z = (Z_k)_k$ is called a semi-Markov chain (SMC) and $(J, T) = (J_n, T_n)_n$ is called a Markov renewal chain (MRC).

Throughout this paper, we assume that the MRC or SMC are homogeneous with respect to the time in the sense that Equation (2) is independent of n. Thus, we will work under the following assumption:

Assumption 1. *The SMC (or, equivalently, the MRC) is assumed to be homogeneous in time.*

It is clear that, if (J, T) is a MRC, then $J = (J_n)_{n\in\mathbb{N}}$ is a Markov chain with state space E, called the *embedded Markov chain* of the MRC (J, T) (or of the SMC Z).

Definition 2. *For a semi-Markov chain, under Assumption 1, we define:*

- *The* semi-Markov core matrix $(q_{ij}(k))_{i,j\in E,k\in\mathbb{N}}$, $q_{ij}(k) = \mathbb{P}(J_{n+1} = j, T_{n+1} - T_n = k | J_n = i)$;
- *The* initial distribution $(\mu_i)_{i\in E}$, $\mu_i = \mathbb{P}(J_0 = i) = \mathbb{P}(Z_0 = i)$;
- *The* transition matrix $(p_{ij})_{i,j\in E}$ *of the embedded Markov chain* $J = (J_n)_n$, $p_{ij} = \mathbb{P}(J_{n+1} = j | J_n = i)$;
- *The* conditional sojourn time distribution $(f_{ij}(k))_{i,j\in E,k\in\mathbb{N}}$, $f_{ij}(k) = \mathbb{P}(T_{n+1} - T_n = k | J_n = i, J_{n+1} = j)$;
- *The* sojourn time distribution in a state $(h_i(k))_{i\in E,k\in\mathbb{N}}$, $h_i(k) = \mathbb{P}(T_{n+1} - T_n = k | J_n = i)$.

Note that:

$$q_{ij}(k) = p_{ij} f_{ij}(k).$$

Remark 1. *We would like to draw the attention to a specific terminological matter that we encountered in the literature of discrete-time semi-Markov processes with finite or countable state space and that may lead to terminological confusion.*

Some authors use the term "semi-Markov kernel" of discrete-time SM processes for $\mathbb{P}(J_{n+1} = j, T_{n+1} - T_n = k | J_n = i)$ *(see, e.g., [1,7,8,10,19,20,22]). Other authors use the term "semi-Markov kernel" of discrete-time SM processes for* $\mathbb{P}(J_{n+1} = j, T_{n+1} - T_n \leq k | J_n = i)$ *(see, e.g., [21,23,26–28]), while the quantity* $\mathbb{P}(J_{n+1} = j, T_{n+1} - T_n = k | J_n = i)$ *can have several names, for instance semi-Markov core matrix. In this article, we used this second terminology.*

In the authors' opinion, this terminological confusion stems from the following reasons:

1. *On the one hand, when working in discrete time and calling* $\mathbb{P}(J_{n+1} = j, T_{n+1} - T_n \leq k | J_n = i), k \in \mathbb{N}$*, the (discrete-time) semi-Markov kernel, this is achieved* by using exactly the same term *as in continuous time, where* $\mathbb{P}(J_{n+1} = j, T_{n+1} - T_n \leq t | J_n = i), t \in \mathbb{R}$ *represents the (continuous-time) semi-Markov kernel.*
2. *On the other hand, when working in discrete time and calling* $\mathbb{P}(J_{n+1} = j, T_{n+1} - T_n = k | J_n = i), k \in \mathbb{N}$*, the (discrete-time) semi-Markov kernel, this is done* by analogy *with continuous time, since for a discrete-time finite/countable state-space, a Markov or sub-Markov kernel is determined by the behaviour on singleton events (the probability mass function defines the distribution).*

In any case, all this discussion is only a matter of notational convenience.

Clearly, a semi-Markov chain is uniquely determined a.s. by an initial distribution $(\mu_i)_{i \in E}$ and a *semi-Markov core matrix* $(q_{ij}(k))_{i,j \in E, k \in \mathbb{N}}$ or, equivalently, by an initial distribution $(\mu_i)_{i \in E}$, a Markov transition matrix $(p_{ij})_{i,j \in E}$ and conditional sojourn time distributions $(f_{ij}(k))_{i,j \in E, k \in \mathbb{N}}$.

Our work will be carried out under the following assumptions:

Assumption 2. *Transitions to the same state are not allowed, i.e.,* $p_{ii} \equiv 0$ *for all* $i \in E$.

Assumption 3. *There are no instantaneous transitions, i.e.,* $q_{ij}(0) \equiv 0$ *for all* $i, j \in E$.

Clearly, Assumption 2 is equivalent to $q_{ii}(k) = 0$ for all $i \in E$, $k \in \mathbb{N}$, and Assumption 3 is equivalent to $f_{ij}(0) \equiv 0$ for all $i, j \in E$; note that this implies that T is a strictly increasing sequence.

For the conditional sojourn time distribution and sojourn time distribution in a state, one can consider the associated cumulative distribution functions defined by

$$\begin{aligned} F_{ij}(k) &:= \mathbb{P}(T_{n+1} - T_n \leq k | J_n = i, J_{n+1} = j) = \sum_{t=1}^{k} f_{ij}(t); \\ H_i(k) &:= \mathbb{P}(T_{n+1} - T_n \leq k | J_n = i) = \sum_{t=1}^{k} h_i(t). \end{aligned}$$

For any distribution function $F(\cdot)$, we can consider the associated survival/reliability function defined by

$$\overline{F}(k) := 1 - F(t).$$

Consequently, we have:

$$\begin{aligned}\overline{F}_{ij}(k) &:= \mathbb{P}(T_{n+1} - T_n > k | J_n = i, J_{n+1} = j) = 1 - \sum_{t=0}^{k} f_{ij}(t) = \sum_{t=k+1}^{\infty} f_{ij}(t);\\ \overline{H}_i(k) &:= \mathbb{P}(T_{n+1} - T_n > k | J_n = i) = 1 - \sum_{t=0}^{k} h_i(t) = \sum_{t=k+1}^{\infty} h_i(t).\end{aligned}$$

To investigate the reliability behaviour of a semi-Markov system, we split the space E into two subsets: U for the up-states and D for the down-states, with $E = U \cup D$ and $E = U \cap D = \varnothing$. For simplicity, we consider $U = \{1, \ldots, s_1\}$ and $D = \{s_1 + 1, \ldots, s\}$.

Two important reliability measures of a system are the reliability (or survival) function at time $k \in \mathbb{N}$, denoted by $R(k)$, and the (instantaneous) availability function at time $k \in \mathbb{N}$, denoted by $A(k)$, defined, respectively, by

$$\begin{aligned}R(k) &:= \mathbb{P}(Z_0 \in U, \ldots, Z_k \in U),\\ A(k) &:= \mathbb{P}(Z_k \in U).\end{aligned}$$

If ever we condition on the initial state, we obtain the corresponding conditional reliability (or survival) function at time $k \in \mathbb{N}$ given that $\{Z_0 = i\}, i \in U$, denoted by $R_i(k)$, and the conditional (instantaneous) availability function at time $k \in \mathbb{N}$ given that $\{Z_0 = i\}, i \in E$, denoted by $A_i(k)$, is defined, respectively, by

$$\begin{aligned}R(k) &:= \mathbb{P}(Z_0 \in U, \ldots, Z_k \in U \mid Z_0 = i), i \in U,\\ A(k) &:= \mathbb{P}(Z_k \in U \mid Z_0 = i), i \in E.\end{aligned}$$

Let us now define the interval reliability, introduced in [20] as the probability that the system is in up-states during a time interval.

Definition 3 (Interval reliability, conditional interval reliability, cf. [20])**.** *For $k, p \in \mathbb{N}$ and $i \in E$, the interval reliability $IR(k, p)$ and conditional interval reliability $IR_i(k, p)$ given the event $\{Z_0 = i\}$ are, respectively, defined by*

$$IR(k, p) := \mathbb{P}(Z_l \in U, l \in [k, k + p]); \tag{3}$$

$$IR_i(k, p) := \mathbb{P}(Z_l \in U, l \in [k, k + p] \mid Z_0 = i). \tag{4}$$

For $k, p \in \mathbb{N}$ and $i \in E$, it is clear that we have the following properties of the interval reliability and conditional interval reliability (cf. Proposition 1 and Remark 1 of [20]):

$$R(k + p) \leq IR(k, p) \leq A(k + p); \tag{5}$$

$$IR_i(0, p) = R_i(p); \tag{6}$$

$$IR_i(k, 0) = A_i(k). \tag{7}$$

3. Sequential Interval Reliability

Let us consider a repairable system. In this section, we introduce a new reliability measure that we will call sequential interval reliability. This will generalise the notion of interval reliability presented before, in the sense that we are looking at the probability that the system is in working mode during two or several non-overlapping intervals.

More precisely, let us consider $\underline{t} := (t_i)_{i=1,\ldots,N}$ and $\underline{p} := (p_i)_{i=1,\ldots,N}$ two time sequences such that:

1. $t_1 > 0, t_i < t_{i+1}$ for all $i = 1, \ldots, N - 1$;
2. $p_i \geq 0$ for all $i = 1, \ldots, N$;
3. $t_i + p_i < t_{i+1}$ for all $i = 1, \ldots, N - 1$.

It is clear that, in this case, $\{[t_i, t_i + p_i]\}_{i=1,\ldots,N}$ is a sequence of non-overlapping real intervals.

For a sequence $\underline{t} := (t_i)_{i=1,\ldots,N}$ indexes $k_1, k_2 \in \mathbb{N}, k_1 \leq k_2$, we will also use the notation $t_{k_1:k_2} := (t_i)_{i=k_1,\ldots,k_2}$.

Definition 4 (Sequential interval reliability). *Let $(Z_k)_{k\in\mathbb{N}}$ be a discrete time semi-Markov system and let $\underline{t} := (t_i)_{i=1,\ldots,N}$ and $\underline{p} := (p_i)_{i=1,\ldots,N}$, $N \in \mathbb{N}^*$, be two time sequences such that $\{[t_i, t_i + p_i]\}_{i=1,\ldots,N}$ is a sequence of non-overlapping real intervals. We assume that Assumptions 1–3 hold true.*

1. *We define the sequential interval reliability, $SIR^{(N)}(\underline{t}, \underline{p})$, as the probability that the system is in the up-states U during the time intervals $\{[t_i, t_i + p_i]\}_{i=1,\ldots,N}$, meaning that:*

$$SIR^{(N)}(\underline{t}, \underline{p}) := \mathbb{P}(Z_l \in U, \text{ for all } l \in [t_i, t_i + p_i], i = 1, \ldots, N); \tag{8}$$

2. *For $v \in \mathbb{N}$ and $k \in E$, we define the conditional sequential interval reliability, $SIR_k^{(N)}(v; \underline{t}, \underline{p})$, as the conditional probability that the system is in the up-states U during the time intervals $\{[t_i, t_i + p_i]\}_{i=1,\ldots,N}$, given the event $(k, v) := \{Z_0 = k, B_0 = v\} = \{J_{N(0)} = k, T_{N(0)} = -v\}$, meaning that:*

$$\begin{aligned} SIR_k^{(N)}(v; \underline{t}, \underline{p}) &:= \mathbb{P}(Z_l \in U, \text{ for all } l \in [t_i, t_i + p_i], i = 1, \ldots, N \mid Z_0 = k, B_0 = v) \\ &= \mathbb{P}_{(k,v)}(Z_l \in U, \text{ for all } l \in [t_i, t_i + p_i], i = 1, \ldots, N), \end{aligned} \tag{9}$$

where $B_t := t - T_{N(t)}$ is the backward time process associated to the semi-Markov process.

Note that we have the obvious relationship between the sequential interval reliability and the conditional sequential interval reliability:

$$SIR^{(N)}(\underline{t}, \underline{p}) = \sum_{k\in E} \mu_k SIR_k^{(N)}(0; \underline{t}, \underline{p}). \tag{10}$$

For notational convenience, we will set:

$$SIR_k^{(N)}(\underline{t}, \underline{p}) \quad := \quad SIR_k^{(N)}(0; \underline{t}, \underline{p}) = \mathbb{P}(Z_l \in U, \text{ for all } l \in [t_i, t_i + p_i], i = 1, \ldots, N \mid Z_0 = k).$$

Remark 2. *Under the previous notation, we have:*

1. *If $t_i + p_i = t_{i+1} - 1$ for all $i = 1, \ldots, N-1$, and $v = 0$, then $SIR_k^{(N)}(0; \underline{t}, \underline{p}) = IR(t_1, t_N + p_N - t_1)$;*
2. *If $t_1 = 0, t_i + p_i = t_{i+1} - 1$ for all $i = 1, \ldots, N-1, k \in U$ and $v = 0$, then $SIR_k^{(N)}(0; \underline{t}, \underline{p}) = R_k(t_N + p_N)$;*
3. *If $p_i = 0$ for all $i = 1, \ldots, N$, and $v = 0$, then $SIR_k^{(N)}(0; \underline{t}, \underline{p}) = \mathbb{P}_k(Z_{t_i} \in U, i = 1, \ldots, N) =: SA_k^{(N)}(\underline{t})$; this function denoted by $SA_k^{(N)}(\underline{t})$ can be called the sequential availability function;*
4. *If there exists a $j \in \{1, \ldots, N\}$ such that $t_i = 0$ for $i < j$ and $t_h = t_j$ for $h \geq j$, $p_i = 0$ for all $i = 1, \ldots, N$, and $v = 0$, then $SIR_k^{(N)}(0; \underline{t}, \underline{p}) = A(t_j)$, the availability function is computed in t_j.*

3.1. Transient Analysis

We will now investigate the recursive formula for computing the sequential interval reliability of a discrete-time semi-Markov system.

Proposition 1. *Let $(Z_k)_{k\in\mathbb{N}}$ be a discrete time semi-Markov system, assuming Assumptions 1–3 hold true and let $\underline{t} := (t_i)_{i=1,\ldots,N}$ and $\underline{p} := (p_i)_{i=1,\ldots,N}$, $N \in \mathbb{N}^*$, be two time sequences such that $\{[t_i, t_i + p_i]\}_{i=1,\ldots,N}$ is a sequence of non-overlapping real intervals. Let $v \in \mathbb{N}$ be the value of the backward process at time $t = 0$ and $k \in E$ be the initial state. Then, the conditional sequential interval reliability, $SIR_k^{(N)}(v; \underline{t}, \underline{p})$, satisfies the following equation:*

$$SIR_k^{(N)}(v;\underline{t},\underline{p}) = g_k^{(N)}(v;\underline{t},\underline{p}) + \sum_{r\in E}\sum_{\theta=1}^{t_1} \frac{q_{kr}(v+\theta)}{\overline{H}_k(v)} SIR_r^{(N)}(0;\underline{t}-\theta\mathbf{1}_{1:N},\underline{p}), \tag{11}$$

where $\mathbf{1}_{1:N}$ *is a vector of* 1*s of length* N*, and* $g_k^{(N)}(v;\underline{t},\underline{p})$ *is given by*

$$\begin{aligned} g_k^{(N)}(v;\underline{t},\underline{p}) &:= \mathbb{1}_{\{k\in U\}}\Bigg[\frac{\overline{H}_k(t_N+p_N+v)}{\overline{H}_k(v)} \\ &+ \sum_{\theta=t_1+1}^{t_1+p_1}\sum_{r\in E}\sum_{m\in U}\sum_{v'=0}^{t_1+p_1-\theta} \frac{q_{kr}(v+\theta)}{\overline{H}_k(v)} R^b_{rm}(v';t_1+p_1-\theta) SIR_m^{(N-1)}(v';t_{2:N}-\mathbf{1}_{2:N}(t_1+p_1),p_{2:N}) \\ &+ \sum_{j=2}^{N}\sum_{\theta=t_j}^{t_j+p_j}\sum_{r\in E} \frac{q_{kr}(v+\theta)}{\overline{H}_k(v)} SIR_r^{(N-j+1)}(0;(\theta,t_{j+1:N}-\mathbf{1}_{j+1:N}\theta),(t_j+p_j-\theta,p_{j+1:N})) \\ &+ \sum_{j=1}^{N-1}\sum_{\theta=t_j+p_j+1}^{t_{j+1}-1}\sum_{r\in E} \frac{q_{kr}(v+\theta)}{\overline{H}_k(v)} SIR_r^{(N-j)}(0;t_{j+1:N}-\mathbf{1}_{j+1:N}\theta,p_{j+1:N})\Bigg], \end{aligned} \tag{12}$$

where $\mathbb{1}_{\{k\in U\}}$ *is the indicator function of the event* $\{k\in U\}$ *and* $R^b_{ij}(v;k)$ *is the reliability with final backward defined by*

$$R^b_{ij}(v;k) := \mathbb{P}(Z_s\in U, \text{ for all } s\in\{0,\ldots,k-v\}, Z_k=j, B_k=v \mid Z_0=i, T_{N(0)}=0). \tag{13}$$

Proof. Before proceeding with the proof, let us introduce the notation:

$$Z(\underline{t},\underline{p}) := \left(Z_{t_1},\ldots,Z_{t_1+p_1},\ldots,Z_{t_2},\ldots,Z_{t_2+p_2},\ldots,Z_{t_N},\ldots,Z_{t_N+p_N}\right).$$

From the definition of the SIR, it is clear that

$$SIR_k^{(N)}(v;\underline{t},\underline{p}) = \mathbb{P}_{(k,v)}\left(Z(\underline{t},\underline{p})\in U^{N+\sum_{i=1}^N p_i}\right). \tag{14}$$

Let us consider now the r.v. T_1 and observe that the events $\{T_1>t_N+p_N\}$, $\{T_1<t_1\}$, $\{T_1\in\cup_{j=1}^N[t_j,t_j+p_j]\}$ and $\{T_1\in\cup_{j=1}^{N-1}[t_j+p_j+1,t_{j+1}-1]\}$ are mutually exclusive. Consequently, we can write (14) as follows:

$$\begin{aligned} SIR_k^{(N)}(v;\underline{t},\underline{p}) &= \mathbb{P}_{(k,v)}\left(Z(\underline{t},\underline{p})\in U^{N+\sum_{i=1}^N p_i}, T_1>t_N+p_N\right) \\ &+\mathbb{P}_{(k,v)}\left(Z(\underline{t},\underline{p})\in U^{N+\sum_{i=1}^N p_i}, T_1<t_1\right)+\mathbb{P}_{(k,v)}\left(Z(\underline{t},\underline{p})\in U^{N+\sum_{i=1}^N p_i}, T_1\in\cup_{j=1}^N[t_j,t_j+p_j]\right) \\ &+\mathbb{P}_{(k,v)}\left(Z(\underline{t},\underline{p})\in U^{N+\sum_{i=1}^N p_i}, T_1\in\cup_{j=1}^{N-1}[t_j+p_j+1,t_{j+1}-1\right). \end{aligned} \tag{15}$$

We need to compute the four terms of the right-hand side of (15); let us denote them by RT_1, RT_2, RT_3 and RT_4, respectively.

First, through a straightforward computation, we obtain

$$RT_1 = \mathbb{1}_{\{k\in U\}}\frac{\overline{H}_k(t_N+p_N+v)}{\overline{H}_k(v)}. \tag{16}$$

Second, using the double expectation formula and conditioning with respect to (J_1,T_1), we immediately obtained the second term given by

$$RT_2 = \sum_{r\in E}\sum_{\theta=1}^{t_1-1}\frac{q_{kr}(v+\theta)}{\overline{H}_k(v)} SIR_r^{(N)}(0;\underline{t}-\theta\mathbf{1}_{1:N},\underline{p}). \tag{17}$$

Third, using the double expectation formula, conditioning with respect to (J_1,T_1), summing over all the possible values of T_1 and splitting the computation according to the interval to which T_1 belongs, a quite long computation yields:

$$RT_3 = \sum_{r\in E} \frac{q_{kr}(v+t_1)}{\overline{H}_k(v)} SIR_r^{(N)}(0;\underline{t}-t_1\mathbf{1}_{1:N},\underline{p})$$
$$+\sum_{\theta=t_1+1}^{t_1+p_1}\sum_{r\in E}\sum_{m\in U}\sum_{v'=0}^{t_1+p_1-\theta} \frac{q_{kr}(v+\theta)}{\overline{H}_k(v)} R^f_{rm}(v';t_1+p_1-\theta) SIR_m^{(N-1)}(v';t_{2:N}-\mathbf{1}_{2:N}(t_1+p_1),p_{2:N})$$
$$+\sum_{j=2}^{N}\sum_{\theta=t_j}^{t_j+p_j}\sum_{r\in E} \frac{q_{kr}(v+\theta)}{\overline{H}_k(v)} SIR_r^{(N-j+1)}(0;(\theta,t_{j+1:N}-\mathbf{1}_{j+1:N}\theta),(t_j+p_j-\theta,p_{j+1:N})). \tag{18}$$

Furthermore, fourth, using the double expectation formula, conditioning with respect to (J_1, T_1) and summing over all the possible values of T_1, we obtain:

$$RT_4 = \mathbb{1}_{\{k\in U\}} \sum_{j=1}^{N-1}\sum_{\theta=t_j+p_j+1}^{t_{j+1}-1}\sum_{r\in E} \frac{q_{kr}(v+\theta)}{\overline{H}_k(v)} SIR_r^{(N-j)}(0;t_{j+1:N}-\mathbf{1}_{j+1:N}\theta,p_{j+1:N}). \tag{19}$$

Substituting these four terms in (15), we obtain the recurrence formula given in (11) and (12). □

If no initial backward is considered, taking $v = 0$ in Equation (11), we immediately obtain the following recursive formula for the sequential interval reliability of a discrete-time semi-Markov system, given the initial state.

Corollary 1. *Let $(Z_k)_{k\in\mathbb{N}}$ be a discrete time semi-Markov system, assuming Assumptions 1–3 hold true and let $\underline{t} := (t_i)_{i=1,\dots,N}$ and $\underline{p} := (p_i)_{i=1,\dots,N}$, $N \in \mathbb{N}^*$, be two time sequences such that $\{[t_i, t_i + p_i]\}_{i=1,\dots,N}$ is a sequence of non-overlapping real intervals. Then, the sequential interval reliability, $SIR^{(N)}(\underline{t},\underline{p})$, satisfies the following equation:*

$$SIR_k^{(N)}(\underline{t},\underline{p}) \quad = \quad g_k^{(N)}(\underline{t},\underline{p}) + \sum_{r\in E}\sum_{\theta=1}^{t_1} q_{kr}(\theta) SIR_r^{(N)}(\underline{t}-\theta\mathbf{1}_{1:N},\underline{p}), \tag{20}$$

where we have set $g_k^{(N)}(\underline{t},\underline{p}) := g_k^{(N)}(0;\underline{t},\underline{p})$.

The next result provides a formula for computing the reliability with the final backward $R^b_{ij}(v;k)$ defined in Equation (13).

Lemma 1. *For a discrete time semi-Markov system $(Z_k)_{k\in\mathbb{N}}$, under Assumptions 1–3, let us define the entrance probabilities $e_{ij}(n)$, $i, j \in E, n \in \mathbb{N}$ by $e_{ij}(n) =$ the probability that the system that entered state i at time 0 will enter state j at time n. Under the previous notations, the reliability with the final backward $R^b_{ij}(v;k)$ defined in Equation (13) is given by*

$$R^b_{ij}(v;k) = \overline{H}_j(v) e^{\widetilde{q}}_{ij}(k-v), \tag{21}$$

where $e^{\widetilde{q}}_{ij}(k-v)$ represents the entrance probabilities for the semi-Markov system (cf. [27]) associated with the semi-Markov core matrix:

$$\widetilde{q}(k) = \begin{pmatrix} q_{UU}(k) & q_{UD}(k)\,\mathbf{1}_{s-s_1} \\ \mathbf{0}_{1s_1} & 0 \end{pmatrix}, \ k \in \mathbb{N},$$

with $q_{UU}(k)$ and $q_{UD}(k)$ being the partitions of the matrix $q(k)$ according to $U \times U$ and $U \times D$.

Proof. First, it can be easily seen that:

$$R^b_{ij}(v;k) = \overline{H}_j(v)\mathbb{P}(Z_{k-v} = j, Z_{k-v-1} \neq j, Z_s \in U, s = 0,1,\dots,k-v \mid Z_0 = i).$$

Second, note that $\mathbb{P}(Z_{k-v} = j, Z_{k-v-1} \neq j, Z_s \in U, s = 0, 1, \ldots, k-v \mid Z_0 = i), i, j \in E$, represent the entrance probabilities for the semi-Markov system associated to the *semi-Markov core matrix* $\widetilde{q}(k)$. See also Proposition 5.1 of [1] for the use of the semi-Markov system associated to the *semi-Markov core matrix* $\widetilde{q}(k)$ in reliability computation.

Third, in order to compute the entrance probabilities, one can use the recurrence formulas (see [27]):

$$e_{ij}(n) = \delta_{ij}\delta(n) + \sum_{r=1}^{s}\sum_{m=1}^{n} p_{ir} f_{ir}(m) e_{rj}(n-m), \tag{22}$$

where $\delta_{ij} := 1$ if $i = j$, $\delta_{ij} := 0$ if $i \neq j$, $\delta(n) := 1$ if $n = 0$, $\delta(n) = 0$ if $n \neq 0$. □

Looking at the recurrence relationship given in Proposition 1 for computing the conditional sequential interval reliability with initial backward $SIR_k^{(N)}(v; \underline{t}, \underline{p})$, and taking into account that, for $N = 1$, we obtain the interval reliability with the initial backward, and see that we need a formula for computing the interval reliability with the initial backward, denoted by $IR_k(v; t, p)$ and defined by

$$IR_k(v; t, p) := \mathbb{P}(Z_l \in U, l \in [t, t+p] \mid Z_0 = k, B_0 = v). \tag{23}$$

The next result provides a formula for computing this quantity.

Lemma 2. *Under the previous notations, the interval reliability with initial backward $IR_k(v; t, p)$ is given by*

$$\begin{aligned} IR_k(v; t, p) &= \frac{1}{\overline{H}_k(v)}\Bigg[\overline{H}_k(v+t+p)\mathbb{1}_{\{k \in U\}} + \sum_{j \in U}\sum_{\theta=t}^{t+p} q_{kj}(v+\theta) R_j(t+p-\theta)\mathbb{1}_{\{k \in U\}} \\ &+ \sum_{j \in E}\sum_{\theta=1}^{t-1} q_{kj}(v+\theta) IR_j(t-\theta, p)\Bigg]. \end{aligned} \tag{24}$$

Proof. The proof is a quite straightforward adaptation of a more general result presented in [21]. □

The next result provides a series of inequalities between sequential interval reliability, sequential availability, conditional reliability and conditional availability.

Proposition 2. *Let $(Z_k)_{k \in \mathbb{N}}$ be a discrete time semi-Markov system, assuming Assumptions 1–3 hold true, and let $k \in E$ and $v \in \mathbb{N}$.*

1. *For any $t_{1:N}$ and $p_{1:N}$, $N \in \mathbb{N}^*$, such that $\{[t_i, t_i + p_i]\}_{i=1,\ldots,N}$ is a sequence of non-overlapping real intervals, and for $s \leq N$ we have:*

$$\begin{aligned} R_k(v, t_N + p_N) &\leq SIR_k^{(N)}(v; t_{1:N}, p_{1:N}) \leq SIR_k^{(N)}(v; t_{1:s}, p_{1:s}) \leq SA_k^{(N)}(v; t_{1:N} + p_{1:N}) \\ &\leq SA_k^{(s)}(v; t_{1:s} + p_{1:s}) \leq A_k(v; t_s + p_s). \end{aligned} \tag{25}$$

2. *For any $t_{1:N}$, $a_{1:N}$ and $b_{1:N}$, $N \in \mathbb{N}^*$, such that $a_i \leq b_i, i = 1, \ldots, N$, and $\{[t_i, t_i + a_i]\}_{i=1,\ldots,N}$ and $\{[t_i, t_i + b_i]\}_{i=1,\ldots,N}$ are two sequences of non-overlapping real intervals, then we have:*

$$SIR_k^{(N)}(v; t_{1:N}, b_{1:N}) \leq SIR_k^{(N)}(v; t_{1:N}, a_{1:N}). \tag{26}$$

3. *For any* $t_{1:N}, p_{1:N}, x_{1:N}, w_{1:N}$, $N \in \mathbb{N}^*$, *such that* $\{[t_i, t_i + p_i]\}_{i=1,\ldots,N}$ *and* $\{[x_i, x_i + w_i]\}_{i=1,\ldots,N}$ *are two sequences of non-overlapping real intervals such that* $t_{1:N} + p_{1:N} = x_{1:N} + w_{1:N}$, *and* $t_{1:N} \geq x_{1:N}$ *(element-wise), then we have:*

$$SIR_k^{(N)}(v; x_{1:N}, w_{1:N}) \leq SIR_k^{(N)}(v; t_{1:N}, p_{1:N}). \tag{27}$$

Proof. For any $k \in E$ and $v \in \mathbb{N}$, let us define the set $\Omega_{(k,v)}$ by

$$\Omega_{(k,v)} := \{\omega \in \Omega \mid Z_0(\omega) = k, B_0(\omega) = v\}.$$

The first point is obtained noticing that, for $s \leq N$, we have:

$$\begin{aligned}
&\left\{\omega \in \Omega_{(k,v)} \mid Z_s \in U, \forall s = 1, \ldots, t_N + p_N\right\} \subseteq \left\{\omega \in \Omega_{(k,v)} \mid Z(t_{1:N}, p_{1:N}) \in U^{N+\sum_{i=1}^N p_i}\right\} \\
&\subseteq \left\{\omega \in \Omega_{(k,v)} \mid Z(t_{1:s}, p_{1:s}) \in U^{s+\sum_{i=1}^s p_i}\right\} \subseteq \left\{\omega \in \Omega_{(k,v)} \mid Z(t_{1:N} + p_{1:N}, 0_{1:N}) \in U^N\right\} \\
&\subseteq \left\{\omega \in \Omega_{(k,v)} \mid Z(t_{1:s} + p_{1:s}, 0_{1:s}) \in U^s\right\} \subseteq \left\{\omega \in \Omega_{(k,v)} \mid Z_{t_s+p_s} \in U\right\}.
\end{aligned}$$

Applying the probability on this chain of inequalities and taking into account the definitions of reliability, sequential reliability, availability and sequential availability, we obtain the inequalities given in (25).

In order to prove the second point, we first observe that $a_i \leq b_i, i = 1, \ldots, N$, implies that the two sequences of non-overlapping real intervals $\{[t_i, t_i + a_i]\}_{i-1,\ldots,N}$ and $\{[t_i, t_i + b_i]\}_{i=1,\ldots,N}$ are such that $[t_i, t_i + a_i] \subseteq [t_i, t_i + b_i], i = 1, \ldots, N$. Thus, we have:

$$\left\{\omega \in \Omega_{(k,v)} \mid Z(t_{1:N}, b_{1:N}) \in U^{N+\sum_{i=1}^N b_i}\right\} \subseteq \left\{\omega \in \Omega_{(k,v)} \mid Z(t_{1:N}, a_{1:N}) \in U^{N+\sum_{i=1}^N a_i}\right\},$$

which implies that $SIR_k^{(N)}(v; t_{1:N}, b_{1:N}) \leq SIR_k^{(N)}(v; t_{1:N}, a_{1:N})$, so we obtain (26).

To prove the last point, since $t_{1:N} \geq x_{1:N}$ and $t_{1:N} + p_{1:N} = x_{1:N} + w_{1:N}$, we have that $[t_i, t_i + p_i] \subseteq [x_i, x_i + w_i], i = 1, \ldots, N$. Consequently, we have:

$$\left\{\omega \in \Omega_{(k,v)} \mid Z(x_{1:N}, w_{1:N}) \in U^{N+\sum_{i=1}^N w_i}\right\} \subseteq \left\{\omega \in \Omega_{(k,v)} \mid Z(t_{1:N}, p_{1:N}) \in U^{N+\sum_{i=1}^N p_i}\right\},$$

which implies that $SIR_k^{(N)}(v; x_{1:N}, w_{1:N}) \leq SIR_k^{(N)}(v; t_{1:N}, p_{1:N})$, so we obtain (27). □

3.2. Asymptotic Analysis

We can investigate the asymptotic analysis of sequential interval reliability $SIR_k^{(N)}(v; t_{1:N}, b_{1:N})$, by letting t_1 tend towards infinity. The next result given in Theorem 1 answers this question.

Let $\underline{t} := (t_i)_{i=1,\ldots,N}$ and $\underline{p} := (p_i)_{i=1,\ldots,N}$, $N \in \mathbb{N}^*$, be two time sequences such that $\{[t_i, t_i + p_i]\}_{i=1,\ldots,andN}$ is a sequence of non-overlapping real intervals. Let us denote by $l_i := t_i - t_{i-1}, i = 2, \ldots, N$.

Theorem 1. *Let us consider an ergodic semi-Markov chain such that Assumptions 1–3 hold true and the mean sojourn times* m_i *in any state i are finite,* $m_i < \infty, i \in E$*, where* m_i *is the mean time of the distribution* $(h_i(k))_{k\in\mathbb{N}}, i \in E$*. Then, under the previous notations, we have:*

$$\lim_{t_1\to\infty} SIR_k^{(N)}(v; \underline{t}, \underline{p}) = \lim_{t_1\to\infty} SIR^{(N)}(\underline{t}, \underline{p}) = \frac{1}{\sum_{i\in E} \nu(i) m_i} \sum_{j\in U} \nu(j) \sum_{t_1\geq 0} g_j^{(N)}(\underline{t}, \underline{p}), \tag{28}$$

where $(\nu(i))_{i\in E}$ *represents the stationary distribution of the embedded Markov chain* $(J_n)_{n\in\mathbb{N}}$.

Before giving the proof of this result, we first need some preliminary notions and results.

First, let us recall some definitions related to the matrix convolution product. Let us denote by $\mathcal{M}_E$ the set of real matrices on $E \times E$ and by $\mathcal{M}_E(\mathbb{N})$ the set of matrix-valued functions defined on $\mathbb{N}$, with values in $\mathcal{M}_E$. For $\mathbf{A} \in \mathcal{M}_E(\mathbb{N})$, we write $\mathbf{A} = (\mathbf{A}(k);\ k \in \mathbb{N})$, where, for $k \in \mathbb{N}$ fixed, $\mathbf{A}(k) = (A_{ij}(k);\ i,j \in E) \in \mathcal{M}_E$. Let $\mathbf{I} \in \mathcal{M}_E$ be the identity matrix and $\mathbf{0} \in \mathcal{M}_E$ be the null matrix. Let us also define $\mathbf{I} := (\mathbf{I}(k);\ k \in \mathbb{N})$ as the constant matrix-valued function whose value for any nonnegative integer k is the identity matrix, that is, $\mathbf{I}(k) := \mathbf{I}$ for any $k \in \mathbb{N}$. Similarly, we set $\mathbf{0} := (\mathbf{0}(k);\ k \in \mathbb{N})$, with $\mathbf{0}(k) := \mathbf{0}$ for any $k \in \mathbb{N}$.

Let $\mathbf{A}, \mathbf{B} \in \mathcal{M}_E(\mathbb{N})$ be two matrix-valued functions. The matrix convolution product $\mathbf{A} * \mathbf{B}$ is the matrix-valued function $\mathbf{C} \in \mathcal{M}_E(\mathbb{N})$ defined by

$$C_{ij}(k) := \sum_{r\in E}\sum_{l=0}^{k} A_{ir}(k-l)\, B_{rj}(l), \quad i,j \in E, \quad k \in \mathbb{N}, \tag{29}$$

or, in matrix form:

$$\mathbf{C}(k) := \sum_{l=0}^{k} \mathbf{A}(k-l)\, \mathbf{B}(l), \quad k \in \mathbb{N}.$$

It can be easily checked whether the identity element for the matrix convolution product in discrete time exists, and whether it is unique and given by $\delta I = (d_{ij}(k);\ i,j \in E) \in \mathcal{M}_E(\mathbb{N})$ defined by

$$d_{ij}(k) := \begin{cases} 1, & \text{if } i = j \text{ and } k = 0, \\ 0, & \text{elsewhere,} \end{cases}$$

or, in matrix form:

$$\delta I(k) := \begin{cases} \mathbf{I}, & \text{if } k = 0, \\ \mathbf{0}, & \text{elsewhere.} \end{cases}$$

The power in the sense of convolution is straightforwardly defined using the previous definition of the matrix convolution product given in (29). For $\mathbf{A} \in \mathcal{M}_E(\mathbb{N})$, a matrix-valued function and $n \in \mathbb{N}$, the n-fold convolution $\mathbf{A}^{(n)}$ is the matrix-valued function recursively defined by

$$\begin{aligned} A_{ij}^{(0)}(k) &:= d_{ij}(k) = \begin{cases} 1, & \text{if } i = j \text{ and } k = 0, \\ 0, & \text{elsewhere,} \end{cases} \\ A_{ij}^{(1)}(k) &:= A_{ij}(k), \end{aligned}$$

$$A_{ij}^{(n)}(k) := \sum_{r\in E}\sum_{l=0}^{k} A_{ir}(l)\, A_{rj}^{(n-1)}(k-l), \quad n \geq 2, k \in \mathbb{N},$$

that is:

$$\mathbf{A}^{(0)} := \delta I,\ \mathbf{A}^{(1)} := \mathbf{A} \text{ and } \mathbf{A}^{(n)} := \mathbf{A} * \mathbf{A}^{(n-1)}, \quad n \geq 2.$$

Second, let us introduce two sets of functions that will be useful for our study. Thus, let us define:

$$\mathcal{A} := \left\{ f : \{(\underline{t}, \underline{p}) \in \mathbb{N}^N \times \mathbb{N}^N \mid t_i \leq t_{i+1}, i = 1, \ldots, N-1\} \to \mathbb{R} \right\} \tag{30}$$

and, for $\underline{l} \in \mathbb{N}^{N-1}$, $\underline{p} \in \mathbb{N}^N$:

$$\mathcal{B}_{\underline{l},\underline{p}} := \left\{ \widetilde{f} : \mathbb{N} \to \mathbb{R} \mid \widetilde{f}(t_1) = \widetilde{f}(t_1; \underline{l}, \underline{p}) \right\}, \tag{31}$$

where, by writing $\widetilde{f}(t_1; \underline{l}, \underline{p})$, we mean that the function $\widetilde{f}$ is a function of the variable t_1, while $\underline{l}, \underline{p}$ are some parameters.

Let us consider a map between the two sets, $\Phi : \mathcal{A} \to \mathcal{B}_{\underline{l},\underline{p}}$ defined by

$$\Phi(f(\underline{t}, \underline{p})) := \widetilde{f}(t_1; \underline{l}, \underline{p}), \tag{32}$$

where $l_i := t_i - t_{i-1}, i = 2, \ldots, N$.

The map Φ allows to represent a function $f(\underline{t}, \underline{p}) \in \mathcal{A}$ as an element of the set $\mathcal{B}_{\underline{l},\underline{p}}$, that is to say as a parametric function of one variable, namely t_1. One can easily check that Φ is bijective and linear.

The last point before giving the proof of Theorem 1 will be to introduce a new matrix convolution product, important for our framework, and to see the relationship with the classical matrix convolution product.

Definition 5. *Let $\mathbf{A} \in \mathcal{M}_E(\mathbb{N})$ be a matrix-valued function and let $\mathbf{b} = (b_1, \ldots, b_s)$ be a vector-valued function such that every component $b_r \in \mathcal{A}, r \in E$. The matrix convolution product $\overline{*}$ is defined by*

$$(\mathbf{A} \overline{*} \mathbf{b})_k(\underline{t}, \underline{p}) := \sum_{r \in E} \sum_{\theta=1}^{t_1} A_{kr}(\theta) b_r(\underline{t} - \theta \mathbf{1}_{1:N}, \underline{p}),$$

or, in matrix form:

$$(\mathbf{A} \overline{*} \mathbf{b})(\underline{t}, \underline{p}) := \sum_{\theta=1}^{t_1} \mathbf{A}(\theta) \mathbf{b}(\underline{t} - \theta \mathbf{1}_{1:N}, \underline{p}).$$

The next result will give a relationship between this new introduced matrix convolution product (cf. Definition 5) and the classical one defined in (29).

Proposition 3. *Let $\mathbf{q} \in \mathcal{M}_E(\mathbb{N})$ be a semi-Markov* semi-Markov core matrix *and let $\mathbf{f} = (f_1, \ldots, f_s)$ be a vector-valued function such that every component $f_r \in \mathcal{A}, r \in E$. Then:*

$$\Phi((\mathbf{q} \overline{*} \mathbf{f})(\underline{t}, \underline{p})) = (\mathbf{q} * \widetilde{\mathbf{f}})(t_1; l_{2:N}, \underline{p}).$$

Proof. From the additivity of the map Φ, we have:

$$\Phi((\mathbf{q} \overline{*} \mathbf{f})(\underline{t}, \underline{p}))_k = \sum_{r \in E} \sum_{\theta=1}^{t_1} \Phi(q_{kr}(\theta) f_r(\underline{t} - \theta \mathbf{1}_{1:N}, \underline{p})) = \sum_{r \in E} \sum_{\theta=1}^{t_1} q_{kr}(\theta) f_r(\underline{t} - \theta \mathbf{1}_{1:N}, \underline{p})$$

$$= \sum_{r \in E} \sum_{\theta=1}^{t_1} q_{kr}(\theta) \widetilde{f}_r(t_1 - \theta; l_{2:N}, \underline{p}) = (\mathbf{q} * \widetilde{\mathbf{f}})_k(t_1; l_{2:N}, \underline{p}).$$

□

Proof of Theorem 1. First of all, it is important to notice that we have:

$$\lim_{t_1 \to \infty} SIR_k^{(N)}(v; t_{1:N}, p_{1:N}) = \lim_{t_1 \to \infty} SIR^{(N)}(t_{1:N}, p_{1:N}),$$

provided that this limit exists. Consequently, since our interest now is in a limiting result, in order to investigate the asymptotic behaviour of $SIR_k^{(N)}(v; t_{1:N}, p_{1:N})$ as t_1 goes to ∞ we can consider the initial backward $v = 0$. Thus, the expression of sequential interval reliability that we will take into account in the next computations will be $SIR^{(N)}(t_{1:N}, p_{1:N})$, that is recurrently obtained through Relation (20).

The main idea of this proof is to consider $SIR_k^{(N)}$ as a function of the variable t_1 and also on other additional parameters; then, we will apply the Markov renewal theory (cf. [1])

to this function of t_1. Using Proposition 3 and applying the function Φ defined in (32) to the left and right hand sides of Equation (20), we obtain:

$$\widetilde{SIR}^{(N)}(t_1; l_{2:N}, \underline{p}) = \widetilde{g}(t_1; l_{2:N}, \underline{p}) + \mathbf{q} * \widetilde{SIR}^{(N)}(t_1; l_{2:N}, \underline{p}), \tag{33}$$

where $\widetilde{SIR}^{(N)} := \Phi(SIR^{(N)})$ and $\widetilde{g} := \Phi(g)$.

It is clear that Equation (33) is an ordinary Markov renewal equation (MRE) in variable t_1, with parameters $(l_{2:N}, \underline{p})$. The solution of this MRE is well known (cf. [1]) and it is given by

$$\widetilde{SIR}^{(N)}(t_1; l_{2:N}, \underline{p}) = (\boldsymbol{\psi} * \widetilde{g})(t_1; l_{2:N}, \underline{p}), \tag{34}$$

or element-wise, using Proposition 3:

$$\widetilde{SIR}_k^{(N)}(t_1; l_{2:N}, \underline{p}) = \sum_{r \in E} \sum_{\theta=1}^{t_1} \psi_{kr}(\theta) g_r(\underline{t} - \theta \mathbf{1}_{1:N}, \underline{p}), \tag{35}$$

where the matrix-valued function $\boldsymbol{\psi} = (\boldsymbol{\psi}(k);\ k \in \mathbb{N})$ is given by

$$\boldsymbol{\psi}(k) = \sum_{n=0}^{k} \mathbf{q}^{(n)}(k),\ k \in \mathbb{N}. \tag{36}$$

Since $\widetilde{SIR}_k^{(N)}(t_1; l_{2:N}, \underline{p}) = \Phi(SIR_k^{(N)}(\underline{t}, \underline{p})) = SIR_k^{(N)}(\underline{t}, \underline{p})$, we obtain:

$$SIR_k^{(N)}(\underline{t}, \underline{p}) = \widetilde{SIR}_k^{(N)}(t_1; l_{2:N}, \underline{p}) = \sum_{r \in E} \sum_{\theta=1}^{t_1} \psi_{kr}(\theta) g_r(\underline{t} - \theta \mathbf{1}_{1:N}, \underline{p}). \tag{37}$$

Consequently:

$$\lim_{t_1 \to \infty} SIR^{(N)}(\underline{t}, \underline{p}) = \lim_{t_1 \to \infty} \widetilde{SIR}_k^{(N)}(t_1; l_{2:N}, \underline{p}).$$

Let us now compute the second limit using the key Markov renewal Theorem (cf. [1]). First, we observe that:

$$\widetilde{g}_k(t_1; l_{2:N}, \underline{p}) = g_k(\underline{t}, \underline{p}) = \mathbb{P}_{(k,0)}(Z_l \in U, \text{ for all } l \in [t_i, t_i + p_i], i = 1, \ldots, N, T_1 > t_1)$$
$$\leq \mathbb{P}_{(k,0)}(Z_l \in U, l \in [t_1, t_1 + p_1], T_1 > t_1) = g_k(t_1, p_1) \leq R_k(t_1 + p_1).$$

Using this result, we have:

$$\sum_{t_1 \geq 0} \mid \widetilde{g}_k(t_1; l_{2:N}, \underline{p}) \mid \leq \sum_{t_1 \geq 0} R_k(t_1 + p_1) = \mathbb{E}_{(k,0)}(T_D),$$

where T_D is the lifetime of the system. Thus, we are under the hypotheses of the key Markov renewal theorem and we obtain:

$$\lim_{t_1 \to \infty} SIR^{(N)}(\underline{t}, \underline{p}) = \lim_{t_1 \to \infty} \mu \psi * \widetilde{g}(t_1; l_{2:N}, \underline{p})$$
$$= \sum_{i \in E} \mu_i \sum_{j \in U} \frac{1}{\mu_{jj}} \sum_{t_1 \geq 0} g_j^{(N)}(\underline{t}, \underline{p}) = \frac{1}{\sum_{i \in E} \nu(i) m_i} \sum_{j \in U} \nu(j) \sum_{t_1 \geq 0} g_j^{(N)}(\underline{t}, \underline{p}),$$

where μ_{jj} is the mean recurrence time to state j for the semi-Markov chain. □

4. A Numerical Example

In this section, we will present a numerical example considering a semi-Markov model that governs a repairable system. The setting is as follows: the state space is $E = \{1, 2, 3\}$, the operational states are the first two, $U = \{1, 2\}$, and the non-working state is the last one, $D = \{3\}$.

The transitions of the repairable semi-Markov model are given in the flowgraph of Figure 1.

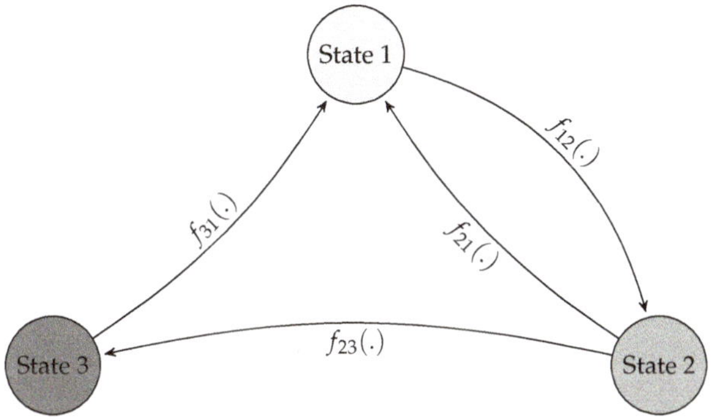

Figure 1. Semi-Markov model.

The transition matrix $\mathbf{p}$ of the EMC J and the initial distribution μ are given by

$$\mathbf{p} = \begin{pmatrix} 0 & 1 & 0 \\ 0.8 & 0 & 0.2 \\ 1 & 0 & 0 \end{pmatrix}, \ \mu = (1, 0, 0).$$

Now, let X_{ij} be the conditional sojourn time of the SMC Z in state i given that the next state is j $(j \neq i)$. The conditional sojourn times are given as follows:

$$\begin{aligned} X_{12} &\sim \text{Geometric}(0.2), \\ X_{21} &\sim \text{discrete Weibull}(0.8, 1.2), \\ X_{23} &\sim \text{discrete Weibull}(0.6, 1.2), \\ X_{31} &\sim \text{discrete Weibull}(0.9, 1.2), \end{aligned}$$

In the following figures, we investigate the semi-Markov repairable system in terms of the proposed reliability measures studied in Section 3.

Figure 2 illustrates the conditional sequential interval reliability for two time intervals moving equally through the time with the same length (one time unit). That is the probability that the system will be operational in the time intervals $(k, k+1)$ and $(k+2, k+3)$ for $k \in \{1, 2, 3, 4, 5, 6, 7, 8\}$. We have to note that, as the time k passes, then the system tends to converge to the asymptotic sequential interval reliability $SIR^{(2)}(0; \underline{t}, \underline{p}) = 0.6603$, as given in Theorem 1. Furthermore, the sequential interval reliability $SIR^{(2)}(0; (k, k+2), (1, 1))$ is equal to the conditional one $SIR_1^{(2)}(0; (k, k+2), (1, 1))$ due to the fact that the only possible initial state is the first one. The point here is to study the probability of a system being operational during different time periods with the same working duration and a fixed time distance between them, equal to one time unit.

Then, Figure 3 examines the probability that the system is working during two time periods, with same working duration (equal to one time unit) and increasing the time distance between them. We considered the first interval to be fixed equal to $(1, 2)$ and the other moving apart with step 1 each time and be $(k+2, k+3)$ for $k \in \{1, 2, 3, 4, 5, 6, 7, 8\}$. It can be easily seen that the probability of the system will be still working in both time intervals and is a decreasing function of time k, which means that the two time intervals are far enough apart for the system to be operational. Note that the probability of the system starting from the non-working state 3 to be in the up-states is sufficiently small.

As in all the previous cases, the concept of the initial backward did not play a role in the simulations (Figures 2 and 3), and in Figure 4, the sequential interval reliability with the initial backward $v = 10$ is examined. The first time interval is considered as fixed, $(1,2)$, and the other one is moving apart with step 1 each time and it is $(k+3, k+4)$ for $k \in \{1,2,3,4,5,6,7,8\}$.

From the point of view of real applications, the proposed reliability measures can be applied to a huge variety of physical phenomena which they characterised from time dependence. In the literature, a lot of research works are presented for modelling a variety of such phenomena via semi-Markov processes, from financial [23] to power demand [29]. D'Amico et al. ([23,26]) proposed a semi-Markov model and associated reliability measures for constructing a credit risk model that solves problems arising from the non-Markovianity nature of the phenomenon. Further developments and recent contributions on this aspect are provided in [28,30,31].

The characteristics of semi-Markov chains which allow for no-memoryless sojourn time distributions, permit considering the duration problem in an effective way. Indeed, it is possible to define and compute different probabilities of changing state, default probabilities included, taking into account the permanence of time in a rating class. This aspect is crucial in credit rating studies because the duration dependence of transition probabilities naturally translates in many financial indicators, which change their values according to the time elapsed in the last rating class, see, e.g., [32].

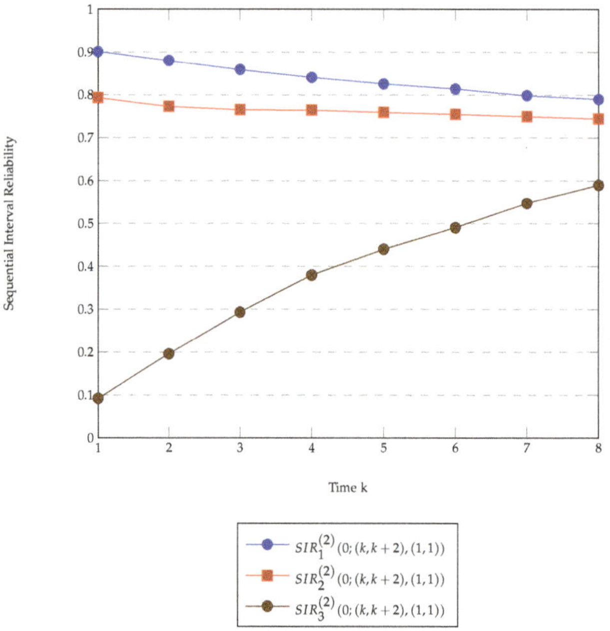

Figure 2. Sequential interval reliability plot with equally moving intervals.

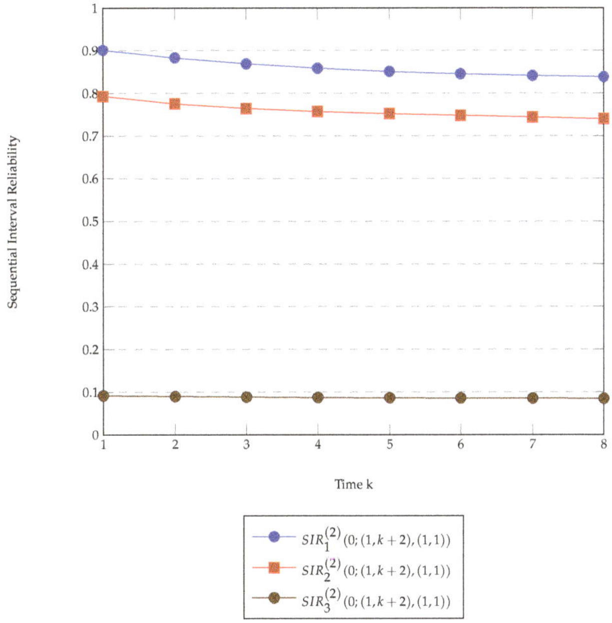

Figure 3. Sequential interval reliability plot with intervals moving apart.

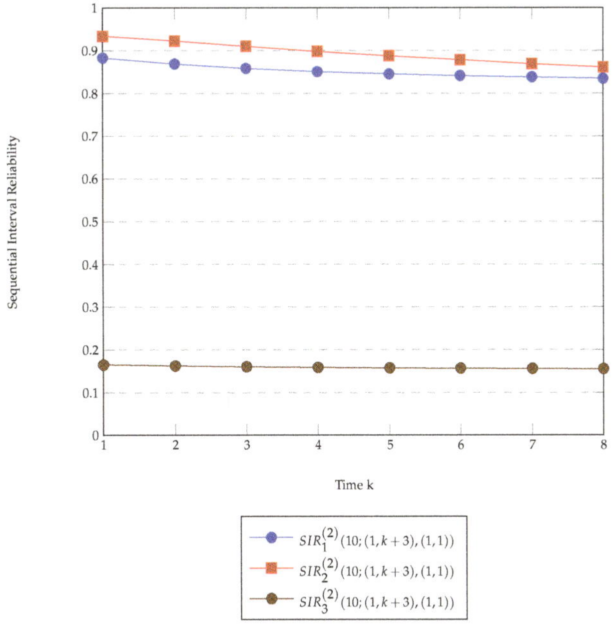

Figure 4. Sequential interval reliability plot with the initial backward ($v = 10$) and intervals moving apart.

In the case of financial modelling (see, e.g., [33]), the presented results could be applied in order to create advanced credit scoring models. A financial asset, similarly to government bond, usually takes a "grade" based on the reliability of the country to pay the debt (also known as creditworthiness). These grades strongly affect the interest rates of the country's debt and they are clearly separated. Let us consider the following set of states, from a simple point of view, as the ratings:

$$E = \{A,\ B,\ C\}.$$

If the country bond receives a rating in the set $U = \{A,\ B\}$, this means that it is thought to be creditworthy and can borrow money from the markets with reasonable interest rates. On the contrary, if the rating is within the set $D = \{C\}$, then the country cannot borrow money from the markets due to very high interest rates caused from its problems for repaying the debt. D'Amico et al. [23] proposed flexible reliability measures based on semi-Markov processes for constructing a credit risk model that solves the problems arising from the non-Markovianity nature of the phenomenon. Following that work, the measures proposed in the present paper can be applied in the same way as follows:

- The conditional sequential interval reliability with final backward $SIR_k^{(N)}(v;\underline{t},\underline{p})$ gives the probability that the bond remains creditworthy in a sequence of different time intervals $\{[t_i, t_i + p_i]\}_{i=1,\ldots,N}$ given its current rating and assuming a secondary process B_t. This measure allows us to estimate the credit risk of the asset for different time periods and at the same time by knowing the complete trajectory of the system due to the backward process B_t.
- The sequential interval reliability $SIR^{(N)}(v;\underline{t},\underline{p})$ which gives the probability that the bond remains creditworthy in a sequence of different time intervals $\{[t_i, t_i + p_i]\}_{i=1,\ldots,N}$ without taking into account the current rating. To implement this measure, we have to know the initial probability of the system being in each state.

It became clear that these measures have applications in a variety of stochastic phenomena due to their flexibility and their ability to significantly extend our knowledge about the process evolution. They can solve problems and provide answers for systems that can shift from failure to operational states, from a probabilistic point of view. Finally, they can be considered as generalisations of the classical measures of reliability for semi-Markov processes.

5. Concluding Remarks

This paper presents a new reliability indicator, called sequential interval reliability (SIR), which is evaluated for a discrete-time homogeneous semi-Markov repairable system. This indicator includes as particular cases several functions that are frequently used in reliability studies, such as the reliability and availability functions, as well as the interval reliability function.

The paper contains new theoretical results on both the transient and the asymptotic cases. More precisely, a recurrent-type equation is established for the calculation of the SIR function in the transient case and a limit theorem establishes its asymptotic behaviour. These results generalise corresponding known results for standard reliability indicators. The possibility to apply our results to real systems is shown by implementing a numerical example where the theoretical results are illustrated from a practical point of view. The paper leaves unresolved several aspects among which an important role is played by the application of the proposed indicator in different applied problems involving the use of real data.

Author Contributions: Conceptualization, G.D.; data curation, T.G.; formal analysis, V.S.B. and G.D.; methodology, G.D.; software, V.S.B. and T.G.; supervision, V.S.B.; validation, V.S.B. and T.G.; writing–original draft, V.S.B. and T.G. All authors have read and agreed to the published version of the manuscript.

Funding: Guglielmo D'Amico acknowledges the financial support by the University G. d'Annunzio of Chieti-Pescara, Italy (FAR2020).

Institutional Review Board Statement: Not applicable.

Informed Consent Statement: Not applicable.

Data Availability Statement: Not applicable.

Acknowledgments: The authors would like to express their gratitude to Professors Panagiotis-Christos Vassiliou and Andreas C. Georgiou for the opportunity to submit the present manuscript to the Special Issue *Markov and Semi-Markov Chains, Processes, Systems and Emerging Related Fields* for possible publication. The authors also wish to express their appreciation to the anonymous referees as well as to the editorial assistance of Devin Zhang for the time dedicated to our paper and for the valuable suggestions and constructive comments that improved both the quality and the presentation of the manuscript.

Conflicts of Interest: The authors declare no conflict of interest.

References

1. Barbu, V.S.; Limnios, N. *Semi-Markov Chains and Hidden Semi-Markov Models toward Applications—Their Use in Reliability and DNA Analysis*; Lecture Notes in Statistics; Springer: New York, NY, USA, 2008; Volume 191.
2. Janssen, J.; Manca, R. *Applied Semi-Markov Processes*; Springer: New York, NY, USA, 2006.
3. Limnios, N.; Oprisan, G. *Semi-Markov Processes and Reliability*; Birkhäuser: Boston, UK, 2001.
4. McClean, S. Semi-Markov models for manpower planning. In *Semi-Markov Models*; Janssen, J., Ed.; Springer: Berlin/Heidelberg, Germany, 1986; pp. 283–300.
5. Papadopoulou, A.; Vassiliou, P.-C.G. Asymptotic behaviour of nonhomogeneous semi-Markov systems. *Linear Algebra Its Appl.* **1994**, *210*, 153–198. [CrossRef]
6. Silvestrov, D.; Silvestrov, S. *Nonlinearly Perturbed Semi-Markov Processes*; Springer: Berlin/Heidelberg, Germany, 2017.
7. Barbu, V.S.; Boussemart, M.; Limnios, N. Discrete-time semi-Markov model for reliability and survival analysis. *Commun. Stat. Theory Methods* **2004**, *33*, 2833–2868. [CrossRef]
8. Barbu, V.S.; Limnios, N. Discrete time semi-Markov processes for reliability and survival analysis—A nonparametric estimation approach. In *Parametric and Semiparametric Models with Applications to Reliability, Survival Analysis and Quality of Life*; Balakrishnan, N., Nikulin, M., Mesbah, M., Limnios, N., Eds.; Birkhäuser: Basel, Switzerland, 2004; pp. 487–502.
9. D'Amico, G. Single-use reliability computation of a semi-Markovian system. *Appl. Math.* **2014**, *59*, 571–588. [CrossRef]
10. Trevezas, S.; Limnios, N. Exact MLE and asymptotic properties for nonparametric semi-Markov models. *J. Nonparametr. Stat.* **2011**, *23*, 719–739. [CrossRef]
11. Limnios, N. Dependability analysis of semi-Markov systems. *Reliab. Eng. Syst. Saf.* **1997**, *55*, 203–207. [CrossRef]
12. Limnios, N.; Ouhbi, B. Nonparametric estimation of some important indicators in reliability for semi-Markov processes. *Stat. Methodol.* **2006**, *3*, 341–350. [CrossRef]
13. Ouhbi, B.; Limnios, N. Nonparametric reliability estimation of semi-Markov processes. *J. Stat. Plan. Inference* **2003**, *109*, 155–165. [CrossRef]
14. Votsi, I.; Gayraud, G.; Barbu, V.S.; Limnios, N. Hypotheses testing and posterior concentration rates for semi-Markov processes. In *Statistical Inference for Stochastic Processes*; Springer: Berlin/Heidelberg, Germany, 2021.
15. Wang, Y.; Shahidehpour, M.; Guo, C. Applications of survival functions to continuous semi-Markov processes for measuring reliability of power transformers. *J. Mod. Power Syst. Clean Energy* **2017**, *5* 959–969. [CrossRef]
16. Blasi, A.; Janssen, J.; Manca, R. Numerical treatment of homogeneous and non-homogeneous semi-Markov reliability models. *Commun. Stat. Theory Methods* **2004**, *33*, 697–714. [CrossRef]
17. Corradi, G.; Janssen, J.; Manca, R. Numerical treatment of homogeneous semi-Markov processes in transient case—A straightforward approach. *Methodol. Comput. Appl. Probab.* **2004**, *6*, 233–246. [CrossRef]
18. Moura, M.D.C.; Droguett, E.L. Mathematical formulation and numerical treatment based on transition frequency densities and quadrature methods for non-homogeneous semi-Markov processes. *Reliab. Eng. Syst. Saf.* **2009**, *94*, 342–349. [CrossRef]
19. Wu, B.; Maya, B.I.G.; Limnios, N. Using semi-Markov chains to solve semi-Markov processes. In *Methodology and Computing in Applied Probability*; Springer: Berlin/Heidelberg, Germany, 2020.
20. Georgiadis, S.; Limnios, N. Interval reliability for semi-Markov systems in discrete time. *J. Soc. Fr. Stat.* **2014**, *153*, 152–166.
21. D'Amico, G.; Manca, R.; Petroni, F.; Selvamuthu, D. On the computation of some interval reliability indicators for semi-Markov systems. *Mathematics* **2021**, *9*, 575. [CrossRef]
22. Georgiadis, S.; Limnios, N. Nonparametric estimation of interval reliability for discrete-time semi-Markov systems. *J. Stat. Theory Pr.* **2015**, *10*, 20–39. [CrossRef]
23. D'Amico, G.; Janssen, J.; Manca, R. Initial and final backward and forward discrete time non-homogeneous semi-Markov credit risk models. *Methodol. Comput. Appl. Probab.* **2010**, *12*, 215–225. [CrossRef]
24. Csenki, A. On the interval reliability of systems modelled by finite semi-Markov processes. *Microelectron. Reliab.* **1994**, *34*, 1319–1335. [CrossRef]

25. Csenki, A. An integral equation approach to the interval reliability of systems modelled by finite semi-Markov processes. *Reliab. Eng. Syst. Saf.* **1995** ,*47*, 37–45. [CrossRef]
26. D'Amico, G.; Jacques, J.; Manca, R. Homogeneous semi-Markov reliability models for credit risk management. *Decis. Econ. Financ.* **2005**, *28*, 79–93. [CrossRef]
27. Howard, R. *Dynamic Probabilistic Systems*; Wiley: New York, NY, USA, 1971; Volume 2.
28. Vassiliou, P.-C.G. Non-homogeneous semi-Markov and Markov renewal processes and change of measure in credit risk. *Mathematics* **2021**, *9*, 55. [CrossRef]
29. Perman, M.; Senegacnik, A.; Tuma, M. Semi-Markov models with an application to power-plant reliability analysis. *IEEE Trans. Reliab.* **1997**, *46*, 526–532. [CrossRef]
30. Vasileiou, A.; Vassiliou, P.-C. G. An inhomogeneous semi-Markov model for the term structure of credit risk spreads. *Adv. Appl. Probab.* **2006**, *38*, 171–198. [CrossRef]
31. Vassiliou, P.-C.G. Semi-Markov migration process in a stochastic market in credit risk. *Linear Algebra Its Appl.* **2014**, *450*, 13–43. [CrossRef]
32. D'Amico, G.; Janssen, J.; Manca, R. Downward migration credit risk problem: a non-homogeneous backward semi-Markov reliability approach. *J. Oper. Res. Soc.* **2016**, *67*, 393–401. [CrossRef]
33. Bulla, J.; Bulla, I. Stylized facts of financial time series and hidden semi-Markov models. *Comput. Stat. Data Anal.* **2006**, *51*, 2192–2209. [CrossRef]

mathematics MDPI

Article

Tails of the Moments for Sums with Dominatedly Varying Random Summands

Mantas Dirma †, Saulius Paukštys † and Jonas Šiaulys *,†

Institute of Mathematics, Vilnius University, Naugarduko 24, LT-03225 Vilnius, Lithuania; m.dirma@lb.lt (M.D.); saulius.paukstys@mif.vu.lt (S.P.)
* Correspondence: jonas.siaulys@mif.vu.lt
† These authors contributed equally to this work.

Abstract: The asymptotic behaviour of the tail expectation $\mathbb{E}\Big(\big(S_n^{\xi}\big)^{\alpha}\mathbb{1}_{\{S_n^{\xi}>x\}}\Big)$ is investigated, where exponent α is a nonnegative real number and $S_n^{\xi}=\xi_1+\ldots+\xi_n$ is a sum of dominatedly varying and not necessarily identically distributed random summands, following a specific dependence structure. It turns out that the tail expectation of such a sum can be asymptotically bounded from above and below by the sums of expectations $\mathbb{E}\Big(\xi_i^{\alpha}\mathbb{1}_{\{\xi_i>x\}}\Big)$ with correcting constants. The obtained results are extended to the case of randomly weighted sums, where collections of random weights and primary random variables are independent. For illustration of the results obtained, some particular examples are given, where dependence between random variables is modelled in copulas framework.

Keywords: tail expectation; asymptotic bound; quasi-asymptotic independence; heavy-tailed distribution; dominated variation; copula

MSC: 91G05; 91G10; 60G70

Citation: Dirma, M.; Paukštys, S.; Šiaulys, J. Tails of the Moments for Sums with Dominatedly Varying Random Summands. *Mathematics* **2021**, *9*, 824. https://doi.org/10.3390/math9080824

Academic Editor: Panagiotis-Christos Vassiliou

Received: 17 March 2021
Accepted: 8 April 2021
Published: 10 April 2021

Publisher's Note: MDPI stays neutral with regard to jurisdictional claims in published maps and institutional affiliations.

Copyright: © 2021 by the authors. Licensee MDPI, Basel, Switzerland. This article is an open access article distributed under the terms and conditions of the Creative Commons Attribution (CC BY) license (https://creativecommons.org/licenses/by/4.0/).

1. Introduction

Let $n\in\mathbb{N}:=\{1,2,\ldots\}$ and let us consider two collections of random variables (r.v.s): heavy-tailed (see definition in Section 2) r.v.s $\{\xi_1,\ldots,\xi_n\}$, called primary r.v.s, and nonnegative, non-degenerate at zero r.v.s $\{\theta_1,\ldots,\theta_n\}$, called random weights. In this paper, we investigate the asymptotic behaviour of the sums of primary r.v.s

$$S_n^{\xi}:=\sum_{k=1}^{n}\xi_k=\xi_1+\ldots+\xi_n, \tag{1}$$

and their weighted counterparts, namely randomly weighted sums

$$S_n^{\theta\xi}:=\sum_{k=1}^{n}\theta_k\xi_k=\theta_1\xi_1+\ldots+\theta_n\xi_n. \tag{2}$$

Asymptotics of (1) and (2) have been studied extensively during recent years in the literature of applied probability under various different assumptions about collections $\{\xi_1,\ldots,\xi_n\}$, $\{\theta_1,\ldots,\theta_n\}$ and their dependence structures. In particular, there are many papers addressing the asymptotic behaviour of the tail probabilities

$$\mathbb{P}\Big(S_n^{\xi}>x\Big) \quad\text{and}\quad \mathbb{P}\Big(S_n^{\theta\xi}>x\Big) \tag{3}$$

expressing them by the sums of tail probabilities of individual summands, i.e., $\sum_{k=1}^{n}\mathbb{P}(\xi_k>x)$ and $\sum_{k=1}^{n}\mathbb{P}(\theta_k\xi_k>x)$, respectively (see, e.g., [1–11]). The main results from the majority of the aforementioned papers are reviewed in detail in Section 3. In line with the tail probabilities $\mathbb{P}\Big(S_n^{\xi}>x\Big)$ and $\mathbb{P}\Big(S_n^{\theta\xi}>x\Big)$, asymptotics of the tail expectations

$\mathbb{E}\Big(S_n^{\xi}\mathbb{1}_{\{S_n^{\xi}>x\}}\Big)$ and $\mathbb{E}\Big(S_n^{\theta\xi}\mathbb{1}_{\{S_n^{\theta\xi}>x\}}\Big)$ are investigated in the literature; however, the number of papers is relatively scarce (see [6,7,12] and the references therein).

In this paper, inspired by the recent results by Leipus et al. [7], we are particularly interested in the asymptotics of the tail expectations

$$\mathbb{E}\Big(\big(S_n^{\xi}\big)^{\alpha}\mathbb{1}_{\{S_n^{\xi}>x\}}\Big) \quad \text{and} \quad \mathbb{E}\Big(\big(S_n^{\theta\xi}\big)^{\alpha}\mathbb{1}_{\{S_n^{\theta\xi}>x\}}\Big), \tag{4}$$

where α is a nonnegative real number. We assume that r.v.s $\xi_1,\ldots,\xi_n$ are not necessarily identically distributed, belong to the class of dominatedly varying distributions (see Section 2.2), a subclass of heavy-tailed distributions, and follow a specific dependence structure, called pairwise quasi-asymptotic independence (see Section 2.3). We seek to asymptotically bound the tail expectations (4) by the sums of individual tail expectations $\mathbb{E}\Big({\xi_k}^{\alpha}\mathbb{1}_{\{\xi_k>x\}}\Big)$ and $\mathbb{E}\Big((\theta_k\xi_k)^{\alpha}\mathbb{1}_{\{\theta_k\xi_k>x\}}\Big)$, respectively, with some specific correcting constants (see Theorems 3 and 4).

Although our paper is more of a theoretical kind, it is worth noting that sums of the form (1) and, especially, (2) are often encountered in the practical applications of probability in financial and actuarial context. One example stems from the so-called discrete time risk model, in which primary r.v. ξ_k could correspond to the net losses (total claim amount minus total premium income) of an insurance company during period $(k-1,k]$, calculated at the moment k, and random weight θ_k could correspond to the stochastic discount factor, from the moment k to the present moment 0, for all $k=1,\ldots,n$. In such a scenario, sum $S_n^{\theta\xi}$ could be treated as the present value of a total discounted net loss of a company in the time interval $(0,n]$ (for more details, see, e.g., [5,8,10,13–16]).

Other insurance related application is based on the individual risk model [17]. Say that an insurance company has a portfolio consisting of n policies. Then, we could interpret $S_n^{\theta\xi}$ as the total claim amount incurred from the whole insurance portfolio. Here, $\theta_k\xi_k$, $k\in\{1,\ldots,n\}$ would correspond to the claim amount from the kth policy. Since there is a possibility that no claim will be incurred, θ_k is an indicatory Bernoulli r.v. which represents the occurrence of the kth claim ($\theta_k=1$ if the claim has occurred and zero otherwise) and r.v. ξ_k corresponds to the claim size of the kth policy given that the payment was made (see [18] and Chapter 4 of [19] as well).

Tang and Yuan [15] considered an example related to the construction of investment portfolio and capital allocation. Suppose that there are n distinct asset classes or lines of business, from which the portfolio is formed. Then, r.v. ξ_k could correspond to the loss incurred from the kth instrument. As for the role of random weights, there could be different viewpoints: θ_k could be treated as a stochastic discount factor of the kth asset class or, for instance, as a weight corresponding to the kth instrument in the portfolio. Then, random sum $S_n^{\theta\xi}$ would correspond to the present value of total loss of a portfolio at the present moment in the former case, and total weighted portfolio loss in the latter.

Highly related to the portfolio construction discussed above are various risk measures quantifying the underlying risk of the portfolio—we list several of them below, which are commonly encountered in the literature of risk management (see, e.g., [20,21]) and in which the asymptotic results concerning tail probabilities and tail expectations could be useful:

- Value-at-Risk (VaR) at level $q\in(0,1)$:

$$\mathrm{VaR}_q\big(S_n^{\xi}\big)=\inf\Big\{x\in\mathbb{R}\mid \mathbb{P}\big(S_n^{\xi}>x\big)\leqslant 1-q\Big\}.$$

- Conditional tail expectation (CTE) at level $q\in(0,1)$:

$$\mathrm{CTE}_q\big(S_n^{\xi}\big)=\mathbb{E}\Big(S_n^{\xi}\mid S_n^{\xi}>\mathrm{VaR}_q(S_n^{\xi})\Big).$$

For the above risk measures, the asymptotic behaviour is mainly considered as $q\uparrow 1$. Nevertheless, as mentioned in [22]: *"as the excessive prudence of the current reg-*

ulatory framework requires a confidence level close to 1, the notion of Extreme Value Theory becomes appropriate". In other words, q being close to 1 results in large $\mathrm{VaR}_q\left(S_n^{\xi}\right)$ values. For more about the estimation of the aforementioned risk measures, see the works of Yang et al. [12], Tang and Yuan [15], Asimit et al. [22], Hua and Joe [23] and Wang et al. [24] (and the references therein).

The rest of the paper is structured as follows. In Section 2, we review the basic definitions of the heavy-tailed distributions and introduce the reader to the specific dependence structure used in this paper. In Section 3, we discuss the related results in literature. In Section 4, our main results, which allow to asymptotically bound the tail expectations (4), are presented and later proved in Section 5. Finally, as applications of our result, in Section 6, we provide three different examples of random sums, for which dependence is controlled via a copula, in a bivariate setting.

2. Definitions and Preliminaries

2.1. Notational Conventions

Before delving into more details, we briefly introduce the notations used throughout the paper. All limiting relationships and asymptotic estimates, unless stated otherwise, are understood as x approaches infinity. For two positive functions f and g, we write:

- $f(x) \lesssim g(x)$ if $\limsup \frac{f(x)}{g(x)} \leqslant 1$
- $f(x) = o(g(x))$ if $\lim \frac{f(x)}{g(x)} = 0$
- $f(x) = O(g(x))$ if $\limsup \frac{f(x)}{g(x)} < \infty$
- $f(x) \sim g(x)$ if $\lim \frac{f(x)}{g(x)} = 1$
- $f(x) \asymp g(x)$ if $0 < \liminf \frac{f(x)}{g(x)} \leqslant \limsup \frac{f(x)}{g(x)} < \infty$

For any r.v. X, by $F_X(x)$, we denote the distribution function (d.f.) of X, i.e., $F_X(x) = \mathbb{P}(X \leqslant x)$. By $\overline{F}(x)$, we denote the tail function of d.f. F, i.e., $\overline{F}(x) = 1 - F(x)$. By $F^{*n}(x)$, we write the n-fold convolution of a d.f. F. That is, if $X_1, \ldots, X_n$ are independent copies of X, then $F_X^{*n}(x) = \mathbb{P}(X_1 + \ldots + X_n \leqslant x)$, and $\overline{F_X^{*n}}(x) = \mathbb{P}(X_1 + \ldots + X_n > x)$.

We say that r.v. X has an infinite right support if $\overline{F}_X(x) > 0$ for all $x \in \mathbb{R}$. In addition, we say that d.f. F is supported on $\mathbb{R}$ if $\overline{F}(x) > 0$ for all $x \in \mathbb{R}$. We write $\mathbb{1}_A$ to denote the indicator function of an event A. For any r.v. X, by X^+ and X^-, we denote its positive and negative parts, respectively: $X^+ = \max\{X, 0\}$, $X^- = \max\{-X, 0\}$. For a given $x \in \mathbb{R}$, by $\lfloor x \rfloor$, we denote the integer part of x and, by $\widehat{x} = x - \lfloor x \rfloor$, we denote the fractional part of x.

2.2. Heavy-Tailed Distributions

In this subsection, we recall the main classes of heavy-tailed distributions. At first, we present the class of dominatedly varying distributions $\mathcal{D}$, which is a central one in this paper.

- *A d.f. F supported on $\mathbb{R}$ is said to be dominatedly varying (belong to class $\mathcal{D}$) if* $\limsup\limits_{x\to\infty} \frac{\overline{F}(xy)}{\overline{F}(x)} < \infty$ *for any (for some)* $y \in (0, 1)$.

As noted in [25], Peter and Paul distribution is an example of a distribution belonging to the class $\mathcal{D}$. We say that r.v. X is distributed according to the generalised Peter and Paul distribution with parameters (a, b), where $b > 1$, $a \in (0, \infty)$, if its tail is characterised by the following equality

$$\overline{F}_X(x) = (b^a - 1) \sum_{k \geqslant 1,\, b^k > x} b^{-ak}.$$

Since $\overline{F}_X(x) = (b^{-a})^{\lfloor \log_b x \rfloor}$ for $x \geqslant 1$, we get that

$$\limsup_{x\to\infty} \frac{\overline{F}_X(xy)}{\overline{F}_X(x)} = \limsup_{x\to\infty} (b^{-a})^{\lfloor \log_b x + \log_b y \rfloor - \lfloor \log_b x \rfloor} \leqslant (b^{-a})^{\lfloor \log_b y \rfloor - 1}.$$

for any $y \in (0,1)$, implying $F_X \in \mathcal{D}$.

Class $\mathcal{D}$ is not the only subclass of heavy-tailed distributions. Below, we briefly recall the other classes of heavy-tailed distributions and describe the relationships between them.

- *A d.f. F is said to be heavy-tailed (belong to class $\mathcal{H}$) if for any $\alpha > 0$*

$$\int_{-\infty}^{\infty} \mathrm{e}^{\alpha x} \mathrm{d}F(x) = \infty.$$

- *A d.f. F is said to be long tailed (belong to class $\mathcal{L}$) if for any $y > 0$ $\overline{F}(x+y) \sim \overline{F}(x)$.*
- *A d.f. F supported on $\mathbb{R}$ is said to be subexponential (belong to class $\mathcal{S}$) if $F \in \mathcal{L}$ and $\overline{F^{*2}}(x) \sim 2\overline{F}(x)$.*
- *A d.f. F is said to be regularly varying with coefficient $\alpha \geqslant 0$ (belong to class $\mathcal{R}_\alpha$) if for any $y > 0$*

$$\lim_{x\to\infty} \frac{\overline{F}(xy)}{\overline{F}(x)} = y^{-\alpha}.$$

- *A d.f. F is said to be consistently varying (belong to class $\mathcal{C}$) if*

$$\lim_{y\uparrow 1} \limsup_{x\to\infty} \frac{\overline{F}(xy)}{\overline{F}(x)} = 1.$$

The class of consistently varying distributions $\mathcal{C}$ is the largest subclass of a class $\mathcal{D}$. The following example of a distribution belonging to $\mathcal{C} \setminus \bigcup_{\alpha\geqslant 0} \mathcal{R}_\alpha$ is given in [26]. Let Y and N be independent r.v.s such that $Y \overset{d}{=} \mathcal{U}([0,1])$ and N is geometric r.v. with parameter $p \in (0,1)$, i.e., $\mathbb{P}(N = k) = (1-p)p^k$ for $k = 0,1,\ldots$). Then, r.v. ξ defined by

$$\xi = (1+Y)2^N \tag{5}$$

belongs to the class $\mathcal{C}$ but not to the class $\bigcup_{\alpha\geqslant 0} \mathcal{R}_\alpha$. This fact can be derived from the expression

$$F_\xi(x) = (1-p)\left(\frac{1-p^{\lfloor \log_2 x\rfloor}}{1-p} + \left(\frac{x}{2^{\lfloor \log_2 x\rfloor}} - 1\right)p^{\lfloor \log_2 x\rfloor}\right)\mathbb{1}_{\{x\geqslant 1\}}. \tag{6}$$

In summary, the interrelationships of the heavy-tailed distribution classes can be expressed by the following relations

$$\mathcal{R} := \bigcup_{\alpha\geqslant 0} \mathcal{R}_\alpha \subsetneq \mathcal{C} \subsetneq \mathcal{L}\cap\mathcal{D}\cap\mathcal{S} \subsetneq \mathcal{L} \subsetneq \mathcal{H}; \quad \mathcal{D} \subsetneq \mathcal{H}; \quad \mathcal{D} \not\subset \mathcal{S}.$$

Some of the above relationships follow directly from the definitions, while proofs of the others can be found in, e.g., [25,27–30], ([31] Sections 6.1 and 6.2) and [32].

The classes $\mathcal{C}$ and $\mathcal{D}$ can be characterised by specific indices. We recall these important indices. The first one is a so-called *L-index*, used in, e.g., [6,7,11,33,34]. The L_F index for d.f. F is

$$L_F := \lim_{y\downarrow 1} \liminf_{x\to\infty} \frac{\overline{F}(xy)}{\overline{F}(x)}.$$

The second important index is the *upper Matuszewska index* introduced in [35]. In this paper, we stick with the slightly different but equivalent formulation given in [36] and used in many other articles (e.g., [2,6,8,11,34]). For a d.f. F, the *upper Matuszewska index* J_F^+ is

$$J_F^+ := \inf_{y>1}\left\{-\frac{1}{\log y}\log\left\{\liminf_{x\to\infty}\frac{\overline{F}(xy)}{\overline{F}(x)}\right\}\right\}.$$

The aforementioned indices give important characterisations for dominatedly varying and consistently varying d.f.s (see, e.g., [33], Proposition 1.1):

$$L_F > 0 \Leftrightarrow F \in \mathcal{D} \Leftrightarrow J_F^+ < \infty; \quad L_F = 1 \Leftrightarrow F \in \mathcal{C}.$$

For a r.v. ξ with d.f. F_ξ, we write for brevity: $L_\xi = L_{F_\xi}$ and $J_\xi^+ = J_{F_\xi}^+$. More information on classes $\mathcal{C}$ and $\mathcal{D}$ can be found in Chapter 2 of [36] and the discussion in Section 3 of [27].

2.3. QAI Dependence Structure

We now introduce the main dependence assumption about r.v.s $\xi_1, \ldots, \xi_n$ used in this paper, so-called *pairwise quasi-asymptotic independence*, which is due to Chen and Yuen [2].

- *R.v.s $\{\xi_1, \ldots, \xi_n\}$ with infinite right supports are called pairwise quasi-asymptotically independent (pQAI) if for any $k, l \in \{1, \ldots, n\}$, $k \neq l$,*

$$\lim_{x\to\infty} \frac{\mathbb{P}(\xi_k^+ > x, \xi_l^+ > x)}{\mathbb{P}(\xi_k^+ > x) + \mathbb{P}(\xi_l^+ > x)} = \lim_{x\to\infty} \frac{\mathbb{P}(\xi_k^+ > x, \xi_l^- > x)}{\mathbb{P}(\xi_k^+ > x) + \mathbb{P}(\xi_l^+ > x)} = 0. \tag{7}$$

Let us construct two examples of r.v.s possessing such a dependence structure using copulas.

Example 1. *Let $\{\xi_1, \ldots, \xi_n\}$ be r.v.s with infinite right supports and corresponding marginal d.f.s $\{F_1, \ldots, F_n\}$. Consider the Farlie–Gumbel–Morgenstein (FGM) copula:*

$$C_\vartheta(u, v) = uv + \vartheta uv(1-u)(1-v),\ u, v \in [0, 1], \quad \vartheta \in [-1, 1].$$

Let r.v.s ξ_i, ξ_j have a joint d.f. $\mathbb{P}(\xi_i \leqslant x_1, \xi_j \leqslant x_2) = C_{\vartheta_i}(F_i(x_1), F_j(x_2))$ with some $\vartheta_i \in [-1, 1]$ if $\max\{i, j\} - \min\{i, j\} = 1$, $\min\{i, j\} = 2k - 1$ for some $k \in \mathbb{N}$ and be independent otherwise. Then, r.v.s $\{\xi_1, \ldots, \xi_n\}$ are pQAI.

It follows from Sklar's theorem (see [37,38] Theorem 2.3.3)) that for any given marginal d.f.s F_1, F_2 and an arbitrary copula $C(u_1, u_2)$, function $F(x_1, x_2) := C(F_1(x_1), F_2(x_2))$ is a bivariate d.f. with marginal d.f.s F_1, F_2. If ξ_i, ξ_j, $i, j = 1, \ldots n$ are independent, then obviously they are pQAI. If $\max\{i, j\} - \min\{i, j\} = 1$, $\min\{i, j\} = 2k - 1$ for some $k \in \mathbb{N}$, then

$$\begin{aligned}\frac{\mathbb{P}(\xi_i > x, \xi_j > x)}{\mathbb{P}(\xi_i > x) + \mathbb{P}(\xi_j > x)} &= \frac{1 - F_i(x) - F_j(x) + C_{\vartheta_i}(F_i(x), F_j(x))}{\overline{F}_i(x) + \overline{F}_j(x)} \\ &= \frac{\overline{F}_i(x)\overline{F}_j(x)(1 + \vartheta_i F_i(x) F_j(x))}{\overline{F}_i(x) + \overline{F}_j(x)} \leqslant 2\overline{F}_i(x).\end{aligned} \tag{8}$$

Similarly, by observing that

$$\mathbb{P}\Big(\xi_i > x, \xi_j^- > x\Big) \leqslant \mathbb{P}\Big(\xi_i > x, \xi_j^- \geqslant x\Big) = \mathbb{P}(\xi_j \leqslant -x) - \mathbb{P}(\xi_i \leqslant x, \xi_j \leqslant -x)$$

for positive x, we get

$$\begin{aligned}\frac{\mathbb{P}(\xi_i > x, \xi_j^- > x)}{\mathbb{P}(\xi_i > x) + \mathbb{P}(\xi_j > x)} &\leqslant \frac{F_j(-x) - C_{\vartheta_i}(F_i(x), F_j(-x))}{\overline{F}_i(x) + \overline{F}_j(x)} \\ &= \frac{\overline{F}_i(x) F_j(-x)(1 - \vartheta_i F_i(x) F_j(-x))}{\overline{F}_i(x) + \overline{F}_j(x)} \leqslant 2F_j(-x).\end{aligned} \tag{9}$$

Estimates (8) and (9) imply (7). Consequently, r.v.s $\{\xi_1, \ldots, \xi_n\}$ in Example 1 are pQAI.

In a more general setting, one can consider n-dimensional ($n \geqslant 2$) Farley–Gumbel–Morgenstein (FGM) distributions (for a detailed treatment on this type of distributions see, e.g., [39]). For r.v.s $\{\xi_1, \dots, \xi_n\}$ with corresponding marginal d.f.s $\{F_1, \dots, F_n\}$, n-variate FGM d.f. is defined as follows:

$$\mathbb{P}(\xi_1 \leqslant x_1, \dots, \xi_n \leqslant x_n) = \prod_{i=1}^{n} F_i(x_i) \left(1 + \sum_{1 \leqslant i < j \leqslant n} \vartheta_{ij} \overline{F}_i(x_i) \overline{F}_j(x_j)\right), \tag{10}$$

where parameters ϑ_{ij}, $i, j \in \{1, \dots, n\}$ should satisfy the following condition

$$1 + \sum_{1 \leqslant i < j \leqslant n} \varepsilon_i \varepsilon_j \vartheta_{ij} \geqslant 0, \tag{11}$$

for all $\varepsilon_i = -\sup_{x \in \mathbb{R}}\{F_i(x)\} \backslash \{0, 1\}$ or $\varepsilon_i = 1 - \inf_{x \in \mathbb{R}}\{F_i(x)\} \backslash \{0, 1\}$ (see [39] Chapter 44, Section 13). Necessary condition (11) is required for (10) to be a well defined d.f..

Note that, if we assume that random vector $(\xi_1, \dots, \xi_n)$ is distributed according to (10) in our example, r.v.s $\xi_1, \dots, \xi_n$ are still pQAI, since bivariate marginal distributions of random vectors (ξ_i, ξ_j), $i, j \in \{1, \dots, n\}$, $i < j$ are distributed according to a bivariate FGM distribution with marginal d.f.s F_i, F_j and FGM copula. Indeed, from (10), we get

$$\begin{aligned} \mathbb{P}(\xi_i \leqslant x_1, \xi_j \leqslant x_2) &= \mathbb{P}(\xi_1 \leqslant \infty, \dots, \xi_i \leqslant x_1, \dots, \xi_j \leqslant x_2, \dots, \xi_n \leqslant \infty) \\ &= F_i(x_1) F_j(x_2) (1 + \vartheta_{ij} \overline{F}_i(x_1) \overline{F}_j(x_2)). \end{aligned}$$

Example 2. *Let ξ_1, ξ_2 be r.v.s with corresponding d.f.s F_1, F_2 and let random vector (ξ_1, ξ_2) have a bivariate d.f. $F(x_1, x_2) := C_\vartheta(F_1(x_1), F_2(x_2))$, where C_ϑ is the Ali–Michail–Haq copula [40]:*

$$C_\vartheta(u, v) = \frac{uv}{1 - \vartheta(1-u)(1-v)}, \quad u, v \in [0, 1], \quad \vartheta \in (-1, 1).$$

Similarly to in Example 1, it can be shown that r.v.s ξ_1, ξ_2 are QAI.

Indeed, for positive x, we have

$$\begin{aligned} \frac{\mathbb{P}(\xi_1 > x, \xi_2 > x)}{\mathbb{P}(\xi_1 > x) + \mathbb{P}(\xi_2 > x)} &= \frac{1 - F_1(x) - F_2(x) + C_\vartheta(F_1(x), F_2(x))}{\overline{F}_1(x) + \overline{F}_2(x)} \\ &= \frac{\overline{F}_1(x)\overline{F}_2(x)(1 + \vartheta(F_2(x) - \overline{F}_1(x)))}{(\overline{F}_1(x) + \overline{F}_2(x))(1 - \vartheta\overline{F}_1(x)\overline{F}_2(x))} \leqslant \frac{2\overline{F}_2(x)}{1 - \vartheta\overline{F}_1(x)\overline{F}_2(x)}. \end{aligned}$$

In the same fashion for positive x, we obtain

$$\begin{aligned} \frac{\mathbb{P}(\xi_1 > x, \xi_2^- > x)}{\mathbb{P}(\xi_1 > x) + \mathbb{P}(\xi_2 > x)} &\leqslant \frac{F_2(-x) - C_\vartheta(F_1(x), F_2(-x))}{\overline{F}_1(x) + \overline{F}_2(x)} \\ &= \frac{F_2(-x)\overline{F}_1(x)(1 - \vartheta\overline{F}_2(-x))}{(\overline{F}_1(x) + \overline{F}_2(x))(1 - \vartheta\overline{F}_1(x)\overline{F}_2(-x))} \leqslant \frac{2F_2(-x)}{1 - \vartheta\overline{F}_1(x)\overline{F}_2(-x)}. \end{aligned}$$

The derived estimates imply that r.v.s ξ_1 and ξ_2 are QAI.

For more about copulas applications in problems related to modelling dependence of heavy-tailed distributions, the reader may refer to the works of Albrechter et al. [41], Asimit et al. [22], Fang et al. [42], Yang et al. [43] and Wang et al. [24] (and the references therein). For a systematic treatment of copulas theory see, for instance, the work of Nelsen [38]. In the next section, we recall briefly more similar dependence structures between r.v.s and examine their relations to pQAI condition (7).

3. Related Results

In this section, we briefly review some of the related results found in the literature, regarding the asymptotic behaviour of the tail probability and tail expectation of random sums in the form of either (1) or (2). Throughout the section, unless mentioned otherwise, we assume that the collections of r.v.s $\{\xi_1, \ldots, \xi_n\}$ and $\{\theta_1, \ldots, \theta_n\}$ are independent.

3.1. Asymptotics of Tail Probabilities

There are many papers in which r.v.s $\xi_1, \ldots, \xi_n$ are assumed to be independent or identically distributed, see, for instance, the works of Tang and Tsitshiashivili [8], Goovaertz et al. [5], Wang et al. [9] and Wang and Tang [10] (and the references therein). In this subsection, however, we concentrate on the results in which such restrictive assumptions have been weakened.

We start with several results in which the exact asymptotic equivalence

$$\mathbb{P}\left(S_n^{\xi} > x\right) \sim \sum_{k=1}^{n} \overline{F}_{\xi_k}(x) \tag{12}$$

was obtained.

Geluk and Tang [4] achieved (12) for distributions $F_{\xi_k} \in \mathcal{D} \cap \mathcal{S}$ (see [4] Theorem 3.1). It was assumed that r.v.s $\xi_1, \ldots, \xi_n$ satisfy the so-called *Assumption A* (as in [4]), which was referred to later as a *strong quasi-asymptotic independence* in other articles (e.g., [6,7,16,44]), as well.

- *R.v.s $\xi_1, \ldots, \xi_n$ with infinite right supports are called pairwise strongly quasi-asymptotically independent (pSQAI) if for any $k, l \in \{1, \ldots n\}$, $k \neq l$,*

$$\lim_{\min\{x_k, x_l\} \to \infty} \mathbb{P}(|\xi_k| > x_k \mid \xi_l > x_l) = 0.$$

Nearly at the same time, Chen and Yuen [2] achieved (12) (see [2] Theorem 3.1) in the smaller class $\mathcal{C}$, but this time the pSQAI condition was replaced by the similar pQAI condition (see Section 2.3). We observe that pSQAI condition implies pQAI. Indeed, by arbitrarily choosing ξ_k, ξ_l, $1 \leqslant k \neq l \leqslant n$, we get that

$$\mathbb{P}(|\xi_k| > x_k \mid \xi_l > x_l) = \mathbb{P}(\xi_k^+ > x_k \mid \xi_l > x_l) + \mathbb{P}(\xi_k^- > x_k \mid \xi_l > x_l) \to 0,$$

as $\min\{x_k, x_l\} \to \infty$. Thus, it follows that

$$\lim_{x\to\infty} \frac{\mathbb{P}(\xi_k^+ > x, \xi_l^+ > x)}{\mathbb{P}(\xi_k^+ > x) + \mathbb{P}(\xi_l^+ > x)} \leqslant \lim_{x\to\infty} \mathbb{P}(\xi_k > x \mid \xi_l > x) = 0,$$

and, in the same way,

$$\lim_{x\to\infty} \frac{\mathbb{P}(\xi_k^- > x, \xi_l^+ > x)}{\mathbb{P}(\xi_k^+ > x) + \mathbb{P}(\xi_l^+ > x)} \leqslant \lim_{x\to\infty} \mathbb{P}(\xi_k^- > x \mid \xi_l > x) = 0.$$

Moreover, in the same article by Chen and Yuen, the results are extended to the case of randomly weighted sums (see [2] Theorem 3.2), resulting in relation

$$\mathbb{P}\left(S_n^{\theta\xi} > x\right) \sim \sum_{k=1}^{n} \overline{F}_{\theta_k\xi_k}(x)$$

under the following moment condition on random weights:

$$\max\{\mathbb{E}\theta_1^p \ldots, \mathbb{E}\theta_n^p\} < \infty \quad \text{for some} \quad p > \max\{J_{\xi_1}^+, \ldots, J_{\xi_n}^+\}. \tag{13}$$

Later, inspired by the results of Chen and Yuen, Yi et al. [11] considered the tail probability asymptotics of the randomly weighted sum $S_n^{\theta\xi}$, when r.v.s $\xi_1, \ldots, \xi_n$ belong to the class $\mathcal{D}$ and follow the same pQAI structure (see [11] Theorems 1 and 2). It was shown that under (13) and additional tail assumption

$$\lim_{x\to\infty} \frac{\overline{F}_{\xi_k^-}(x)}{\overline{F}_{\xi_k}(x)} = 0 \qquad \text{for all} \qquad k \in \{1,\ldots,n\}, \tag{14}$$

the following asymptotic bounds hold:

$$L_n^{\xi} \sum_{k=1}^{n} \overline{F}_{\theta_k \xi_k}(x) \lesssim \mathbb{P}\left(S_n^{\theta\xi} > x\right) \lesssim \frac{1}{L_n^{\xi}} \sum_{k=1}^{n} \overline{F}_{\theta_k \xi_k}(x), \tag{15}$$

where $L_n^{\xi} := \min\{L_{\xi_1}, \ldots, L_{\xi_n}\}$. Cheng [3] managed to tighten the bounds in (15) (see [3] Theorems 1.1 and 1.2) by putting the L-indices inside the sums and obtaining

$$\sum_{k=1}^{n} L_{\xi_k} \overline{F}_{\theta_k \xi_k}(x) \lesssim \mathbb{P}\left(S_n^{\theta\xi} > x\right) \lesssim \sum_{k=1}^{n} \frac{1}{L_{\xi_k}} \overline{F}_{\theta_k \xi_k}(x),$$

where $F_{\xi_k} \in \mathcal{D}$ for all $k \in \{1,\ldots,n\}$. The assumption (13) was substituted by a weaker condition (see [3] Assumption C and Remark 1.1):

$$\lim_{x\to\infty} \frac{F_{\theta_k \xi_k}(x)}{F_{\xi_k}(x)} = 0 \qquad \text{for all} \qquad k \in \{1,\ldots,n\}.$$

Moreover, instead of pQAI, two other dependence structures, namely *pairwise tail quasi-asymptotic independence* (see [3] Assumption B) and *pairwise asymptotic independence*, together with condition (14) (see [3] Assumption A) were considered.

- *R.v.s $\xi_1, \ldots, \xi_n$ with infinite right supports are called pairwise tail quasi-asyptotically independent (pTQAI) if for any $k, l = 1, \ldots, n$, $k \neq l$,*

$$\lim_{\min\{x_k, x_l\}\to\infty} \frac{\mathbb{P}(\xi_k^+ > x_k, \xi_l^+ > x_l)}{\mathbb{P}(\xi_k^+ > x_k) + \mathbb{P}(\xi_l^+ > x_l)} = \lim_{\min\{x_k, x_l\}\to\infty} \frac{\mathbb{P}(\xi_k^- > x_k, \xi_l^+ > x_l)}{\mathbb{P}(\xi_k^+ > x_k) + \mathbb{P}(\xi_l^+ > x_l)} = 0.$$

- *R.v.s $\xi_1, \ldots, \xi_n$ with infinite right supports are called pairwise asymptotically independent (pAI) if for any $k, l = 1, \ldots, n$, $k \neq l$,*

$$\lim_{x\to\infty} \frac{\mathbb{P}(\xi_k > x, \xi_l > x)}{\mathbb{P}(\xi_k > x)} = \lim_{x\to\infty} \frac{\mathbb{P}(\xi_k > x, \xi_l > x)}{\mathbb{P}(\xi_l > x)} = 0.$$

As noted in [3], implication pTQAI $\Rightarrow$ pQAI follows trivially, by allowing x_k and x_l to attain the same value x in the definition of pTQAI. It is easy to see that pAI implies pQAI if r.v.s $\xi_1, \ldots, \xi_n$ are nonnegative. Nonetheless, (14) is a sufficient condition for pAI $\Rightarrow$ pQAI to hold in the general case because, for any $1 \leqslant k \neq l \leqslant n$,

$$\lim_{x\to\infty} \frac{\mathbb{P}(\xi_k > x, \xi_l^- > x)}{\mathbb{P}(\xi_k > x) + \mathbb{P}(\xi_l > x)} \leqslant \lim_{x\to\infty} \frac{F_{\xi_l}(-x)}{\overline{F}_{\xi_l}(x)} = 0.$$

Quite recently, Jaunė et al. [6] reconsidered the asymptotic behaviour of tail probability $\mathbb{P}\left(S_n^{\theta\xi} > x\right)$ under the pQAI condition on r.v.s $\xi_1, \ldots, \xi_n$ in the class $\mathcal{D}$. The statement of Lemma 1 from [6] extends mainly the results of Yi et al. [11], resulting in (15) under the moment condition (13), but without the additional assumption (14).

3.2. Asymptotics of Tail Expectations

Having reviewed the main results about the asymptotic behaviour of tail probabilities (3), we now turn to the asymptotics of tail expectation of random sums S_n^{ξ} and $S_n^{\theta\xi}$ which is the main object of this paper. Tang and Yuan [15] obtained the relation

$$\mathbb{E}\Big(\theta_1\xi_1\mathbb{1}_{\{S_n^{\theta\xi}>x\}}\Big) \sim \mathbb{E}\Big(\theta_1\xi_1\mathbb{1}_{\{\theta_1\xi_1>x\}}\Big)$$

for i.i.d. r.v.s $\xi_1,\dots,\xi_n$ from the class $\mathcal{D}\cap\mathcal{S}$ and random weights $\theta_1,\dots,\theta_n$, satisfying $\mathbb{E}\theta_k^{p_k}<\infty$, $p_k>J_{\xi_k}^+$, $\overline{F}_{\theta_k\xi_k}(x)=O(\overline{F}_{\theta_1\xi_1}(x))$ for all $k\in\{1,\dots,n\}$ (see [15] Theorem 4). It was noted by Yang et al. [12] that, under additional condition $\overline{F}_{\theta_k\xi_k}(x)\asymp\overline{F}_{\theta_1\xi_1}(x)$ for all $k\in\{1,\dots,n\}$, relation

$$\mathbb{E}\Big(S_n^{\theta\xi}\mathbb{1}_{\{S_n^{\theta\xi}>x\}}\Big) \sim \sum_{k=1}^{n}\mathbb{E}\Big(\theta_k\xi_k\mathbb{1}_{\{\theta_k\xi_k>x\}}\Big) \tag{16}$$

holds.

Jaunė et al. [6] later weakened the i.i.d. condition of the previous result, allowing pQAI or pSQAI dependence structures among primary r.v.s $\xi_1,\dots,\xi_n$, at the cost of exact asymptotics in (16).

Now, we turn to the recent result by Leipus et al. [7], which inspired our investigation. Before stating the relevant theorems, we note that, in [7], the new dependence structure called *Assumption* $\mathcal{B}$, regarding r.v.s $\xi_1,\dots,\xi_n$, is used.

Assumption $\mathcal{B}$. *R.v.s $\xi_1,\dots,\xi_n$ have infinite right supports and, for all $k,l=1,\dots,n$, $k\neq l$ satisfy*

$$\begin{aligned}\lim_{x\to\infty}\sup_{u\geqslant x}\mathbb{P}(\xi_k^+>x\mid\xi_l^+>u) &= \lim_{x\to\infty}\sup_{u\geqslant x}\mathbb{P}(\xi_k^->x\mid\xi_l^+>u)\\ &= \lim_{x\to\infty}\sup_{u\geqslant x}\mathbb{P}(\xi_k^+>x\mid\xi_l^->u)=0.\end{aligned}$$

Similarly, as in the case pSQAI $\Rightarrow$ pQAI, we can show that assumption $\mathcal{B}$ implies the pQAI condition because for any ξ_k,ξ_l, $1\leqslant k\neq l\leqslant n$,

$$\begin{aligned}\lim_{x\to\infty}\frac{\mathbb{P}(\xi_k^+>x,\xi_l^+>x)}{\mathbb{P}(\xi_k^+>x)+\mathbb{P}(\xi_l^+>x)} &\leqslant \lim_{x\to\infty}\sup_{u\geqslant x}\mathbb{P}(\xi_k^+>x\mid\xi_l^+>u)=0,\\ \lim_{x\to\infty}\frac{\mathbb{P}(\xi_k^->x,\xi_l^+>x)}{\mathbb{P}(\xi_k^+>x)+\mathbb{P}(\xi_l^+>x)} &\leqslant \lim_{x\to\infty}\sup_{u\geqslant x}\mathbb{P}(\xi_k^->x\mid\xi_l^+>u)=0.\end{aligned}$$

The following assertion is the main result in [7].

Theorem 1 (See [7] Theorem 4)**.** *Let $\xi_1,\dots,\xi_n$ be r.v.s satisfying assumption $\mathcal{B}$ such that $F_{\xi_1}\in\mathcal{D}$, $\mathbb{E}|\xi_1|^m<\infty$ for some $m\in\mathbb{N}$ and $\overline{F}_{\xi_k}(x)\asymp\overline{F}_{\xi_1}(x)$, $\overline{F}_{\xi_k^-}(x)=O(\overline{F}_{\xi_1}(x))$, for all $k=2,\dots,n$. Then,*

$$L_n^{\xi}\sum_{k=1}^{n}\mathbb{E}\Big(\xi_k^m\mathbb{1}_{\{\xi_k>x\}}\Big)\lesssim\mathbb{E}\Big(\big(S_n^{\xi}\big)^m\mathbb{1}_{\{S_n^{\xi}>x\}}\Big)\lesssim\frac{1}{L_n^{\xi}}\sum_{k=1}^{n}\mathbb{E}\Big(\xi_k^m\mathbb{1}_{\{\xi_k>x\}}\Big).$$

Moreover, the results were extended to the general case of weighted sums. This time, however, a quite restrictive assumption about random weights was made; namely, it was supposed that random weights $\theta_1,\dots,\theta_n$ are bounded.

Theorem 2 (See [7] Theorem 5)**.** *Let $\xi_1,\dots,\xi_n$ be r.v.s satisfying assumption $\mathcal{B}$ such that $F_{\xi_1}\in\mathcal{D}$, $\mathbb{E}|\xi_1|^m<\infty$ for some $m\in\mathbb{N}$. Let $\theta_1,\dots,\theta_n$ be nonnegative, non-degenerate at zero,* *bounded r.v.s, independent of* $\theta_1, \ldots, \theta_n$. *If* $\overline{F}_{\theta_k\xi_k}(x) \asymp \overline{F}_{\theta_1\xi_1}(x)$, $\overline{F}_{\theta_k\xi_k^-}(x) = O(\overline{F}_{\theta_1\xi_1}(x))$, *for all* $k = 2, \ldots, n$. *Then,*

$$L_n^{\xi} \sum_{k=1}^{n} \mathbb{E}\Big((\theta_k\xi_k)^m \mathbb{1}_{\{\theta_k\xi_k > x\}} \Big) \lesssim \mathbb{E}\Big(\Big(S_n^{\theta\xi}\Big)^m \mathbb{1}_{\{S_n^{\theta\xi} > x\}} \Big) \lesssim \frac{1}{L_n^{\xi}} \sum_{k=1}^{n} \mathbb{E}\Big((\theta_k\xi_k)^m \mathbb{1}_{\{\theta_k\xi_k > x\}} \Big).$$

4. Main Results

In this section we present the main results of this paper. Theorem 3 states the asymptotic bounds for the tail expectation of a random sum S_n^{ξ} and Theorem 4 is mainly a generalisation to the case of randomly weighted sums $S_n^{\theta\xi}$.

We note that Theorems 3 and 4 improve previous results in several ways. For instance, compared to Theorems 1 and 2, we put individual L-indices inside the sums in (17) and (18), thus obtaining more accurate asymptotic bounds. Moreover, we weaken the condition for exponent, from being a nonnegative integer to any nonnegative real number. In addition, assumption $\mathcal{B}$ considered in [7] is substituted by a weaker pQAI structure and random weights $\theta_1, \ldots, \theta_n$ need not to be bounded as in Theorem 2. In addition, it is worth noting that, by setting $\alpha = 0$ in Theorems 3 and 4, we obtain asymptotics for the tail probabilities (3) (see Remark 2 as well), thus our results can be compared with those discussed in Section 3.1.

Theorem 3. *Let* $\xi_1, \ldots, \xi_n$ *be pQAI real-valued r.v.s. If* $\mathbb{E}|\xi_k|^{\alpha} < \infty$, $F_{\xi_k} \in \mathcal{D}$ *for all* $k \in \{1, \ldots, n\}$ *and some* $\alpha \in [0, \infty)$, *then*

$$\sum_{k=1}^{n} L_{\xi_k} \mathbb{E}\Big(\xi_k^{\alpha} \mathbb{1}_{\{\xi_k > x\}} \Big) \lesssim \mathbb{E}\Big(\Big(S_n^{\xi}\Big)^{\alpha} \mathbb{1}_{\{S_n^{\xi} > x\}} \Big) \lesssim \sum_{k=1}^{n} \frac{1}{L_{\xi_k}} \mathbb{E}\Big(\xi_k^{\alpha} \mathbb{1}_{\{\xi_k > x\}} \Big). \tag{17}$$

Theorem 4. *Let* $\xi_1, \ldots, \xi_n$ *be pQAI real valued r.v.s, such that* $F_{\xi_k} \in \mathcal{D}$ *for all* $k \in \{1, \ldots, n\}$, *and let* $\theta_1, \ldots, \theta_n$ *be arbitrarily dependent, nonnegative, non-degenerate at zero r.v.s with*

$$\max\{\mathbb{E}\theta_1^p, \ldots, \mathbb{E}\theta_n^p\} < \infty \quad \textit{for some} \quad p > \max\{J_{\xi_1}^+, \ldots, J_{\xi_n}^+\}.$$

If collections $\{\xi_1, \ldots, \xi_n\}$ *and* $\{\theta_1, \ldots, \theta_n\}$ *are independent and* $\mathbb{E}(\theta_k|\xi_k|)^{\alpha} < \infty$ *for all* $k \in \{1, \ldots, n\}$ *and some* $\alpha \in [0, \infty)$, *then*

$$\sum_{k=1}^{n} L_{\xi_k} \mathbb{E}\Big((\theta_k\xi_k)^{\alpha} \mathbb{1}_{\{\theta_k\xi_k > x\}} \Big) \lesssim \mathbb{E}\Big(\Big(S_n^{\theta\xi}\Big)^{\alpha} \mathbb{1}_{\{S_n^{\theta\xi} > x\}} \Big) \lesssim \sum_{k=1}^{n} \frac{1}{L_{\xi_k}} \mathbb{E}\Big((\theta_k\xi_k)^{\alpha} \mathbb{1}_{\{\theta_k\xi_k > x\}} \Big). \tag{18}$$

Remark 1. *By narrowing the class* $\mathcal{D}$ *to the class* $\mathcal{C}$ *of consistently varying distributions (for which the L-index is unit), we get the exact asymptotic equivalence in* (18). *That is, if* $F_{\xi_k} \in \mathcal{C}$ *for all* $k \in \{1, \ldots, n\}$ *and all other conditions of Theorem* 4 *hold, then*

$$\sum_{k=1}^{n} \mathbb{E}\Big((\theta_k\xi_k)^{\alpha} \mathbb{1}_{\{\theta_k\xi_k > x\}} \Big) \sim \mathbb{E}\Big(\Big(S_n^{\theta\xi}\Big)^{\alpha} \mathbb{1}_{\{S_n^{\theta\xi} > x\}} \Big).$$

Remark 2. *When* $\alpha = 0$, *from* (18), *we obtain asymptotic bounds for tail probabilities:*

$$\sum_{k=1}^{n} L_{\xi_k} \overline{F}_{\theta_k\xi_k}(x) \lesssim \mathbb{P}\Big(S_n^{\theta\xi} > x \Big) \lesssim \sum_{k=1}^{n} \frac{1}{L_{\xi_k}} \overline{F}_{\theta_k\xi_k}(x). \tag{19}$$

Remark 3. *Under the same conditions as in Theorem 4, we can obtain asymptotic bounds for the conditional expectation $\mathbb{E}\Big(\big(S_n^{\theta\xi}\big)^\alpha \mid S_n^{\theta\xi} > x\Big)$. Namely, combining (18) with (19) we obtain the following asymptotic bounds:*

$$\frac{\sum_{k=1}^n L_{\xi_k}\mathbb{E}((\theta_k\xi_k)^\alpha \mathbb{1}_{\{\theta_k\xi_k>x\}})}{\sum_{k=1}^n \frac{1}{L_{\xi_k}}\overline{F}_{\theta_k\xi_k}(x)} \lesssim \mathbb{E}\Big(\big(S_n^{\theta\xi}\big)^\alpha \mid S_n^{\theta\xi} > x\Big) \lesssim \frac{\sum_{k=1}^n \frac{1}{L_{\xi_k}}\mathbb{E}((\theta_k\xi_k)^\alpha \mathbb{1}_{\{\theta_k\xi_k>x\}})}{\sum_{k=1}^n L_{\xi_k}\overline{F}_{\theta_k\xi_k}(x)}. \tag{20}$$

In addition, by using the min–max inequality (22), we can express (20) fully in conditional expectations at the cost of tightness of the initial bounds:

$$\min_{1\leqslant k\leqslant n}\Big\{L_{\xi_k}^2\mathbb{E}((\theta_k\xi_k)^\alpha \mid \theta_k\xi_k > x)\Big\} \lesssim \mathbb{E}\Big(\big(S_n^{\theta\xi}\big)^\alpha \mid S_n^{\theta\xi} > x\Big) \lesssim \max_{1\leqslant k\leqslant n}\Big\{\frac{1}{L_{\xi_k}^2}\mathbb{E}((\theta_k\xi_k)^\alpha \mid \theta_k\xi_k > x)\Big\}.$$

5. Proofs of Main Results

To prove Theorem 3, we need some auxiliary assertions. The lemma below is proved in [7].

Lemma 1. *Let ξ be a real-valued r.v. If $\mathbb{E}(\xi^+)^\alpha < \infty$ for some $\alpha \in [0,\infty)$, then for all $x \geqslant 0$.*

$$\mathbb{E}(\xi^\alpha \mathbb{1}_{\{\xi>x\}}) = x^\alpha\mathbb{P}(\xi > x) + \alpha\int_x^\infty u^{\alpha-1}\mathbb{P}(\xi > u)\mathrm{d}u.$$

The next lemma is crucial for the proof of Theorem 3.

Lemma 2. *Let $\xi_1,\ldots,\xi_n$ be pQAI real-valued r.v.s, such that $F_{\xi_k} \in \mathcal{D}$ for all $k \in \{1,\ldots,n\}$. Then,*

$$\sum_{k=1}^n L_{\xi_k}\overline{F}_{\xi_k}(x) \lesssim \mathbb{P}(S_n^\xi > x) \lesssim \sum_{k=1}^n \frac{1}{L_{\xi_k}}\overline{F}_{\xi_k}(x). \tag{21}$$

Proof. The case $n = 1$ in (21) follows trivially from the definition of coefficient L_{ξ_1}. Let $n \geqslant 2$. First, let us consider the upper asymptotic bound in (21).

For an arbitrary $\delta \in (0,1)$,

$$\mathbb{P}(S_n^\xi > x) \leqslant \sum_{k=1}^n \overline{F}_{\xi_k}((1-\delta)x) + \mathbb{P}\Big(S_n^\xi > x, \bigcap_{k=1}^n\{\xi_k \leqslant (1-\delta)x\}\Big) =: \sum_{k=1}^n \overline{F}_{\xi_k}((1-\delta)x) + \mathcal{A}(x,\delta).$$

By observing that for all $k \in \{1,\ldots,n\}$

$$\Big\{S_n^\xi > x, \xi_k \leqslant (1-\delta)x\Big\} \subseteq \Big\{S_n^\xi - \xi_k > \delta x\Big\},$$

we can estimate the term $\mathcal{A}(x,\delta)$ as follows:

$$\mathcal{A}(x,\delta) \leqslant \sum_{k=1}^n \mathbb{P}\Big(\xi_k > \frac{x}{n}, S_n^\xi - \xi_k > \delta x\Big) \leqslant \sum_{k=1}^n \mathbb{P}\Big(\xi_k > \frac{x}{n}, \bigcup_{l=1,l\neq k}^n\Big\{\xi_l > \frac{\delta x}{n-1}\Big\}\Big)$$

$$\leqslant \sum_{k=1}^{n} \sum_{l=1,l\neq k}^{n} \mathbb{P}(\xi_k > \delta_1 x, \xi_l > \delta_1 x),$$

where in the last inequality $\delta_1 = \delta_1(\delta) = \min\{1/n, \delta/(n-1)\}$.

Consequently,

$$\frac{\mathbb{P}\left(S_n^{\xi} > x\right)}{\sum_{k=1}^{n} \frac{1}{L_{\xi_k}} \overline{F}_{\xi_k}(x)} \leqslant \frac{\sum_{k=1}^{n} \overline{F}_{\xi_k}((1-\delta)x)}{\sum_{k=1}^{n} \frac{1}{L_{\xi_k}} \overline{F}_{\xi_k}(x)} + \frac{\sum_{k=1}^{n} \sum_{l=1,l\neq k}^{n} \mathbb{P}(\xi_k > \delta_1 x, \xi_l > \delta_1 x)}{\sum_{k=1}^{n} \frac{1}{L_{\xi_k}} \overline{F}_{\xi_k}(x)}$$
$$=: \mathcal{I}_1(x,\delta) + \mathcal{I}_2(x,\delta).$$

Using the min–max inequality,

$$\min\left\{\frac{a_1}{b_1}, \ldots, \frac{a_m}{b_m}\right\} \leqslant \frac{a_1 + \ldots + a_m}{b_1 + \ldots + b_m} \leqslant \max\left\{\frac{a_1}{b_1}, \ldots, \frac{a_m}{b_m}\right\}, \tag{22}$$

provided that $m \in \mathbb{N}$ and $a_i \geqslant 0$, $b_i > 0$ for $i = 1, \ldots, m$, we get

$$\mathcal{I}_1(x,\delta) \leqslant \max_{1\leqslant k\leqslant n} \left\{ L_{\xi_k} \frac{\overline{F}_{\xi_k}((1-\delta)x)}{\overline{F}_{\xi_k}(x)} \right\}.$$

Taking into account (22) and observing that

$$\sum_{k=1}^{n} \sum_{l=1,l\neq k}^{n} \left(\overline{F}_{\xi_k}(\delta_1 x) + \overline{F}_{\xi_l}(\delta_1 x)\right) \leqslant 2(n-1) \sum_{k=1}^{n} \overline{F}_{\xi_k}(\delta_1 x),$$

we similarly obtain

$$\mathcal{I}_2(x,\delta) = \frac{\sum_{k=1}^{n} \sum_{l=1,l\neq k}^{n} \mathbb{P}(\xi_k > \delta_1 x, \xi_l > \delta_1 x)}{\sum_{k=1}^{n} \sum_{l=1,l\neq k}^{n} \left(\overline{F}_{\xi_k}(\delta_1 x) + \overline{F}_{\xi_l}(\delta_1 x)\right)}$$
$$\times \frac{\sum_{k=1}^{n} \sum_{l=1,l\neq k}^{n} \left(\overline{F}_{\xi_k}(\delta_1 x) + \overline{F}_{\xi_l}(\delta_1 x)\right)}{\sum_{k=1}^{n} \frac{1}{L_{\xi_k}} \overline{F}_{\xi_k}(x)}$$
$$\leqslant \max_{1\leqslant k\neq l\leqslant n} \left\{ \frac{\mathbb{P}(\xi_k > \delta_1 x, \xi_l > \delta_1 x)}{\overline{F}_{\xi_k}(\delta_1 x) + \overline{F}_{\xi_l}(\delta_1 x)} \right\} \times 2(n-1) \max_{1\leqslant k\leqslant n} \left\{ L_{\xi_k} \frac{\overline{F}_{\xi_k}(\delta_1 x)}{\overline{F}_{\xi_k}(x)} \right\}. \tag{23}$$

The fact that $F_{\xi_k} \in \mathcal{D}$ for all $k \in \{1, \ldots, n\}$ and condition of pQAI for r.v.s $\{\xi_1, \ldots, \xi_n\}$ implies:

$$\limsup_{x\to\infty} \mathcal{I}_1(x,\delta) \leqslant \max_{1\leqslant k\leqslant n} \left\{ L_{\xi_k} \limsup_{x\to\infty} \frac{\overline{F}_{\xi_k}((1-\delta)x)}{\overline{F}_{\xi_k}(x)} \right\}, \tag{24}$$

$$\limsup_{x\to\infty} \mathcal{I}_2(x,\delta) \leqslant 2(n-1) \max_{1\leqslant k\neq l\leqslant n} \left\{ \limsup_{x\to\infty} \frac{\mathbb{P}(\xi_k > \delta_1 x, \xi_l > \delta_1 x)}{\overline{F}_{\xi_k}(\delta_1 x) + \overline{F}_{\xi_l}(\delta_1 x)} \right\}$$
$$\times \max_{1\leqslant k\leqslant n} \left\{ L_{\xi_k} \limsup_{x\to\infty} \frac{\overline{F}_{\xi_k}(\delta_1 x)}{\overline{F}_{\xi_k}(x)} \right\} = 0. \tag{25}$$

Therefore, by letting $\delta \downarrow 0$, from estimates (24), (25) and definition of indices L_{ξ_k}, we get the upper bound in (21):

$$\limsup_{x\to\infty} \frac{\mathbb{P}(S_n^{\xi} > x)}{\sum_{k=1}^{n} \frac{1}{L_{\xi_k}} \overline{F}_{\xi_k}(x)} \leqslant 1.$$

Let us consider the lower asymptotic bound in (21). Again, choose arbitrary $\delta \in (0,1)$. By the Bonferroni inequality, for this δ, we get

$$\begin{aligned}
&\mathbb{P}\Big(S_n^{\xi} > x\Big) \geqslant \mathbb{P}\left(S_n^{\xi} > x, \bigcup_{k=1}^{n}\{\xi_k > (1+\delta)x\}\right) \\
&\geqslant \sum_{k=1}^{n}\mathbb{P}\Big(S_n^{\xi} > x, \xi_k > (1+\delta)x\Big) - \sum_{k=1}^{n}\sum_{l=1, l\neq k}^{n}\mathbb{P}(\xi_k > (1+\delta)x, \xi_l > (1+\delta)x) \\
&=: \mathcal{A}_1(x,\delta) - \mathcal{A}_2(x,\delta). \qquad (26)
\end{aligned}$$

For the first summand in (26), we obtain

$$\begin{aligned}
\mathcal{A}_1(x,\delta) &\geqslant \sum_{k=1}^{n}\mathbb{P}\Big(S_n^{\xi} - \xi_k > -\delta x, \xi_k > (1+\delta)x\Big) \\
&= \sum_{k=1}^{n}\overline{F}_{\xi_k}((1+\delta)x) - \sum_{k=1}^{n}\mathbb{P}\Big(S_n^{\xi} - \xi_k \leqslant -\delta x, \xi_k > (1+\delta)x\Big) \\
&=: \mathcal{A}_{11}(x,\delta) - \mathcal{A}_{12}(x,\delta). \qquad (27)
\end{aligned}$$

For the second term in (27), we get

$$\begin{aligned}
\mathcal{A}_{12}(x,\delta) &\leqslant \sum_{k=1}^{n}\mathbb{P}\left(\bigcup_{l=1,l\neq k}^{n}\left\{\xi_l \leqslant -\frac{\delta x}{n-1}\right\}, \xi_k > (1+\delta)x\right) \\
&\leqslant \sum_{k=1}^{n}\sum_{l=1,l\neq k}^{n}\mathbb{P}\left(\xi_k > (1+\delta)x, \xi_l^- \geqslant \frac{\delta x}{n-1}\right) \\
&\leqslant \sum_{k=1}^{n}\sum_{l=1,l\neq k}^{n}\mathbb{P}(\xi_k > \delta_2 x, \xi_l^- > \delta_2 x), \qquad (28)
\end{aligned}$$

where $\delta_2 = \delta_2(\delta) = \delta/2(n-1)$ in the last inequality.

We have from (26), (27) and (28) that

$$\begin{aligned}
\frac{\mathbb{P}(S_n^{\xi} > x)}{\sum_{k=1}^{n} L_{\xi_k}\overline{F}_{\xi_k}(x)} \geqslant &\frac{\mathcal{A}_{11}(x,\delta)}{\sum_{k=1}^{n} L_{\xi_k}\overline{F}_{\xi_k}(x)} \\
&- \frac{\sum_{k=1}^{n}\sum_{l=1,l\neq k}^{n}\mathbb{P}(\xi_k > \delta_2 x, \xi_l^- \geqslant \delta_2 x)}{\sum_{k=1}^{n} L_{\xi_k}\overline{F}_{\xi_k}(x)} \\
&- \frac{\mathcal{A}_2(x,\delta)}{\sum_{k=1}^{n} L_{\xi_k}\overline{F}_{\xi_k}(x)} \\
=: &\mathcal{J}_1(x,\delta) - \mathcal{J}_2(x,\delta) - \mathcal{J}_3(x,\delta).
\end{aligned}$$

Now, we estimate each term $\mathcal{J}_i(x,\delta)$, $i \in \{1,2,3\}$, separately. For the case $i = 1$, using inequality (22), we get

$$\mathcal{J}_1(x,\delta) \geqslant \min_{1\leqslant k\leqslant n}\left\{\frac{\overline{F}_{\xi_k}((1+\delta)x)}{L_{\xi_k}\overline{F}_{\xi_k}(x)}\right\}.$$

For $\mathcal{J}_2(x,\delta)$, similarly to in the derivation of (23), we obtain

$$\mathcal{J}_2(x,\delta) \leqslant 2(n-1)\max_{1\leqslant k\neq l\leqslant n}\left\{\frac{\mathbb{P}(\xi_k > \delta_2 x, \xi_l^- > \delta_2 x)}{\overline{F}_{\xi_k}(\delta_2 x) + \overline{F}_{\xi_l}(\delta_2 x)}\right\}\max_{1\leqslant k\leqslant n}\left\{\frac{\overline{F}_{\xi_k}(\delta_2 x)}{L_{\xi_k}\overline{F}_{\xi_k}(x)}\right\}.$$

Finally,

$$\mathcal{J}_3(x,\delta) \leqslant 2(n-1) \max_{1\leqslant k\neq l\leqslant n} \left\{ \frac{\mathbb{P}(\xi_k > (1+\delta)x, \xi_l > (1+\delta)x)}{\overline{F}_{\xi_k}((1+\delta)x) + \overline{F}_{\xi_l}((1+\delta)x)} \right\}$$
$$\times \max_{1\leqslant k\leqslant n} \left\{ \frac{\overline{F}_{\xi_k}((1+\delta)x)}{L_{\xi_k}\overline{F}_{\xi_k}(x)} \right\}$$
$$\leqslant \max_{1\leqslant k\leqslant n} \left\{ \frac{2(n-1)}{L_{\xi_k}} \right\} \max_{1\leqslant k\neq l\leqslant n} \left\{ \frac{\mathbb{P}(\xi_k > (1+\delta)x, \xi_l > (1+\delta)x)}{\overline{F}_{\xi_k}((1+\delta)x) + \overline{F}_{\xi_l}((1+\delta)x)} \right\}.$$

From the fact that $F_{\xi_k} \in \mathcal{D}$ for all $k \in \{1,\ldots,n\}$ and condition of pQAI for r.v.s $\{\xi_1,\ldots,\xi_n\}$, we get the following estimates:

$$\liminf_{x\to\infty} \mathcal{J}_1(x,\delta) \geqslant \min_{1\leqslant k\leqslant n} \left\{ \frac{1}{L_{\xi_k}} \liminf_{x\to\infty} \frac{\overline{F}_{\xi_k}((1+\delta)x)}{\overline{F}_{\xi_k}(x)} \right\}, \tag{29}$$

$$\limsup_{x\to\infty} \mathcal{J}_2(x,\delta) \leqslant 2(n-1) \max_{1\leqslant k\neq l\leqslant n} \left\{ \limsup_{x\to\infty} \frac{\mathbb{P}(\xi_k > \delta_2 x, \xi_l^- > \delta_2 x)}{\overline{F}_{\xi_k}(\delta_2 x) + \overline{F}_{\xi_l}(\delta_2 x)} \right\}$$
$$\times \max_{1\leqslant k\leqslant n} \left\{ \frac{1}{L_{\xi_k}} \limsup_{x\to\infty} \frac{\overline{F}_{\xi_k}(\delta_2 x)}{\overline{F}_{\xi_k}(x)} \right\} = 0, \tag{30}$$

$$\limsup_{x\to\infty} \mathcal{J}_3(x,\delta) \leqslant \max_{1\leqslant k\leqslant n} \left\{ \frac{2(n-1)}{L_{\xi_k}} \right\}$$
$$\times \max_{1\leqslant k\neq l\leqslant n} \left\{ \limsup_{x\to\infty} \frac{\mathbb{P}(\xi_k > (1+\delta)x, \xi_l > (1+\delta)x)}{\overline{F}_{\xi_k}((1+\delta)x) + \overline{F}_{\xi_l}((1+\delta)x)} \right\} = 0. \tag{31}$$

Thus, letting $\delta \downarrow 0$ from the estimates (29), (30), (31) and definition of indices L_{ξ_k}, we obtain the lower asymptotic bound in (21):

$$\limsup_{x\to\infty} \frac{\mathbb{P}(S_n^\xi > x)}{\sum_{k=1}^n L_{\xi_k}\overline{F}_{\xi_k}(x)} \geqslant 1.$$

This finish the proof of Lemma 2. □

Proof of Theorem 3. The special case, when $\alpha = 0$, is covered by Lemma 2. Consider $\alpha > 0$. The case $n = 1$ follows trivially from the definition of index L_{ξ_1}. Let $n \geqslant 2$. First, observe that, by Lemma 1 and the min–max inequality (22), we have

$$\frac{\mathbb{E}\left(\left(S_n^\xi\right)^\alpha \mathbb{1}_{\{S_n^\xi > x\}}\right)}{\sum_{k=1}^n \frac{1}{L_{\xi_k}} \mathbb{E}\left(\xi_k^\alpha \mathbb{1}_{\{\xi_k > x\}}\right)} = \frac{x^\alpha \mathbb{P}(S_n^\xi > x) + \alpha \int_x^\infty u^{\alpha-1}\mathbb{P}(S_n^\xi > u)\mathrm{d}u}{\sum_{k=1}^n \frac{1}{L_{\xi_k}} \left(x^\alpha \overline{F}_{\xi_k}(x) + \alpha \int_x^\infty u^{\alpha-1}\overline{F}_{\xi_k}(u)\mathrm{d}u\right)}$$
$$\leqslant \max\left\{ \frac{\mathbb{P}(S_n^\xi > x)}{\sum_{k=1}^n \frac{1}{L_{\xi_k}}\overline{F}_{\xi_k}(x)}, \frac{\int_x^\infty u^{\alpha-1}\mathbb{P}(S_n^\xi > u)\mathrm{d}u}{\int_x^\infty u^{\alpha-1}\sum_{k=1}^n \frac{1}{L_{\xi_k}}\overline{F}_{\xi_k}(u)\mathrm{d}u} \right\}$$
$$=: \max\{\mathcal{C}_1(x), \mathcal{C}_2(x)\}. \tag{32}$$

By Lemma 2, we obtain $\limsup_{x\to\infty} \mathcal{C}_1(x) \leqslant 1$, and, for the term $\mathcal{C}_2(x)$, we have that

$$\limsup_{x\to\infty} \mathcal{C}_2(x) = \limsup_{x\to\infty} \frac{\int_x^\infty u^{\alpha-1}\mathbb{P}(S_n^\xi > u)\mathrm{d}u}{\int_x^\infty u^{\alpha-1}\mathbb{P}(S_n^\xi > u)\frac{\sum_{k=1}^n \frac{1}{L_{\xi_k}}\overline{F}_{\xi_k}(u)}{\mathbb{P}(S_n^\xi > u)}\,\mathrm{d}u}$$

$$\leqslant \limsup_{x\to\infty} \sup_{u\geqslant x} \frac{\mathbb{P}(S_n^{\xi} > u)}{\sum_{k=1}^{n} \frac{1}{L_{\xi_k}} \overline{F}_{\xi_k}(u)}$$

$$= \limsup_{x\to\infty} \frac{\mathbb{P}(S_n^{\xi} > x)}{\sum_{k=1}^{n} \frac{1}{L_{\xi_k}} \overline{F}_{\xi_k}(x)} \leqslant 1,$$

where the last estimate follows from Lemma 2 as well. The desired upper estimate in (17) follows now from (32).

The asymptotic lower bound in (17) follows similarly. Indeed, in the same fashion, we obtain

$$\frac{\mathbb{E}\Big((S_n^{\xi})^{\alpha}\mathbb{1}_{\{S_n^{\xi}>x\}}\Big)}{\sum_{k=1}^{n} L_{\xi_k}\mathbb{E}\Big(\xi_k^{\alpha}\mathbb{1}_{\{\xi_k>x\}}\Big)} \geqslant \min\left\{\frac{\mathbb{P}(S_n^{\xi} > x)}{\sum_{k=1}^{n} L_{\xi_k}\overline{F}_{\xi_k}(x)}, \frac{\int_x^{\infty} u^{\alpha-1}\mathbb{P}(S_n^{\xi} > u)\mathrm{d}u}{\int_x^{\infty} u^{\alpha-1}\sum_{k=1}^{n} L_{\xi_k}\overline{F}_{\xi_k}(u)\mathrm{d}u}\right\}$$
$$=: \min\{C_3(x), C_4(x)\}. \tag{33}$$

Using Lemma 2, we have that $\liminf_{x\to\infty} C_3(x) \geqslant 1$ and

$$\liminf_{x\to\infty} C_4(x) = \liminf_{x\to\infty} \frac{\int_x^{\infty} u^{\alpha-1}\mathbb{P}(S_n^{\xi} > u)\,\mathrm{d}u}{\int_x^{\infty} u^{\alpha-1}\mathbb{P}(S_n^{\xi} > u)\frac{\sum_{k=1}^{n} L_{\xi_k}\overline{F}_{\xi_k}(u)}{\mathbb{P}(S_n^{\xi}>u)}\,\mathrm{d}u}$$

$$\geqslant \liminf_{x\to\infty} \inf_{u\geqslant x} \frac{\mathbb{P}(S_n^{\xi} > u)}{\sum_{k=1}^{n} L_{\xi_k}\overline{F}_{\xi_k}(u)}$$

$$= \liminf_{x\to\infty} \frac{\mathbb{P}(S_n^{\xi} > x)}{\sum_{k=1}^{n} L_{\xi_k}\overline{F}_{\xi_k}(x)} \geqslant 1,$$

which implies the lower estimate in (17) due to (33). Theorem 3 is proved. □

To prove Theorem 4, we need the following two additional lemmas from [2,6,27].

Lemma 3 (See Lemma 3.1 of [27] and Lemma 3 of [6])**.** *If ξ and θ are two independent r.v.s such that $F_{\xi} \in \mathcal{D}$ and θ is nonnegative, non-degenerate at zero r.v., then d.f. $F_{\theta\xi}$ of product $\theta\xi$ belongs to the class $\mathcal{D}$. If, in addition, $\mathbb{E}\theta^p < \infty$ for some $p > J_{\xi}^{+}$, then the inequality $L_{\theta\xi} \geqslant L_{\xi}$ holds for L-indices.*

Lemma 4 (See [6] Lemma 4)**.** *Let two pairs of r.v.s $\{\xi_1, \xi_2\}$ and $\{\theta_1, \theta_2\}$ be independent. Let ξ_1, ξ_2 be QAI r.v.s such that $F_{\xi_k} \in \mathcal{D}$, $k \in \{1,2\}$, and let θ_1, θ_2 be two arbitrarily dependent, nonnegative, non-degenerate at zero r.v.s with $\max\{\mathbb{E}\theta_1^p, \mathbb{E}\theta_2^p\} < \infty$ for some $p > \max\{J_{\xi_1}^{+}, J_{\xi_2}^{+}\}$. Then, r.v.s $\theta_1\xi_1$ and $\theta_2\xi_2$ are QAI as well.*

Proof. Although the proof of this lemma can be found in [6], we present a more detailed derivation based on the proof of Lemma 3.1 from [2]. Firstly, we need one result from [8]. Namely, by Lemma 3.7 of [8], we have that

$$\mathbb{P}\Big(\theta_i > x^{1-\varepsilon}\Big) = o(\overline{F}_{\xi_i}(x)) \tag{34}$$

for $i \in \{1,2\}$ and $\varepsilon \in (0, 1 - \max\{J_{\xi_1}^{+}, J_{\xi_2}^{+}\}/p)$.

It is obvious that, for a given $\widehat{\varepsilon} = (1 - \max\{J_{\xi_1}^{+}, J_{\xi_2}^{+}\}/p)/2$,

$$\frac{\mathbb{P}(\theta_1\xi_1 > x, \theta_2\xi_2 > x)}{\overline{F}_{\theta_1\xi_1}(x) + \overline{F}_{\theta_2\xi_2}(x)} = \frac{\mathbb{P}(\theta_1\xi_1 > x, \theta_2\xi_2 > x, \max\{\theta_1, \theta_2\} > x^{1-\widehat{\varepsilon}})}{\overline{F}_{\theta_1\xi_1}(x) + \overline{F}_{\theta_2\xi_2}(x)}$$

$$+\frac{\mathbb{P}(\theta_1\xi_1 > x, \theta_2\xi_2 > x, \max\{\theta_1, \theta_2\} \leqslant x^{1-\widehat{\varepsilon}})}{\overline{F}_{\theta_1\xi_1}(x) + \overline{F}_{\theta_2\xi_2}(x)}$$
$$:= \mathcal{L}_1(x, \widehat{\varepsilon}) + \mathcal{L}_2(x, \widehat{\varepsilon}). \quad (35)$$

Using (22), we estimate the first term in the following way:

$$\mathcal{L}_1(x, \widehat{\varepsilon}) \leqslant \frac{\mathbb{P}(\theta_1 > x^{1-\widehat{\varepsilon}}) + \mathbb{P}(\theta_2 > x^{1-\widehat{\varepsilon}})}{\overline{F}_{\theta_1\xi_1}(x) + \overline{F}_{\theta_2\xi_2}(x)} \leqslant \max_{i\in\{1,2\}}\left\{\frac{\mathbb{P}(\theta_i > x^{1-\widehat{\varepsilon}})}{\overline{F}_{\xi_i}(x)}\frac{\overline{F}_{\xi_i}(x)}{\overline{F}_{\theta_i\xi_i}(x)}\right\}.$$

Therefore,

$$\limsup_{x\to\infty} \mathcal{L}_1(x, \widehat{\varepsilon}) = 0 \quad (36)$$

because of (34) and

$$\limsup_{x\to\infty}\frac{\overline{F}_{\xi_i}(x)}{\overline{F}_{\theta_i\xi_i}(x)} \leqslant \limsup_{x\to\infty}\frac{\overline{F}_{\xi_i}(x)}{\mathbb{P}(\xi_i a > x, \theta_i > a)} \leqslant \frac{1}{\mathbb{P}(\theta_i > a)}\limsup_{x\to\infty}\frac{\overline{F}_{\xi_i}(x)}{\overline{F}_{\xi_i}(x/a)} < \infty, \quad (37)$$

provided that $F_{\xi_i} \in \mathcal{D}$ and $\mathbb{P}(\theta_i > a) > 0$ for some positive a.

For the second term of (35), using (22) once again, we get

$$\mathcal{L}_2(x, \widehat{\varepsilon}) = \frac{\iint\limits_{\{0<u_1,u_2\leqslant x^{1-\widehat{\varepsilon}}\}} \mathbb{P}\Big(\xi_1 > \frac{x}{u_1}, \xi_2 > \frac{x}{u_2}\Big)\mathrm{d}\mathbb{P}(\theta_1 \leqslant u_1, \theta_2 \leqslant u_2)}{\overline{F}_{\theta_1\xi_1}(x) + \overline{F}_{\theta_2\xi_2}(x)}$$
$$\leqslant \frac{\iint\limits_{\{0<u_1,u_2\leqslant x^{1-\widehat{\varepsilon}}\}} \mathbb{P}\Big(\xi_1 > \frac{x}{\max\{u_1,u_2\}}, \xi_2 > \frac{x}{\max\{u_1,u_2\}}\Big)\mathrm{d}\mathbb{P}(\theta_1 \leqslant u_1, \theta_2 \leqslant u_2)}{\overline{F}_{\theta_1\xi_1}(x) + \overline{F}_{\theta_2\xi_2}(x)}$$
$$\leqslant \frac{\mathbb{P}(\max\{\theta_1, \theta_2\}\xi_1 > x) + \mathbb{P}(\max\{\theta_1, \theta_2\}\xi_2 > x)}{\overline{F}_{\theta_1\xi_1}(x) + \overline{F}_{\theta_2\xi_2}(x)}$$
$$\times \sup_{\{0<u_1,u_2\leqslant x^{1-\widehat{\varepsilon}}\}} \frac{\mathbb{P}\Big(\xi_1 > \frac{x}{\max\{u_1,u_2\}}, \xi_2 > \frac{x}{\max\{u_1,u_2\}}\Big)}{\overline{F}_{\xi_1}\Big(\frac{x}{\max\{u_1,u_2\}}\Big) + \overline{F}_{\xi_2}\Big(\frac{x}{\max\{u_1,u_2\}}\Big)}$$
$$\leqslant \max_{i\in\{1,2\}}\left\{\frac{\mathbb{P}(\max\{\theta_1, \theta_2\}\xi_i > x)}{\overline{F}_{\xi_i}(x)}\frac{\overline{F}_{\xi_i}(x)}{\overline{F}_{\theta_i\xi_i}(x)}\right\}$$
$$\times \sup_{z\geqslant x^{\varepsilon}}\frac{\mathbb{P}(\xi_1 > z, \xi_2 > z)}{\overline{F}_{\xi_1}(z) + \overline{F}_{\xi_2}(z)}.$$

Since ξ_1, ξ_2 are QAI r.v.s and $\mathbb{E}(\max\{\theta_1, \theta_2\})^p < \infty$, the last estimate and relations (34), (37) imply that

$$\limsup_{x\to\infty} \mathcal{L}_2(x, \widehat{\varepsilon}) = 0. \quad (38)$$

By substituting relations (36) and (38) into (35), we get

$$\lim_{x\to\infty}\frac{\mathbb{P}(\theta_1\xi_1 > x, \theta_2\xi_2 > x)}{\overline{F}_{\theta_1\xi_1}(x) + \overline{F}_{\theta_2\xi_2}(x)} = 0.$$

The equality

$$\lim_{x\to\infty}\frac{\mathbb{P}((\theta_1\xi_1)^- > x, \theta_2\xi_2 > x)}{\overline{F}_{\theta_1\xi_1}(x) + \overline{F}_{\theta_1\xi_1}(x)} = 0$$

follows analogously, by observing that

$$\frac{\mathbb{P}((\theta_1\xi_1)^- > x, \theta_2\xi_2 > x)}{\overline{F}_{\theta_1\xi_1}(x) + \overline{F}_{\theta_1\xi_1}(x)} = \frac{\mathbb{P}(\theta_1\xi_1^- > x, \theta_2\xi_2 > x)}{\overline{F}_{\theta_1\xi_1}(x) + \overline{F}_{\theta_1\xi_1}(x)}$$

and replacing ξ_1 by ξ_1^- in the given proof. The lemma is proved. □

Proof of Theorem 4. Since we have that $\max\{\mathbb{E}\theta_1^p \ldots, \mathbb{E}\theta_n^p\} < \infty$ for some $p > \max\{J_{\xi_1}^+, \ldots, J_{\xi_n}^+\}$, Lemma 3 implies that $F_{\theta_k\xi_k} \in \mathcal{D}$ and $L_{\theta_k\xi_k} \geqslant L_{\xi_k}$ for all $k \in \{1, \ldots, n\}$. Additionally, by Lemma 4, we have that for any $k, l \in \{1, \ldots, n\}$, $k \neq l$, r.v.s $\theta_k\xi_k$, $\theta_l\xi_l$ are QAI. In other words, r.v.s $\theta_1\xi_1, \ldots, \theta_n\xi_n$ are pQAI. Thus, we only need to apply Theorem 3 for r.v.s $\theta_1\xi_1, \ldots, \theta_n\xi_n$ to obtain the desired result. □

6. Examples

In this section, we present three examples illustrating Theorem 3. For the sake of simplicity, in this section, we consider sums consisting of exactly two summands, i.e., we only consider bivariate distributions (ξ_1, ξ_2). Furthermore, we assume that their dependence structure is defined by the FGM copula described in Example 1 of Section 2.3. To illustrate the behaviour of dominatedly varying summands better, we consider three different cases of marginal distributions from the disjoint subclasses of $\mathcal{D}$.

Example 3. *Let the vector (ξ_1, ξ_2) coordinates follow a bivariate FGM copula with a parameter ϑ, and let ξ_1 and ξ_2 be distributed according to the Pareto distribution with parameters $\{\gamma_1, \varkappa_1\}$ and $\{\gamma_2, \varkappa_2\}$ (case of class $\mathcal{R}$), i.e.,*

$$F_{\xi_1}(x) = \left(1 - \left(\frac{\varkappa_1}{x}\right)^{\gamma_1}\right)\mathbb{1}_{\{x \geqslant \varkappa_1\}},\; F_{\xi_2}(x) = \left(1 - \left(\frac{\varkappa_2}{x}\right)^{\gamma_2}\right)\mathbb{1}_{\{x \geqslant \varkappa_2\}}.$$

For the parameter values $\gamma_1 = 4, \varkappa_1 = 5, \gamma_2 = 2, \varkappa_2 = 5$ and $\vartheta \in \{-0.8, 0, 0.8\}$, we compare simulated values of the moment tail $\mathbb{E}\Big((\xi_1 + \xi_2)^{1/2}\mathbb{1}_{\{\xi_1+\xi_2>x\}}\Big)$ with its asymptotic values obtained via Theorem 3.

Example 4. *Let ξ_1, ξ_2 be dependent r.v.s which dependence is controlled by the bivariate FGM copula as in the previous example. In addition, let ξ_1 and ξ_2 be distributed according to the generalised Peter and Paul distribution described in Section 2.2 with parameters $\{a_1, b_1\}$ and $\{a_2, b_2\}$ (case of class $\mathcal{D} \setminus \mathcal{L}$), i.e.,*

$$\overline{F}_{\xi_1}(x) = \mathbb{1}_{\{x<1\}} + \left(b_1^{-a_1}\right)^{\lfloor \log_{b_1} x \rfloor}\mathbb{1}_{\{x \geqslant 1\}},\; \overline{F}_{\xi_2}(x) = \mathbb{1}_{\{x<1\}} + \left(b_1^{-a_1}\right)^{\lfloor \log_{b_1} x \rfloor}\mathbb{1}_{\{x \geqslant 1\}}.$$

For the parameter values $a_1 = 1, b_1 = 2, a_2 = 1/2, b_2 = 2$ and $\vartheta \in \{-0.8, 0, 0.8\}$, we compare simulated values of the moment tail $\mathbb{E}\Big((\xi_1 + \xi_2)^{0.06}\mathbb{1}_{\{\xi_1+\xi_2>x\}}\Big)$ with its asymptotic bounds derived from Theorem 3.

Example 5. *Let us suppose that r.v.s ξ_1, ξ_2 is dependent with the dependence structure generated by the bivariate FGM copula as in the previous examples, and let ξ_1 and ξ_2 be distributed according to the Tang distribution described in Section 2.2 with parameters p_1 and p_2 case of class $\mathcal{C}$), i.e.,*

$$\xi_1 = (1+Y)2^{N_1},\; \xi_2 = (1+Y)2^{N_2},\; Y \stackrel{d}{=} \mathcal{U}([0,1]),$$
$$\mathbb{P}(N_1 = k) = (1-p_i)p_i^k,\; k \in \mathbb{N}_0,\; i = \{1, 2\}.$$

For the parameter values $p_1 = 0.2$, $p_2 = 0.3$ and $\vartheta \in \{-0.8, 0, 0.8\}$, we compare simulated values of the moment tail $\mathbb{E}\Big((\xi_1 + \xi_2)^{0.8}\mathbb{1}_{\{\xi_1+\xi_2>x\}}\Big)$ with its asymptotic values derived from Theorem 3.

Even though we can usually derive the analytic expression of an expectation $\mathbb{E}(\xi^\alpha \mathbb{1}_{\{\xi>x\}})$ knowing the distribution of a r.v. ξ, finding an analytic expression for $\mathbb{E}((\xi_1 + \xi_2)^\alpha \mathbb{1}_{\{\xi_1+\xi_2>x\}})$ might be unfeasible if we assume that r.v.s ξ_1 and ξ_2 are not independent. For this reason, in all the examples of this section, we derive the exact analytic expressions for the asymptotic bounds in (17) and find values of the moments tails of sums of r.v.s using the Monte-Carlo simulation method. Before turning to the final results for the Examples 3–5, which are stated in Section 6.3, we present some preliminaries.

6.1. Sampling Procedure

To obtain samples of r.v.s having arbitrary distributions using pseudo-random numbers generator, we use the so-called inverse probability integral transform property.

Lemma 5 (Inverse probability integral transform). *Let $U \overset{d}{=} \mathcal{U}([0,1])$ and X be an arbitrary r.v. with d.f. F. Then, $F^{\leftarrow}(U) \overset{d}{=} X$, where, by $F^{\leftarrow}(y)$, we denote the generalised inverse function (g.i.f.) of a d.f. F*

$$F^{\leftarrow}(y) := \inf\{x \mid F(x) \geqslant y\}.$$

The proof of this lemma, as well as some additional properties of g.i.f.s, can be found in [45]. Further, in this subsection, we derive the expressions of g.i.f.s of d.f.s in Examples 3–5.

- *G.i.f. of the Pareto d.f.* Consider the regularly varying Pareto d.f. F with parameters $\{\gamma, \varkappa\}$, i.e.,

$$F(x) = \left(1 - \left(\frac{\varkappa}{x}\right)^\gamma\right) \mathbb{1}_{\{x \geqslant \varkappa\}}. \tag{39}$$

Since F is strictly monotone and increasing on interval $[\varkappa, \infty)$, one can derive that $F^{\leftarrow}(y) = F^{-1}(y)$ and, therefore, for all $y \in [0,1)$

$$F^{\leftarrow}(y) = \varkappa(1-y)^{1/\gamma}.$$

- *G.i.f. of the Peter and Paul d.f.* Recall that Peter and Paul distribution with parameters $\{a, b\}$, $b > 1$, $a \in (0, \infty)$ is defined by the following d.f.

$$F(x) = (b^a - 1) \sum_{k \geqslant 1,\, b^k \leqslant x} b^{-ak} = \left(1 - (b^{-a})^{\lfloor \log_b x \rfloor}\right) \mathbb{1}_{\{x \geqslant 1\}}. \tag{40}$$

To find the g.i.f. $F^{\leftarrow}(y)$ we need to find the smallest x, for which $F(x) \geqslant y$. Since

$$1 - (b^{-a})^{\lfloor \log_b x \rfloor} \geqslant y \quad \Leftrightarrow \quad \lfloor \log_b x \rfloor \geqslant -\frac{1}{a} \log_b(1-y),$$

we get that

$$F^{\leftarrow}(y) = b^{\lceil -\frac{1}{a} \log_b(1-y) \rceil},$$

for all $y \in [0,1)$, where symbol $\lceil .. \rceil$ denotes the ceiling function.

- *G.i.f. of d.f. of the Cai–Tang (5) distribution.* In Section 2.2, we show that the d.f. of the r.v. $(1+Y)2^N$ with independent $Y \overset{d}{=} \mathcal{U}[0,1]$ and geometric N with parameter $p \in (0,1)$, is the following

$$F(x) = (1-p)\left(\frac{1 - p^{\lfloor \log_2 x \rfloor}}{1-p} + \left(\frac{x}{2^{\lfloor \log_2 x \rfloor}} - 1\right) p^{\lfloor \log_2 x \rfloor}\right) \mathbb{1}_{\{x \geqslant 1\}}.$$

To find the g.i.f. $F^{\leftarrow}$, we observe that the d.f. F is continuous, strictly monotone on the interval $[1, \infty)$ and linearly increasing on intervals $[2^k, 2^{k+1})$, $k \in \{0, 1, \ldots\}$. Hence, g.i.f. $F^{\leftarrow}$ coincides with F^{-1}.

Suppose that, for a given $y \in (0,1)$, variable $x \in [2^k, 2^{k+1})$ is such that $F(x) = y$. In such situation, we have

$$F(x) = (1-p)\left(\frac{1-p^k}{1-p} + \left(\frac{x}{2^k} - 1\right)p^k\right) = y \quad \Leftrightarrow \quad x = \frac{2^k\left(y - 1 + p^k(2-p)\right)}{p^k(1-p)}. \tag{41}$$

Since $F(2^n) = 1 - p^n$ for all $n \in \{0, 1, \ldots\}$, we obtain

$$F^{\leftarrow}(y) = \frac{2^{\lfloor \log_p(1-y) \rfloor}\left(y - 1 + p^{\lfloor \log_p(1-y) \rfloor}(2-p)\right)}{p^{\lfloor \log_p(1-y) \rfloor}(1-p)}$$

for all $y \in [0,1)$.

Similarly to the case described in Lemma 5, one can draw samples from multivariate distributions which marginals are not necessarily mutually independent. The procedure is mainly based on the so-called *Rosenblatt* transformation presented in [46]. According to the results of Rosenblatt [46], Brockwell [47], for an arbitrary random vector $(X_1, \ldots, X_n)$ with absolutely continuous distribution, the collection

$$\{F_1(X_1), F_2(X_2 \mid X_1), \ldots, F_n(X_n \mid X_{n-1}, \ldots, X_1)\}$$

consists of independent r.v.s which are uniformly distributed on interval $[0,1]$, where

$$F_1(x_1) = \mathbb{P}(X_1 \leqslant x_1),\ F_2(x_2 \mid x_1) = \mathbb{P}(X_2 \leqslant x_2 \mid X_1 = x_1),$$
$$F_k(x_k \mid x_{k-1}, \ldots, x_1) = \mathbb{P}(X_k \leqslant x_k \mid X_{k-1} = x_{k-1}, \ldots, X_1 = x_1), k \in \{3, \ldots, n\}.$$

For any copula $C(u,v) = \mathbb{P}(U \leqslant u, V \leqslant v)$, the conditional distribution function of U for the given event $\{V = v\}$ is defined by equality

$$C_v(u) := \mathbb{P}(U \leqslant u \mid V = v) = \lim_{\delta \downarrow 0} \frac{C(u, v+\delta) - C(u,v)}{\delta} = \frac{\partial}{\partial v} C(u,v).$$

By Theorem 2.2.7 of [38], it follows that the partial derivative in the last expression exists for almost all v in the interval $[0,1]$. To sample from a bivariate copula, we follow the algorithm presented in Section 2.9 of [38].

- *Algorithm $\mathcal{N}$. Generation of samples from a bivariate distribution characterised by marginal d.f.s F_1, F_2 and copula $C(u,v)$.*

  Step **1**: Generate two independent realisations $\{t^*, v^*\}$ of distribution $\mathcal{U}([0,1])$.

  Step **2**: To induce the copula implied dependence, transform t^* into $u^* = C_{v^*}^{\leftarrow}(t^*)$, where $C_v^{\leftarrow}(t)$ is the g.i.f. of the conditional distribution $C_v(u)$. In such a way, we obtain the realisation (u^*, v^*) from copula $C(u,v)$.

  Step **3**: Obtain the realisation of the desired distribution using Lemma 5 by transforming (u^*, v^*) into $\left(F_1^{\leftarrow}(u^*), F_2^{\leftarrow}(v^*)\right)$.

  In what follows, we derive the conditional distribution function $C_v(u)$ and its generalised inverse $C_v^{\leftarrow}(t)$ for the bivariate FGM copula which is used in Examples 3–5.

- *Inverse conditional distribution of bivariate FGM copula.*

  FGM copula is described in Section 2.3. Since

$$C_\vartheta(u,v) = uv + \vartheta uv(1-u)(1-u), \quad \theta \in [-1,1],$$

for $u, v \in [0,1]$, we get that

$$C_{\vartheta,v}(u) = u(1 - \vartheta + 2v\vartheta) + u^2(\vartheta - 2v\vartheta).$$

To obtain the inverse $C^{\leftarrow}_{\vartheta,v}(t)$, we observe that equation

$$x^2(\vartheta - 2v\vartheta) + x(1 - \vartheta + 2v\vartheta) - t = 0, \qquad t \in (0,1),$$

has two roots

$$x_{1,2} = \frac{-(1 - \vartheta + 2v\vartheta) \pm \sqrt{(1 - \vartheta + 2v\vartheta)^2 - 4t(\vartheta - 2v\vartheta)}}{2(\vartheta - 2v\vartheta)}.$$

We are interested in $C^{\leftarrow}_{\vartheta,v}(t) \in (0,1)$. Consequently,

$$C^{\leftarrow}_{\vartheta,v}(t) = \frac{-(1 - \vartheta + 2v\vartheta) + \sqrt{(1 - \vartheta + 2v\vartheta)^2 - 4t(\vartheta - 2v\vartheta)}}{2(\vartheta - 2v\vartheta)}.$$

6.2. *Analytic Expressions of Individual Summands' Tail Expectations*

To obtain exact analytic expressions of the bounding functions in (17), we need to find the tail expectations $\mathbb{E}(\xi^\alpha \mathbb{1}_{\{\xi > x\}})$ for all marginal distributions considered in Examples 3–5 together with L-indices in the case of the generalised Peter and Paul r.v.s. Note that both the Pareto distribution and the Cai–Tang distribution (5) defined in Section 2.2 belong to the class $\mathcal{C}$. Hence, the L-indices for both distributions are equal to units, and we obtain the exact asymptotic equivalences in (17).

- *Truncated expectation of the Pareto distribution.* Let us consider r.v. ξ having the Pareto distribution with parameters $\{\gamma, \varkappa\}$ presented in Equation (39). If $\gamma > \alpha$, then it is obvious that

$$\mathbb{E}\Big(\xi^\alpha \mathbb{1}_{\{\xi > x\}}\Big) = \frac{\gamma \varkappa^\gamma \max\{x, \varkappa\}^{\alpha - \gamma}}{\gamma - \alpha},$$

- *Truncated expectation and L-index of Peter and Paul distribution.* If r.v. ξ has the generalised Peter and Paul distribution (40) with parameters $\{a, b\}$, $\alpha < a$, then

$$\mathbb{E}\Big(\xi^\alpha \mathbb{1}_{\{\xi > x\}}\Big) = \sum_{k=1}^{\infty} (b^a - 1) b^{k(\alpha - a)} \mathbb{1}_{\{b^k > x\}} = \sum_{k = \lfloor \log_b x \rfloor + 1}^{\infty} (b^a - 1) b^{k(\alpha - a)}$$
$$= (b^a - 1) \frac{b^{\lfloor \log_b x \rfloor (\alpha - a)}}{b^{a - \alpha} - 1}.$$

In addition, for any $y > 1$,

$$\liminf_{x \to \infty} \frac{\overline{F}_\xi(xy)}{\overline{F}_\xi(x)} = \liminf_{x \to \infty} \big(b^{-a}\big)^{\lfloor \log_b y \rfloor + \lfloor \widehat{\log_b x} + \widehat{\log_b y} \rfloor} = \big(b^{-a}\big)^{\lfloor \log_b y \rfloor + 1},$$

where the symbol $\widehat{z}$ denotes the fractional part of z. Hence, L-index of r.v. ξ

$$L_\xi = \lim_{y \downarrow 1} \liminf_{x \to \infty} \frac{\overline{F}_\xi(xy)}{\overline{F}_\xi(x)} = \lim_{y \downarrow 1} \big(b^{-a}\big)^{\lfloor \log_b y \rfloor + 1} = b^{-a}.$$

- *Truncated expectation of the Cai–Tang distribution.* Let ξ be r.v. defined by Equation (5). If $\alpha < \log_2(1/p)$, then according to (6) and (41), we get

$$\mathbb{E}\Big(\xi^\alpha \mathbb{1}_{\{\xi > x\}}\Big) = \frac{1-p}{\alpha+1}\left(\frac{2^{\alpha+1} - 1}{1 + 2^\alpha p}\right) \mathbb{1}_{\{x < 1\}} + \frac{1-p}{\alpha+1}\left(\left(\frac{p}{2}\right)^{\lfloor \log_2 x \rfloor}\left(2^{\lceil \log_2 x \rceil (\alpha+1)} - x^{\alpha+1}\right)\right.$$
$$\left. + \frac{(2^{\alpha+1} - 1)(2^\alpha p)^{\lceil \log_2 x \rceil}}{1 - 2^\alpha p}\right) \mathbb{1}_{\{x \geqslant 1\}},$$

because for $x < 1$

$$\begin{aligned}
\mathbb{E}\Big(\xi^{\alpha}\mathbb{1}_{\{\xi>x\}}\Big) &= \mathbb{E}(\xi^{\alpha}) = \int_{[1,\infty)} u^{\alpha}\mathrm{d}F_{\xi}(u) \\
&= (1-p)\int_{[1,\infty)} u^{\alpha}\mathrm{d}\left(\frac{1-p^{\lfloor \log_2 u\rfloor}}{1-p} + \left(\frac{u}{2^{\lfloor \log_2 u\rfloor}} - 1\right)p^{\lfloor \log_2 u\rfloor}\right) \\
&= (1-p)\sum_{k=0}^{\infty}\int_{[2^k,2^{k+1})} u^{\alpha}\mathrm{d}\left(\frac{1-p^{\lfloor \log_2 u\rfloor}}{1-p} + \left(\frac{u}{2^{\lfloor \log_2 u\rfloor}} - 1\right)p^{\lfloor \log_2 u\rfloor}\right) \\
&= (1-p)\sum_{k=0}^{\infty}\int_{[2^k,2^{k+1})} u^{\alpha}\mathrm{d}\left(\frac{1-p^{k}}{1-p} + \left(\frac{u}{2^{k}} - 1\right)p^{k}\right) \\
&= (1-p)\sum_{k=0}^{\infty}\left(\frac{p}{2}\right)^{k}\int_{[2^k,2^{k+1})} u^{\alpha}\mathrm{d}u \\
&= (1-p)\sum_{k=0}^{\infty}\left(\frac{p}{2}\right)^{k}\frac{\left(2^k\right)^{\alpha+1}\left(2^{\alpha+1}-1\right)}{\alpha+1} \\
&= \frac{\left(2^{\alpha+1}-1\right)p(1-p)}{(\alpha+1)(1-2^{\alpha}p)},
\end{aligned}$$

and for $x \geqslant 1$

$$\mathbb{E}\Big(\xi^{\alpha}\mathbb{1}_{\{\xi>x\}}\Big) = \int_{\left(x,2^{\lceil \log_2 x\rceil}\right)} u^{\alpha}\mathrm{d}F_{\xi}(u) + \int_{\left[2^{\lceil \log_2 x\rceil},\infty\right)} u^{\alpha}\mathrm{d}F_{\xi}(u) =: \mathcal{K}_1 + \mathcal{K}_2,$$

with

$$\begin{aligned}
\mathcal{K}_1 &= \int_{\left(x,2^{\lceil \log_2 x\rceil}\right)} u^{\alpha}\mathrm{d}\left((1-p)\left(\frac{1-p^{\lfloor \log_2 u\rfloor}}{1-p} + \left(\frac{u}{2^{\lfloor \log_2 u\rfloor}} - 1\right)\right)p^{\lfloor \log_2 u\rfloor}\right) \\
&= \int_{\left(x,2^{\lceil \log_2 x\rceil}\right)} u^{\alpha}\mathrm{d}\left((1-p)\left(\frac{1-p^{\lfloor \log_2 x\rfloor}}{1-p} + \left(\frac{u}{2^{\lfloor \log_2 x\rfloor}} - 1\right)\right)p^{\lfloor \log_2 x\rfloor}\right) \\
&= (1-p)\left(\frac{p}{2}\right)^{\lfloor \log_2 x\rfloor}\int_{\left(x,2^{\lceil \log_2 x\rceil}\right)} u^{\alpha}\mathrm{d}u \\
&= (1-p)\left(\frac{p}{2}\right)^{\lfloor \log_2 x\rfloor}\frac{2^{\lceil \log_2 x\rceil(\alpha+1)} - x^{\alpha+1}}{\alpha+1}
\end{aligned}$$

and

$$\begin{aligned}
\mathcal{K}_2 &= \int_{\left[2^{\lceil \log_2 x\rceil},\infty\right)} u^{\alpha}\mathrm{d}\left((1-p)\left(\frac{1-p^{\lfloor \log_2 u\rfloor}}{1-p} + \left(\frac{u}{2^{\lfloor \log_2 u\rfloor}} - 1\right)\right)p^{\lfloor \log_2 u\rfloor}\right) \\
&= (1-p)\sum_{k=0}^{\infty}\left(\frac{p}{2}\right)^{\lceil \log_2 x\rceil+k}\int_{\left[2^{\lceil \log_2 x\rceil+k},2^{\lceil \log_2 x\rceil+k+1}\right)} u^{\alpha}\mathrm{d}u \\
&= (1-p)\sum_{k=0}^{\infty}\left(\frac{p}{2}\right)^{\lceil \log_2 x\rceil+k}\frac{2^{(\lceil \log_2 x\rceil+k)(\alpha+1)}\left(2^{\alpha+1}-1\right)}{\alpha+1} \\
&= \frac{(1-p)\left(2^{\alpha+1}-1\right)\left(2^{\alpha}p\right)^{\lceil \log_2 x\rceil}}{(\alpha+1)}\sum_{k=0}^{\infty}\left(2^{\alpha}p\right)^{k} \\
&= \frac{(1-p)\left(2^{\alpha+1}-1\right)\left(2^{\alpha}p\right)^{\lceil \log_2 x\rceil}}{(\alpha+1)(1-2^{\alpha}p)}.
\end{aligned}$$

6.3. Simulation Procedure and Results

We performed three different simulation studies described in Examples 3–5. We specified the concrete d.f.s of r.v.s ξ_1, ξ_2 and exponent α in (17). For every case, we considered three different scenarios defined by parameter ϑ of the FGM copula. In particular, we chose $\theta = 0$ to include independent case of ξ_1, ξ_2 and two other cases, namely $\theta = -0.8$ and $\theta = 0.8$, to reflect how imposed dependence affect the overall asymptotic behaviour. For all three cases of bivariate distributions, we calculated asymptotic bounds in (17) for various x values to see how quickly the theoretical asymptotics are attained as x tends to infinity.

- Under the conditions of Example 3, we get from Theorem 3 that

$$\mathbb{E}\Big((\xi_1+\xi_2)^{1/2}\mathbb{1}_{\{\xi_1+\xi_2>x\}}\Big) \sim \sum_{k=1}^{2}\mathbb{E}\Big(\xi_k^{1/2}\mathbb{1}_{\xi_k>x}\Big)\left[=\frac{100}{3x^{3/2}}\left(1+\frac{150}{7x^2}\right), x>5\right]$$

for all $\vartheta \in \{-0.8, 0, 0.8\}$, according to the expressions of truncated moments derived in Section 6.2. The results of simulated values of $\mathbb{E}\Big((\xi_1+\xi_2)^{1/2}\mathbb{1}_{\{\xi_1+\xi_2>x\}}\Big)$ together with the values of the derived asymptotic formula are presented in Figure 1.

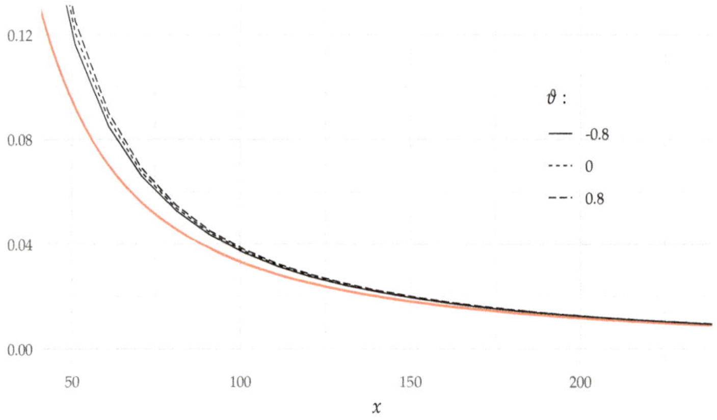

Figure 1. Simulated and asymptotic values for the truncated expectation of Example 3. Solid red line represents the exact asymptotic values of $\sum_{k=1}^{2}\mathbb{E}\Big(\xi_k^{1/2}\mathbb{1}_{\{\xi_k>x\}}\Big)$.

- The conditions of Example 4 and Theorem 3 imply that

$$\mathbb{E}\Big((\xi_1+\xi_2)^{0.06}\mathbb{1}_{\{\xi_1+\xi_2>x\}}\Big) \lesssim \sum_{k=1}^{2}\frac{1}{L_{\xi_k}}\mathbb{E}\Big(\xi_k^{0.06}\mathbb{1}_{\{\xi_k>x\}}\Big)$$
$$\left[=\frac{2}{2^{0.94}-1}2^{-0.94\lfloor\log_2 x\rfloor}+\frac{2-\sqrt{2}}{2^{0.44}-1}2^{-0.44\lfloor\log_2 x\rfloor}, x>1\right],$$
$$\mathbb{E}\Big((\xi_1+\xi_2)^{0.06}\mathbb{1}_{\{\xi_1+\xi_2>x\}}\Big) \gtrsim \sum_{k=1}^{2}L_{\xi_k}\mathbb{E}\Big(\xi_k^{0.06}\mathbb{1}_{\{\xi_k>x\}}\Big)$$
$$\left[=\frac{1}{2(2^{0.94}-1)}2^{-0.94\lfloor\log_2 x\rfloor}+\frac{(\sqrt{2}-1)}{\sqrt{2}(2^{0.44}-1)}2^{-0.44\lfloor\log_2 x\rfloor}, x>1\right]$$

due to the formulas derived in Section 6.2. The results of the simulated values of $\mathbb{E}\Big((\xi_1+\xi_2)^{0.06}\mathbb{1}_{\{\xi_1+\xi_2>x\}}\Big)$ together with the asymptotic values are presented in Figure 2.

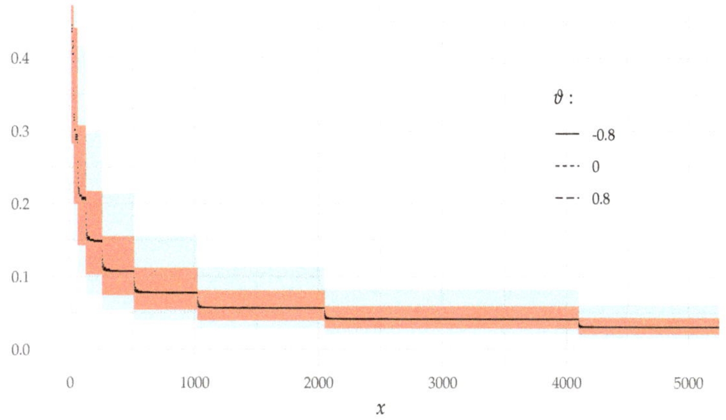

Figure 2. Simulated and asymptotic values for the truncated expectation of Example 4. Red area represents the region bounded by $\sum_{k=1}^{2} L_{\xi_k} \mathbb{E}\left(\xi_k^{0.06} \mathbb{1}_{\{\xi_k>x\}}\right)$ and $\sum_{k=1}^{2} L_{\xi_k}^{-1} \mathbb{E}\left(\xi_k^{0.06} \mathbb{1}_{\{\xi_k>x\}}\right)$. Cyan area reflects the additional error using bounding coefficients $L_2^{\xi} = \min\{L_{\xi_1}, L_{\xi_2}\}$ and $(L_2^{\xi})^{-1}$.

- Under the conditions of Example 5, Theorem 3 implies that $\mathbb{E}\left((\xi_1+\xi_2)^{0.8}\mathbb{1}_{\{\xi_1+\xi_2>x\}}\right)$ can be approximated by sum

$$\sum_{k=1}^{2} \mathbb{E}\left(\xi_k^{0.8}\mathbb{1}_{\xi_k>x}\right) = \frac{4}{9}\left(10^{-\lfloor \log_2 x\rfloor}\left(2^{1.8\lceil \log_2 x\rceil} - x^{1.8}\right) + \frac{(2^{1.8}-1)(2^{4/5}/5)^{\lceil \log_2 x\rceil}}{1-2^{4/5}/5}\right)$$
$$+\frac{7}{18}\left[\left(\frac{3}{10}\right)^{\lfloor \log_2 x\rfloor}\left(2^{\lceil \log_2 x\rceil} - x^{1.8}\right) + \frac{(2^{1.8}-1)(2^{4/5}3/10)^{\lceil \log_2 x\rceil}}{1-2^{4/5}3/10}\right], x>1,$$

for large x and for all three parameter ϑ values. The simulated values of $\mathbb{E}\left((\xi_1+\xi_2)^{0.8}\mathbb{1}_{\{\xi_1+\xi_2>x\}}\right)$ and its asymptotic values are presented in Figure 3.

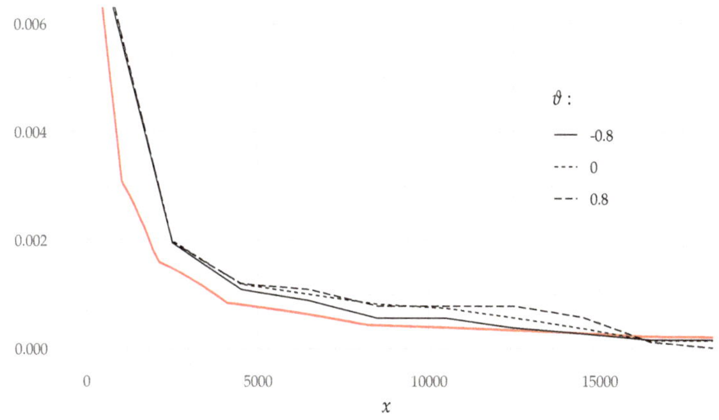

Figure 3. Simulated and asymptotic values for the truncated expectation of Example 5. Solid red line represents the exact asymptotic values of $\sum_{k=1}^{2} \mathbb{E}\left(\xi_k^{0.8}\mathbb{1}_{\{\xi_k>x\}}\right)$.

From the presented graphs, we can see that the tail expectations $\mathbb{E}((\xi_1+\xi_2)^\alpha \mathbb{1}_{\{\xi_1+\xi_2>x\}})$ are approximated quite accurately by their asymptotic values in all three examples. In addition, the effect of the copula dependence implied by the parameter ϑ quickly becomes negligible, as x attains larger values. In addition, we observe that the scale of the horizontal axis is different in each of the graphs, which reflects the different rate of convergence in the three examples.

To perform Monte Carlo simulations evaluating $\mathbb{E}((\xi_1+\xi_2)^\alpha \mathbb{1}_{\{\xi_1+\xi_2>x\}})$ for each of the three examples, we firstly generated two samples consisting of $M = 2 \times 10^7$ uniform random variates, namely vectors $(t_1, \ldots, t_M)$ and $(v_1, \ldots, v_M)$, and transformed them according to *Algorithm* $\mathcal{N}$ to induce the FGM copula dependence. Then, we modified each of the resulting vectors $(u_1, \ldots, u_M)$ and $(v_1, \ldots, v_M)$ according to the inverse probability integral transform described in Section 6.1, obtaining samples $\left\{F_{\xi_1}^{\leftarrow}(u_k), F_{\xi_2}^{\leftarrow}(v_k)\right\}_{k=1}^{M}$ from distributions F_{ξ_1} and F_{ξ_2}. Finally, we replaced elements of the collection $\left\{(F_{\xi_1}^{\leftarrow}(u_k) + F_{\xi_2}^{\leftarrow}(v_k))^\alpha\right\}_{k=1}^{M}$ exceeding given threshold x by zeroes and calculated the empirical mean of the resulting vector.

All simulations were computed in the statistical programming package **R** [48]. Apart from the base **R** functions, several others from the *tictoc* [49], *tikzDevice* [50], *furrr* [51] and *tidyverse* [52] libraries were used.

7. Conclusions

In this paper, we investigate the asymptotic behaviour of tails of the moments for randomly weighted sums with possibly dependent dominatedly varying summands. Our results improve and generalise other related findings in the literature. Firstly, by putting the L-indices of individual summands inside the bounding sums, we achieve sharper asymptotic bounds under pQAI dependence structure. Moreover, we relax the condition for the exponent, allowing it to be any fixed nonnegative real number. Finally, in the case of randomly weighted sums, we substitute the boundedness condition on random weights by a less restrictive moments condition.

To illustrate and further validate the obtained results, we performed a Monte Carlo simulation study in which we considered three concrete examples of random sums from disjoint subclasses of dominatedly varying distributions. The simulations confirmed our derived asymptotic relations.

Author Contributions: Conceptualization, S.P. and J.Š.; methodology, M.D.; software, M.D.; validation, M.D., S.P. and J.Š.; formal analysis, S.P.; investigation, M.D.; writing-original draft preparation, M.D. and S.P.; writing-review and editing, J.Š; visualization, M.D. and S.P.; supervision, J.Š; project administration, J.Š; funding acquisition, J.Š. All authors have read and agreed to the published version of the manuscript.

Funding: This research was funded by grant No. S-MIP-20-16 from the Research Council of Lithuania.

Institutional Review Board Statement: Not applicable.

Informed Consent Statement: Not applicable.

Data Availability Statement: Not applicable.

Acknowledgments: The authors would like to express deep gratitude to two anonymous referees for their valuable suggestions and comments which have helped to improve the previous version of the paper.

Conflicts of Interest: The authors declare no conflict of interest.

References

1. Chen, Y. A renewal shot noise process with subexponential shot marks. *Risks* **2019**, *7*, 63. [CrossRef]
2. Chen, Y.; Yuen, K.C. Sums of pairwise quasi-asymptotically independent random variables with consistent variation. *Stoch. Models* **2009**, *25*, 76–89. [CrossRef]

3. Cheng, D. Randomly weighted sums of dependent random variables with dominated variation. *J. Math. Anal. Appl.* **2014**, *420*, 1617–1633. [CrossRef]
4. Geluk, J.; Tang, Q. Asymptotic tail probabilities of sums of dependent subexponential random variables. *J. Theoret. Probab.* **2009**, *22*, 871–882. [CrossRef]
5. Goovaerts, M.J.; Kaas, R.; Laeven, R.J.A.; Tang, Q.; Vernic, R. The tail probability of discounted sums of Pareto-like losses in insurance. *Scand. Actuar. J.* **2005**, *2005*, 446–461. [CrossRef]
6. Jaunė, E.; Ragulina, O.; Šiaulys, J. Expectation of the truncated randomly weighted sums with dominatedly varying summands. *Lith. Math. J.* **2018**, *58*, 421–440. [CrossRef]
7. Leipus, R.; Šiaulys, J.; Vareikaitė, I. Tails of higher-order moments with dominatedly varying summands. *Lith. Math. J.* **2019**, *59*, 389–407. [CrossRef]
8. Tang, Q.; Tsitsiashvili, G. Precise estimates for the ruin probability in finite horizon in a discrete-time model with heavy-tailed insurance and financial risks. *Stoch. Process. Appl.* **2003**, *108*, 299–325. [CrossRef]
9. Wang, D.; Su, C.; Zeng, Y. Uniform estimate for maximum of randomly weighted sums with applications to insurance risk theory. *Sci. China Ser. A* **2005**, *48*, 1379–1394. [CrossRef]
10. Wang, D.; Tang, Q. Tail probabilities of randomly weighted sums of random variables with dominated variation. *Stoch. Models* **2006**, *22*, 253–272. [CrossRef]
11. Yi, L.; Chen, Y.; Su, C. Approximation of the tail probability of randomly weighted sums of dependent random variables with dominated variation. *J. Math. Anal. Appl.* **2011**, *376*, 365–372. [CrossRef]
12. Yang, Y.; Ignatavičiūtė, E.; Šiaulys, J. Conditional tail expectation of randomly weighted sums with heavy-tailed distributions. *Stat. Probab. Lett.* **2015**, *105*, 20–28. [CrossRef]
13. Fougeres, A.L.; Mercadier, C. Risk measures and multivariate extensions of Breiman's theorem. *J. Appl. Probab.* **2012**, *49*, 364–384. [CrossRef]
14. Nyrhinen, H. On the ruin probabilities in a general economic environment. *Stoch. Process. Appl.* **1999**, *83*, 319–330. [CrossRef]
15. Tang, Q.; Yuan, Z. Randomly weighted sums of subexponential random variables with application to capital allocation. *Extremes* **2014**, *17*, 467–493. [CrossRef]
16. Wang, S.; Chen, C.; Wang, X. Some novel results on pairwise quasi-asymptotical independence with applications to risk theory. *Comm. Stat. Theory Methods* **2017**, *46*, 9075–9085. [CrossRef]
17. Verrall, R.J. The individual risk model: A compound distribution. *J. Inst. Actuaries* **1989**, *116*, 101–107. [CrossRef]
18. Dhaene, J.; Goovaerts, M.J. On the dependency of risks in the individual life model. *Insur. Math. Econom.* **1997**, *19*, 243–253. [CrossRef]
19. Dickson, D.C.M. *Insurance Risk and Ruin*; Cambridge University Press: Cambridge, UK, 2005.
20. Acerbi, C. Spectral measures of risks: A coherent representation of subjective risk aversion. *J. Bank. Financ.* **2002**, *26*, 1505–1518. [CrossRef]
21. Artzner, P.; Delbaen, F.; Eber, J.M.; Heath, D. Coherent measures of risk. *Math. Financ.* **1999**, *9*, 203–228. [CrossRef]
22. Asimit, A.V.; Furman, E.; Tang, Q.; Vernic, R. Asymptotics for risk capital allocations based on conditional tail expectation. *Insur. Math. Econom.* **2011**, *49*, 310–324. [CrossRef]
23. Hua, L.; Joe, H. Strength of tail dependence based on conditional tail expectation. *J. Multivar. Anal.* **2014**, *123*, 143–159. [CrossRef]
24. Wang, S.; Hu, Y.; Yang, L.; Wang, W. Randomly weighted sums under a wide type of dependence structure with application to conditional tail expectation. *Commun. Stat. Theory Methods* **2018**, *47*, 5054–5063. [CrossRef]
25. Goldie, C.M. Subexponential distributions and dominated-variation tails. *J. Appl. Probab.* **1978**, *15*, 440–442. [CrossRef]
26. Cai, J.; Tang, Q. On max-sum equivalence and convolution closure of heavy-tailed distributions and their applications. *J. Appl. Probab.* **2004**, *41*, 117–130. [CrossRef]
27. Cline, D.B.H.; Samorodnitsky, G. Subexponentiality of the product of independent random variables. *Stoch. Process. Appl.* **1994**, *49*, 75–98. [CrossRef]
28. Embrechts, P.; Goldie, C.M. On closure and factorization properties of subexponential and related distributions. *J. Austral. Math. Soc. Ser. A* **1980**, *29*, 243–256. [CrossRef]
29. Embrechts, P.; Omey, E. A property of longtailed distributions. *J. Appl. Probab.* **1984**, *21*, 80–87. [CrossRef]
30. Foss, S.; Korshunov, D.; Zachary, S. *An Introduction to Heavy-Tailed and Subexponential Distributions*, 2nd ed.; Springer: New York, NY, USA, 2013.
31. Konstantinides, D.G. *Risk Theory. A Heavy Tail Approach*; World Scientific Publishing: Singapore, 2018.
32. Pitman, E.J.G. Subexponential distribution functions. *J. Austral. Math. Soc. Ser. A* **1980**, *29*, 337–347. [CrossRef]
33. Wang, Y.B.; Wang, K.Y.; Cheng, D.Y. Precise large deviations for sums of negatively associated random variables with common dominatedly varying tails. *Acta Math. Sin. (Engl. Ser.)* **2006**, *22*, 1725–1734. [CrossRef]
34. Yang, Y.; Wang, Y. Asymptotics for ruin probability of some negatively dependent risk models with a constant interest rate and dominatedly-varying-tailed claims. *Statist. Probab. Lett.* **2010**, *80*, 143–154. [CrossRef]
35. Matuszewska, W. On a generalization of regularly increasing functions. *Studia Math.* **1964**, *24*, 271–279. [CrossRef]
36. Bingham, N.H.; Goldie, C.M.; Teugels, J.L. *Regular Variation*; Cambridge University Press: Cambridge, UK, 1987.
37. Sklar, M. Fonctions de répartition à *n* dimensions et leurs marges. *Publ. Inst. Stat. Univ. Paris* **1959**, *8*, 229–231.
38. Nelsen, R.B. *An introduction to Copulas*, 2nd ed.; Springer: New York, NY, USA, 2006.

39. Kotz, S.; Balakrishnan, N.; Johnson, N.L. *Continuous Multivariate Distributions*, 2nd ed.; John Wiley and Sons: New York, NY, USA, 2000; Volume 1.
40. Ali, M.M.; Mikhail, N.N.; Haq, M.S. A class of bivariate distributions including the bivariate logistic. *J. Multivar. Anal.* **1978**, *8*, 405–412. [CrossRef]
41. Albrecher, H.; Asmussen, S.; Kortschak, D. Tail asymptotics for the sum of two heavy-tailed dependent risks. *Extremes* **2006**, *9*, 107–130. [CrossRef]
42. Fang, H.; Ding, S.; Li, X.; Yang, W. Asymptotic approximations of ratio moments based on dependent sequences. *Mathematics* **2020**, 8, 361. [CrossRef]
43. Yang, H.; Gao, W.; Li, J. Asymptotic ruin probabilities for a discrete-time risk model with dependent insurance and financial risks. *Scand. Actuar. J.* **2016**, *2016*, 1–17. [CrossRef]
44. Li, J. On pairwise quasi-asymptotically independent random variables and their applications. *Stat. Probab. Lett.* **2013**, *83*, 2081–2087. [CrossRef]
45. Embrechts, P.; Hofert, M. A note on generalized inverses. *Math. Methods Oper. Res. (Heidelb)* **2013**, *77*, 423–432. [CrossRef]
46. Rosenblatt, M. Remarks on a multivariate transformation. *Ann. Math. Stat.* **1952**, *23*, 470–472. [CrossRef]
47. Brockwell, A.E. Universal residuals: A multivariate transformation. *Stat. Probab. Lett.* **2007**, *77*, 1473–1478. [CrossRef] [PubMed]
48. R Core Team. *R: A Language and Environment for Statistical Computing*; R Foundation for Statistical Computing: Vienna, Austria, 2018. Available online: https://www.R-project.org/ (accessed on 15 December 2020).
49. Izrailev, S. Tictoc: Functions for Timing R Scripts, as Well as Implementations of Stack and List Structures. R Package Version 1.0. 2014. Available online: https://CRAN.R-project.org/package=tictoc (accessed on 23 December 2020).
50. Sharpsteen, C.; Bracken, C. tikzDevice: R Graphics Output in LaTeX Format. R Package Version 0.12.3.1. 2020. Available online: https://CRAN.R-project.org/package=tikzDevice (accessed on 9 January 2021).
51. Vaughan, D.; Dancho, M. Furrr: Apply Mapping Functions in Parallel Using Futures. R Package Version 0.1.0. 2018. Available online: https://CRAN.R-project.org/package=furrr (accessed on 17 January 2021).
52. Wickham, H.; Averick, M.; Bryan, J.; Chang, W.; McGowan, L.D.; François, R.; Grolemund, G.; Hayes, A.; Henry, L.; Hester, J.; et al. Welcome to the tidyverse. *J. Open Source Softw.* **2019**, *4*, 1686. [CrossRef]

mathematics

MDPI

Article

Particle Filtering: A Priori Estimation of Observational Errors of a State-Space Model with Linear Observation Equation †

Rodi Lykou *,‡ and George Tsaklidis ‡

Department of Statistics and Operational Research, School of Mathematics, Aristotle University of Thessaloniki, GR54124 Thessaloniki, Greece; tsaklidi@math.auth.gr
* Correspondence: lykourodi@math.auth.gr
† This paper is an extended version of our published in the 6th Stochastic Modeling Techniques and Data Analysis International Conference with Demographics Workshop. 2–5 June 2020, Turned into an Online Conference.
‡ These authors contributed equally to this work.

Abstract: Observational errors of Particle Filtering are studied over the case of a state-space model with a linear observation equation. In this study, the observational errors are estimated prior to the upcoming observations. This action is added to the basic algorithm of the filter as a new step for the acquisition of the state estimations. This intervention is useful in the presence of missing data problems mainly, as well as sample tracking for impoverishment issues. It applies theory of Homogeneous and Non-Homogeneous closed Markov Systems to the study of particle distribution over the state domain and, thus, lays the foundations for the employment of stochastic control against impoverishment. A simulating example is quoted to demonstrate the effectiveness of the proposed method in comparison with existing ones, showing that the proposed method is able to combine satisfactory precision of results with a low computational cost and provide an example to achieve impoverishment prediction and tracking.

Keywords: particle filter; missing data; single imputation; impoverishment; Markov Systems

MSC: 60G35; 60G20; 60J05; 62M20

Citation: Lykou, R.; Tsaklidis, G Particle Filtering: A Priori Estimation of Observational Errors of a State-Space Model with Linear Observation Equation. *Mathematics* **2021**, *9*, 1445. https://doi.org/10.3390/math9121445

Academic Editor: Panagiotis-Christos Vassiliou and Andreas C. Georgiou

Received: 20 May 2021
Accepted: 17 June 2021
Published: 21 June 2021

Publisher's Note: MDPI stays neutral with regard to jurisdictional claims in published maps and institutional affiliations.

Copyright: © 2021 by the authors. Licensee MDPI, Basel, Switzerland. This article is an open access article distributed under the terms and conditions of the Creative Commons Attribution (CC BY) license (https://creativecommons.org/licenses/by/4.0/).

1. Introduction

Particle Filter (PF) methodology deals with the estimation of latent variables of stochastic processes taking into consideration noisy observations generated by the latent variables [1]. This technique mainly consists of Monte-Carlo (MC) simulation [2] of the hidden variables and the weight assignment to the realizations of the random trials during simulation, the particles. This procedure is repeated sequentially, at every time step of a stochastic process. The involvement of sequential MC simulation in the method is accompanied by a heavy computational cost. However, the nature of the MC simulation makes the PF estimation methodology suitable for a wide variety of state-space models, including non-linear models with non-Gaussian noise. The weights are defined according to observations, which are received at every time step. The weight assignment step constitutes an evaluation process of the existing particles, which are created at the simulation step.

As weight assignment according to an observation dataset is a substantial part of PF, missing observations hinder the function of the filter. Wang et al. [3] wrote a review concerning PF on target tracking, wherein they mentioned cases of missing data and measurement uncertainties within multi-target tracking, as well as methods that deal with this problem (see, e.g., [4]). Techniques that face the problem of missing data focus mainly on substitution of the missing data. In recent decades, Expectation-Maximization algorithm [5] and Markov-Chain Monte Carlo methods [6] became popular for handling missing data problems. These algorithms have been constructed independently of PF. Gopaluni [7] proposed a combination of Expectation-Maximization algorithm with PF for parameter estimation with missing data. Housfater et al. [8] devised Multiple Imputations Particle Filter (MIPF), wherein missing data are substituted by multiple imputations from a proposal distribution and these imputations are evaluated with an additional weight assignment according to their proposal distribution. Xu et al. [9] involved uncertainty on data availability in the observations with the form of additional random variables in the subject state-space model. All the aforementioned approaches are powerful, although computationally costly.

This paper focuses on state-space models with linear observation equations and provides an estimation of the errors of missing observations (in cases of missing data), aiming at the approximation of weights, under a Missing At Random (MAR) assumption [10]. Linearity in an observation equation permits sequential replacements of missing values with equal quantities of known distributions. Although this method is applicable to a smaller set of models than the former ones, it is much faster as it leads to a single imputation process. A simulating example is provided for the comparison of the suggested method with existing techniques for the advantages of the proposed algorithm to be highlighted. The contribution of the a priori estimation step to the study of impoverishment phenomena is also exhibited through Markov System (MS) framework (see, e.g., [11]). The substitution of future weights renders the estimation of future distribution of particles in the state domain feasible. The significance of this initiative lays on the possible estimation of the sample condition concerning impoverishment, in future steps, based on the suggested theory. Such a practice permits the coordinated application of stochastic control [12] instead of the mostly empirical approaches that been proposed so far [13].

The present article is based on the work of Lykou and Tsaklidis [14]. Further mathematical propositions are formed by the sparse remarks exhibited in [14], and the incorporation procedure of MS-theory in the study of particle distribution is explained in detail. The presentation of the initial results of the simulation example is reformulated for the example to be more easily understandable, as well as a new application of MS-theory for the quantitative prior estimation of the particle distribution one time step forward is added to the initial example. In Section 2, PF algorithm is presented analytically. In Section 3, the new weight estimation step is introduced and its connection with the study of degeneracy and impoverishment is explained. In Section 4, a simulating example is provided, where the results of the current method are compared with those of MIPF and the results of the basic PF algorithm in the case when all data are available. An example for the estimation of the particle distribution one step ahead after the current is also presented in the direction of impoverishment tracking and prediction. In Section 5, the discussion and concluding remarks are quoted.

2. Particle Filter Algorithm

Let $\{x_t\}_{t\in\mathbb{N}}$ be a stochastic process described by m-dimensional latent vectors, $x_t \in \mathbb{R}^m$, and $\{y_t\}_{t\in\mathbb{N}}$ be the k-dimensional process of noisy observations, $y_t \in \mathbb{R}^k$. The states and observations are inter-related according to the state-space model,

$$x_t = f(x_{t-1}) + v_t \tag{1}$$

$$y_t = c + Ax_t + u_t \Leftrightarrow y_t - c - Ax_t = u_t. \tag{2}$$

In the system of Equations (1) and (2), f is a known deterministic function of x_t, v_t stands for the process noise, and u_t symbolizes the observation noise. Each sequence $\{v_t\}_{t\in\mathbb{N}}$ and $\{u_t\}_{t\in\mathbb{N}}$ consists of independent and identically distributed (iid) random vectors, while $c \in \mathbb{R}^k$ is a constant vector and $A \in \mathbb{R}^{k\times m}$ is a constant matrix.

PF methodology employs Bayesian inference for state estimation. The Bayesian approach aims at the construction of the posterior probability distribution function $p(x_t|y_{1:t})$, where $y_{1:t} = (y_1, y_2, ..., y_t)$, resorting to the recursive equations,

$$\begin{aligned} p(x_t|y_{1:t-1}) &= \int p(x_t|x_{t-1})p(x_{t-1}|y_{1:t-1})dx_{t-1} \text{ (prediction)} \\ p(x_t|y_{1:t}) &= \frac{p(y_t|x_t)p(x_t|y_{1:t-1})}{\int p(y_t|x'_t)p(x'_t|y_{1:t-1})dx'_t} \text{ (update).} \end{aligned}$$

These equations are analytically solvable in cases of linear state-space models with Gaussian noises. However, for more general models, analytical solutions are usually infeasible. For this reason, PF can be applied by utilizing MC simulation and integration to represent the state posterior probability distribution function, $p(x_t|y_{1:t})$, with the help of a set of $N \in \mathbb{N}$ particles x_t^i, $i = 1, 2, ..., N$, with corresponding weights w_t^i, $i = 1, 2, ..., N$. Then, $p(x_t|y_{1:t})$ can be approximated by the discrete mass probability distribution of the weighted particles $\{x_t^i\}_{i=1}^N$ as

$$\hat{p}(x_t|y_{1:t}) \approx \sum_{i=1}^{N} w_t^i \delta(x_t - x_t^i),$$

where δ is the Dirac delta function and weights are normalized, so that $\sum_i w_k^i = 1$. As $p(x_t|y_{1:t})$ is usually unknown; this MC simulation is based on importance sampling, namely a known probability (importance) density $q(x_t|y_{1:t})$ is chosen in order for the set of particles to be produced. Then, the state posterior distribution is approximated by

$$\hat{p}(x_t|y_{1:t}) \approx \sum_{i=1}^{N} \tilde{w}_t^i \delta(x_t - \tilde{x}_t^i),$$

with $\tilde{x}_t^i \sim q(x_t|y_{1:t})$, while

$$\tilde{w}_t^i \propto \tilde{w}_{t-1}^i \frac{p(y_t|\tilde{x}_t^i)p(\tilde{x}_t^i|\tilde{x}_{t-1}^i)}{q(\tilde{x}_t^i|\tilde{x}_{t-1}^i, y_t)} \tag{3}$$

are the normalized particle weights for $i = 1, 2, ..., N$.

As PF is applied successively for several time steps, it happens that increasing weights are assigned to the most probable particles, while the weights of the other particles become negligible progressively. Thus, only a very small proportion of particles is finally used for the state estimation. This phenomenon is known as PF degeneracy. In order to face this problem, a resampling with replacement step according to the calculated weights has been incorporated into the initial algorithm, resulting in the Sampling Importance Resampling (SIR) algorithm. Nevertheless, sequential resampling leads the particles to take values from a very small domain and exclude many other less probable values. This problem is called impoverishment. A criterion over the weight variability has been introduced for a decision to be made at every time step, whether existing particles should be resampled or not, to reach the middle ground between degeneracy and impoverishment. In this criterion, the Effective Sample Size measure of degeneracy, defined as

$$N_{eff}(t) = \frac{N}{1 + Var_{p(\bullet|y_{1:t})}(w(x_t))},$$

is involved (see, e.g., [15], pp. 35–36). As this quantity cannot be calculated directly, it can be estimated as

$$\hat{N}_{eff}(t) = \frac{1}{\sum_{i=1}^{N} (\tilde{w}_t^i)^2}.$$

The decision on resampling is positive whenever $\hat{N}_{eff}(t) < N_T$, where $N_T = c_1 N, c_1 \in \mathbb{R}$ is a fixed threshold. A usually selected value for N_T is 75%N. Establishing a criterion for resampling slows down sample impoverishment of the sample but does not prevent it.

Algorithm 1 summarizes PF steps. The sampling part corresponds to the prior (prediction) step of Bayesian inference, while weight assignment and possible resampling constitute the posterior (update) step.

Algorithm 1 SIR Particle Filter

Require: N, q, N_{eff}, T
  Initialize $\{x_0^i, w_0^i\}$
  for $t = 1, 2, ..., T$ **do**
    1. Importance Sampling
    Sample $\tilde{x}_t^i \sim q(x_t | x_{0:t-1}^i, y_t)$
    Set $\tilde{x}_{0:t}^i = (x_{1:t-1}^i, \tilde{x}_t^i)$,
    Calculate importance weights
    $\tilde{w}_t^i \propto w_{t-1}^i \frac{p(y_t|\tilde{x}_t^i) p(\tilde{x}_t^i | x_{t-1}^i)}{q(\tilde{x}_t^i | x_{t-1}^i, y_t)}$.
  end for
  for $i = 1, 2, ..., N$ **do**
    Normalize weights $w_t^i = \frac{\tilde{w}_t^i}{\sum_{i=1}^N \tilde{w}_t^i}$
    2. Resampling
    if $\hat{N}_{eff}(t) \geq N_T$ **then**
      for $i = 1, 2, ..., N$ **do**
        $x_{0:t}^i = \tilde{x}_{0:t}^i$
      end for
    else
      for $i = 1, 2, ..., N$ **do**
        Sample with replacement index $j(i)$ according to the discrete weight distribution $P(j(i) = d) = w_t^d, d = 1, ..., N$
        Set $x_{0:t}^i = \tilde{x}_{0:t}^{j(i)}$ and $w_t^i = \frac{1}{N}$
      end for
    end if
  end for

3. The Missing Data Case—Estimation of Weights

We now proceed to the addition of a new step to Algorithm 1 for the case of missing data. For that purpose, some new definitions need be quoted. As the missing data mechanism is usually unknown, its possible dependence on the missing data themselves could introduce bias to the statistical inference. For this reason, a Missing at Random (MAR) assumption is adopted: let a random indicator variable $R_{t,j}, j = 1, ..., k$, indicate whether the jth component of the tth observation is available or not. That is,

$$R_{t,j} = \begin{cases} 0 & \text{the } j\text{th component is available at time } t \\ 1 & \text{otherwise} \end{cases}.$$

Additionally, sets Z_t and W_t are defined as the collections of missing and available components of observations y_t, respectively, for every time step $t \in \mathbb{N}$.

According to the MAR assumption, the missing data mechanism does not depend on the missing observations, given the available ones:

$$P(R_{t,j} | Z_t, W_t) = P(R_{t,j} | W_t), t \in \mathbb{N}, j = 1, ..., k.$$

Proposition 1. *Let $\{x_{t-1}^i\}_{i=1}^N$ be the set of particles produced for the posterior estimation of latent vector x_{t-1}, while whole observation y_t is missing. In addition, let u_t^i, $i = 1 \dots N$, be the observational errors for corresponding candidate particles $\tilde{x}_t^i$, so that $u_t^i = y_t - c - A\tilde{x}_{t-1}^i$,* *according to (2). Then, the conditional distribution of every observational error u_t^i on the particle set $\{x_{t-1}^i\}_{i=1}^N$ is approximated as*

$$p(u_t^i|\{x_{t-1}^i\}_{i=1}^N) \approx p(A(f(\hat{x}_{t-1}) - f(x_{t-1}^i)) + Av_t - Av_t^i + u_t), \tag{4}$$

where $\hat{x}_{t-1}$ is a point estimation of x_{t-1} and v_t^i represents the process noise for the generation of x_t^i.

Proof. Let x_{t-1}^i be a particle for the posterior estimation of the hidden state x_{t-1} of the state-space model (1)–(2). Then, according to Algorithm 1 and Equation (1), the ith prior estimation of the hidden state x_t is produced by equation

$$\tilde{x}_t^i = f(x_{t-1}^i) + v_t^i. \tag{5}$$

According to Equation (2), the observational error of the particle is calculated as

$$u_t^i = y_t - c - A\tilde{x}_t^i. \tag{6}$$

If the (whole) observation y_t is unavailable, sequential replacements of u_t^i and y_t from Equations (6) and (2), respectively, contribute to the creation of the formula,

$$\begin{aligned} u_t^i &= c + Ax_t + u_t - c - A\tilde{x}_t^i \\ &= A(x_t - \tilde{x}_t^i) + u_t. \end{aligned}$$

As observation y_t is considered missing, particles $\tilde{x}_t^i$ cannot be evaluated. Thus, both x_t and $\tilde{x}_t^i$ are replaced according to Equations (1) and (5),

$$\begin{aligned} u_t^i &= Af(x_{t-1}) + Av_t - Af(x_{t-1}^i) - Av_t^i + u_t \\ &= A(f(x_{t-1}) - f(x_{t-1}^i)) + Av_t - Av_t^i + u_t. \end{aligned}$$

The hidden state x_{t-1} is unknown, but its posterior distribution is available, so that a point estimation of it, $\hat{x}_{t-1}$, can be calculated. Then,

$$u_t^i \approx A(f(\hat{x}_{t-1}) - f(x_{t-1}^i)) + Av_t - Av_t^i + u_t. \tag{7}$$

Therefore, since the quantity $A(f(\hat{x}_{t-1}) - f(x_{t-1}^i))$ is a constant at time t, the distribution of u_t^i can be approximated as

$$p(u_t^i|\{x_{t-1}^i\}_{i=1}^N) \approx p(A(f(\hat{x}_{t-1}) - f(x_{t-1}^i)) + Av_t - Av_t^i + u_t). \tag{8}$$

□

Remark 1. *Given that the distributions of the random vectors v_t, v_t^i, and u_t are known, the distribution of $Av_t - Av_t^i + u_t$ is also known, as it is the convolution of linear functions of the initial components v_t, v_t^i, and u_t. Calculation of such convolutions is not always an easy task, as analytical solutions may not be feasible, leading to numerical approximation options ([16]). However, given that each noise process consists of iid vectors and matrix A is constant, the distribution of this sum needs to be calculated only once.*

Remark 2. *The replacement of $\tilde{x}_t^i$ can be avoided, if MC simulation has been implemented at this time point.*

The weight assigned to $\tilde{x}_t^i$ depends on u_t^i, according to Equations (3) and (6), because

$$p(y_t|\tilde{x}_t^i) = p(u_t^i).$$

Then, as the two variables ($\tilde{w}_t^i$ and u_t^i) are closely associated, knowledge of the distribution of u_t^i leads to the derivation of the distribution of $\tilde{w}_t^i$. Even if the distribution of $\tilde{w}_t^i$ may not be exactly calculated, in cases where $\tilde{w}_t^i$ are complicated functions of u_t^i, knowledge on the distribution of u_t^i will suffice to provide information on the weight distribution. Thereby, calculation of $p(u_t^i \in D), D \subset \mathbb{R}^k$, as it is suggested in Remark 1, is of interest for the concomitant estimation of weights.

Proposition 2. *If the conditions of Proposition 1 hold, while y_t is partially observed, the conditional distribution of every observational error u_t^i on particle set $\{x_{t-1}^i\}_{i=1}^N$ and collection W_t of available components of y_t is approximated as*

$$p(u_t^i|\{x_{t-1}^i\}_{i=1}^N, W_t) \approx p(A(f(\hat{x}_{t-1}) - f(x_{t-1}^i)) + Av_t - Av_t^i + u_t|W_t).$$

Proof. Estimation of u_t^i implies the estimation of y_t, according to Equation (6). If observation y_t is partially available, its available components, say W_t collection, can be placed into the above equations. Thus, some components of the observational error will also be available, while the rest of the components, say Z_t collection, possibly dependent on the available ones, is estimated in the same rationale. In this case, (4) takes the form

$$p(u_t^i|\{x_{t-1}^i\}_{i=1}^N, W_t) \approx p(A(f(\hat{x}_{t-1}) - f(x_{t-1}^i)) + Av_t - Av_t^i + u_t|W_t).$$

□

In any case, missing (parts of) observational errors u_t^i along with their weights can be substituted by single values, as expected values or modes. Consequently, the initial PF algorithm undergoes a slight change, as presented in Algorithm 2. Further to Remark 2, the substitution of observational errors u_t^i is implemented after the Importance Sampling step in Algorithm 2 for the sake of simplicity.

3.1. Connection to Markov Systems and Contribution to the Study over Impoverishment

Impoverishment over the particle samples can be studied in connection with certain Markov models, the Homogeneous or Non-homogeneous Markov Systems (denoted as HMSs or NHMSs, respectively) , which have their roots in [17]. With the consideration of a grid of $d \in \mathbb{N}$ cells over the state domain, problems of impoverishment reduce to a problem concerning the derivation of the distribution of the particle population over the grid cells. Term "grid" denotes here a single partition over the whole state domain. The cells of this grid represent the states of the MS. At every time step, a particle moves from cell i ($i = 1, \ldots, d$) to cell j ($j = 1, \ldots, d$) with (time-dependent) transition probability $p_{ij,h}(t)$, ($h = 1, \ldots, N$) in the general case. However, MS consideration is based on the hypothesis that population members which are situated in the same state move to any cell at the next time step according to a joint transition probability. Thus, for the introduction of MS-theory in the study of particle distribution over the grid, probabilities $p_{ij,h}(t)$ are approximated by single quantities $p_{ij}(t)$ for all particles in cell i at time point t. The fact that PF is applied to dynamical systems entails that different areas of the state domain become of particular interest at different time steps. Therefore, it is preferable for the grid lines not to remain constant over time. A simple time-varying grid is constructed within the simulating example in Section 4, while a more complex structure is provided in [18]. In the simple case that the PF algorithm comprises constant parameters and excludes the resampling step, the corresponding MS can be considered homogeneous, as the particles only move according to a state equation with constant approximating transition probability values. Resampling causes the redistribution of particles over the grid. The probability vectors for this redistribution are defined by the observational errors at every time step. Thus, steps of changing probability vectors are introduced in the MS rendering this MS non-homogeneous. Moreover, the results over the grid of both production of weighted particles on the basis of system (1)–(2) and resampling, at the end of every time step, derive the results of sums of multinomial trials with varying probability vectors (see also [19], p. 28); this procedure corresponds to the transitions of population members between the state of a MS. In general, the sums of multinomial trials can be considered to follow generalized multinomial distribution [20] or, more generally, Poisson Binomial distribution (which has its roots in [21], §14). As the number of particles remains constant, according to Algorithm 1, the MS is considered to be closed. The difficulty in making predictions on the MS lies in the fact that observational errors are not a priori acquired during the filtering procedure.

Algorithm 2 SIR Particle Filter for missing data with observational error estimation

Require: N, q, N_{eff}, T
  Initialize $\{x_0^i, w_0^i\}$
  for $t = 1, 2, ..., T$ **do**
    1. Importance Sampling
    Sample $\tilde{x}_t^i \sim q(x_t | x_{0:t-1}^i, W_t)$
    Set $\tilde{x}_{0:t}^i = (x_{1:t-1}^i, \tilde{x}_t^i)$,
    Produce observational error estimations $\hat{u}_t^i$ for the missing components Z_t and calculate importance weights
    $\tilde{w}_t^i \propto w_{t-1}^i \frac{p(y_t|\tilde{x}_t^i) p(\tilde{x}_t^i | x_{t-1}^i)}{q(\tilde{x}_t^i | x_{t-1}^i, y_t)}$.
  end for
  for $i = 1, 2, ..., N$ **do**
    Normalize weights $w_t^i = \frac{\tilde{w}_t^i}{\sum_{i=1}^N \tilde{w}_t^i}$
    2. Resampling
    if $\hat{N}_{eff}(t) \geq N_T$ **then**
      for $i = 1, 2, ..., N$ **do**
        $x_{0:t}^i = \tilde{x}_{0:t}^i$
      end for
    else
      for $i = 1, 2, ..., N$ **do**
        Sample with replacement index $j(i)$ according to the discrete weight distribution $P(j(i) = d) = w_t^d, d = 1, ..., N$
        Set $x_{0:t}^i = \tilde{x}_{0:t}^{j(i)}$ and $w_t^i = \frac{1}{N}$
      end for
    end if
  end for

In this study, observational errors are substituted by single values for one time step, so that weights can be estimated one step ahead. The set of weights configures the probability vectors of the resampling step. Thus, the distribution of particles over the grid cells can be approximated during the upcoming step of resampling and new Importance Sampling. Thus, the distribution of the particle population can be estimated for the next step on the basis of the estimated weights for the dispersion of the future particles over the grid to be assessed and impoverishment phenomena to be predicted. Such a practice paves the way to the involvement of stochastic control theory [12] (leading to control of asymptotic variability [22]) into the matter of the avoidance of impoverishment.

4. Simulating Example

4.1. Contribution to the Missing Data Case

A simulation example is presented in this section to emphasize the benefits of the proposed algorithm. The proposed method is compared with the typical PF algorithm, when the complete dataset is available, and the multiple imputation particle filter (MIPF) for $n = 5$ imputations [8]. The reduction step proposed in [23] is incorporated in the initial

MIPF algorithm for the best possible results to be achieved. The data simulation is based on the state-space model of Equations (1) and (2), with two-dimensional vectors

$$x_t = \begin{bmatrix} x_{1,t} \\ x_{2,t} \end{bmatrix} = \begin{bmatrix} cos(x_{1,t-1} - x_{1,t-1}/x_{2,t-1}) \\ cos(x_{2,t-1} - x_{2,t-1}/x_{1,t-1}) \end{bmatrix} + \begin{bmatrix} v_{1,t} \\ v_{2,t} \end{bmatrix} \tag{9}$$

$$y_t = x_t + u_t,$$

where $v_t = \begin{bmatrix} v_{1,t} \\ v_{2,t} \end{bmatrix} \sim N(\mu, S_1)$, $\mu = \begin{bmatrix} 0 \\ 0 \end{bmatrix}$, $S_1 = \begin{pmatrix} 0.05 & 0 \\ 0 & 0.05 \end{pmatrix}$, $u_t \sim N(\mu, S_2)$, $S_2 = \begin{pmatrix} 0.03 & 0 \\ 0 & 0.03 \end{pmatrix}$ and N symbolizes Gaussian distribution. Let $x_0 = \begin{bmatrix} 1 \\ 0.5 \end{bmatrix}$ be considered known. Next, concerning missing data, we let $R_{t,j} \sim Bernoulli(0.15)$, $j = 1,2$, that is the data are missing completely at random. $N = 100$ particles have been used for every filter. The distributions of noises are also considered known. The weighted mean is used as a point estimator of a hidden state and missing observational errors are substituted by their expected values. All the filters have been repeated for 100 times and their performance concerning their precision and consumed time has been recorded. (The code was written in R project [24]. Packages *mvnorm* [25], with its corresponding reference book [26], ggplot [27], and ggforce [28] were also used. Simulations were performed on an AMD A8-7600 3.10 GHz processor with 8 GB of RAM.)

The results of the three methods are shown in Table 1. In the first two columns, the means over the simulations of Root-Mean-Square Errors (RMSE) of the estimators (weighted means) for each component of the hidden states are presented. The mean of the two aforementioned columns is also calculated, as well as the mean time consumed in each approach. In the table, it is shown that the weight estimation with the suggested method outperforms MIPF concerning both precision and time elapsed. The precision of the suggested method supersedes that of MIPF slightly, while the mean required computational time is about 50% less than the corresponding mean time required for MIPF. The proposed method is also compared with the results of the standard PF algorithm, for which all observations are available, and it seems that, even if the precision is inevitably reduced in the case of missing data, the computational time remains nearly the same. The small differentiation in the mean elapsed time is probably connected with the resampling decision. That is, in this example, the precision of the suggested method slightly supersedes its competitor, while its computational cost is much lower than the cost of its competitor, reaching the levels of the basic filter (which is practically infeasible in the missing data case). In Figures 1 and 2, the performances of the proposed method and MIPF are depicted for the two components of the state process, respectively, for one iteration of each filter. The estimators (weighted means) of the two approaches are close to each other, tracking the hidden vector satisfactorily. Therefore, in this example, the suggested method appears to provide the best option between the available ones in the missing data case.

Table 1. Comparison of the results over three methods: the basic PF algorithm, when all observations are available; the weight estimation method, which is proposed in this study; and MIPF for $n = 5$ imputations. The methods are compared through the mean of RMSE and the time consumed over the 100 repeated implementations.

Method	Mean RMSE for ($x_{1,t}$)	Mean RMSE for ($x_{2,t}$)	Overall Mean Precision	Mean Time Elapsed (s)
Basic PF	0.1610253	0.1566881	0.1588567	2.5570
Weight est.	0.2065578	0.2102287	0.2083933	2.5491
MIPF	0.2267527	0.2173670	0.2220598	4.9137

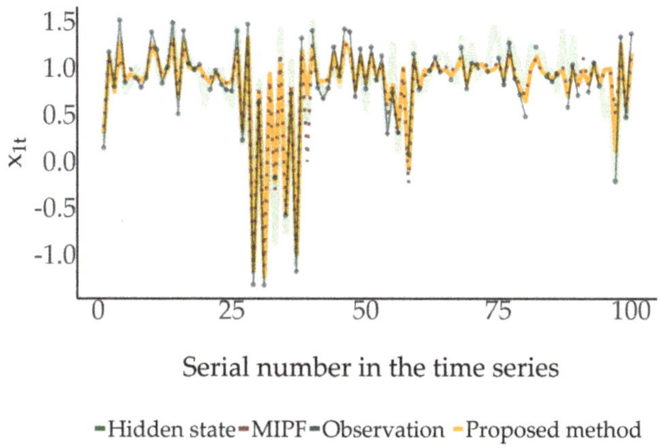

Figure 1. Time-series of the hidden values, the observations, and the corresponding point estimations of the proposed method and MIPF imputations for the first component $x_{1,t}$ of the state process.

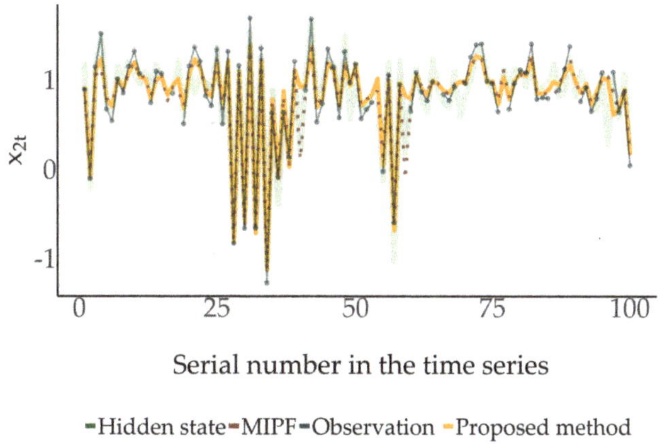

Figure 2. Time-series of the hidden values, the observations, and the corresponding point estimations of the proposed method and MIPF imputations for the second component $x_{2,t}$ of the state process.

4.2. Contribution to Impoverishment Prediction

As far as estimation of particle distribution one step ahead is concerned, an application for the transition of the particles from time point $t = 9$ to $t = 10$ is presented during one implementation of the suggested PF with single imputation for missing values on the available dataset. In the time interval (0,10], only the first component of observation y_6 is unavailable. In the end of time step $t = 9$, resampling is implemented and the histograms of the particle sets are exhibited in Figure 3. The sample mean of the particles is $m = [m_1 = 0.947 \quad m_2 = 1.07]^T$ and the standard deviations of the corresponding components are $s_1 = 0.126$ and $s_2 = 0.127$. According to Equation (4) and the given parameters of the problem, the random factor needed to be estimated for every particle at the next time step is

$$z_t^i = Av_{t+1} - Av_{t+1}^i + u_{t+1} \sim N(0, S_z), \quad S_z = 2S_1 + S_2 = \begin{pmatrix} \sigma_z^2 & 0 \\ 0 & \sigma_z^2 \end{pmatrix},$$

where $\sigma_z = 0.36$. Thus, in both dimensions, the following partitions are considered,

$$\Pi_j = \{(-\infty, m_j - \sigma_z/2), [m_j - \sigma_z/2, m_j + \sigma_z/2), [m_j + \sigma_z/2, +\infty)\}, \quad j = 1,2,$$

so that a grid of nine cells is configured over the two dimensions. The frequency table (Table 2) exhibits the particle distribution over the grid.

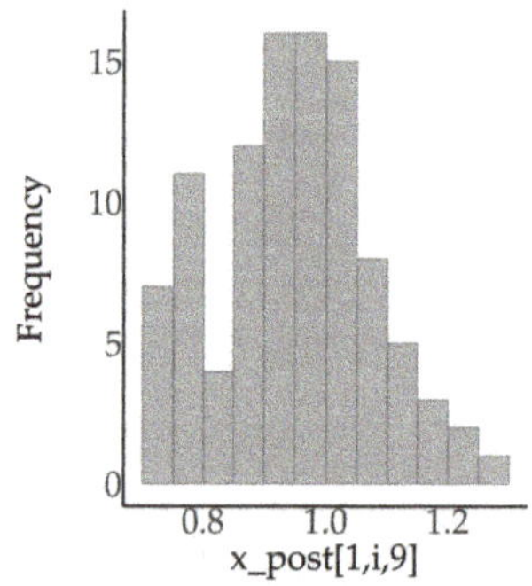

(**a**) Histogram of the particle sample ($i = 1, \ldots, N$) for the posterior estimation of first component $x_{1,9}$.

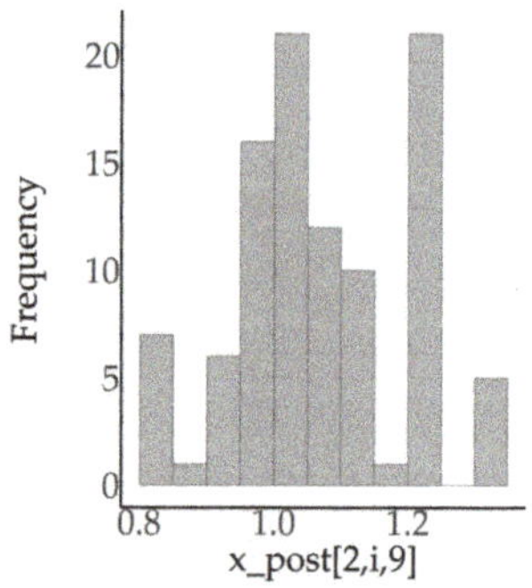

(**b**) Histogram of the particle sample ($i = 1, \ldots, N$) for the posterior estimation of second component $x_{2,9}$.

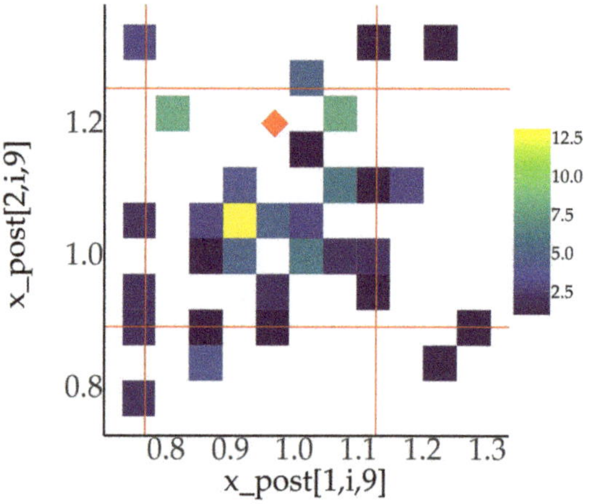

(**c**) Joint histogram of the particle sample ($i = 1, \ldots, N$) for the posterior estimation of both components of the whole hidden vector x_9. Red lines delimit the suggested grid cells.The red diamond stands for the hidden state.

Figure 3. Histograms of particle samples for the posterior estimation of hidden variable x_9.

The selected time period was chosen because there is a considerable number of preceding steps that permits a relatively good adaptation of the particle samples over the hidden states and the samples have not yet collapsed to a tiny neighborhood around a single point (utter impoverishment). This argument is evinced in Figure 3c and Table 2, where the distribution of the particles is presented in connection with the hidden state and the suggested grid. The produced particles are both close to the hidden state, as most of them are less than one standard deviation σ_z from it, and sparse enough for the existence of particles outside the central cell of the grid. Thus, the condition of the sample during these time points configures a typical example of filter implementation before its collapse. Such time points may be suitable starting points for the introduction of control (which exceeds the limits of this study) as the subject sample is in a good condition concerning both impoverishment and accuracy over hidden variable estimation.

Table 2. Frequency table of the particle distribution over the suggested grid at the end of $t = 9$.

1st Component \ 2nd Component	$(-\infty, 0.89)$	$[0.89, 1.25)$	$[1.25, +\infty)$
$(-\infty, 0.767)$	2	6	3
$[0.767, 1.13)$	5	77	1
$[1.13, +\infty)$	1	4	1

For the next time step, a prior estimation for hidden vector x_{10} is implemented. For the formation of the new grid, the existing particles are moved according to the deterministic part of Equation (9), resulting in $x_{10}(-) = [0.993 \quad 0.984]^T$, the mean of the new particle set. This quantity constitutes a prior point estimator of the hidden state. Thus, the grid of $t = 9$ is shifted by $x_{10}(-) - m$ to a new grid, as shown it Table 3, the central cell of which is

$$[x_{10}(-)[1] - 0.18, x_{10}(-)[1] + 0,18) \times [x_{10}(-)[2] - 0.18, x_{10}(-)[2] + 0,18) = \\ [0.813, 1.17) \times [0.804, 1.16).$$

The movement of all the particles according to the deterministic part of Equation (9) results in the frequency table in Table 3, where it is shown that all the new particles belong to the central cell. Even though the particles are identically distributed, with the addition of the process noise to the particles, the probabilities for particles to move from the central cell to random ones defer from particle to particle, as the particles have different distances form the grid lines initially. This fact is in contrast to the theoretical background of MS, according to which population members have a common transition probability matrix P to move during a time step. For this reason, the probabilities of particles to move to a cell with the addition of the random noise are approximated by the probabilities of the point estimation $x_{10}(-)$ to move to a random cell with the addition of noise. These probabilities (rounded values) are provided in Table 4. Thus, the expected numbers of particles over the grid cells are

$$N * P = 100 * \begin{pmatrix} 0.044 & 0.122 & 0.044 \\ 0.122 & 0.335 & 0.122 \\ 0.044 & 0.122 & 0.044 \end{pmatrix}$$

and the expected distribution of the particles over the grid is presented in Table 5. Concerning the expected posterior distribution of the particles, the expected observational errors are zero, so that particle weights are expected to remain the same. Thus, no further change is expected in their distribution in cells even if resampling is decided to take place, as all weights are equal after resampling in the previous time step.

Remark 3. *The expected values of observational errors are zero. Nevertheless, the prior estimation of their distribution according to relation (4) and model parameters, where the variances of the errors are presented, evinces the increased uncertainty for them, as* $\sigma_u^2 = 0.03$ *while* $\sigma_z^2 = 0.13$.

Table 3. Frequency table of the particle distribution over the suggested grid at $t = 10$ when the particles $x_9^i, i = 1, \ldots, 100$, move only according to the deterministic part of Equation (9).

2nd Component / 1st Component	$(-\infty, 0.804)$	$[0.804, 1.16)$	$[1.16, +\infty)$
$(-\infty, 0.813)$	0	0	0
$[0.813, 1.17)$	0	100	0
$[1.17, +\infty)$	0	0	0

Table 4. Transition probability table for $x_{10}(-)$ to move with the addition of process noise v_{10} .

2nd Component / 1st Component	$(-\infty, 0.804)$	$[0.804, 1.16)$	$[1.16, +\infty)$
$(-\infty, 0.813)$	0.044	0.122	0.044
$[0.813, 1.17)$	0.122	0.335	0.122
$[1.17, +\infty)$	0.044	0.122	0.044

Table 5. Frequency table for the expected numbers of the particles over the grid cells after the addition of process noise realizations at $t = 10$.

2nd Component / 1st Component	$(-\infty, 0.804)$	$[0.804, 1.16)$	$[1.16, +\infty)$
$(-\infty, 0.813)$	4.4	12.2	4.4
$[0.813, 1.17)$	12.2	33.5	12.2
$[1.17, +\infty)$	4.4	12.2	4.4

The results of the implementation of PF at time step $t = 10$ are also exhibited. Resampling has taken place at this time step as well. The joint histogram of the posterior sample over both dimensions (Figure 4) indicates that the majority of the particles do not belong to the central cell. This fact is reasonable, as the length of the sides of the central cell equals only one standard deviation σ_z, so that prior probabilities for the particles to be placed outside of the central cell at this time point are considerably big according to the Empirical Rule (68-95-99.7) for normal distribution. For the consolidation of these results towards this rule, the orange squares are drawn in Figure 4 for the corresponding areas of the rule to be defined for each separate dimension, while the orange circles are the corresponding standard deviation circles (and not ellipses, generally, as the two components have the same variance $\sigma_z^2 = 0.13$) of the whole vector. Thus, questions on the suitability of the proposed grid structure are raised for future study. Nevertheless, it should be mentioned that a grid with a central cell of double side length would have classified all particles to the central cell during time step $t = 9$, rendering further study on the issue meaningless. Additionally, a new grid of nine cells is also constructed around the mean of this posterior particle set, the central cell of which also has length σ_z. The distribution of the particles in the new grid is quoted in Table 6. In comparison with Table 2, it seems that the number of particles in the central cell is increased in Table 6.

Table 6. Frequency table of the particle distribution over the grid around the mean of the sample in the end of $t = 10$.

2nd Component / 1st Component	$(-\infty, 1.01)$	$[1.01, 1.37)$	$[1.37, +\infty)$
$(-\infty, 1.06)$	1	3	0
$[1.06, 1.42)$	6	90	0
$[1.42, +\infty)$	0	0	0

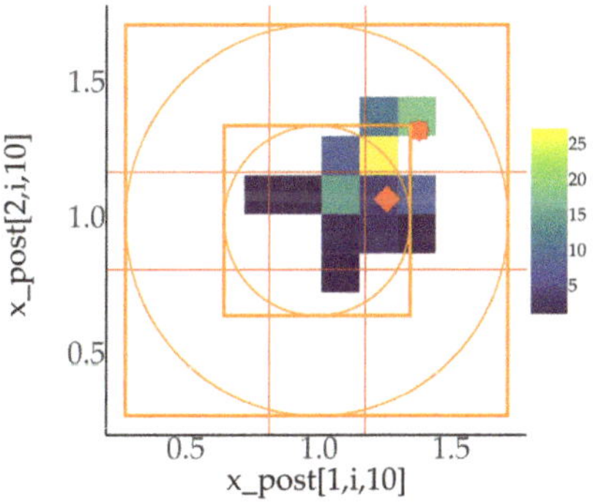

Figure 4. Joint histogram of the particle sample ($i = 1, \ldots, N$) for the posterior estimation of both components of the whole hidden vector x_{10}. Red lines delimit the suggested grid cells of Table 3. The red diamond stands for the hidden state. The red star stands for the observation at this time point. The sides of the two squares are correspondingly one and two standard deviations σ_z from the center of the diagram. The circles are inscribed in the corresponding squares.

Remark 4. *In the present example, the transitions of the particles according to the deterministic function led all particles to a single cell (Table 3), so that the result of the addition of process noise was handled as a result of a multinomial trial. In the case that the deterministic function leads the particles to more than one cell, then it is suggested that different means be found for each cell as well as corresponding transition probabilities, so that the final result can be considered the sum of results of multinomial trials for the transitions to every cell.*

5. Discussion and Conclusions

In this study, single substitution (in contrast to MIPF) of observational errors is proposed for missing data cases, when PF is implemented and MAR assumption is adopted. This method is a single imputation procedure. Acuña et al. [23] argued against single imputation, as it is rather simplistic and it cannot attribute to a single value the distributional characteristics that can be approached and described by a sample of multiple imputation. Nevertheless, the primary target of the proposed technique is the minimization of the computational cost that is added to the initial PF algorithm, in the case of missing data. For this purpose, interventions in the PF algorithm are slight. Moreover, in the provided simulation, the suggested method outperforms the multiple imputation approach even for a considerable number of imputations, whereas Acuña et al. [23] noticed that MIPF with $n = 3, 4, 5$ imputations produces very satisfactory results, according to the approximation of multiple imputation estimator of efficiency provided by Rubin [29]. As a result, in this example, estimation of observational errors performs better with respect to both the computational time it requires and the precision it achieves. Besides, knowledge on the distribution of the observational errors contributes to the quantification possibility of the uncertainty over the point estimations. Thus, the suggested method takes advantage of the low computational cost of the single imputation option, while the study of more general distributional characteristics of the observational noise can also be taken into account at the same time (see Remark 3).

The contribution of such a method in order to cope with impoverishment problems is also worth mentioning. This method permits the estimation of observational errors and their corresponding weights one time step forward. The evolution of weight distribution has not been a priori estimated for multiple time steps yet, to the best of the authors' knowledge, but this is feasible at least for one step ahead. As the weights of the next step can be estimated, the probabilities that a particle will be chosen at the resampling step can also be estimated. As explained in Section 3.1, the assessment of weight distribution for the forthcoming time steps could be very interesting, as far as it is connected with impoverishment issues. Concerning future perspectives over this issue, the study of impoverishment problems can be implemented with the use of input control [12], in order for the impoverishment to be controlled; laws of large numbers [30], as MC approximation employs large samples; state capacity restrictions [31]; for the existence of a population limit at every grid cell; literature on the evolution of attainable structures [32]; the evolution of the distribution of particles [33] or of the corresponding moments [34] in the direction of HMSs and NHMSs; and for the estimation of the future behaviour of the sample, possibly reaching continuous-time models [35]. Research on automatic optimal control [36] could be combined with the suggested methodology, possibly leading to interesting joint applications of PF [37] along with artificial intelligence [38]. The performance of the method could also be tested when data are missing for longer time periods [39], while more sophisticated grid structures could also be examined [18]. Correspondingly, in a broader sense, the main idea of the proposed method could be implemented in the errors-in-variables signal processing for missing data cases [40], or it could be involved in more complex models that require MC simulation for the prior estimation of variables [41].

Author Contributions: Conceptualization, R.L. and G.T.; methodology, R.L. and G.T.; software, R.L.; validation, R.L. and G.T.; formal analysis, R.L. and G.T.; writing—original draft preparation, R.L.; writing—review and editing, G.T.; supervision, G.T.; and funding acquisition, R.L. and G.T. All authors have read and agreed to the published version of the manuscript.

Funding: This research was funded by A.G. Leventis foundation (Grant for Doctoral Studies).

Institutional Review Board Statement: Not applicable.

Informed Consent Statement: Not applicable.

Data Availability Statement: Not applicable.

Acknowledgments: The authors would like to express their gratitude to Panagiotis-Christos Vassiliou and Andreas C. Georgiou for the opportunity to submit to this Special Issue for possible publication. The constructive comments and the valuable suggestions of the three anonymous reviewers as well as the editorial assistance of Bella Chen are highly appreciated.

Conflicts of Interest: The funders had no role in the design of the study; in the collection, analyses, or interpretation of data; in the writing of the manuscript, or in the decision to publish the results.

Abbreviations

The following abbreviations are used in this manuscript:

PF	Particle Filter
MC	Monte Carlo
MIPF	Multiple Imputation Particle Filter
MAR	Missing At Random
MS	Markov System

iid	independent and identically distributed
SIR	Sampling Importance Resampling
HMSs	Homogeneous Markov Systems
NHMSs	Non-homogeneous Markov Systems
RMSE	Root Mean Square Error

References

1. Gordon, N.; Salmond, D.; Smith, A. Novel approach to nonlinear/non-Gaussian Bayesian state estimation. *IEE Proc. Radar Signal Process.* **1993**, *140*, 107. [CrossRef]
2. Metropolis, N.; Ulam, S. The Monte Carlo Method. *J. Am. Stat. Assoc.* **1949**. [CrossRef]
3. Wang, X.; Li, T.; Sun, S.; Corchado, J.M. A Survey of Recent Advances in Particle Filters and Remaining Challenges for Multitarget Tracking. *Sensors* **2017**, *17*, 2707. [CrossRef]
4. Degen, C.; Govaers, F.; Koch, W. Track maintenance using the SMC-intensity filter. In Proceedings of the 2012 Workshop on Sensor Data Fusion: Trends, Solutions, Applications (SDF), Bonn, Germany, 4–6 September 2012; pp. 7–12. [CrossRef]
5. Dempster, A.P.; Laird, N.M.; Rubin, D.B. Maximum Likelihood from Incomplete Data Via the EM Algorithm. *J. R. Stat. Soc. Ser. B (Methodol.)* **1977**, *39*, 1–22. [CrossRef]
6. Metropolis, N.; Rosenbluth, A.W.; Rosenbluth, M.N.; Teller, A.H.; Teller, E. Equation of State Calculations by Fast Computing Machines. *J. Chem. Phys.* **1953**, *21*, 1087–1092. [CrossRef]
7. Gopaluni, R.B. A particle filter approach to identification of nonlinear processes under missing observations. *Can. J. Chem. Eng.* **2008**, *86*, 1081–1092. [CrossRef]
8. Housfater, A.S.; Zhang, X.P.; Zhou, Y. Nonlinear fusion of multiple sensors with missing data. In Proceedings of the 2006 IEEE International Conference on Acoustics Speech and Signal Processing Proceedings, Toulouse, France, 14–19 May 2006. [CrossRef]
9. Xu, L.; Ma, K.; Li, W.; Liu, Y.; Alsaadi, F.E. Particle filtering for networked nonlinear systems subject to random one-step sensor delay and missing measurements. *Neurocomputing* **2018**, *275*, 2162–2169. [CrossRef]
10. Rubin, D.B. Inference and Missing Data. *Biometrika* **1976**, *63*, 581. [CrossRef]
11. Vassiliou, P.C.G. Asymptotic behavior of Markov systems. *J. Appl. Probab.* **1982**, *19*, 851–857. [CrossRef]
12. Vassilliou, P.C.G.; Tsantas, N. Stochastic control in non- homogeneous markov systems. *Int. J. Comput. Math.* **1984**, *16*, 139–155. [CrossRef]
13. Li, T.; Sun, S.; Sattar, T.P.; Corchado, J.M. Fight sample degeneracy and impoverishment in particle filters: A review of intelligent approaches. *Expert Syst. Appl.* **2014**, *41*, 3944–3954. doi:10.1016/j.eswa.2013.12.031. [CrossRef]
14. Lykou, R.; Tsaklidis, G. A priori estimation methodology on observation errors of a state space model with linear observation equation using Particle Filtering. InProceedings of the 6th Stochastic Modeling Techniques and Data Analysis International Conference, Skiadas, C.H., Ed.; Barcelona, Spain, 2–5 June 2020; pp. 333–341.
15. Liu, J.S. *Monte Carlo Strategies in Scientific Computing*; Springer Series in Statistics; Springer: New York, NY, USA, 2004. [CrossRef]
16. Lykou, R.; Tsaklidis, G. Prior estimation of observational errors of Particle Filter. In Proceedings of the 32nd Panhellenic Statistics Conference, Ioannina, Greece, 30 May–1 June 2019; pp. 195–204. (In Greek)
17. Bartholomew, D.J. A Multi-Stage Renewal Process. *J. R. Stat. Soc. Ser. B (Methodol.)* **1963**, *25*, 150–168. [CrossRef]
18. Li, T.; Sattar, T.P.; Sun, S. Deterministic resampling: Unbiased sampling to avoid sample impoverishment in particle filters. *Signal Process.* **2012**, *92*, 1637–1645. [CrossRef]
19. Bartholomew, D.J. *Stochastic Models for Social Processes*, 3rd ed.; Wiley: New York, NY, USA, 1982.
20. Beaulieu, N. On the generalized multinomial distribution, optimal multinomial detectors, and generalized weighted partial decision detectors. *IEEE Trans. Commun.* **1991**, *39*, 193–194. [CrossRef]
21. Poisson, S.D. *Recherches sur la Probabilité des Jugements en Matière Criminelle et en Matière Civile*; Bachelier: Paris, France, 1837.
22. Vassiliou, P.C.G.; Georgiou, A.C.; Tsantas, N. Control of Asymptotic Variability in Non-Homogeneous Markov Systems. *J. Appl. Probab.* **1990**, *27*, 756–766. [CrossRef]
23. Acuña, D.E.; Orchard, M.E.; Silva, J.F.; Pérez, A. Multiple-imputation-particle-filtering for uncertainty characterization in battery state-of-charge estimation problems with missing measurement data: Performance analysis and impact on prognostic algorithms. *Int. J. Progn. Health Manag.* **2015**, *6*, 1–12.
24. R Core Team. R: A Language and Environment for Statistical Computing. 2014. Available online: https://www.gbif.org/tool/81287/r-a-language-and-environment-for-statistical-computing (accessed on 20 June 2021)
25. Genz, A.; Bretz, F.; Miwa, T.; Mi, X.; Leisch, F.; Scheipl, F.; Hothorn, T. {mvtnorm}: Multivariate Normal and t Distributions. 2020. Available online: http://mvtnorm.r-forge.r-project.org/ (accessed on 20 June 2021)
26. Genz, A.; Bretz, F. *Computation of Multivariate Normal and t Probabilities*; Lecture Notes in Statistics; Springer: Heidelberg, Germany, 2009.
27. Wickham, H. *ggplot2: Elegant Graphics for Data Analysis*; Springer: New York, NY, USA, 2016.
28. Pedersen, T.L. ggforce: Accelerating 'ggplot2'; R Package Version 0.3.3. 2021. Available online: https://cran.r-project.org/web/packages/ggforce/index.html (accessed on 20 June 2021)
29. Rubin, D.B. (Ed.) *Multiple Imputation for Nonresponse in Surveys*; Wiley Series in Probability and Statistics; John Wiley & Sons, Inc.: Hoboken, NJ, USA, 1987. [CrossRef]

30. Vassiliou, P.C. Laws of Large Numbers for Non-Homogeneous Markov Systems. *Methodol. Comput. Appl. Probab.* **2020**, *22*, 1631–1658. [CrossRef]
31. Vasiliadis, G.; Tsaklidis, G. On the Distributions of the State Sizes of the Closed Discrete-Time Homogeneous Markov System with Finite State Capacities (HMS/c). *Markov Process. Relat. Fields* **2011**, *17*, 91–118.
32. Tsaklidis, G.M. The evolution of the attainable structures of a homogeneous Markov system by fixed size. *J. Appl. Probab.* **1994**, *31*, 348–361. [CrossRef]
33. Kipouridis, I.; Tsaklidis, G. The size order of the state vector of discrete-time homogeneous Markov systems. *J. Appl. Probab.* **2001**, *38*, 357–368. [CrossRef]
34. Vasiliadis, G.; Tsaklidis, G. On the moments of the state sizes of the discrete time homogeneous Markov system with a finite state capacity. In *Recent Advances in Stochastic Modeling and Data Analysis*; Skiadas, C.H., Ed.; 2007; pp. 190–197. [CrossRef]
35. Vasiliadis, G.; Tsaklidis, G. On the Distributions of the State Sizes of Closed Continuous Time Homogeneous Markov Systems. *Methodol. Comput. Appl. Probab.* **2009**, *11*, 561–582. [CrossRef]
36. Sands, T. Optimization Provenance of Whiplash Compensation for Flexible Space Robotics. *Aerospace* **2019**, *6*, 93. [CrossRef]
37. Bachmann, A.; Williams, S.B. Outlier handling when using particle filters in terrain-aided navigation. *IFAC Proc. Vol.* **2004**, *37*, 358–363. [CrossRef]
38. Sands, T. Development of Deterministic Artificial Intelligence for Unmanned Underwater Vehicles (UUV). *J. Mar. Sci. Eng.* **2020**, *8*, 578. [CrossRef]
39. Orchard, M.E.; Cerda, M.A.; Olivares, B.E.; Silva, J.F. Sequential monte carlo methods for discharge time prognosis in lithium-ion batteries. *Int. J. Progn. Health Manag.* **2012**, *3*, 90–101.
40. Maciak, M.; Pešta, M.; Peštová, B. Changepoint in dependent and non-stationary panels. *Stat. Pap.* **2020**, *61*, 1385–1407. [CrossRef]
41. Maciak, M.; Okhrin, O.; Pešta, M. Infinitely stochastic micro reserving. *Insur. Math. Econ.* **2021**, *100*, 30–58. [CrossRef]

mathematics

MDPI

Article

State Space Modeling with Non-Negativity Constraints Using Quadratic Forms

Ourania Theodosiadou *,† and George Tsaklidis *,†

Department of Mathematics, Aristotle University of Thessaloniki, 54124 Thessaloniki, Greece
* Correspondence: outheod@math.auth.gr (O.T.); tsaklidi@math.auth.gr (G.T.); Tel.: +30-2310-997964 (O.T.)
† These authors contributed equally to this work.

Abstract: State space model representation is widely used for the estimation of nonobservable (hidden) random variables when noisy observations of the associated stochastic process are available. In case the state vector is subject to constraints, the standard Kalman filtering algorithm can no longer be used in the estimation procedure, since it assumes the linearity of the model. This kind of issue is considered in what follows for the case of hidden variables that have to be non-negative. This restriction, which is common in many real applications, can be faced by describing the dynamic system of the hidden variables through non-negative definite quadratic forms. Such a model could describe any process where a positive component represents "gain", while the negative one represents "loss"; the observation is derived from the difference between the two components, which stands for the "surplus". Here, a thorough analysis of the conditions that have to be satisfied regarding the existence of non-negative estimations of the hidden variables is presented via the use of the Karush–Kuhn–Tucker conditions.

Keywords: state space model; Kalman filter; constrained optimization; two-sided components

Citation: Theodosiadou, O.; Tsakilids, G. State Space Modeling with Non-Negativity Constraints Using Quadratic Forms. *Mathematics* **2021**, *9*, 1908. https://doi.org/10.3390/math9161908

Academic Editors: Panagiotis-Christos Vassiliou and Andreas C. Georgiou

Received: 11 July 2021
Accepted: 5 August 2021
Published: 10 August 2021

Publisher's Note: MDPI stays neutral with regard to jurisdictional claims in published maps and institutional affiliations.

Copyright: © 2021 by the authors. Licensee MDPI, Basel, Switzerland. This article is an open access article distributed under the terms and conditions of the Creative Commons Attribution (CC BY) license (https://creativecommons.org/licenses/by/4.0/).

1. Introduction

State space modeling is used for estimating—revealing the dynamic evolution of hidden variables' processes. In some cases, the state vector, which includes the hidden components, is subject to constraints, which are derived either due to the physical meaning of the states or because of the mathematical properties that have to be satisfied. For example, state space models with constraints are used in camera surveillance [1,2], navigation issues [3], and biological systems [4]. Especially, in finance, the hidden variables are often subject to non-negative constraints or in general have to be bounded. For example, in the Vasicek model [5] and its extension [6], the interest rates are considered to be hidden random variables subject to non-negative constraints, while in [7,8], the eigenvalues of the VAR process were restricted within the unit circle. Considering the use of state space models in the domain of finance, a discrete state space model could be implemented for the estimation of the hidden jump components of asset returns [9,10]. The use of jumps has been proposed for the description of the dynamics of asset prices since they can explain some of the empirical characteristics of the asset prices, e.g., the lack of a normal distribution or the existence of leptokurticity (see for example [11]).

When dealing with state space models that are subject to constraints, the Kalman filtering algorithm [12] can no longer be used, since it assumes linearity in the model. In the domain of nonlinear filters, the particle filtering approach (see for example [13–16]) has wide applicability, and it adopts resampling techniques for the estimation of the state vector at every time t. However, the use of resampling techniques adds considerable computational cost in the estimation procedure.

In this work, the observation is defined as the difference between the two-sided components under noise inclusion. The components are considered to be hidden random variables, and therefore, a state space model is established, where the state equation describes the dynamic evolution of the two hidden components. This equation represents a first-order Markov process, i.e., all the information needed for the estimation of the components at time t is derived by the components at time $t-1$, and no other information from past times is needed. Moreover, the state vector is subject to non-negative constraints that have to be taken into account for its estimation in time. Such a model could describe, for example, the evolution of a system where the positive component represents "gain", while the negative one represents "loss"; the observation is derives from the difference between the two components, which stands for the "surplus", under noise inclusion. In asset pricing, an asset return can be defined as the difference between the two-sided non-negative return jump components under noise inclusion, and the jump components are considered to be hidden variables. Another example could be the one-dimensional random walk, where a positive jump could represent (the measure of) a move to the right and a negative jump (the measure of) a move to the left, while the observation could be a function of the two jump components given at discrete times. To handle such kinds of problems, non-negative definite quadratic forms are adopted in the state equation for the dynamic evolution of the two-sided components. In this case, the recursive equations of the Kalman filter cannot be used for the estimation of the state vector, since this filter assumes linearity in the measurement and state equation. To this end, this work first derives the recursive equations for the estimation of the state vector based on the state space model representation with non-negative definite quadratic forms in the state equation and their Taylor expansions. Then, a thorough analysis of the necessary conditions that have to be satisfied in order to obtain the non-negative estimations at every time t is provided. In Proposition 1, the stationary points of the optimization problem with the non-negative constraints are given by using the Karush–Kuhn–Tucker conditions, while in Proposition 2, the necessary conditions for the existence of feasible solutions in the constrained optimization problem are provided.

Overall, this work proposes a method in state space modeling representation, which can be used when dealing with hidden components that are subject to non-negativity constraints. The method results in the formulation of a constrained optimization problem for which the stationary points are derived via Proposition 1, and the necessary conditions for the existence of feasible solutions in this optimization problem are provided via Proposition 2; to that end, the iterative formulas for the minimum variance a posteriori estimators for the (hidden) state vector are illustrated. Moreover, the proposed method has a low computational burden compared to other nonlinear filtering methods that can be used in state space modeling with inequality constraints and are based on resampling techniques (e.g., particle filtering).

The paper is organized as follows. In Section 2, the state space model proposed for the estimation of the two jump components is established. Two non-negative quadratic forms are adopted to describe the dynamic evolution of the two-sided components subject to their non-negative restrictions. In Section 3, the recursive equations of the second-order Kalman filter are presented, while in Section 4, a thorough analysis of the conditions that have to be fulfilled so as to have non-negative estimations is presented. The results of this analysis are summarized in Propositions 1 and 2. In Section 5, an illustrative example concerning the evolution of positive and negative jumps of asset returns is presented to demonstrate the theoretical results. Finally, Section 6 concludes on the findings and provides suggestions for future work.

2. State Space Model

In this section, a state space model representation is illustrated considering the case where there are two hidden processes subject to non-negativity constraints. The state equation that describes the dynamic evolution of the hidden components adopts the use of non-negative definite quadratic forms, while the measurement equation is linear.

The state equation is given by:

$$\left.\begin{aligned} X_t &= (\mathbf{z}_{t-1} + \mathbf{w}_{t-1})^\top \mathbf{G}^{(1)} (\mathbf{z}_{t-1} + \mathbf{w}_{t-1}) = f_1(\mathbf{z}_{t-1}, \mathbf{w}_{t-1}) \\ Y_t &= (\mathbf{z}_{t-1} + \mathbf{w}_{t-1})^\top \mathbf{G}^{(2)} (\mathbf{z}_{t-1} + \mathbf{w}_{t-1}) = f_2(\mathbf{z}_{t-1}, \mathbf{w}_{t-1}) \end{aligned}\right\} \tag{1}$$

or equivalently:

$$\mathbf{z}_t = \sum_{k=1}^{2} \boldsymbol{\phi}_k (\mathbf{z}_{t-1} + \mathbf{w}_{t-1})^\top \mathbf{G}^{(k)} (\mathbf{z}_{t-1} + \mathbf{w}_{t-1}) \tag{2}$$

where:

- $\mathbf{z}_t = (X_t, Y_t)^\top = (z_{t,1}, z_{t,2})^\top$ stands for the state vector;
- $\mathbf{w_t}$ stands for the noise, and it is assumed that $\mathbf{w_t} \sim N(\mathbf{0}, \mathbf{Q})$, where $\mathbf{Q} = \begin{bmatrix} \sigma_x^2 & 0 \\ 0 & \sigma_y^2 \end{bmatrix}$;
- $\mathbf{G}^{(k)}, k = 1, 2$, is a (symmetric, 2×2) non-negative definite matrix, i.e.,

$$g_{11}^{(k)} > 0 \quad \text{and} \quad g_{11}^{(k)} g_{22}^{(k)} - (g_{12}^{(k)})^2 > 0 \,, \; k = 1, 2.$$

The vector $\boldsymbol{\phi}_k$ is a (2×1) column vector, where the k-th element equals 1, and the other element equals 0. The measurement equation is given by the relation:

$$R_t = \mathbf{H}\mathbf{z_t} + e_t \,, \tag{3}$$

where $\mathbf{H} = \begin{bmatrix} 1 & -1 \end{bmatrix}$ and $e_t \sim N(0, V)$. Moreover, it is assumed that $\mathbb{E}(e_k \mathbf{w}_j^T) = \mathbf{0}$.

Apparently, state Equation (2) describes a (nonobservable) first-order non-negative valued Markovian process, the evolution of which and its characteristics (e.g., periodicity, convergence etc.) depend on the structure (values) of the associated noisy observation sequence. The aim of our study here was to estimate (reveal) the Markovian process (2) (i.e., the matrices $\mathbf{G}^{(k)}, k = 1, 2$, and $\mathbf{Q}$), through the observation Equation (3), if the components of the state vector have to be non-negative. For this purpose, Model (2) and (3) adopts the use of non-negative definite quadratic forms to describe the dynamic evolution of the hidden two-sided components; that is, to ensure that the estimations of the components will be non-negative. To that end, the extended Kalman filter of second order is proposed in order to estimate at every time t the state vector $\mathbf{z}_t$ that incorporates the hidden jump components. It is noticed here that the noise component in Relation (2) is multiplicative and not additive.

Next, the extended Kalman filter of second order is described and its iterative equations for the estimation of the state vector are presented.

3. Extended Kalman Filter of Second Order

Model (2) and (3) presented in Section 2 is nonlinear, and subsequently, the recursive standard algorithm of the Kalman filter cannot be used for the estimation of the state vector. Aiming to derive the recursive equations for the estimation of the hidden states taking into consideration that the state Equation (2) is a quadratic form, the following notation is used:

- $\hat{\mathbf{z}}_t^-$: the a priori estimation of the state vector $\mathbf{z}_t$, i.e., without taking into consideration the measurement at time t;
- $\hat{\mathbf{z}}_t^+$: the a posteriori estimation of the state vector $\mathbf{z}_t$, i.e., by considering the measurement at time t;
- $\mathbf{P}_t^-, \mathbf{P}_t^+$: the variance–covariance matrices of the a priori and a posteriori error estimations of $\mathbf{z}_t$, respectively, i.e.,

$$\mathbf{P}_t^- = \mathbb{E}[(\mathbf{z}_t - \hat{\mathbf{z}}_t^-)(\mathbf{z}_t - \hat{\mathbf{z}}_t^-)^\top] \quad \text{and} \quad \mathbf{P}_t^+ = \mathbb{E}[(\mathbf{z}_t - \hat{\mathbf{z}}_t^+)(\mathbf{z}_t - \hat{\mathbf{z}}_t^+)^\top] \,.$$

According to (2), $z_{t,k}$, $k = 1,2$ is a function of the random variables $\mathbf{z}_{t-1}$, and $\mathbf{w}_{t-1}$, i.e., $z_{t,k} = z_{t,k}(\mathbf{z}_{t-1}, \mathbf{w}_{t-1})$. Then, using the Taylor expansion of second order of $z_{t,k}$ at $(\hat{\mathbf{z}}_{t-1}^{+}, \mathbf{0})$, it is derived that:

$$\begin{aligned} z_{t,k} =& f_k(\hat{\mathbf{z}}_{t-1}^{+}, \mathbf{0}) \\ &+ (\frac{\partial f_k(\hat{\mathbf{z}}_{t-1}^{+}, \mathbf{0})}{\partial \mathbf{z}_{t-1}})^{\top} (\mathbf{z}_{t-1} - \hat{\mathbf{z}}_{t-1}^{+}) + (\frac{\partial f_k(\hat{\mathbf{z}}_{t-1}^{+}, \mathbf{0})}{\partial \mathbf{w}_{t-1}})^{\top} \mathbf{w}_{t-1} \\ &+ \frac{1}{2} (\mathbf{z}_{t-1} - \hat{\mathbf{z}}_{t-1}^{+})^{\top} \frac{\partial^2 f_k(\hat{\mathbf{z}}_{t-1}^{+}, \mathbf{0})}{\partial \mathbf{z}_{t-1}^2} (\mathbf{z}_{t-1} - \hat{\mathbf{z}}_{t-1}) \\ &+ \frac{1}{2} \mathbf{w}_{t-1}^{\top} \frac{\partial^2 f_k(\hat{\mathbf{z}}_{t-1}^{+}, \mathbf{0})}{\partial \mathbf{w}_{t-1}^2} \mathbf{w}_{t-1} \\ &+ (\mathbf{z}_{t-1} - \hat{\mathbf{z}}_{t-1}^{+})^{\top} \frac{\partial^2 f_k(\hat{\mathbf{z}}_{t-1}^{+}, \mathbf{0})}{\partial \mathbf{z}_{t-1} \partial \mathbf{w}_{t-1}} \mathbf{w}_{t-1}, \quad k = 1,2 \end{aligned} \tag{4}$$

where functions $f_k = f_k(\mathbf{z}_{t-1}, \mathbf{w}_{t-1})$, $k = 1,2$, are given in (1). By equating the mean values in Relation (4), the a priori estimation of $\mathbf{z}_t$ (*prediction stage*) is derived, that is:

$$\begin{aligned} \hat{z}_{t,k}^{-} =& f_k(\hat{\mathbf{z}}_{t-1}^{+}, \mathbf{0}) + \frac{1}{2} \operatorname{tr}(\frac{\partial^2 f_k(\hat{\mathbf{z}}_{t-1}^{+}, \mathbf{0})}{\partial \mathbf{z}_{t-1}^2} \mathbf{P}_{t-1}^{+}) \\ &+ \frac{1}{2} \operatorname{tr}(\frac{\partial^2 f_k(\hat{\mathbf{z}}_{t-1}^{+}, \mathbf{0})}{\partial \mathbf{w}_{t-1}^2} \mathbf{Q}), \quad k = 1,2 \end{aligned} \tag{5}$$

and the entries of the respective variance–covariance matrix $\mathbf{P}_t^{-}$ are given by the relation,

$$\begin{aligned} (\mathbf{P}_t^{-})_{k,m} =& (\frac{\partial f_k(\hat{\mathbf{z}}_{t-1}^{+}, \mathbf{0})}{\partial \mathbf{z}_{t-1}})^{\top} \mathbf{P}_{t-1}^{+} \frac{\partial f_m(\hat{\mathbf{z}}_{t-1}^{+}, \mathbf{0})}{\partial \mathbf{z}_{t-1}} \\ &+ (\frac{\partial f_k(\hat{\mathbf{z}}_{t-1}^{+}, \mathbf{0})}{\partial \mathbf{w}_{t-1}})^{\top} \mathbf{Q} \frac{\partial f_m(\hat{\mathbf{z}}_{t-1}^{+}, \mathbf{0})}{\partial \mathbf{w}_{t-1}} \\ &+ \frac{1}{2} \operatorname{tr}(\frac{\partial^2 f_k(\hat{\mathbf{z}}_{t-1}^{+}, \mathbf{0})}{\partial \mathbf{z}_{t-1}^2} \mathbf{P}_{t-1}^{+} \frac{\partial^2 f_m(\hat{\mathbf{z}}_{t-1}^{+}, \mathbf{0})}{\partial \mathbf{z}_{t-1}^2} \mathbf{P}_{t-1}^{+}) \\ &+ \frac{1}{2} \operatorname{tr}(\frac{\partial^2 f_k(\hat{\mathbf{z}}_{t-1}^{+}, \mathbf{0})}{\partial \mathbf{w}_{t-1}^2} \mathbf{Q} \frac{\partial^2 f_m(\hat{\mathbf{z}}_{t-1}^{+}, \mathbf{0})}{\partial \mathbf{w}_{t-1}^2} \mathbf{Q}), \quad k,m = 1,2 \end{aligned} \tag{6}$$

where $(\mathbf{P}_t^{-})_{k,m}$ denotes the (k,m)-element of matrix $\mathbf{P}_t^{-}$ and $\operatorname{tr}(.)$ denotes the trace of the respective matrix. Taking into consideration the properties of the trace of a matrix, it is derived after some algebraic manipulations on Relations (5) and (6) that:

$$\hat{z}_{t,k}^{-} = \hat{\mathbf{z}}_{t-1}^{+\,T} \mathbf{G}^{(k)} \hat{\mathbf{z}}_{t-1}^{+} + \operatorname{tr}(\mathbf{G}^{(k)} \mathbf{P}_{t-1}^{+}) + \operatorname{tr}(\mathbf{G}^{(k)} \mathbf{Q}), \quad k = 1,2 \tag{7}$$

$$\begin{aligned} (\mathbf{P}_t^{-})_{k,m} =& 4\hat{\mathbf{z}}_{t-1}^{+\,T} \mathbf{G}^{(k)} \mathbf{P}_{t-1}^{+} \mathbf{G}^{(m)} \hat{\mathbf{z}}_{t-1}^{+} + 4\hat{\mathbf{z}}_{t-1}^{+\,T} \mathbf{G}^{(k)} \mathbf{Q} \mathbf{G}^{(m)} \hat{\mathbf{z}}_{t-1}^{+} \\ &+ 2\operatorname{tr}(\mathbf{G}^{(k)} \mathbf{P}_{t-1}^{+} \mathbf{G}^{(m)} \mathbf{P}_{t-1}^{+}) + 2\operatorname{tr}(\mathbf{G}^{(k)} \mathbf{Q} \mathbf{G}^{(m)} \mathbf{Q}), \; k,m = 1,2\,. \end{aligned} \tag{8}$$

Regarding the a posteriori estimations of $\mathbf{z}_t$, it is taken into account that the joint distribution of $\mathbf{z}_t$ and R_t is normal, based on the relation:

$$\begin{bmatrix} \mathbf{z}_t \\ R_t \end{bmatrix} \sim N(\begin{bmatrix} \hat{\mathbf{z}}_t^{-} \\ \mathbf{H}\hat{\mathbf{z}}_t^{-} \end{bmatrix}, \begin{bmatrix} \mathbf{P}_t^{-} & \mathbf{P}_t^{-}\mathbf{H}^T \\ \mathbf{H}\mathbf{P}_t^{-\,T} & \mathbf{H}\mathbf{P}_t^{-}\mathbf{H}^T + V \end{bmatrix}).$$

Then, we make use of the following Lemma (see for example [17]):

Lemma 1. *Let* $\mathbf{x}, \mathbf{y}$ *be two random variables that are jointly normally distributed with:*

$$\mathbb{E}(\begin{bmatrix} \mathbf{x} \\ \mathbf{y} \end{bmatrix}) = \begin{bmatrix} \mu_x \\ \mu_y \end{bmatrix} \quad \text{and} \quad \boldsymbol{\Sigma} = \begin{bmatrix} \boldsymbol{\Sigma}_{11} & \boldsymbol{\Sigma}_{12} \\ \boldsymbol{\Sigma}_{21} & \boldsymbol{\Sigma}_{22} \end{bmatrix}.$$

Then, $(\mathbf{x}/\mathbf{y}) \sim N(\boldsymbol{\mu}', \boldsymbol{\Sigma}')$, *where:*

$$\boldsymbol{\mu}' = \boldsymbol{\mu}_x + \boldsymbol{\Sigma}_{11}\boldsymbol{\Sigma}_{22}^{-1}(\mathbf{y} - \boldsymbol{\mu}_x) \quad \text{and} \quad \boldsymbol{\Sigma}' = \boldsymbol{\Sigma}_{11} - \boldsymbol{\Sigma}_{12}\boldsymbol{\Sigma}_{22}^{-1}\boldsymbol{\Sigma}_{21}\,.$$

Based on Lemma 1, the a posteriori estimation of $\mathbf{z}_t$ (*update stage*) and the related variance–covariance matrix $\mathbf{P}_t(+)$ are given by,

$$\hat{\mathbf{z}}_t^+ = \hat{\mathbf{z}}_t^- + \mathbf{K}_t(R_t - \mathbf{H}\hat{\mathbf{z}}_t^-)\,, \tag{9}$$

$$\mathbf{P}_t^+ = (\mathbf{I} - \mathbf{K}_t\mathbf{H})\mathbf{P}_t^-\,, \tag{10}$$

where $\mathbf{K}_t = \mathbf{P}_t^-\mathbf{H}^T(\mathbf{H}\mathbf{P}_t^-\mathbf{H}^T + V)^{-1}$. By using the recursive Relations (7)–(10), we can estimate the hidden components at every time t.

Next, a detailed investigation regarding the existence of non-negative solutions (i.e., non-negative a posteriori estimations of $\mathbf{z}_t$) derived from (9) is presented.

4. Investigation of the State Space Model

In what follows, we present an investigation concerning the conditions that have to be satisfied so as to derive non-negative a posteriori estimations of the state vector $\mathbf{z}_t$. Obviously, Relation (7) ensures the existence of non-negative a priori estimations of $\mathbf{z}_t$ at every time t. However, the a posteriori estimations of $\mathbf{z}_t$ given by (9) may not fulfil the non-negativity condition. We note that the solutions depend on the term $\mathbf{K}_t(R_t - \mathbf{H}\hat{\mathbf{z}}_t^-)$, the sign of which is not time invariant. To this end, in order to ensure that the a posteriori unbiased estimator $\hat{\mathbf{z}}_t^+$ will be a minimum variance estimator under the non-negativity restrictions that its components must satisfy, the following optimization problem arises,

$$min_{\hat{\mathbf{z}}_t^+}\{\text{tr}(\mathbf{P}_t^+) = \mathbb{E}[(\mathbf{z}_t - \hat{\mathbf{z}}_t^+)(\mathbf{z}_t - \hat{\mathbf{z}}_t^+)^T]\} \tag{11}$$

$$\text{where } \hat{\mathbf{z}}_t^+ \succeq \mathbf{0}.$$

Symbol $\succeq$ (or $\preceq$) is used for the elementwise inequality, while $\mathbf{z}_t = (X_t, Y_t)^T$ is given by Equation (1) (or (2)). The following Proposition 1 provides the set of *stationary points* related to the optimization problem (11), subject to the non-negativity restrictions. This set includes the optimal solution, i.e., the unbiased minimum variance estimator $\hat{\mathbf{z}}_t^+$. In what follows, we use the following notations:

$$a_t = R_t - \mathbf{H}\hat{\mathbf{z}}_t^-, b_t = \mathbf{H}\mathbf{P}_t^-\mathbf{H}^T \text{ and } \mathbf{K}_t = (K_{t,1}, K_{t,2})^\top.$$

Remark 1. *Notice that, if* $a_t = 0$, *then Relation* (9) *leads to* $\hat{\mathbf{z}}_t^+ = \hat{\mathbf{z}}_t^- \succeq \mathbf{0}$, *and consequently, the solution is acceptable.*

Taking into consideration Remark 1, it is assumed in the sequel that $a_t \neq 0$ for every t.

Proposition 1. *The weight matrix* $\mathbf{K}_t$ *and the stationary points related to the optimization problem* (11) *are given by the relations:*

(i) $\mathbf{K}_t = (b_t + R_t)^{-1}\mathbf{P}_t^-\mathbf{H}^T$, *which leads to the solution:*

$$\hat{\mathbf{z}}_t^+ = \hat{\mathbf{z}}_t^- + a_t(b_t + R_t)^{-1}\mathbf{P}_t^-\mathbf{H}^T;$$

(ii) $\mathbf{K}_t = \begin{pmatrix} (b_t + R_t)^{-1}(\mathbf{P}_t^-\mathbf{H}^T)_1 \\ -a_t^{-1}\hat{z}_{t,2}^- \end{pmatrix}$ *which leads to the solution:*

$$\hat{z}^+_{t,1} = \hat{z}^-_{t,1} + a_t(b_t + R_t)^{-1}(\mathbf{P}^-_t\mathbf{H}^T)_1,$$

and:

$$\hat{z}^+_{t,2} = 0\,;$$

(iii) $\mathbf{K}_t = \begin{pmatrix} -a_t^{-1}\hat{z}^-_{t,1} \\ (b_t + R_t)^{-1}(\mathbf{P}^-_t\mathbf{H}^T)_2 \end{pmatrix}$, *which leads to the solution:*

$$\hat{z}^+_{t,1} = 0$$

and:

$$\hat{z}^+_{t,2} = \hat{z}^-_{t,2} + a_t(b_t + R_t)^{-1}(\mathbf{P}^-_t\mathbf{H}^T)_2;$$

(iv) $\mathbf{K}_t = -a_t^{-1}\hat{\mathbf{z}}^-_t$, *which leads to the solution:*

$$\hat{\mathbf{z}}^+_t = \mathbf{0},$$

where $(\mathbf{P}^-_t\mathbf{H}^T)_i$ *denotes the ith-row of matrix* $\mathbf{P}^-_t\mathbf{H}^T$, $i = 1,2$.

Proof. The Lagrangian function related to the optimization problem (11) is defined as:

$$\begin{aligned} \Lambda &= \mathrm{tr}(\mathbf{P}^+_t) + \lambda_1(-\hat{z}^+_{t,1}) + \lambda_2(-\hat{z}^+_{t,2}) \\ &= \mathrm{tr}(\mathbb{E}[(\mathbf{z}_t - \hat{\mathbf{z}}^+_t)(\mathbf{z}_t - \hat{\mathbf{z}}^+_t)^T]) + \lambda_1(-\hat{z}^+_{t,1}) + \lambda(-\hat{z}^+_{t,2})\,, \quad \lambda_1, \lambda_2 \geq 0. \end{aligned} \tag{12}$$

Based on (10), it is derived that:

$$\mathrm{tr}(\mathbf{P}^+_t) = \mathrm{tr}[(\mathbf{I} - \mathbf{K}_t\mathbf{H})\mathbf{P}^-_t(\mathbf{I} - \mathbf{K}_t\mathbf{H})^T + \mathbf{K}_t V \mathbf{K}^T_t]$$

while (by assuming the dependence of $\hat{z}^+_{t,i}$ on R_t and $\hat{z}^-_{t,i}$, $i = 1,2$, as provided in Kalman filtering):

$$\hat{z}^+_{t,1} = \hat{z}^-_{t,1} + K_{t,1}(R_t - \mathbf{H}\hat{\mathbf{z}}^-_t) = \hat{z}^-_{t,1} + a_t K_{t,1}\,,$$

$$\hat{z}^+_{t,2} = \hat{z}^-_{t,2} + K_{t,1}(R_t - \mathbf{H}\hat{\mathbf{z}}^-_t) = \hat{z}^-_{t,2} + a_t K_{t,2}\,.$$

By calculating the first derivative of the Lagrangian function and equating it to 0, it is derived that:

$$\begin{aligned} \frac{d\Lambda}{d\mathbf{K}_t} &= \frac{d}{d\mathbf{K}_t}[\mathrm{tr}[(\mathbf{I} - \mathbf{K}_t\mathbf{H})\mathbf{P}^-_t(\mathbf{I} - \mathbf{K}_t\mathbf{H})^T + \mathbf{K}_t V\mathbf{K}^T_t] - \lambda_1(\hat{z}^-_{t,1} + a_t K_{t,1}) \\ &\quad - \lambda_2(\hat{z}^-_{t,2} + a_t K_{t,2})] \\ &= -2\mathbf{P}^{-\,T}_t\mathbf{H}^T + 2\mathbf{K}_t\mathbf{H}\mathbf{P}^-_t\mathbf{H}^T + 2\mathbf{K}_t V - a_t\boldsymbol{\lambda} \\ &= 0 \end{aligned} \tag{13}$$

where $\boldsymbol{\lambda} = (\lambda_1, \lambda_2)^T$. Thus, matrix $\mathbf{K}_t$ has to satisfy the following condition (by noticing that $\mathbf{P}^-_t$ is symmetric):

$$-2\mathbf{P}^-_t\mathbf{H}^T + 2\mathbf{K}_t\mathbf{H}\mathbf{P}^-_t\mathbf{H}^T + 2\mathbf{K}_t V = a_t\boldsymbol{\lambda} \tag{14}$$

based on the constraints [18]:

$$\begin{aligned} \lambda_1(\hat{z}^-_{t,1} + a_t K_{t,1}) &= 0, \\ \lambda_2(\hat{z}^-_{t,2} + a_t K_{t,2}) &= 0 \\ \lambda_1, \lambda_2 &\geq 0. \end{aligned}$$

The following cases have to be considered:

(i) **The two constraint conditions are inactive**. Then, $\lambda_1 = \lambda_2 = 0$, and the optimization problem, leading to (14), is transformed into the unconstrained one considered in the case of the Kalman filter. It is derived that:

$$\mathbf{K}_t = \mathbf{P}_t^- \mathbf{H}^T (\mathbf{H}\mathbf{P}_t^- \mathbf{H}^T + V)^{-1}, \tag{15}$$

which is the well-known *Kalman gain matrix*. The related solution in terms of the a posteriori estimator $\hat{\mathbf{z}}_t^+$ is:

$$\hat{\mathbf{z}}_t^+ = \hat{\mathbf{z}}_t^- + a_t (b_t + R_t)^{-1} \mathbf{P}_t^- \mathbf{H}^T \,. \tag{16}$$

Relation (16) constitutes a possible solution of the optimization problem (11), and it has to satisfy the constraint $\hat{\mathbf{z}}_t^+ \succeq \mathbf{0}$;

(ii) **The first constraint condition is inactive** (i.e., $\lambda_1 = 0$), **while the second one is active**. Then, the following two cases are considered:

(a) If $\lambda_2 = 0$, then we are led to the unconstrained optimization problem presented in Case (i), and the solution must satisfy the non-negative restrictions, i.e., $\hat{\mathbf{z}}_t^+ \succeq \mathbf{0}$;

(b) If $\hat{z}_{t,2}^- + a_t K_{t,2} = 0$, it is derived via the active constraint condition that:

$$K_{t,2} = -a_t^{-1} \hat{z}_{t,2}^- \,. \tag{17}$$

By using (17), Relation (14) is transformed into:

$$\begin{pmatrix} (\mathbf{P}_t^- \mathbf{H}^T)_1 \\ (\mathbf{P}_t^- \mathbf{H}^T)_2 \end{pmatrix} + \begin{pmatrix} K_{t,1} \\ -a_t^{-1} \hat{z}_{t,2}^- \end{pmatrix} \mathbf{H}\mathbf{P}_t^- \mathbf{H}^T + \begin{pmatrix} K_{t,1} \\ -a_t^{-1} \hat{z}_{t,2}^- \end{pmatrix} V = a_t \boldsymbol{\lambda} \,.$$

Consequently,

$$K_{t,1} = (b_t + R_t)^{-1} (\mathbf{P}_t^- \mathbf{H}^T)_1, \tag{18}$$

where $b_t = \mathbf{H}\mathbf{P}_t^- \mathbf{H}^T \geq 0$. By using (17) and (18), it is derived that:

$$K_t = \begin{pmatrix} (b_t + R_t)^{-1} (\mathbf{P}_t^- \mathbf{H}^T)_1 \\ -a_t^{-1} \hat{z}_{t,2}^- \end{pmatrix} .$$

Thus,

$$\hat{z}_{t,1}^+ = \hat{z}_{t,1}^- + a_t (b_t + R_t)^{-1} (\mathbf{P}_t^- \mathbf{H}^T)_1$$

and:

$$\hat{z}_{t,2}^+ = 0 \,;$$

(iii) **The first constraint condition is active, while the second one is inactive** (i.e., $\lambda_2 = 0$). The following two cases are considered:

(a) If $\lambda_1 = 0$, then we obtain the unconstrained optimization problem presented in Case (i), and the solution must fulfil the nonnegative restrictions, i.e., $\hat{\mathbf{z}}_t^+ \succeq \mathbf{0}$;

(b) If $\hat{z}_{t,1}^- + a_t K_{t,1} = 0$ and $\lambda_1 = 0$, then it is derived that:

$$K_{t,1} = -a_t^{-1} \hat{z}_{t,1}^- \,, \tag{19}$$

and Relation (14) is transformed into:

$$\begin{pmatrix} (\mathbf{P}_t^- \mathbf{H}^T)_1 \\ (\mathbf{P}_t^- \mathbf{H}^T)_2 \end{pmatrix} + \begin{pmatrix} K_{t,1} \\ -a_t^{-1} \hat{z}_{t,2}^- \end{pmatrix} \mathbf{H}\mathbf{P}_t^- \mathbf{H}^T + \begin{pmatrix} K_{t,1} \\ -a_t^{-1} \hat{z}_{t,2}^- \end{pmatrix} V = a_t \boldsymbol{\lambda} \,.$$

Then,

$$K_{t,2} = (b_t + R_t)^{-1}(\mathbf{P}_t^- \mathbf{H}^T)_2 \tag{20}$$

where $b_t = \mathbf{H}\mathbf{P}_t^- \mathbf{H}^T \geq 0$. By using (19) and (20), it is derived that:

$$K_t = \begin{pmatrix} -a_t^{-1}\hat{z}_{t,1}^- \\ (b_t + R_t)^{-1}(\mathbf{P}_t^- \mathbf{H}^T)_2 \end{pmatrix},$$

and consequently:

$$\hat{z}_{t,1}^+ = 0$$

and:

$$\hat{z}_{t,2}^+ = \hat{z}_{t,2}^- + a_t(b_t + R_t)^{-1}(\mathbf{P}_t^- \mathbf{H}^T)_2\,;$$

(iv) **The two constraint conditions are active**, i.e., $\hat{z}_{t,1}^- + a_t K_{t,1} = 0$ and $\hat{z}_{t,2}^- + a_t K_{t,2} = 0$. In this case, we have to seek solutions such that $\lambda_1, \lambda_2 \geq 0$.

Based on the active constraint conditions, it is derived that:

$$K_{t,1} = -a_t^{-1}\hat{z}_{t,1}^- \quad \text{and} \quad K_{t,2} = -a_t^{-1}\hat{z}_{t,2}^-$$

i.e., $\mathbf{K}_t = -a_t^{-1}\hat{\mathbf{z}}_t^-$, resulting in the relation,

$$\begin{aligned} \hat{\mathbf{z}}_t^+ &= \hat{\mathbf{z}}_t^- + a_t \mathbf{K}_t \\ &= \hat{\mathbf{z}}_t^- - a_t a_t^{-1}\hat{\mathbf{z}}_t^- \\ &= \mathbf{0}. \end{aligned}$$

The state vector $\hat{\mathbf{z}}_t^+ = \mathbf{0}$ constitutes a feasible solution, and it has to be checked whether Relation (14) is satisfied with $\lambda_1, \lambda_2 \geq 0$. □

In what follows, Proposition 2 provides the necessary conditions for the existence of feasible solutions regarding the constrained filter.

Proposition 2. *The solutions given in Proposition 1 regarding the optimization problem (11) are feasible upon the following conditions (necessary conditions):*

(i)

$$\hat{\mathbf{z}}_t^+ = \hat{\mathbf{z}}_t^- + a_t(b_t + R_t)^{-1}\mathbf{P}_t^- \mathbf{H}^\top$$

constitutes a feasible solution, if:

$$a_t \mathbf{P}_t^- \mathbf{H}^T \succeq -(b_t + R_t)\hat{\mathbf{z}}_t^-\,;$$

(ii)

$$\hat{\mathbf{z}}_t^+ = \begin{pmatrix} \hat{z}_{t,1}^- + a_t(b_t + R_t)^{-1}(\mathbf{P}_t^- \mathbf{H}^T)_1 \\ 0 \end{pmatrix}$$

constitutes a feasible solution, if:

$$a_t(\mathbf{P}_t^- \mathbf{H}^\top)_1 \geq -(b_t + R_t)\hat{\mathbf{z}}_{t,1}^-$$

and:

$$a_t(\mathbf{P}_t^- \mathbf{H}^\top)_2 < -(b_t + R_t)\hat{\mathbf{z}}_{t,2}^-\,;$$

(iii)

$$\hat{\mathbf{z}}_t^+ = \begin{pmatrix} 0 \\ \hat{z}_{t,2}^- + a_t(b_t + R_t)^{-1}(\mathbf{P}_t^- \mathbf{H}^T)_2 \end{pmatrix}$$

constitutes a feasible solution, if:

$$a_t(\mathbf{P}_t^- \mathbf{H}^T)_1 < -(b_t + R_t)\hat{\mathbf{z}}_{t,1}^-$$

and:

$$a_t(\mathbf{P}_t^- \mathbf{H}^T)_2 \geq -(b_t + R_t)\hat{\mathbf{z}}_{t,2}^- ;$$

(iv) $\hat{\mathbf{z}}_t^+ = \mathbf{0}$ *constitutes a feasible solution, if:*

$$a_t(\mathbf{P}_t^- \mathbf{H}^T)_1 < -(b_t + R_t)\hat{\mathbf{z}}_{t,1}^-$$

and:

$$a_t(\mathbf{P}_t^- \mathbf{H}^T)_2 < -(b_t + R_t)\hat{\mathbf{z}}_{t,2}^- .$$

Proof. Similar to the proof of Proposition 1, four cases are considered:

(i) **The two constraint conditions are inactive**. Then, $\lambda_1 = \lambda_2 = 0$, and the optimization problem is transformed into the unconstrained one that is met in the case of the Kalman filter. In this case, based on Proposition 1, we obtain that $\mathbf{K}_t = (b_t + R_t)^{-1}\mathbf{P}_t^- \mathbf{H}^T$, resulting in the estimation:

$$\hat{\mathbf{z}}_t^+ = \hat{\mathbf{z}}_t^- + a_t(b_t + R_t)^{-1}\mathbf{P}_t^- \mathbf{H}^T ,$$

where $\hat{\mathbf{z}}_t^+$ is a feasible solution of the optimization problem with the nonnegative constraints, if:

$$\hat{\mathbf{z}}_t^- + a_t(b_t + R_t)^{-1}\mathbf{P}_t^- \mathbf{H}^T \succeq \mathbf{0}.$$

Consequently, the necessary condition is formulated as follows:

$$a_t\mathbf{P}_t^- \mathbf{H}^T \succeq -a_t(b_t + R_t)\hat{\mathbf{z}}_t^- ;$$

(ii) **The first constraint condition is inactive, while the second one is active**, i.e, $\lambda_1 = 0$ and $\hat{z}_{t,2}^- + a_t K_{t,2} = 0$, respectively. The following two cases are considered:

(a) If $\lambda_2 = 0$, then based on (14), the solution is given by (15), which is related to the Kalman filter and the unconstrained optimization problem. This solution is acceptable if it is aligned with the active constraint condition. Otherwise, it is rejected;

(b) If $\lambda_2 > 0$, matrix $\mathbf{K}_t$ has to be in such a form so that $\hat{z}_{t,1}^+ \geq 0$.
It is derived via the active constraint condition that $K_{t,2} = -a_t^{-1}\hat{z}_{t,2}^-$ where $a_t \neq 0$ based on Remark 1. Then, (14) results in:

$$\begin{aligned} a_t\boldsymbol{\lambda} &= \begin{pmatrix} (\mathbf{P}_t^- \mathbf{H}^T)_1 \\ (\mathbf{P}_t^- \mathbf{H}^T)_2 \end{pmatrix} + \begin{pmatrix} K_{t,1} \\ -a_t^{-1}\hat{z}_{t,2}^- \end{pmatrix} \mathbf{H}\mathbf{P}_t^- \mathbf{H}^T + \begin{pmatrix} K_{t,1} \\ -a_t^{-1}\hat{z}_{t,2}^- \end{pmatrix} V \\ \begin{pmatrix} \lambda_1 \\ \lambda_2 \end{pmatrix} &= \begin{pmatrix} -a_t^{-1}(\mathbf{P}_t^- \mathbf{H}^T)_1 \\ -a_t^{-1}(\mathbf{P}_t^- \mathbf{H}^T)_2 \end{pmatrix} + \begin{pmatrix} a_t^{-1}K_{t,1} \\ -a_t^{-2}\hat{z}_{t,2}^- \end{pmatrix} \mathbf{H}\mathbf{P}_t^- \mathbf{H}^T + \begin{pmatrix} a_t^{-1}K_{t,1} \\ -a_t^{-2}\hat{z}_{t,2}^- \end{pmatrix} V \\ &= \begin{pmatrix} -a_t^{-1}(\mathbf{P}_t(-)\mathbf{H}^T)_1 + a_t^{-1}b_t K_{t,1} + a_t^{-1}K_{t,1}V \\ -a_t^{-1}(\mathbf{P}_t^- \mathbf{H}^T)_2 + a_t^{-2}b_t\hat{z}_{t,2}^- + a_t^{-2}\hat{z}_{t,2}^- V \end{pmatrix} . \end{aligned} \quad (21)$$

Since $\lambda_1 = 0$ and for $a_t \neq 0$, Relation (21) implies:

$$\begin{pmatrix} 0 \\ \lambda_2 \end{pmatrix} = \begin{pmatrix} -(\mathbf{P}_t^- \mathbf{H}^T)_1 + b_t K_{t,1} + K_{t,1}V \\ -a_t^{-1}(\mathbf{P}_t^- \mathbf{H}^T)_2 - a_t^{-2}b_t\hat{z}_{t,2}^- - a_t^{-2}\hat{z}_{t,2}^- V \end{pmatrix} \quad (22)$$

It is derived via (22) that if $\lambda_2 > 0$, then:

$$\lambda_2 = -a_t(\mathbf{P}_t^- \mathbf{H}^T)_2 - b_t\hat{z}_{t,2}^- - \hat{z}_{t,2}^- V > 0 .$$

Consequently,

$$a_t(\mathbf{P}_t^{-}\mathbf{H}^T)_2 < -(b_t+V)\hat{z}_{t,2}^{-}\,, \tag{23}$$

resulting in $a_t(\mathbf{P}_t^{-}\mathbf{H}^T)_2 < 0$. Moreover, taking into consideration that:

$$\hat{z}_{t,1}^{+} = \hat{z}_t^{-} + a_t K_{t,1} = \hat{z}_{t,1}^{-} - a_t(b_t+V)(\mathbf{P}_t^{-}\mathbf{H}^T)_1$$

and $\hat{z}_{t,1}^{+} \geq 0$, it is derived that:

$$a_t(\mathbf{P}_t^{-}\mathbf{H}^T)_1 < -(b_t+V)\hat{z}_{t,1}^{-}\,; \tag{24}$$

(iii) **The first constraint condition is active, while the second one is inactive**, i.e., $\hat{z}_{t,1}^{-} + a_t K_{t,1} = 0$ and $\lambda_2 = 0$, respectively. The following two cases are considered:

(a) $\lambda_1 = 0$ and $\lambda_2 = 0$;

(b) $\lambda_1 > 0$ and $\lambda_2 = 0$.

Similar to Case (ii), the third part of Proposition 2 can be derived;

(iv) **The two constraint conditions are active**, i.e., $\hat{z}_{t,1}^{-} + a_t K_{t,1} = 0$ and $\hat{z}_{t,2}^{-} + a_t K_{t,2} = 0$. In this case, we have to search for solutions where $\lambda_1, \lambda_2 \geq 0$.
It is derived via Proposition 1 that:

$$\mathbf{K}_t = a_t^{-1}\hat{\mathbf{z}}_t^{-}\,, \tag{25}$$

which leads to the solution $\hat{\mathbf{z}}_t^{+} = \mathbf{0}$.
The following subcases are considered:

(a) If $\lambda_1 = \lambda_2 = 0$, Solution (15) is derived via Relation (14), and it is accepted if it coincides with Relation (25). Otherwise, it is rejected;

(b) If $\lambda_1 = 0$ and $\lambda_2 > 0$, then by taking into consideration Case (iib), it is concluded that the solution $\mathbf{z}_t(+) = \mathbf{0}$ is accepted, if:

$$a_t(\mathbf{P}_t^{-}\mathbf{H}^T)_2 < -(b_t+V)\hat{z}_{t,2}^{-}\,.$$

Otherwise, it is rejected since it is not aligned with the conditions of the considered case (i.e., $\lambda_1 = 0$ $\lambda_2 > 0$);

(c) If $\lambda_1 > 0$ and $\lambda_2 = 0$, similar to Case (*iv*)-c, the solution $\mathbf{z}_t^{+} = \mathbf{0}$ is accepted, if:

$$a_t(\mathbf{P}_t^{-}\mathbf{H}^T)_1 < -(b_t+V)\hat{z}_{t,1}^{-}\,;$$

(d) If $\lambda_1 > 0$ and $\lambda_2 > 0$, then by taking into consideration Relations (23) and (25), it turns out that the necessary condition in order for $\mathbf{z}_t^{+} = \mathbf{0}$ to be accepted as a feasible solution (which satisfies the conditions of the considered case, i.e., $\lambda_1 > 0$ $\lambda_2 > 0$) is:

$$a_t(\mathbf{P}_t^{-}\mathbf{H}^T)_1 < -(b_t+V)\hat{z}_{t,1}^{-} \quad \text{and} \quad a_t(\mathbf{P}_t^{-}\mathbf{H}^T)_2 < -(b_t+V)\hat{z}_{t,2}^{-}.$$

In conclusion, in Case (*iv*), the vector $\mathbf{z}_t^{+} = \mathbf{0}$ is a possible optimal solution, if at least one of the following conditions holds:

$$a_t(\mathbf{P}_t(-)\mathbf{H}^T)_1 < -(b_t+V)\hat{z}_{t,1}^{-} \quad \text{or} \quad a_t(\mathbf{P}_t^{-}\mathbf{H}^T)_2 < -(b_t+V)\hat{z}_{t,2}^{-}.$$

□

Remark 2. *Based on the low computational cost, the four possible solutions of the constrained optimization problem (11) can be examined one-to-one, aiming to find the optimal solution. In any case, the necessary conditions presented in Proposition 2 can be examined simultaneously to have a more comprehensive view in the process of searching for the optimal solution.*

Next, an illustrative application of the described methodology is presented regarding the estimation (revelation) of the two-sided jump components of asset returns.

5. Application; Estimation of the Two-Sided Jump Components of the NASDAQ Index

In this section, an application example of the proposed methodology analyzed in Section 4 is illustrated concerning the estimation of the hidden two-sided jump components of the NASDAQ index for the 3 y period 2006–2008. To estimate the parameters of the model, i.e., the parameter set $\phi = (\mathbf{G}^{(1)}, \mathbf{G}^{(2)}, \sigma_x^2, \sigma_y^2, V)$, the maximum likelihood estimation method is used taking into consideration that the distribution of R_t conditioned on $\mathbf{z}_t$ is normal, i.e.,

$$R_t|\mathbf{z}_t \sim N(\mathbf{H}\hat{\mathbf{z}}_t^-, \mathbf{H}\mathbf{P}_t^-\mathbf{H}^T + V)\,.$$

Therefore, the log-likelihood function, LogL, is of the form:

$$LogL(R_1, \ldots, R_n) = -n/2\log(2\pi) - 0.5\sum_{t=1}^{n}(log(|\omega_t|) + u_t^T\omega_t^{-1}u_t) \tag{26}$$

where,

$$u_t = R_t - \mathbf{H}\hat{\mathbf{z}}_t^- \quad \text{and} \quad \omega_t = \mathbf{H}\mathbf{P}_t^-\mathbf{H}^T + V\,.$$

The estimations derived by maximizing $LogL$, given in (26), are as follows:

$$\mathbf{G}^{(1)} = \begin{bmatrix} 5.4741 & -2.8498 \\ -2.8498 & 7.3474 \end{bmatrix}, \mathbf{G}^{(2)} = \begin{bmatrix} 7.4368 & 1.4909 \\ 1.4909 & 2.8304 \end{bmatrix}$$

and:

$$\sigma_x^2 = 0.9897 \times 10^{-3}, \quad \sigma_y^2 = 0.86281 \times 10^{-3}, \quad V = 4.961 \times 10^{-11},$$

with $LogL = 995.9854$. Based on the estimated parameters, the estimated two-sided jump components of the NASDAQ index are showcased in Figure 1.

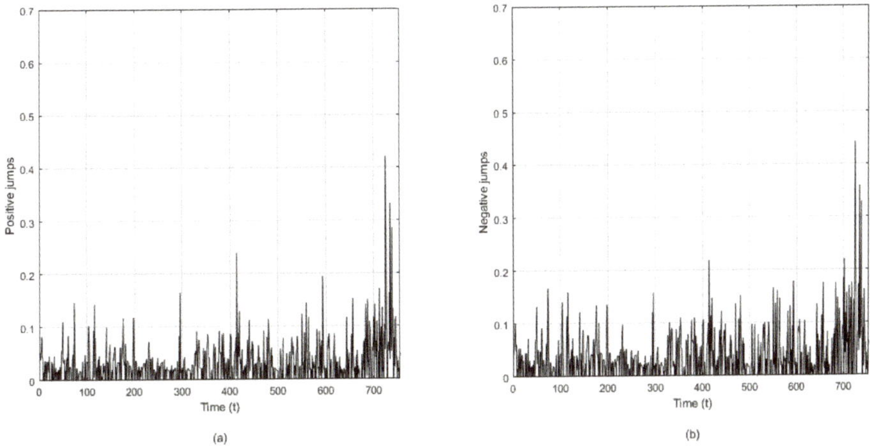

Figure 1. (**a**) Estimated positive return jumps of the NASDAQ index during 2006–2008. (**b**) Estimated negative return jumps of the NASDAQ index during 2006–2008.

6. Conclusions

In this work, the topic of state space modeling with non-negative constraints was considered. For that purpose, a state space model was constructed where the state equation that describes the dynamic evolution of the components of the hidden state vector was expressed via non-negative definite quadratic forms and represents a non-negative valued Markovian stochastic process of order one. Due to the inequality conditions, a constrained optimization problem arises to derive estimators for the states, which are unbiased and of minimum variance. Towards this direction, a thorough analysis was illustrated via Propositions 1 and 2, concerning the stationary points of the optimization problem along with the special conditions that have to be satisfied in order to derive non-negative estimations for the state vectors at every time. Thus, in Proposition 2, necessary conditions were derived for a stationary point to constitute a feasible solution. The proposed method constitutes an alternative for handling state space models with non-negativity constraints, and it has a low computational burden compared to resampling methods for the estimation procedure.

Regarding future work, the generalization of the proposed method for the case of an n-dimensional non-negative state vector, $n > 2$, could be examined. This is a challenging problem in many applications. For example, in navigation problems, for $n = 3$, state space models with non-negativity constraints are suitable to describe the distance covered during the motion of a vehicle, if we let the three non-negative components of the state vector represent the measures of the velocities (speeds) along the axes in R^3.

Author Contributions: Methodology, O.T. and G.T.; Writing—original draft, O.T. and G.T.; Writing—review & editing, O.T. and G.T. Both authors have read and agreed to the published version of the manuscript.

Funding: This research received no external funding.

Institutional Review Board Statement: Not applicable.

Informed Consent Statement: Not applicable.

Data Availability Statement: The dataset used to illustrate the applicability of the proposed method is publicly available at www.finance.yahoo.com accessed on 11 June 2021.

Conflicts of Interest: The authors declare no conflict of interest.

References

1. Julier, S.; LaViola, J. On kalman filtering with nonlinear equality constraints. *IEEE Trans. Signal Process.* **2007**, *55*, 2774–2784. [CrossRef]
2. Loumponias, K.; Vretos, N.; Tsaklidis, G.; Daras, P. An Improved Tobit Kalman Filter with Adaptive Censoring Limits. *Circuits Syst. Signal Process.* **2020**, *39*, 5588–5617. [CrossRef]
3. Wang, L.; Chiang, Y.; Chang, F. Filtering method for nonlinear systems with constraints. *Control. Theory Appl. IEE Proc.* **2002**, *149*, 525–531. [CrossRef]
4. Chia, T.; Chow, P.; Chizeck, H. Recursive parameter identification of constrained systems: An application to electrically stimulated muscle. *IEEE Trans. Biomed. Eng.* **1991**, *38*, 429–442. [CrossRef] [PubMed]
5. Vasicek, O. An equilibrium characterization of the term structure. *J. Financ. Econ.* **1997**, *5*, 177–188. [CrossRef]
6. Hull, J.; White, A. Pricing interest rate derivatives. *Rev. Financ. Stud.* **1990**, *3*, 573–592. [CrossRef]
7. Cogleya, T.; Sargent, T. Drifts and volatilities: monetary policies and outcomes in the post WWII US. *Rev. Econ. Dyn.* **2005**, *8*, 262–302. [CrossRef]
8. Cogley, T.; Sargent, T. Evolving post World War II US inflation dynamics. *Nber Macroecon. Annu.* **2001**, *16*, 331–373. [CrossRef]
9. Theodosiadou, O.; Tsaklidis, G. Estimating the Positive and Negative Jumps of Asset Returns via Kalman Filtering: The Case of NASDAQ Index. *J. Methodol. Comput. Appl. Probab.* **2017**, *19*, 1123–1134. [CrossRef]
10. Theodosiadou, O.; Skaperas, S.; Tsaklidis, G. Change Point Detection and Estimation of the Two-Sided Jumps of Asset Returns using a Modified Kalman Filter. *J. Risks* **2017**, *5*, 15. [CrossRef]
11. Cont, R. Empirical properties of asset returns: stylized facts and statistical issues. *Quant. Financ.* **2001**, *1*, 223–236. [CrossRef]
12. Kalman, R. A new approach to linear filtering and prediction problems. *J. Basic Eng. Trans. ASME Ser.* **1960**, *82*, 35–45. [CrossRef]
13. Gordon, N.; Salmond, D.; Smith, A. Novel Approach to Nonlinear/Non-Gaussian Bayesian State Estimation. *Radar Signal Process. IEE Proc. F* **1993**, *140*, 107–113. [CrossRef]

14. Doucet, A.; Johansen, A. A Tutorial on Particle Filtering and Smoothing: Fifteen years later. In *Handbook of Nonlinear Filtering*; Oxford University Press: Oxford, UK, 2008.
15. Doucet, A.; de Freitas, N.; Gordon, N. (Eds.) *Sequential Monte Carlo Methods in Practice*; Springer: New York, NY, USA, 2001.
16. Urteaga, I.; Bugallo, M.F.; Djurić, P.M. Sequential Monte Carlo methods under model uncertainty. In Proceedings of the 2016 IEEE Statistical Signal Processing Workshop (SSP), Palma de Mallorca, Spain, 26–29 June 2016; pp. 1–5.
17. Kotz, S.; Balakrinshnan, N.; Johnson, N. *Continouous Multivariate Distributions*; Wiley: New York, NY, USA, 1995; Volume 1.
18. Griva, I.; Nash, S.; Sofer, A. *Linear and Nonlinear Optimization*, 3rd ed.; SIAM: Philadelphia, PA, USA, 2008.

MDPI
St. Alban-Anlage 66
4052 Basel
Switzerland
Tel. +41 61 683 77 34
Fax +41 61 302 89 18
www.mdpi.com

Mathematics Editorial Office
E-mail: mathematics@mdpi.com
www.mdpi.com/journal/mathematics

Lightning Source UK Ltd.
Milton Keynes UK
UKHW050257231122
412584UK00004BA/135

9 783036 523989